FOLDING AND FRACTURING OF ROCKS

AGER   *Principles of paleoecology*

BELOUSSOV (TRANS. BRONEER)   *Basic problems in geotectonics*

CAGNIARD (TRANS. FLINN AND DIX)   *Reflection and refraction of progressive seismic waves*

DE SITTER   *Structural geology*

EWING, JARDETSKY, AND PRESS   *Elastic waves in layered media*

GRANT AND WEST   *Interpretation theory in applied geophysics*

GRIFFITHS   *Scientific method in analysis of sediments*

GRIM   *Applied clay mineralogy*

GRIM   *Clay mineralogy*

HEINRICH   *Mineralogy and geology of radioactive raw materials*

HEISKANEN AND VENING MEINESZ   *The earth and its gravity field*

HOWELL   *Introduction to geophysics*

JACOBS, RUSSELL, AND WILSON   *Physics and geology*

KRAUSKOPF   *Introduction to geochemistry*

KRUMBEIN AND GRAYBILL   *An introduction to statistical models in geology*

KUZNETSOV, IVANOV, AND LYALIKOVA (TRANS. BRONEER)   *Introduction to geological microbiology*

LEGGET   *Geology and engineering*

MENARD   *Marine geology of the pacific*

MILLER   *Photogeology*

OFFICER   *Introduction to the theory of sound transmission*

RAMSAY   *Folding and fracturing of rocks*

SHAW   *Time in stratigraphy*

SHROCK AND TWENHOFEL   *Principles of invertebrate paleontology*

TURNER AND VERHOOGEN   *Igneous and metamorphic petrology*

TURNER AND WEISS   *Structural analysis of metamorphic tectonites*

WHITE   *Seismic waves: radiation, transmission, and attenuation*

Folding and fracturing of rocks

McGRAW-HILL BOOK COMPANY

New York   *San Francisco*
St. Louis   *Toronto*
London   *Sydney*

JOHN G. RAMSAY

Professor of Geology
Imperial College of
Science and Technology
London

Preface

Through geological time the earth's crust has suffered a sequence of spasmodic deformations. Since this phenomenon is now well established, structural investigations have entered a second phase concerned with a rigorous analysis of the form and formation of the structures and with the nature and origin of the forces which led to the deformation. This book outlines the basic theories of stress, strain, and the properties of rock materials, and considers some of their implications in geological processes.

Existing texts relevant to these matters are written primarily for engineers and physicists; and although it may be profitable for the geologist to seek information from sources not specifically geological, much is of limited relevance to problems confronting him. For example, discussion of strain in engineering textbooks concentrates mainly on an analysis of strains of less than 2 percent and may omit any study of the large strains which frequently occur in deformed rock strata. While this book may be useful to civil and mechanical engineers, and to miners and others concerned with the mechanical properties of material under stress, it has been written with the needs of the geologist and the geophysicist in mind.

I have tried to explain in elementary mathematical terms how the various stress and strain formulas are developed from first principles and how they may be applied to rock deformation. Where more advanced techniques enable a speedier or more complete analysis, I have included some as alternatives, but the science student with a knowledge of simple algebra and elementary differential and integral calculus should be able to follow the mathematical development of these subjects as presented here. Intermediate steps in the reasoning have generally been included; and while this will be superfluous for the experienced mathematician, it should prove useful to many students.

The basic groundwork leads to a study of fracture and flow and the geological aspects of fracturing and folding of rock materials which, it is hoped, will supplement the textbooks of descriptive structural geology and lead to an understanding of the fundamentals of these processes. This book is not meant to be a study of the structural development of specific regions, but many real examples have been chosen to illustrate the particular structural phenomena being discussed. Diagrammatic aids have been included to show the visual significance of stress and strain in three dimensions, and the geological significance of the theoretical analyses is constantly indicated.

Chapter 1 introduces the methods used to analyze the orientation of structures. Chapters 2 to 6 develop the analysis of stress and strain from first principles. The basic mathematical methods are developed here and the standard techniques for manipulating geological data are described. The various types of folds to be seen in rocks are described in Chap. 7 with an analysis of their formation and a discussion of their general tectonic environment. After the geometry and significance of folding has been discussed, the geometry of originally rectilinear structures deformed during folding is described (Chap. 8) and the effects of folding plane surfaces which were not initially parallel are discussed (Chap. 9). These subjects have important geological applications: deformed unconformities and disconformities, repeated folding of strata, deformed sedimentation structures, etc. Chapter 10 deals with some of the structural complexities developed when two systems of folds are superimposed one on the other, and much of this analysis builds on the principles established in Chaps. 7, 8, and 9.

Because limitations of space have not permitted an adequate account of microfabric techniques and the use of the universal stage for fabric analysis, it was decided that discussions of these topics were best omitted completely.

I should like to thank my colleagues and friends who gave me much help by correcting and commenting on the text, in particular, D. Elliott, M. J. Fleuty, N. J. Price, F. J. Turner, and J. Watson. I am greatly indebted to Gilbert Wilson who kindled my enthusiasm for structural geology. Also, thanks are due J. Gee for his expert advice and assistance in the preparation of the photographic material used for the illustrations. Finally, but not least, it was only because of the constant help and encouragement of my wife that the book ever reached its present form.

JOHN G. RAMSAY

Contents

Preface   vii
Nomenclature   xv

1   ORIENTATION ANALYSIS   1

1-1   *Development of the stereographic projection   1*
1-2   *Planar structures   4*
1-3   *Linear structures   4*
1-4   *Directional properties of lines   5*
1-5   *Intersection of planes   6*
1-6   *Angle between lines and angle between planes   6*
1-7   *Presentation of structural data   7*
1-8   *Determination of density distribution using contour methods   8*
1-9   *Analysis of the form of curving surfaces   9*
    π diagrams   *9*
    β diagrams   *12*
1-10   *Statistical methods applied to structural analysis   14*
    Unimodal pole distribution   *15*
    The plane of best fit (π circle) to cylindrically disposed s-plane poles   *18*
    The cone of best fit to conically disposed s-plane poles   *20*

2   STRESS   23

2-1   *Stress notation   23*
2-2   *Stress components   24*
2-3   *Analysis of plane stress   26*

2-4   *Values of the principal stresses   27*
2-5   *Invariants in plane stress   28*
2-6   *Stresses acting on a plane whose normal makes an angle ϕ with the principal axis   29*
2-7   *Stress in three dimensions   31*
2-8   *Orientation of the principal axes of stress   34*
2-9   *Stress invariants in three dimensions   34*
2-10   *Stresses acting on any planar surface   35*
2-11   *Planes of maximum shearing stress   36*
2-12   *Hydrostatic stress   39*
2-13   *Deviatoric stress   39*
2-14   *Variation in stress through a body   41*
2-15   *Stress trajectories   43*
2-16   *Superposition of stress systems   44*
2-17   *Photoelastic techniques for determining stress distribution   47*

3   STRAIN IN TWO DIMENSIONS   50

3-1   *The measurement of strain   51*
3-2   *Homogeneous and inhomogeneous strain   53*
3-3   *Finite and infinitesimal strain   55*
3-4   *Finite strain in two dimensions   55*
3-5   *Change in length of lines   65*
        Lines of no finite elongation   66
3-6   *Changes in angles   67*
3-7   *Shear strain   67*
        Maximum value of the finite shear strain   68
3-8   *The Mohr diagram   69*
3-9   *Strain invariants   81*
3-10   *Simple shear   83*
3-11   *The superposition of two finite strains   91*
3-12   *Methods of graphically recording the shape of the finite-strain ellipse   94*
3-13   *Infinitesimal strain in two dimensions   96*
3-14   *Changes in length of lines   99*
        Lines of no infinitesimal longitudinal strain   100
3-15   *Infinitesimal shear strain   102*
3-16   *Infinitesimal area change   103*
3-17   *Geological implications of deformation in two dimensions   103*

4   STRAIN IN THREE DIMENSIONS 121

4-1   *Finite strain in three dimensions* *121*
4-2   *The finite-strain tensor* *123*
4-3   *Changes in length of lines* *124*
         Lines of no finite longitudinal strain *127*
4-4   *Shear strain* *128*
4-5   *Changes in angle during deformation* *129*
4-6   *Methods of graphically recording the components of finite strain* *134*
         Strain components (1), (2), and (3) *134*
         Strain components (4) through (9) *139*
         Limitations of applications to data from deformed rocks *140*
4-7   *Determining the principal axes of the strain ellipsoid from two-dimensional data* *142*
         Measurements of strain made on any three mutually perpendicular planes *142*
         Measurements of strain made on any three nonparallel plane surfaces *147*
         From a knowledge of the state of strain on a single plane surface when the orientation of the principal axes of the ellipsoid is known *148*
4-8   *Mohr's construction for representing states of strain in three dimensions* *149*
4-9   *Finite-strain properties of the five types of constant-volume ellipsoids* *154*
         Strain-ellipse sections of the five types of ellipsoids *158*
         Changes in volume during deformation *161*
4-10   *Dispersion of lines and planes as a result of strain* *162*
4-11   *Superposition of two finite strains* *165*
4-12   *Infinitesimal strain in three dimensions* *167*
4-13   *Infinitesimal longitudinal strain* *172*
4-14   *Infinitesimal shear strain* *173*
4-15   *Mohr diagram for representing infinitesimal strain in three dimensions* *173*
4-16   *Infinitesimal volume change* *174*
4-17   *Progressive deformation in three dimensions* *174*
4-18   *Relationship between progressive deformation in two and three dimensions* *175*
4-19   *The significance of cleavage and schistosity* *177*

5   DETERMINATION OF FINITE STRAIN IN ROCKS 185

5-1   *Deformation of initially spherical objects* *187*
5-2   *Determination of the strain ellipse* *193*

5-3   *Calculation of the strain components from the strain
       ellipse   199*

5-4   *Measurement of the individual ellipsoids   200*

5-5   *Deformation of nonspherical objects   202*

5-6   *Deformation of elliptical objects with random fabric   209*

5-7   *Deformation of ellipsoidal objects with random fabric   211*

5-8   *Deformation of ellipsoidal objects with an original fabric   216*

5-9   *The behavior of rigid or competent objects in a ductile
       matrix   221*

5-10   *Pressure solution and its effects in deformed conglomerates   226*

5-11   *Determination of strain from deformed fossils   228*

5-12   *Fossils which deform homogeneously with their matrix   229*
       Originally circular disks   *229*
       Originally circular tubes   *229*
       Fossils with bilateral symmetry   *230*
       Fossils without original bilateral symmetry   *243*

5-13   *Fossils which deform inhomogeneously with their matrix   247*

5-14   *Determination of strain from folded and boudinaged veins and
       other features   250*

6   **RELATIONSHIP BETWEEN STRESS AND STRAIN   255**

6-1   *Behavior of rocks under experimental conditions   256*

6-2   *Deformation of polycrystalline aggregates   265*

6-3   *Rock behavior in terms of mechanical analogs   268*

6-4   *The Kelvin or Voigt viscoelastic model   270*

6-5   *The Maxwell elastoviscous model   274*

6-6   *The standard linear solid (viscoelastic)   277*

6-7   *Elastoviscous and plastic models for creep strain   279*

6-8   *Relationships between the stress and strain tensors   281*

6-9   *Elastic solids   283*

6-10   *Brittle failure; the development of faults and fractures   289*

6-11   *Stress functions and their application to faulting problems   297*

6-12   *Viscous fluids   306*

6-13   *Stream functions and the solution of problems of
       viscous flow   312*

6-14   *Plastic flow   313*

6-15   *Progressive deformation   322*

6-16   *The concept of symmetry and tectonic axes   333*

7 FOLDS AND FOLDING 342

7-1   *Description and classification of folds*   345
7-2   *Description of single folded surfaces*   345
7-3   *Relations of adjacent surfaces in the fold*   355
7-4   *Terms used to describe the attitude of folds*   358
7-5   *The geometrical classification of folds*   359
7-6   *Folds developed by buckling*   372
         Fold wavelength   *373*
7-7   *Buckling: the shape of the folded layers*   386
7-8   *States of strain within the buckled layer*   391
         Shearing parallel to layer boundaries   *392*
         Longitudinal strains parallel to layer boundaries   *397*
         Combinations of layer boundary slip and tangential longitudinal strain   *403*
         Modifications of shape by superimposed homogeneous strain   *411*
7-9   *States of strain outside the buckled layers*   ·415
7-10   *Similar folds and the problem of shear folding*   421
7-11   *The shear components*   423
7-12   *The compressive strain components*   427
7-13   *The states of finite strain within similar folds*   429
7-14   *The mechanism of formation of similar folds*   430
7-15   *Kink bands, chevron folds, and conjugate folds*   437

8 DEFORMATION OF LINEAR STRUCTURES 461

8-1   *Folds formed by buckling*   463
8-2   *Conical folds*   468
8-3   *Similar folds*   469
8-4   *Restoration of linear structures to their undeformed state*   486

9 FOLDING OF OBLIQUELY INCLINED SURFACES 491

9-1   *Flexural-slip folding*   492
9-2   *Variations in dihedral angle produced by flexural slip*   496
9-3   *Flattened flexural-slip folds*   498
9-4   *Buckle folds with internal deformation by tangential longitudinal strain*   500
9-5   *Similar folds formed by shear*   504
9-6   *Variation in dihedral angle in shear folds*   506
9-7   *Similar folds formed by shear and a homogeneous strain*   506

10 SUPERIMPOSED FOLDING 518

10-1 Environments of superimposed folding 518
10-2 Interference patterns produced by two successive foldings 520
10-3 Stratigraphic order 537
10-4 Geometric control of new folds in superimposed fold systems 538
10-5 Geometric forms of deformed first folds 546
10-6 Reactivation of old folds 548
10-7 Initial orientation of first-fold axes 549
10-8 Geometric analysis of superimposed folds using projection techniques 551

Author index 557
Subject index 563

Nomenclature

x, y, z	cartesian coordinate axes
l, m, n	direction cosines of a line with respect to x, y, and z axes
A	area
δA	area increase
V	volume
δV	volume increase
Δ	area or volume dilation $\delta A/A$ or $\delta V/V$
e, e_x, e_y, e_z	extension, and extension in the direction of x, y, and z axes
e_1, e_2, e_3	principal extensions
$\dot{e}$	rate of change of extension with respect to time, $\partial e/\partial t$
λ	quadratic extension
λ'	reciprocal quadratic extension $= 1/\lambda$
$\lambda_1, \lambda_2, \lambda_3$	principal quadratic extensions
$\omega, \omega_1, \omega_2, \omega_3$	rigid-body rotation, and rigid rotations around x, y, and z axes
X, Y, Z	measured ellipsoid axes from deformed sphere of radius r, $X = r(1 + e_1)$, etc.
a	ratio of principal strains $(1 + e_1)/(1 + e_2) = X/Y$
b	ratio of principal strains $(1 + e_2)/(1 + e_3) = Y/Z$
ε	logarithmic or natural strain $= \log_e (1 + e)$
ψ	angular shear; plastic proportionality factor; stream function
γ	shear strain $= \tan \psi$
γ'	strain parameter γ/λ

$\dot{\gamma}$	rate of change of shear strain with respect to time $\partial\gamma/\partial t$
J_1, J_2, J_3	invariants of the finite-strain tensor
K_1, K_2, K_3	invariants of the infinitesimal-strain tensor
$\sigma, \sigma_x, \sigma_y, \sigma_z$	normal stress, and normal stresses parallel to x, y, and z axes
$\sigma_1, \sigma_2, \sigma_3$	principal stresses
$\bar{\sigma}$	mean or hydrostatic stress
σ'	deviatoric stress
$\sigma'_1, \sigma'_2, \sigma'_3$	principal deviatoric stresses
σ_Y	yield stress
I_1, I_2, I_3	invariants of stress tensor
I'_1, I'_2, I'_3	invariants of deviatoric stress tensor
ϕ	stress function
E	Young's modulus of elasticity
v	Poisson's ratio
m	Poisson's number $= 1/v$
G	rigidity or shear modulus of elasticity
K	bulk or volume modulus of elasticity
μ, λ	Lamé's elastic moduli
μ	viscosity shear modulus
η	viscosity tensile modulus
a, b, c	axes defining direction of translation by simple shear
A, B, C	axes of a monoclinic symmetry fabric

1 ‖ Orientation analysis

MANY of the problems that are encountered in structural geology require careful geometric analysis of three-dimensional forms. The stereographic projection affords one of the most useful methods of representing the angular relationships of planar and linear features of geological structures. It is a graphical method of solving these geometrical problems that is easy to use and quick to produce results. The accuracy of the solutions is generally to within half a degree; that is more precise than is generally contained in the data obtained from field study. In any detailed analysis of an area, it may be necessary to collect a vast amount of information on the orientations of planes and lines in the deformed rocks. It is possible to incorporate thousands of such primary observations onto a single diagram, apply statistical methods to the data, and arrive at results which have a high degree of precision.

1-1 DEVELOPMENT OF THE STEREOGRAPHIC PROJECTION

The development of the stereographic projection may be compared directly to that employed in constructing certain types of map projection used to represent the global distribution of land and sea masses on the earth's surface. If we take a plane surface in space and project it into the *lower hemisphere*[1] of

---

[1] In crystallography it is usual to develop the projections of crystal facies onto an *upper hemisphere*. In most published works on the analysis of geological structures, the lower hemisphere is generally used. This rule has not always been followed, and it is always best to state somewhere in the diagram which hemisphere has been used for the constructions.

a sphere through the center of the sphere (Fig. 1-1*A*), the plane intersects the hemispherical surface in a semicircular arc known as the *spherical projection* of the plane. This projection is a *great circle*, and it is now projected onto the horizontal surface of the hemisphere from a point on the apex of the upper hemisphere. The curved line obtained in this way is known as the *stereographic projection* of the plane (Fig. 1-1*B*, 1-1*C*). If the inclina-

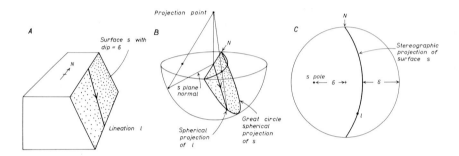

Figure 1-1

The development of the stereographic projection of a plane surface s with dip δ and with lineation l.

tion of the plane surface is vertical, its projection is a diameter of the circle; if it is horizontal, the projection coincides with the periphery of the circle. To develop the complete graph known as a *stereographic net* or *Wulff net*, a complete angular grid of small and great circles drawn on the hemisphere at 10° intervals (cf. meridians of longitude and parallels of latitude) is projected onto the horizontal plane (Fig. 1-2). This grid is a graph which makes

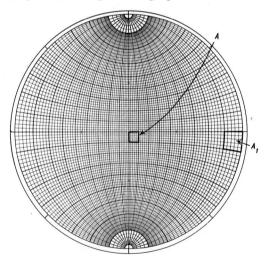

Figure 1-2

The stereographic net.

it possible to represent the three-dimensional orientation of any plane or line. The surface areas of all 10° great-circle intercepts of any zone between two given small circles on the hemisphere surface are equal. On the stereographic net, however, they are not of equal area (Fig. 1-2, *A* and *A*$_1$). For the statistical interpretation of the frequency of distribution of data, the position of the grid lines can be adjusted so that the areas are equal. This adjustment is directly analogous to the corrections for area distortion that are made by some map projections. The *equal-area Schmidt* or *Lambert net* (Fig. 1-3) that results from this correction is the most useful graph for plotting and interpreting structural data. Equal-area projection has been used throughout this book. Another useful change may be made in the grid to produce a net analogous to the polar map projection. This grid, known as the *polar* or *Billing equal-area net* (Fig. 1-4) is particularly useful where large numbers of observations of lineation orientations are to be plotted.

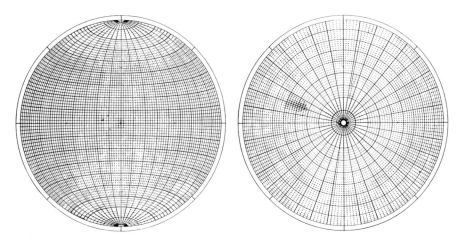

Figure 1-3

The equal-area net.

Figure 1-4

The polar equal-area net.

All the measurements we make which refer to the geometrical orientation of rock structures fall into one of two groups: they refer either to planes or to lines. Curving surfaces may be analyzed by imagining that the surface is made up of an infinitely large number of tangent planes. For example, when the strike and the angle of dip of the layer on the flank of some fold are recorded, we are in reality measuring the attitude of the tangential plane to the fold surface at this point. Lines, such as axial directions of folds, which may have a curving form in space can also be envisaged as the locus traced out by an infinitely large number of straight or rectilinear sections. From these two ideas we deduce that *any* curved surface or curved line may be represented in the projection by plotting its plane or rectilinear component parts. Because all structural forms can be represented as a series of either varying planes or lines, there are only a few basic plotting techniques. The plotting of structural data is generally carried out on a sheet of tracing paper placed over the net and held in position by a pin at the center of the circle so that the overlay may be rotated about this point.

1-2   PLANAR STRUCTURES

Any plane may be represented on the net as a great-circle projection. The method of drawing this great circle is most conveniently illustrated with an actual example. To plot the great circle representing a plane with a strike of 38° and a dip of 71° to the southeast:

1. Place the tracing paper over the net and mark the north-point azimuth.
2. Rotate the overlay 38° in an anticlockwise direction and trace the great circle which has a dip of 71°. The angle is measured from the *right-hand edge* of the net toward the center along an E-W diameter.

If large numbers of planar structures are to be plotted, it is not always practicable to record each one as a great circle, the diagram becomes overloaded with lines which may produce a very confusing effect. Where large amounts of data are to be recorded, the diagrams are usually simplified by plotting the line which is normal or perpendicular to each surface. Each plane is then represented by a single point or *pole* in the projection. The pole always falls into the opposite part of the net to that of the dip direction, and its distance from the edge of the net is 90° minus the angle of dip of the surface. To plot the pole of the same surface as in the previous example, the second step is modified as follows:

2*a*. Rotate the overlay 38° in an anticlockwise direction and locate the point on the *left-hand side* of the net (i.e., the NW quadrant) at a distance of 71° from the *center* of the net on the W-E diameter.

1-3   LINEAR STRUCTURES

To record the position of a lineation that has a plunge of 35° toward 77°:

1. Place the tracing paper over the net and mark the north azimuth.
2. Rotate the paper anticlockwise through 77° and locate the point 35° from the north point of the underlying net on the N-S diameter.

If a large number of lineations have to be plotted, it is more convenient to use the polar net than the Lambert net since the points may be located without recourse to rotation of the overlay.

1-4 DIRECTIONAL PROPERTIES OF LINES

Linear structures and poles to plane surfaces are recorded on the stereogram by points which represent their projections onto the lower hemisphere. The lines recorded in this way may or may not have a directional significance. For example, neither a pole to a cleavage surface nor the direction of a hinge line of a fold has a significant sense of direction, whereas the pole to a bedding surface may point either toward or away from the direction of "younging" of the strata, and a line parallel to the current flow depositing a sediment has a significant directional sense. It may sometimes be important to indicate the directional sense of a line by distinguishing with different symbols those lines which have a direction passing into the lower hemisphere from those emerging out of the lower hemisphere. Figure 1-5 shows the use of such symbols referred to bedding surfaces with the same planar orientation. One is inverted and the other is in normal order. This type of directional symbol

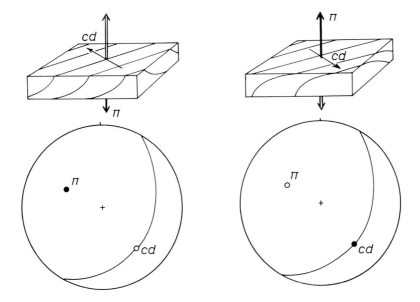

Figure 1-5

The directional significance of lines. Both pole to the bedding surface π and the current direction cd have directional sense and are plotted into the lower hemisphere as downward-directed (solid circles) or upward-directed (open circles) poles.

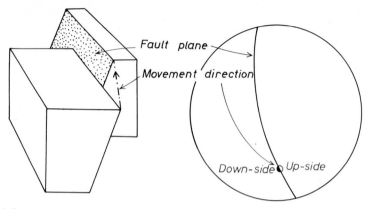

Figure 1-6

Method of using the split-circle symbol to indicate relative displacements on faults.

should also be used to indicate the relative displacements on adjacent sides of fault surfaces (Fig. 1-6).

1-5   INTERSECTION OF PLANES

Two nonparallel-plane surfaces always intersect in a line. The orientation of this line of intersection can easily be determined by a construction: the two great circles that represent the two planes cross at a point that gives the linear intersection. Another problem that can be solved rapidly by these methods is the determination of true dip from any two apparent dips. The two apparent dips are the lines of intersection of the bedding surface on the various surfaces of erosion. If the two points that represent these two lines of intersection are plotted onto a net, it will be found by rotating the overlay that there is only one great circle on the underlying net which passes through these two points; this great circle gives the true orientation of the surface.

1-6   ANGLE BETWEEN LINES AND ANGLE BETWEEN PLANES

To measure the angle between any two nonparallel lines (or skew lines) the points representing these lines are first plotted onto the net. The paper overlay is now rotated until the two points fall on a single great circle. The angular intercept made on this great circle by the crosscutting small circles (Fig. 1-7) is the angle between the lines. This process may be directly compared with that of counting the angular distance between any two parallels of latitude on a global surface.

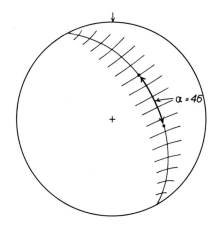

Figure 1-7

Measurement of angle α between two lines.

The smallest or *dihedral angle* between any two nonparallel planes is determined by first constructing the two normals (poles) to the plane surfaces. The angle between these lines is measured by the technique described above and gives the true angle between the planes. The bisectors of two inclined surfaces can be determined by finding the bisectors of the two poles, but it should be remembered that the bisector of the *acute* angle between the poles is the bisector of the *obtuse* angle between the planes.

It is always important with these constructions to keep a clear visual representation of the actual physical form of the structures and not to work by the routine application of rules. When the north point on the overlay is restored to its correct orientation, it is always advisable to convert mentally the two-dimensional projection into a lower hemisphere and check that the structure that has been plotted has the correct orientation.

1-7 PRESENTATION OF STRUCTURAL DATA

Although the projection often affords an excellent practical method for determining the angular relationships that exist between different structures, it is not always the most convenient way to present the conclusions of a structural synthesis. Published descriptions of structures of an area some-times lean too heavily on the stereographic or equal-area diagram as a method of presentation. It is often difficult to determine the significance of the data recorded in large numbers of stereograms, and descriptions of these diagrams often make very dull reading. Stereograms can never indicate the spatial distribution of the observations, only their overall angular relationships. The results of this analysis can often be clarified by abstracting the significant information from each net (axial directions of folds, axial planes of folds, lineation trends) and replotting it onto a map.

Stereographic and equal-area diagrams in reports or publications should always indicate the north-point azimuth and the center of the circle. The number and types of data plotted should always be recorded and if the stereograms refer to particular subareas within a region, there should always be a clear key map to indicate their location. Data which are plotted as poles may be left as individual points on an equal-area net, but sometimes certain features of the distribution of the poles will stand out more clearly if the intensity of the pole concentration is contoured.

1-8   DETERMINATION OF DENSITY DISTRIBUTION USING CONTOUR METHODS

Although the intensity of distribution of structural elements plotted on an equal-area net can often be approximately evaluated by inspection, it is sometimes necessary to measure the density more accurately. If the density of points in certain zones of the net is high, then it is best to use a contouring counter (Knopf and Ingerson, 1938, pp. 23, 145-151; Turner and Weiss, 1963, pp. 58-64). The contouring counter is a card with a circular hole 1/100 of the area of the complete equal-area net. This counter is placed on the net and the number of points which can be seen within the circle are counted and the percentage concentration calculated (Fig. 1-8*A*). For example, if there

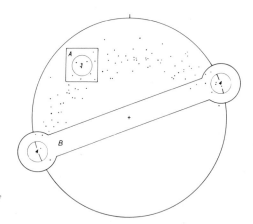

Figure 1-8

Contouring counters for determination of distribution in center (A) and near periphery (B) of net.

were 300 poles on the net, and at one position of the counter nine occurred within the hole, then the percentage at the center of the counting circle would be $(9 \times 100)/300 = 3$ percent. These percentages are recorded from place to place on the net and the density distribution contoured by normal methods. At the periphery of the net the percentages are calculated with a special counter (Fig. 1-8*B*) so that a complete one percent area is used.

If the poles show a more uniform distribution, the contouring is more accurately and conveniently done by constructing circles of constant radius around each of the poles. For example, if 100 poles were plotted, then 100 circles each of radius 1/10 of that of the net are drawn with the poles as centers (Fig. 1-9). The overlapping areas between two circles have a 2-percent concentration, those between three circles have a 3-percent concentration, etc.

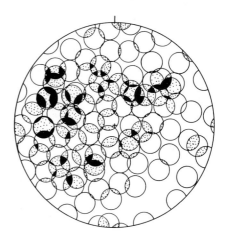

Figure 1-9

Density evaluation using 1 percent circles. Concentrations of 1 to 2 percent stippled, greater than 2 percent solid black.

1-9 ANALYSIS OF THE FORM OF CURVING SURFACES

A curved surface may be termed either *cylindrical* or *noncylindrical*. Cylindrical surfaces can be envisaged as the form traced out or generated when a straight line (the rectilinear generator) moves parallel to itself in space. Noncylindrical surfaces may be generated when a rectilinear generator moves in a nonparallel manner through space (e.g., a cone surface is produced when the generator always passes through a single point) or they may have a more complex form and not be developable in this way (e.g., a paraboloid). Certain geometrical features of all cylindrical surfaces may be determined from the angular relationships of their components, but the geometrical forms of nondevelopable surfaces are indeterminate by stereographic or equal-area projection methods since their properties are dependent on the spatial distribution of the elements, features which cannot be recorded on these diagrams.

Two graphical methods, the π *diagram* and the β *diagram*, are most frequently employed for the analysis of cylindrical and conical surfaces.

π **diagrams** The π-diagram method of analysis utilizes the plots of the poles (*s* poles) or normals to the measured tangential surfaces (*s* surfaces) through the structure. In cylindrical folds each *s* pole is normal to the fold generator (axial direction) and therefore, when plotted onto a stereogram or equal-area net, the *s* poles fall on (or about) one particular great circle known as the π circle. The normal to the π circle is known as the π axis and is parallel to the axial direction of the folds (Fig. 1-10).

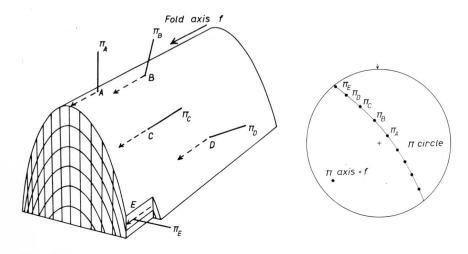

Figure 1-10

The perpendicular relationship of surface normals (π_A, π_B, etc.) to the axial direction and fold axis f, and the construction of a π diagram.

If the surface has the form of a cone of apical angle α, each *s* pole makes a constant angle of $(90-\alpha/2)$ degrees with the axis of the cone. If α is constant the *s* poles fall on part of a small circle making an angle of $(90-\alpha/2)$ degrees with the point representing the cone axis (Fig. 9-15).

The special value of the π-diagram method of structural analysis is that a great number of observations may be recorded on one diagram and therefore the geometric analysis may be made with a high degree of precision, particularly if statistical methods are employed (Sec. 1-10).

The axial surface (or axial plane) of a cylindrical fold cannot be determined from the *s* poles alone. If, however, the axial trace (see p. 357) can be found, then it is possible to define exactly the orientation of the axial surface. This is done by finding the one great circle which passes through the π axis (fold axis) and the point representing the plot of the axial trace (Fig. 1-11).

The *s* poles generally fall on only part of the π circle; the angular spread of the *s* poles gives the range of variation of the planes in the structure and the angle between the limbs of the fold. The *s* poles also frequently show a variation in concentration within the π circle. If the measured planes represent a reasonably uniform sample of the folds, this density variation indicates that there are certain preferred orientations of the *s* planes related to the

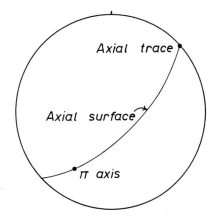

Figure 1-11

The determination of the axial surface orientation.

fold shape and gives a measure of the comparative development of the fold limbs and hinge zones. By finding the mean values of these concentrations, it is sometimes possible to determine the "best fit" orientation of the fold limbs.

It is always important to include in a single plot only data drawn from a homogeneously folded region where the axial direction of the folds does not depart more than about 5° from the mean value. If the region shows greater variation than this, it is usual to divide it into smaller units or subareas and make separate π diagrams for each area.

Minor structures (hinge lines and axial surfaces of small folds, lineations,

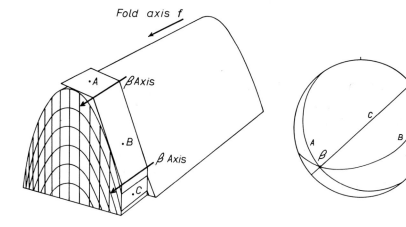

Figure 1-12

The intersection of surfaces A, B, and C, to establish β axes.

Figure 1-13

The β diagram prepared from Figure 1-12.

cleavage, joints, etc.) may also be plotted onto the same net as the *s* poles to determine whether the small-scale structures are geometrically related to the form of the major folds.

β **diagrams** Surfaces within a cylindrical fold contain a line parallel to the rectilinear generator and therefore the intersections of any two observations of measured tangent planes to the folded surface intersect in a line parallel to this generator. This line is known as a β axis and is parallel to the fold axis (Fig. 1-12). The stereogram or equal-area projection offers a very convenient way for graphic determination of the orientation of such β axes (Fig. 1-13). All the β axes computed in this manner should have a parallel orientation (i.e., all the great circles representing individual surfaces in the fold should pass through a single point). In practice, however, folds rarely have a perfectly cylindrical form and the measurements of the fold surfaces are always subject to a certain error. This means that the computed

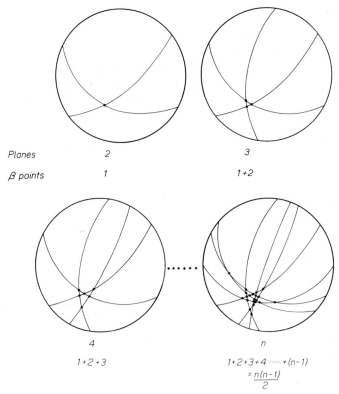

Planes	2	3
β points	1	1+2

4

1+2+3

n

$1+2+3+4 \cdots + (n-1)$

$= \dfrac{n(n-1)}{2}$

Figure 1-14

The relationship between number of β axes and number of planes n plotted.

β axes do not generally coincide, but they are unimodally grouped around a point which gives the "best fit" orientation for the mean β axis or fold axis. The number of β intersections (S) produced by the intersection of n observed planes (tangent planes) in the fold (Fig. 1-14) is given by the arithmetic progression:

$$S = 0 + 1 + 2 + 3 + \cdots + (n - 1) = \frac{n(n - 1)}{2} \tag{1-1}$$

If $n > 3$, the number of β intersections outnumbers the observations, and in practice it may be almost impossible to determine their position in a single diagram (for example, if $n = 500$, $S = 124{,}750$). Where n is large it is usual to subdivide the region into a number of smaller areas. The large number of β points and β diagrams that can be obtained from relatively few primary observations can be overimpressive and lead to a false idea of analytical rigor. The mean of the β axes is statistically parallel to the fold axis and can be determined by inspection, by using a contour counter, or by mathematical methods (Sec. 1-10).

Because the field measurements are always subject to a certain error, the positions of the β axes are also subject to certain positional errors. The maximum error in the positions of the β points (max e_β) is dependent on the initial measurement error (e_π) and the angle between the two plane surfaces (d). The β axis is located in the zone of intersection of two cones with apical angles $2(90 - e_\pi)$ degrees and axes at angular distance d (Fig. 1-15). The

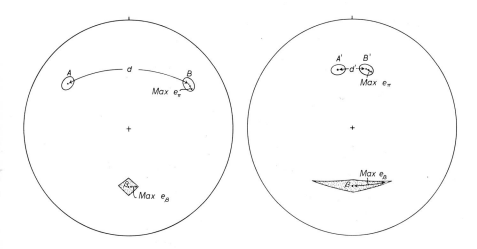

Figure 1-15

The maximum error in position of the β axis (e_β) resulting from plotting two planes with positional error e_π and angular distance d apart.

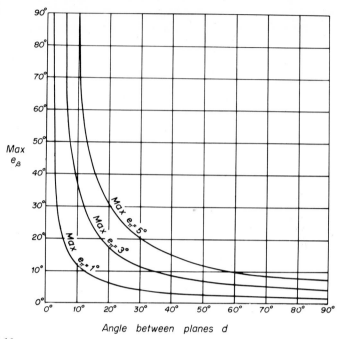

Figure 1-16

Values of maximum error e_β for variations in e_π and d.

maximum error is

$$\max e_\beta = \cos^{-1}\left[1 - \frac{2\sin^2 e_\pi}{\sin^2 d}(1 + \cos d)\right]^{\frac{1}{2}} \qquad (1\text{-}2)$$

Values of this maximum error for various values of e_π and d are graphically illustrated in Fig. 1-16. These errors may become very large if the angle between the surfaces is small and this means that the β-diagram method for determining the position of the axial direction of cylindrical folds is unsatisfactory where the folds have an open or tight cross-section form (where $d < 40°$ or $> 140°$) (Ramsay, 1963).

If the folding is noncylindrical, or if the axial direction of the folds is variable, the β-diagram method of analysis is mathematically unsound, and many geologically meaningless β axes will appear on the plot (Ramsay, 1963).

1-10   STATISTICAL METHODS APPLIED TO STRUCTURAL ANALYSIS

The method of plotting poles to surfaces on the equal-area net enables a large amount of data to be incorporated into one diagram. Geometrical interpretations of these data can often be simply done by inspection, assisted by the use of various contouring methods to determine graphically the distribution density and its significance. Such simple methods often make it possible for the geometrical analysis to be carried out with a fair degree of precision. It is also possible, however, to employ statistical techniques to give a more exact analysis of the data and, in consequence, the reliability of the deductions made from the data can be assessed more accurately. Some of the more common types of pole distribution will now be examined to illustrate the way in which statistical methods may be used in this type of analysis.

Unimodal pole distribution   If the poles are distributed fairly uniformly around a central mode (e.g., a single β-point maxima), then it is possible to determine the orientation of the mean vector defining the central point of density of the poles.

Three mutually perpendicular directions are chosen for coordinate axes. The most convenient for purposes of measurement are x axis E-W horizontal; y axis N-S horizontal, and z axis vertical. The angles α, β, and γ between each pole (unit vector) and the x, y, and z directions are measured on the net (Fig. 1-17). The x, y, and z components of the side of the vector "box" for

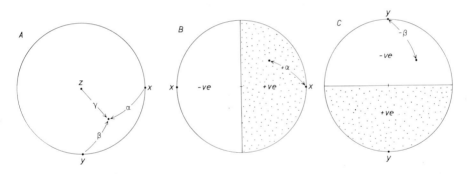

Figure 1-17

The method of defining orientation of a pole using the angles α, β, and γ from three mutually perpendicular directions X, Y, and Z. The sign convention for α and β is shown in B and C; γ is always negative if directed into the lower hemisphere, or positive if directed out of the lower hemisphere.

each measurement are then given by cos α, cos β, and cos γ, respectively (Fig. 1-18). These measurements are repeated for each of the other poles and the sums of the vector components are computed (Σ cos α, Σ cos β, and Σ cos γ). These sums give the dimensions of the x, y, and z components of the total vector sum and the diagonal of this box gives the strength of the total vector sum, that is, $[(\Sigma \cos \alpha)^2 + (\Sigma \cos \beta)^2 + (\Sigma \cos \gamma)^2]^{\frac{1}{2}}$. The direc-

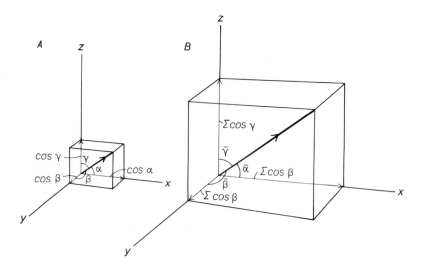

Figure 1-18

Method of summing vectors.

tions of this mean vector with the x, y, and z axes are given by

$$\cos \bar{\alpha} = \frac{\Sigma \cos \alpha}{[(\Sigma \cos \alpha)^2 + (\Sigma \cos \beta)^2 + (\Sigma \cos \gamma)^2]^{\frac{1}{2}}} \tag{1-3}$$

$$\cos \bar{\beta} = \frac{\Sigma \cos \beta}{[(\Sigma \cos \alpha)^2 + (\Sigma \cos \beta)^2 + (\Sigma \cos \gamma)^2]^{\frac{1}{2}}} \tag{1-4}$$

$$\cos \bar{\gamma} = \frac{\Sigma \cos \gamma}{[(\Sigma \cos \alpha)^2 + (\Sigma \cos \beta)^2 + (\Sigma \cos \gamma)^2]^{\frac{1}{2}}} \tag{1-5}$$

As an example, Fig. 1-19 shows a pole distribution which has been analyzed in this way to illustrate the method of setting out the data for computation of the mean using a desk calculating machine (Table 1-1).

Using the data set out in Table 1-1, in Eqs. (1-3) to (1-5) calculate

$$\cos \bar{\alpha} = \frac{5.738}{9.901} = 0.579 \qquad \bar{\alpha} = 54.6°$$

$$\cos \bar{\beta} = \frac{4.359}{9.901} = 0.440 \qquad \bar{\beta} = 63.9°$$

$$\cos \bar{\gamma} = \frac{6.791}{9.901} = 0.686 \qquad \bar{\gamma} = 46.7°$$

As a check, $\cos^2 \bar{\alpha} + \cos^2 \bar{\beta} + \cos^2 \bar{\gamma} = 0.335 + 0.193 + 0.470 = 1.000$.

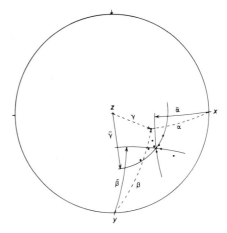

Figure 1-19

Determination of the mean vector of 10 plotted poles.

The angles $\bar{\alpha}$, $\bar{\beta}$, and $\bar{\gamma}$ can be set off on the stereogram to locate the mean vector (Fig. 1-19).

It is now possible to give values to the extent of dispersion of the poles around their mean to determine the intensity and significance of the concentration, e.g., by recording the standard deviation, variance, etc.

Table 1-1   *Computation of a mean vector*

α	β	γ	$\cos \alpha$	$\cos \beta$	$\cos \gamma$
	degrees				
58	78	35	0.530	0.208	0.819
47	72	47	0.682	0.309	0.682
57	68	41	0.545	0.375	0.755
56	64	45	0.559	0.438	0.707
52	65	47	0.616	0.443	0.682
51	62	51	0.629	0.470	0.629
55	60	49	0.574	0.500	0.656
61	62	42	0.485	0.470	0.743
68	52	46	0.375	0.616	0.695
42	58	65	0.743	0.530	0.423
	Σ		5.738	4.359	6.791
	Σ^2		32.925	19.001	46.118

$$(\Sigma \cos \alpha)^2 + (\Sigma \cos \beta)^2 + (\Sigma \cos \gamma)^2 = 98.044$$
$$(98.044)^{\frac{1}{2}} = 9.901$$

The plane of best fit (π circle) to cylindrically disposed s-plane poles   The projection method used to determine the axial direction of cylindrically folded surfaces relies on finding the normal to the plane of best fit to the s poles (π). The statistical procedure for finding this plane of best fit is based on the usual technique of minimizing the squares of the deviations of the observed s poles from this surface.

Let the normal to the π circle make angles of α, β, and γ, respectively with any three mutually perpendicular directions, x, y, and z. If the angles between an s pole on the same coordinate directions are a, b, and c, respectively, then the s poles and the π-circle normal should be perpendicular. The condition for this is that the product of their cosines should equal zero, that is,

$$\cos a \cos \alpha + \cos b \cos \beta + \cos c \cos \gamma = \cos 90° = 0 \qquad (1\text{-}6)$$

Replacing $\cos a$ by l, $\cos b$ by m, $\cos c$ by n, $\cos \alpha / \cos \gamma$ by A, $\cos \beta / \cos \gamma$ by B, then the condition expressed in (1-6) simplifies to

$$Al + Bm + n = 0 \qquad (1\text{-}7)$$

If actual values of l, m, and n are substituted in (1-7), the value will generally differ from zero by some small deviation e. The standard statistical procedure for finding the values of A and B which have the closest fit to the observed values is to minimize the squares of these deviations, that is, $\Sigma e^2 = \Sigma(Al + Bm + n)^2$ has a minimum value.

$$\Sigma e^2 = A^2\Sigma l^2 + B^2\Sigma m^2 + \Sigma n^2 + 2AB\Sigma lm + 2A\Sigma ln + 2B\Sigma mn \qquad (1\text{-}8)$$

Differentiating (1-8) with respect to A and equating to zero,

$$A\Sigma l^2 + B\Sigma lm + \Sigma ln = 0 \qquad (1\text{-}9)$$

Differentiating (1-8) with respect to B and equating to zero,

$$A\Sigma lm + B\Sigma m^2 + \Sigma mn = 0 \qquad (1\text{-}10)$$

Solving the simultaneous equations (1-9) and (1-10) for A and B,

$$A = \frac{\Sigma lm \Sigma mn - \Sigma ln \Sigma m^2}{\Sigma l^2 \Sigma m^2 - (\Sigma lm)^2} \qquad (1\text{-}11)$$

$$B = \frac{\Sigma lm \Sigma ln - \Sigma mn \Sigma l^2}{\Sigma l^2 \Sigma m^2 - (\Sigma lm)^2} \qquad (1\text{-}12)$$

Because $\cos^2 \alpha + \cos^2 \beta + \cos^2 \gamma = 1$,

$$A^2 \cos^2 \gamma + B^2 \cos^2 \gamma + \cos^2 \gamma = 1$$

or

$$\cos \gamma = (1 + A^2 + B^2)^{-\frac{1}{2}} \qquad (1\text{-}13)$$

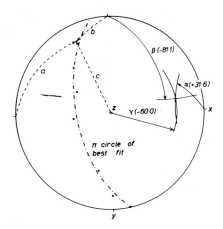

Figure 1-20

Determination of the best-fit π circle to seven s poles.

also

$$\cos \alpha = A(1 + A^2 + B^2)^{-\frac{1}{2}} \qquad (1\text{-}14)$$

$$\cos \beta = B(1 + A^2 + B^2)^{-\frac{1}{2}} \qquad (1\text{-}15)$$

As an example of the use of this method, Fig. 1-20 shows a series of *s* poles distributed about a great circle. To find the best fit for the π circle the values of *a*, *b*, and *c* are measured for each pole. These data are tabulated in Table 1-2 for computation with a desk calculating machine. It should be noted how care must be taken in assigning +*ve* or −*ve* values to the angles *a*, *b*, and *c* according to the convention shown in Fig. 1-17.

Using the data of Table 1-2 in Eqs. (1-11) through (1-14), we can calculate that

$$A = \frac{0.717 \times 0.086 - 1.375 \times 3.669}{0.940 \times 3.669 - (0.717)^2} = -1.698$$

$$B = \frac{0.717 \times 1.375 - 0.940 \times 0.086}{0.940 \times 3.669 - (0.717)^2} = +0.308$$

$$1 + A^2 + B^2 = 3.978$$

$$\cos \gamma = (3.978)^{-\frac{1}{2}} = 0.502 \qquad\qquad \gamma = +60.0°$$

$$\cos \alpha = 0.502 \times -1.698 = -0.852 \qquad \alpha = -31.6°$$

$$\cos \beta = 0.502 \times 0.308 = +0.154 \qquad \beta = +81.1°$$

Check, $\cos^2 \alpha + \cos^2 \beta + \cos^2 \gamma = 1.002$.

Because the calculation leads to a +*ve* value for *γ* (upper hemisphere), all signs are changed to give the correct lower hemisphere location of the pole to the π circle, that is, $\alpha = +31.6°$, $\beta = -81.1°$, $\gamma = -60.0°$. These are now plotted onto the net and the best-fit π circle can be drawn (Fig. 1-20).

Table 1-2   *Computation of the plane of best fit to several s-plane poles*

a	b	c	l	m	n	l²	m²	xy	xz	yz
−67	−31	−71	−0.391	−0.857	−0.326	0.153	0.734	+0.335	+0.127	+0.279
−62	−53	−50	−0.469	−0.602	−0.643	0.220	0.362	+0.282	+0.302	+0.387
−62	−75	−34	−0.469	−0.259	−0.829	0.220	0.067	+0.121	+0.389	+0.215
−58	+85	−34	−0.530	+0.087	−0.829	0.281	0.008	−0.046	+0.439	−0.072
−79	+40	−52	−0.191	+0.766	−0.616	0.036	0.587	−0.146	+0.118	−0.472
+90	+14	−75	0.000	+0.970	−0.259	0.000	0.941	0.000	0.000	−0.251
+80	+10	+90	+0.174	+0.985	0.000	0.030	0.970	+0.171	0.000	0.000
	Σ		−1.876	+1.090	−3.502	0.940	3.669	+0.717	+1.375	+0.086

The cone of best fit to conically disposed s-plane poles   If a series of measured surfaces is believed to form part of a circular cone surface, it is possible to determine the orientation of the axis of this cone and its apical angle.

Let the cone axis be situated at angular distances α, β, and γ from three mutually perpendicular direction axes x, y, and z, respectively. If the angles between each measured s pole and the x, y, and z direction axes are a, b, and c, respectively, then for these surfaces to be arranged on a cone they should be situated at a constant angle K from the cone axis, where K is half the apical angle. The condition for this is given by

$$\cos a \cos \alpha + \cos b \cos \beta + \cos c \cos \gamma = \cos K \qquad (1\text{-}16)$$

Replacing $\cos a$, $\cos b$, and $\cos c$ by l, m, and n, respectively, $\cos \alpha/\cos \gamma$ by A, $\cos \beta/\cos \gamma$ by B, and $\cos K/\cos \gamma$ by C, the condition of (1-16) simplifies to

$$Al + Bm + n + C = 0$$

Substituting the measured values of l, m, and n into this equation generally produces a small deviation e from its true zero value. Minimizing the sums of the squares of these deviations, we obtain the condition that

$$\Sigma e^2 = \Sigma (Al + Bm + n + C)^2$$

$$= A^2 \Sigma l^2 + B^2 \Sigma m^2 + \Sigma n^2 + NC^2 + 2AB\Sigma lm + 2B\Sigma mn$$

$$+ 2AC\Sigma l + 2BC\Sigma m + 2C\Sigma n \qquad (1\text{-}17)$$

has a minimum value where N is the number of s poles measured. Differentiating (1-17) with respect to A, to B, and then to C, and equating each to zero, we obtain three linear equations for A, B, and C:

$$A\Sigma l^2 + B\Sigma lm + C\Sigma l + \Sigma ln = 0 \qquad (1\text{-}18)$$

$$A\Sigma lm + B\Sigma m^2 + C\Sigma m + \Sigma mn = 0 \qquad (1\text{-}19)$$

$$A\Sigma l + B\Sigma m + CN + \Sigma n = 0 \qquad (1\text{-}20)$$

To solve these equations the determinants D, D_A, D_B, and D_C are computed using the standard technique known as the *Rule of Sarrus*.

$$D = \begin{vmatrix} \Sigma l^2 & \Sigma lm & \Sigma l \\ \Sigma lm & \Sigma m^2 & \Sigma m \\ \Sigma l & \Sigma m & N \end{vmatrix} \qquad D_A = \begin{vmatrix} -\Sigma ln & \Sigma lm & \Sigma l \\ -\Sigma mn & \Sigma m^2 & \Sigma m \\ -\Sigma n & \Sigma m & N \end{vmatrix}$$

$$(1\text{-}21)$$

$$D_B = \begin{vmatrix} \Sigma l^2 & -\Sigma ln & \Sigma l \\ \Sigma lm & -\Sigma mn & \Sigma m \\ \Sigma l & -\Sigma n & N \end{vmatrix} \qquad D_C = \begin{vmatrix} \Sigma l^2 & \Sigma lm & -\Sigma ln \\ \Sigma lm & \Sigma m^2 & -\Sigma mn \\ \Sigma l & \Sigma m & -\Sigma n \end{vmatrix}$$

For example,

$$D = \Sigma l^2 \Sigma m^2 N + \Sigma lm \Sigma m \Sigma l + \Sigma l \Sigma lm \Sigma m - \Sigma l^2 (\Sigma m)^2 - N(\Sigma lm)^2 - \Sigma m^2 (\Sigma l)^2$$

The values of A, B, and C are then given by

$$A = \frac{D_A}{D} \qquad B = \frac{D_B}{D} \qquad C = \frac{D_C}{D} \qquad (1\text{-}22)$$

From $\cos^2 \alpha + \cos^2 \beta + \cos^2 \gamma = 1$,

$$\cos \gamma = (1 + A^2 + B^2)^{-\frac{1}{2}} \qquad (1\text{-}23)$$

Also

$$\cos \alpha = A(1 + A^2 + B^2)^{-\frac{1}{2}} \qquad (1\text{-}24)$$

$$\cos \beta = B(1 + A^2 + B^2)^{-\frac{1}{2}} \qquad (1\text{-}25)$$

$$\cos K = -C(1 + A^2 + B^2)^{-\frac{1}{2}} \qquad (1\text{-}26)$$

As an example of the use of this method, the data from Fig. 1-20 and Table 1-2 will be used to see if the assumption that the s poles lay on a plane was reasonably correct, for a plane may be considered as a cone with apical angle equal to $180°$ (that is, $K = 90°$).

Calculation of the determinants using a desk calculating machine gives

$$D = +3.583 \qquad D_B = +1.151$$

$$D_A = -6.234 \qquad D_C = +0.057$$

where $A = -1.739$, $B = +0.321$, and $C = +0.016$. Thus $\cos \gamma = +0.492$, $\cos \alpha = -0.856$, $\cos \beta = +0.157$, and $\cos K = -0.008$. This shows that the data fit a cone with axial direction given by $\alpha = +31.1°$, $\beta = -81.0°$, and

$\gamma = -60.5°$, and with a half apical angle of 89.5°. The assumption was justified; the cone axis almost exactly corresponds with the normal to the π circle, and the half apical angle is close to a right angle.

REFERENCES AND FURTHER READING

Badgley, P. C.: "Structural Methods for the Exploration Geologist," pp. 187-242, Harper & Brothers, New York, 1959.

Bucher, W. H.: The Stereographic Projection, a Handy Tool for the Practical Geologist, *J. Geol.*, **52**: 191-212 (1944).

Chayes, F.: Application of the Coefficient of Correlation to Fabric Diagrams, *Am. Geophys. Union Trans.*, **27**: 400-405 (1946).

Dahlstrom, C. D. A.: Statistical Analysis of Cylindroidal Folds, *Can. Inst. Min. Met. Bull.*, **47**: 234-239 (1954).

Donn, W. L., and J. A. Shiner: "Graphic Methods in Structural Geology," Appleton-Century-Crofts, Inc., New York, 1958.

Fairbairn, H. W.: "Structural Petrology of Deformed Rocks," Addison-Wesley Publishing Company, Inc., Reading, Mass., 1949.

Flinn, D.: On Tests of Significance of Preferred Orientation in Three Dimensional Fabric Diagrams, *J. Geol.*, **66**: 526-539 (1958).

Haman, P. J.: Manual of Stereographic Projection, *West Can. Res. Pub., Geology Related Sciences*, **1**: 1-67 (1961).

Kamb, W. B.: Petrofabric Observations from Blue Glacier, Washington, in Relation to Theory and Experiment, *J. Geophys. Res.*, **64**: 1908-1909 (1959).

Knopf, E. B., and E. Ingerson: Structural Petrology, *Geol. Soc. Am., Mem.*, **6** (1938).

Noble, D. C., and S. W. Eberly: A Digital Computer Procedure for Preparing Beta Diagrams, *Am. J. Sci.*, **262**: 1124-1129 (1964).

Phillips, F. C.: "The Use of Stereographic Projection in Structural Geology," Edward Arnold (Publishers) Ltd., London, 1954.

Pincus, H. J.: The Analysis of Aggregates of Orientation Data in the Earth Sciences, *J. Geol.*, **61**: 482-509 (1953).

Ramsay, J. G.: The Uses and Limitations of Beta-Diagrams and Pi-Diagrams in the Geometrical Analysis of Folds, *Quart. J. Geol. Soc.*, **120**: 435-454 (1963).

Robinson, P., R. Robinson, and S. J. Garland: Preparation of Beta-Diagrams in Structural Geology by a Digital Computer, *Am. J. Sci.*, **261**: 913-928 (1963).

Sander, B.: Geologie des Tauern-Westendes. Ueber Flächen-und Achsen Gefuge, *Mitt. Reichsamts Bodenforsch., Wien*, **4**: 1-94 (1942).

Sander, B.: "Einführung in die Gefügekunde der geologischen Körper," Springer-Verlag OHG, Vienna and Innsbruck, 1948.

den Tex, E.: in "Elements of Structural Geology" by E. S. Hills, pp. 392-406, John Wiley & Sons, Inc., New York, 1963.

Turner, F. J., and L. Weiss: "Structural Analysis of Metamorphic Tectonites," pp. 46-75, McGraw-Hill Book Company, New York, 1963.

Winchell, H.: A new Method of Interpretation of Petrofabric Diagrams, *Am. Mineralogist*, **22**: 15-36 (1937).

2 ‖ Stress

THE forces which act on an element of material can be divided into two types. The first arise in the bulk of the material and are proportional to the mass of the substance (e.g., gravity, centrifugal force, magnetic force). These forces are known as *body forces* and are measured in units of force per unit volume. The second type, known as *surface forces*, act over the surface of the body and are measured in units of force per unit area. This force per unit area is termed *stress* and gives a measure of the intensity of the reaction of the material which lies on one side of the surface or that which lies on the other. The stress acting on imaginary surfaces inside the body can be investigated without actual physical division of the body to form free surfaces. These surface stresses generally arise from externally applied forces, but they can develop from body forces if the component parts of the body are in equilibrium.

2-1 STRESS NOTATION

The forces acting on a surface can be resolved into *normal stresses* or *direct stresses* denoted by the Greek letter sigma (σ), and *shearing stresses* denoted by the letter tau (τ) (Fig. 2-1). Direct stresses can be either tensile or compressive; the convention is to use a positive sign for tensile and a negative sign for compressive stresses. In many geological texts compressive normal stresses are considered positive, but this nomenclature is not consistent with that generally employed in works on

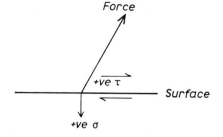

Figure 2-1

The resolution of the forces acting on a surface of area a into normal (σ) and shearing (τ) stress components. If the force vector F makes an angle θ with the surface, then σ = F sin θ/a, and τ = F cos θ/a.

elasticity and plasticity. Also, it leads to difficulties in notation when stress-strain relationships are discussed. The sign of the shearing stresses is defined as follows: If the upper side of the surface is sheared to the right (dextral or right-handed shear), the sign is positive; if it is sheared to the left, the sign is negative.

2-2 STRESS COMPONENTS

In order to develop a description of the state of stress in a mathematical way, three mutually perpendicular coordinate axes x, y, and z are first set up. We now consider the stresses that act on the surfaces of a cubic element of the substance when a force is applied (Fig. 2-2), assuming that the state of stress is perfectly homogeneous throughout the element and that the body is in equilibrium. The stresses on each face can be resolved into three parts, one normal stress and a shearing stress which itself can be resolved into two

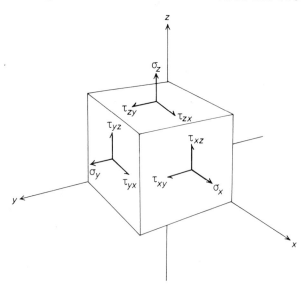

Figure 2-2

The stress components acting on the faces of a cube.

components parallel to the directions of two of the coordinates. There are nine quantities in all acting on the faces of the cube; these are known as the *stress components*:

$$\begin{matrix} \sigma_x & \tau_{xy} & \tau_{xz} \\ \tau_{yx} & \sigma_y & \tau_{yz} \\ \tau_{zx} & \tau_{zy} & \sigma_z \end{matrix}$$

If they are all known, they completely define the state of stress in the elemental cube.

Great variations in notation for the shearing stress components occur in the many texts on stress analysis. The one used here is most common; the first suffix refers to the normal to the plane on which the shear stress acts and the second suffix refers to the direction of shear on this plane.

Although nine stress components are derived in this way, not all of these are independent quantities. Since the elemental cube is in equilibrium, the forces must balance and there is no rotational motion. Consider first the stresses which are acting to cause the cube to rotate about a central axis parallel to the z axis. There are two which do this, the shearing stresses τ_{xy} and τ_{yx}. If we take moments (stress × area on which it acts × length of arm) about the z axis and equate them to zero, then

$$\tau_{xy} \times a^2 \times \frac{a}{2} - \tau_{yx} \times a^2 \times \frac{a}{2} = 0$$

where a is the length of the side of the cube; therefore,

$$\tau_{xy} = \tau_{yx} \tag{2-1}$$

Similarly, by taking moments about the x and y axes, $\tau_{yz} = \tau_{zy}$ and $\tau_{xz} = \tau_{zx}$. Therefore, of the nine stress components only six are independent, and any stress system and state of stress at a point can be expressed in terms of the *six independent stress parameters*:

$$\begin{matrix} \sigma_x & \tau_{xy} & \tau_{zx} \\ \tau_{xy} & \sigma_y & \tau_{yz} \\ \tau_{zx} & \tau_{yz} & \sigma_z \end{matrix}$$

Before a general analysis of stress is developed, it is initially instructive to carry out a simplified analysis where the stresses parallel to one of the coordinate axes (say z) are of zero value and where only the relationships of the stress components which are independent of z are determined. This analysis of "plane stress" is only applicable to two-dimensional problems but it takes the same form as that used for the general analysis of stress and as such forms a simple introduction to the more complicated general case. It must be emphasized, however, that the state of stress is always a three-dimensional quantity and there is really no such thing as "two-dimensional" stress.

2-3   ANALYSIS OF PLANE STRESS

In plane stress where the relationships of only those parameters independent of z are investigated (Fig. 2-3), there are four stress components:

$$\sigma_x \qquad \tau_{yx}$$

$$\tau_{xy} \qquad \sigma_y$$

of which, from (2-1), only three are independent ($\tau_{yx} = \tau_{xy}$). Let us investigate the nature of the stresses which act on a surface AB of unit size whose normal is inclined at an angle ϕ to the x axis. In general, these stresses can be resolved into a shear stress τ and a normal stress σ. If the angle ϕ varies, then the values of τ and σ will also vary. At certain values of ϕ the shear stress will be zero, and the stresses which act on the surface AB will be a single direct stress perpendicular to the surface. At this position, if Sx and Sy are the components of stress parallel to the x and y axes, respectively, then

$$Sx = \sigma \cos \phi \qquad Sy = \sigma \sin \phi \tag{2-2}$$

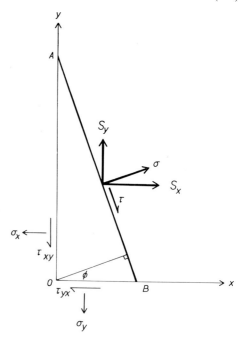

Figure 2-3

The stresses σ and τ which act on any face AB in plane-stress conditions.

The stresses Sx and Sy can themselves be resolved into the three stress components

$$Sx = \sigma_x \cos \phi + \tau_{xy} \sin \phi \qquad Sy = \tau_{xy} \cos \phi + \sigma_y \sin \phi \qquad (2\text{-}3)$$

Substituting (2-2) in (2-3)

$$0 = (\sigma_x - \sigma) \cos \phi + \tau_{xy} \sin \phi \qquad (2\text{-}4a)$$

$$0 = \tau_{xy} \cos \phi + (\sigma_y - \sigma) \sin \phi \qquad (2\text{-}4b)$$

By eliminating σ from these functions an equation can be obtained which gives values of ϕ determining the orientation of planes of no shearing stress. Dividing (2-4a) by $\cos \phi$ and (2-4b) by $\sin \phi$, we get

$$0 = \sigma_x - \sigma + \tau_{xy} \tan \phi$$

$$0 = \frac{\tau_{xy}}{\tan \phi} + \sigma_y - \sigma$$

Therefore

$$\sigma_x + \tau_{xy} \tan \phi = \sigma_y + \frac{\tau_{xy}}{\tan \phi}$$

$$\tan^2 \phi + \tan \phi \, \frac{\sigma_x - \sigma_y}{\tau_{xy}} - 1 = 0 \qquad (2\text{-}5)$$

As this is a quadratic in $\tan \phi$, there are two directions of AB making angles of ϕ_1 and ϕ_2 with the y axis on which shear stresses are zero. Because the product of the two roots $\tan \phi_1$, $\tan \phi_2$ of this equation is -1, the two directions are at right angles.

The normal stresses which act on these two planes are known as the *principal stresses* σ_1 and σ_2; and the directions in which they act, as the *principal directions*. Their values can be found by solving (2-4a) and (2-4b) for σ.

2-4 VALUES OF THE PRINCIPAL STRESSES

Solving (2-4) for σ by eliminating ϕ from (2-4a)

$$\tan \phi = \frac{-(\sigma_x - \sigma)}{\tau_{xy}}$$

and from (2-4b)

$$\tan \phi = \frac{-\tau_{xy}}{\sigma_y - \sigma}$$

Therefore

$$\tau_{xy}^2 = \sigma^2 - \sigma(\sigma_x + \sigma_y) + \sigma_x \sigma_y$$

$$0 = \sigma^2 - \sigma(\sigma_x + \sigma_y) + \sigma_x \sigma_y - \tau_{xy}^2 \qquad (2\text{-}6)$$

The two roots of this equation give the values of the two principal stresses in terms of the stress components referred to any perpendicular coordinate axis.

$$2\sigma_1 = \sigma_x + \sigma_y + [(\sigma_x - \sigma_y)^2 + 4\tau_{xy}^2]^{\frac{1}{2}}$$

$$2\sigma_2 = \sigma_x + \sigma_y - [(\sigma_x - \sigma_y)^2 + 4\tau_{xy}^2]^{\frac{1}{2}}$$

$$(2\text{-}7)$$

2-5 INVARIANTS IN PLANE STRESS

From the theory of equations it is known that the sum and product of the roots of the equation $ax^2 + bx + c = 0$ are b/a and c/a, respectively; therefore,

$$\sigma_1 + \sigma_2 = \sigma_x + \sigma_y$$

$$\sigma_1\sigma_2 = \sigma_x\sigma_y - \tau_{xy}^2$$

In describing the stress system we made no special choice of the x and y coordinate axes. Had we chosen rectangular coordinate axes u and v (Fig. 2-4) with different orientations to describe the same stress system, we should have obtained a similar quadratic equation in σ

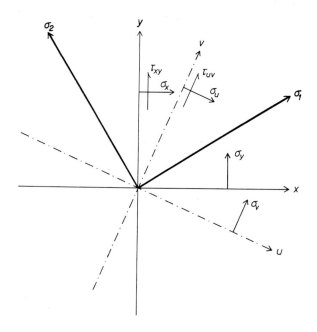

Figure 2-4

The principal stresses σ_1 and σ_2 referred to xy and uv coordinate systems.

$$0 = \sigma^2 - \sigma(\sigma_u + \sigma_v) + \sigma_u\sigma_v - \tau_{uv}^{2}$$

The roots of this must also be σ_1 and σ_2, the principal stresses, and therefore

$$\sigma_1 + \sigma_2 = \sigma_x + \sigma_y = \sigma_u + \sigma_v = I_1 \qquad (2\text{-}8)$$

$$\sigma_1\sigma_2 = \sigma_x\sigma_y - \tau_{xy}^{2} = \sigma_u\sigma_v - \tau_{uv}^{2} = I_2 \qquad (2\text{-}9)$$

The stress system is an entity which is not dependent on the reference axes used to describe it. Such mathematical entities which have properties that are independent of the coordinate frame used to describe them are known as *tensors*. In equations which describe a tensor quantity within a particular reference frame, there are certain components of the equation which always have constant values (I_1 and I_2 above) known as *invariants*. The *invariants of the stress tensor* are important in certain theories of rock failure and yield and their significance will be discussed in more detail when the general theory of stress has been developed.

2-6 STRESSES ACTING ON A PLANE WHOSE NORMAL MAKES AN ANGLE ϕ WITH THE PRINCIPAL AXIS

Equations (2-5) and (2-7) have established that any plane-stress state can be resolved into two perpendicular principal normal stresses. To determine the state of stress on any plane we first rotate the coordinate axes x and y

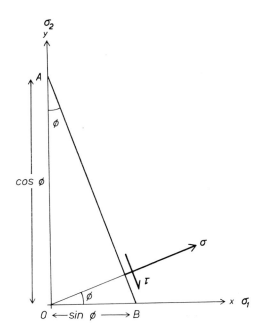

Figure 2-5

The stresses σ and τ which act on any face AB inclined at an angle ϕ to the principal axes of stress in plane-stress conditions.

so that they coincide with the directions of two principal normal stresses σ_1 and σ_2, respectively. We can now determine fairly simply the normal stress σ and the shear stress τ which act on any plane whose normal is inclined at an angle ϕ to the principal axis of stress σ_1 (x axis).

If AB (Fig. 2-5) is of unit length, then $AO = \cos \phi$ and $OB = \sin \phi$, and resolving from normal stresses σ_1 and σ_2 on to AB, we find that

$$\sigma = \cos \phi(\sigma_1 \cos \phi) + \sin \phi(\sigma_2 \sin \phi)$$

$$= \sigma_1 \cos^2 \phi + \sigma_2 \sin^2 \phi \tag{2-10}$$

This can be put into terms of an angle 2ϕ by replacing $\cos^2 \phi$ by $(1 + \cos 2\phi)/2$ and $\sin^2 \phi$ by $(1 - \cos 2\phi)/2$, and we obtain

$$\sigma = \frac{\sigma_1 + \sigma_2}{2} + \frac{\sigma_1 - \sigma_2}{2} \cos 2\phi \tag{2-11}$$

Similarly, resolving the same principal stresses into shearing components along AB, remembering that the sense of shear from stress σ_2 is opposite in sign to that from σ_1,

$$\tau = \cos \phi(\sigma_1 \sin \phi) - \sin \phi(\sigma_2 \cos \phi)$$

$$= (\sigma_1 - \sigma_2) \sin \phi \cos \phi \tag{2-12}$$

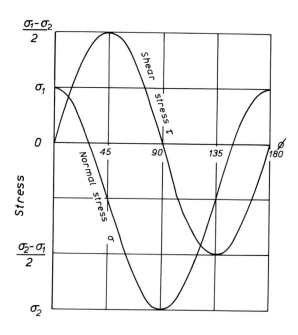

Figure 2-6

Variations of normal stress σ and shear stress τ with changes in ϕ.

or using double angles,

$$\tau = \frac{\sigma_1 - \sigma_2}{2} \sin 2\phi \qquad (2\text{-}13)$$

The variations in σ and τ with change in angle ϕ are most simply expressed graphically (Fig. 2-6). The maximum shearing stresses of $\pm(\sigma_1 - \sigma_2)/2$ are always developed on planes which make an angle of $45°$ with the principal axes of stress no matter what values are taken by these principal stresses. This fact is of importance in the theories of rock failure by mechanisms of shearing and faulting. The position of the planes where the normal stresses are zero does vary with the absolute values of the principal stresses, they occur where $\cos 2\phi = (\sigma_1 + \sigma_2)/(\sigma_2 - \sigma_1)$. As far as is known, the planes which have no normal stresses acting across them are of no significance in controlling or localizing the actual fractures which develop in deformed substances.

2-7 STRESS IN THREE DIMENSIONS

In the previous section on plane stress it was found that no matter what forces are applied to a body, there always exist two mutually perpendicular directions where the shear stresses are zero. We will now investigate the stress relationships in three dimensions to see if this principle can be extended and to see whether there are three mutually perpendicular directions across which there is no shearing stress.

First it will be necessary to describe the angular orientation of a plane in terms of three cartesian coordinates x, y and z. This is most conveniently done by referring to what are known as the three *direction cosines l, m, and n* of the normal to the plane. If, in Fig. 2-7, P is the point on ABC through which the normal to the plane through the origin O passes, then the direction cosines are defined as

$$\cos POx = l$$

$$\cos POy = m$$

$$\cos POz = n$$

If P has coordinates x, y, and z, then $x = lOP$, $y = mOP$, and $z = nOP$ and it follows by Pythagoras that $x^2 + y^2 + z^2 = OP^2$

or
$$l^2 + m^2 + n^2 = 1 \qquad (2\text{-}14)$$

Another important property of direction cosines is that if two lines have orientations (l_1, m_1, n_1) and (l_2, m_2, n_2) respectively, the angle between them is given by

$$\cos \alpha = l_1 l_2 + m_1 m_2 + n_1 n_2 \qquad (2\text{-}15)$$

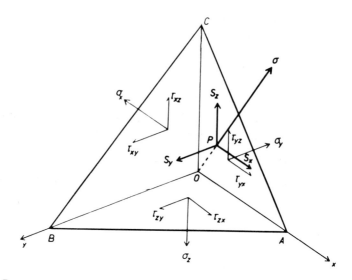

Figure 2-7

The stresses Sx, Sy, and Sz acting on unit area ABC where the normal stress σ is a principal stress.

and that if these lines are perpendicular then

$$0 = l_1 l_2 + m_1 m_2 + n_1 n_2 \tag{2-16}$$

We will now proceed to discuss the stresses that act on the plane ABC in terms of the six stress components. Let us suppose that the orientation of the plane is varied until at one position of ABC oblique to the x, y, and z axes no shearing stresses act and the stress consists only of the normal stress. Such a plane will be a principal plane, and its normal will be a principal axis of stress. Because there are no shearing stresses, the normal stress σ can be resolved into three stress components S_x, S_y, and S_z parallel to the coordinates such that

$$S_x = \sigma l$$
$$S_y = \sigma m \tag{2-17}$$
$$S_z = \sigma n$$

These components can also be expressed in terms of direction cosines and the main stress components

$$S_x = l\sigma_x + m\tau_{yx} + n\tau_{zx}$$
$$S_y = l\tau_{xy} + m\sigma_y + n\tau_{zy} \tag{2-18}$$
$$S_z = l\tau_{xz} + m\tau_{yz} + n\sigma_z$$

Combining (2-17) and (2-18),

$$l(\sigma_x - \sigma) + m\tau_{yx} + n\tau_{zx} = 0 \qquad (2\text{-}19a)$$

$$l\tau_{xy} + m(\sigma_y - \sigma) + n\tau_{zy} = 0 \qquad (2\text{-}19b)$$

$$l\tau_{xz} + m\tau_{yz} + n(\sigma_z - \sigma) = 0 \qquad (2\text{-}19c)$$

By eliminating l, m, and n we can obtain values for σ, the principal stresses. Multiplying (2-19a) by τ_{zy} and (2-19b) by τ_{zx} and subtracting, we get

$$l(\sigma_x\tau_{zy} - \sigma\tau_{zy} - \tau_{xy}\tau_{zx}) + m(\tau_{yx}\tau_{zy} - \sigma_y\tau_{zx} + \sigma\tau_{zx}) = 0 \qquad (2\text{-}20)$$

Multiplying (2-19b) by $(\sigma_z - \sigma)$ and (2-19c) by τ_{zy} and subtracting, we get

$$l(\tau_{xy}\sigma_z - \tau_{xy}\sigma - \tau_{zy}\tau_{xz}) + m(\sigma_y\sigma_z - \sigma_y\sigma - \sigma_z\sigma + \sigma^2 - \tau_{yz}\tau_{zy}) = 0 \quad (2\text{-}21)$$

Rearranging and using $\tau_{xy} = \tau_{yx}$, etc.,

$$\frac{l}{m} = \frac{\tau_{xy}\tau_{yz} - \sigma_y\tau_{zx} + \sigma\tau_{zx}}{\sigma\tau_{yz} - \sigma_x\tau_{yz} + \tau_{xy}\tau_{zx}} = \frac{\sigma_y\sigma_z - \sigma_y\sigma - \sigma_z\sigma + \sigma^2 - \tau_{yz}^2}{\sigma\tau_{xy} + \tau_{zx}\tau_{yz} - \tau_{xy}\sigma_z}$$

Cross multiplying and collecting terms, we obtain a cubic equation for σ

$$\sigma^3 - (\sigma_x + \sigma_y + \sigma_z)\sigma^2 + (\sigma_x\sigma_y + \sigma_y\sigma_z + \sigma_z\sigma_x - \tau_{xy}^2 - \tau_{yz}^2 - \tau_{zx}^2)\sigma$$

$$- (\sigma_x\sigma_y\sigma_z + 2\tau_{xy}\tau_{yz}\tau_{zx} - \sigma_x\tau_{yz}^2 - \sigma_y\tau_{zx}^2 - \sigma_z\tau_{xy}^2) = 0 \quad (2\text{-}22)$$

For all practical values of the six stress components, all three roots of this equation are real and are the three principal stresses $\sigma_1 \geqslant \sigma_2 \geqslant \sigma_3$. These principal stresses are proved mutually perpendicular in the following way. First Eqs. (2-19) are rewritten in terms of the principal stress σ_1 and its direction cosines l_1, m_1, and n_1; then multiplying (2-19a) by l_2, (2-19b) by m_2, and (2-19c) by n_2 (l_2, m_2, and n_2 are the directions of the principal stress σ_2)

$$l_1l_2(\sigma_x - \sigma_1) + m_1l_2\tau_{xy} + n_1l_2\tau_{zx} = 0$$

$$l_1m_2\tau_{xy} + m_1m_2(\sigma_y - \sigma_1) + n_1m_2\tau_{yz} = 0 \qquad (2\text{-}23)$$

$$l_1n_2\tau_{zx} + m_1n_2\tau_{yz} + n_1n_2(\sigma_z - \sigma_1) = 0$$

Rewriting (2-19) in terms of σ_2 and its direction cosines l_2, m_2, and n_2, multiplying (2-19a) by l_1, (2-19b) by m_1, and (2-19c) by n_1, we get

$$l_1l_2(\sigma_x - \sigma_2) + l_1m_2\tau_{xy} + l_1n_2\tau_{zx} = 0$$

$$m_1l_2\tau_{xy} + m_1m_2(\sigma_y - \sigma_2) + m_1n_2\tau_{yz} = 0 \qquad (2\text{-}24)$$

$$n_1l_2\tau_{zx} + n_1m_2\tau_{yz} + n_1n_2(\sigma_z - \sigma_2) = 0$$

Subtracting the sum of Eqs. (2-23) from the sum of Eqs. (2-24), we find that

$$(\sigma_1 - \sigma_2)(l_1l_2 + m_1m_2 + n_1n_2) = 0 \qquad (2\text{-}25)$$

Because Eq. (2-25) is true for all stress states and because σ_1 is not generally equal to σ_2, then from (2-16) the lines (l_1, m_1, n_1) and (l_2, m_2, n_2) must be perpendicular. Similarly it can be proved that $l_2 l_3 + m_2 m_3 + n_2 n_3 = 0$ and that the three principal stresses are therefore mutually perpendicular.

2-8   ORIENTATION OF THE PRINCIPAL AXES OF STRESS

From Eqs. (2-19) we can obtain a value for l_1/n_1 in the same way as Eq. (2-20) defines l_1/m_1.

$$\frac{l_1}{m_1} = \frac{\sigma_y \sigma_z - \sigma_y \sigma_1 - \sigma_1 \sigma_z + \sigma_1^2 - \tau_{yz}^2}{\sigma_1 \tau_{xy} + \tau_{zx} \tau_{yz} - \tau_{xy} \sigma_z}$$

$$\frac{l_1}{n_1} = \frac{\sigma_y \sigma_z - \sigma_y \sigma_1 - \sigma_1 \sigma_z + \sigma_1^2 - \tau_{yz}^2}{\sigma_1 \tau_{zx} + \tau_{xy} \tau_{yz} - \tau_{zx} \sigma_y} \qquad (2\text{-}26)$$

It therefore follows that

$$\frac{l_1}{\sigma_y \sigma_z - \sigma_y \sigma_1 - \sigma_1 \sigma_z + \sigma_1^2 - \tau_{yz}^2} = \frac{m_1}{\sigma_1 \tau_{xy} + \tau_{zx} \tau_{yz} - \tau_{xy} \sigma_z}$$

$$= \frac{n_1}{\sigma_1 \tau_{zx} + \tau_{xy} \tau_{yz} - \tau_{zx} \sigma_y} \qquad (2\text{-}27)$$

By substituting the values of the stress components and σ_1 in these equations together with the relationship $l_1^2 + m_1^2 + n_1^2 = 1$ the actual values of l_1, m_1 and n_1 may be determined.

2-9   STRESS INVARIANTS IN THREE DIMENSIONS

The cubic equation (2-22) which gives the three principal stresses is of fundamental importance in stress analysis. We have seen that its three roots are the three principal stresses, σ_1, σ_2, and σ_3, and it therefore follows that whatever choice of the coordinates x, y, and z we make to describe the stress system, this equation and its roots must remain constant. This means that the coefficients of the terms in σ^2, σ, together with the constant must always have a fixed value no matter what the orientation of the coordinate axes, and these coefficients are known as the *three stress invariants*. The cubic equation whose roots are σ_1, σ_2, and σ_3 has the form

$$(\sigma - \sigma_1)(\sigma - \sigma_2)(\sigma - \sigma_3) = 0$$

or $$\sigma^3 - (\sigma_1 + \sigma_2 + \sigma_3)\sigma^2 + (\sigma_1 \sigma_2 + \sigma_2 \sigma_3 + \sigma_3 \sigma_1) - \sigma_1 \sigma_2 \sigma_3 = 0 \qquad (2\text{-}28)$$

By comparing the coefficients of (2-22) with (2-28), we obtain the stress invariants

$$\sigma_1 + \sigma_2 + \sigma_3 = \sigma_x + \sigma_y + \sigma_z = I_1 \tag{2-29}$$

$$\sigma_1\sigma_2 + \sigma_2\sigma_3 + \sigma_3\sigma_1 = \sigma_x\sigma_y + \sigma_y\sigma_z + \sigma_z\sigma_x - \tau_{xy}^2 - \tau_{yz}^2 - \tau_{zx}^2$$
$$= I_2 \tag{2-30}$$

$$\sigma_1\sigma_2\sigma_3 = \sigma_x\sigma_y\sigma_z + 2\tau_{xy}\tau_{yz}\tau_{zx} - \sigma_x\tau_{yx}^2 - \sigma_y\tau_{zx}^2 - \sigma_z\tau_{xy}^2 = I_3 \tag{2-31}$$

These invariants are of great importance in the theories of rock yield and plastic flow. Their determination is also a first step in defining the principal stress axes from measurements of stress made along any three perpendicular directions. The method for calculating the principal stresses from six known stress components depends first on the calculation of the invariants I_1, I_2, and I_3 to establish the cubic equation $\sigma^3 - I_1\sigma^2 + I_2\sigma - I_3 = 0$. This equation is most easily solved in practice by first finding one of its roots (the intermediate principal stress σ_2) by trial and error, knowing that it lies between the turning points of the cubic and occurs where $3\sigma^2 - 2I_1\sigma - I_2 = 0$, that is, between $\sigma = [I_1 + (I_1^2 - 3I_2)^{\frac{1}{2}}]/3$, and $\sigma = [I_1 - (I_1^2 - 3I_2)^{\frac{1}{2}}]/3$. When this root is found with sufficient accuracy, the cubic is divided by $(\sigma - \sigma_2)$, and σ_1 and σ_3 are then determined from the roots of the quadratic equation

$$\sigma^2 + (\sigma_2 - I_1)\sigma + (\sigma_2^2 - I_1\sigma_2 - I_2) = 0 \tag{2-32}$$

2-10 STRESSES ACTING ON ANY PLANAR SURFACE

To determine the stresses which act on any plane of known orientation we first rotate the coordinate axes x, y, and z into directions which correspond to the mutually perpendicular axes of principal stress σ_1, σ_2, and σ_3, respectively. Consider now any plane ABC of unit area which is oblique to these axes (Fig. 2-8B) and has a normal with direction cosines l, m, and n. The normal stress which acts on this plane is σ, and the shearing stress τ. If the three components of stress which act parallel to the axes x, y, and z are S_x, S_y, and S_z, then

$$S_x = \sigma_1 l$$
$$S_y = \sigma_2 m \tag{2-33}$$
$$S_z = \sigma_3 n$$

and the normal stress σ on ABC can be found by resolving these components.

$$\sigma = \sigma_1 l^2 + \sigma_2 m^2 + \sigma_3 n^2 \tag{2-34}$$

Having obtained the normal stress in terms of the principal stresses, we now discover the shearing stress τ on ABC. The total stress S acting on ABC

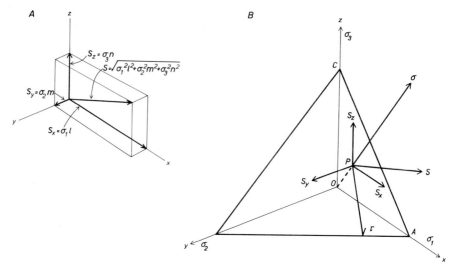

Figure 2-8

The normal stress σ and shear stress τ which act on unit area ABC inclined to the three principal axes of stress. Part A illustrates the relationships between the total stress S acting on ABC and the principal stresses σ_1, σ_2, and σ_3.

(Fig. 2-8*A*) is given by

$$S^2 = \sigma_1^2 l^2 + \sigma_2^2 m^2 + \sigma_3^2 n^2 \tag{2-35}$$

We know that for the general resolution of stresses on plane *ABC* the total stress *S* is compounded of a normal stress and a shearing stress with the simple relationship

$$S^2 = \sigma^2 + \tau^2 \tag{2-36}$$

As we have already determined σ (2-34), we can discover the shearing stress

$$\tau^2 = S^2 - \sigma^2$$

$$= \sigma_1^2 l^2 + \sigma_2^2 m^2 + \sigma_3^2 n^2 - (\sigma_1 l^2 + \sigma_2 m^2 + \sigma_3 n^2)^2 \tag{2-37}$$

Multiplying out, putting $1 - l^2 = m^2 + n^2$, $1 - m^2 = l^2 + n^2$ and $1 - n^2 = l^2 + m^2$, and collecting terms,

$$\tau^2 = (\sigma_1 - \sigma_2)^2 l^2 m^2 + (\sigma_2 - \sigma_3)^2 m^2 n^2 + (\sigma_3 - \sigma_1)^2 n^2 l^2 \tag{2-38}$$

2-11   PLANES OF MAXIMUM SHEARING STRESS

The planes of maximum shearing stress are determined by finding the maximum values of (2-38), and they occur at the stationary values of the function.

To discover these values we must first rearrange (2-38) to eliminate the dependent variable n, $(n^2 = 1 - l^2 - m^2)$

$$\tau^2 = (\sigma_1 - \sigma_2)^2 l^2 m^2 + (\sigma_2 - \sigma_3)^2(1 - l^2 - m^2)m^2$$
$$+ (\sigma_3 - \sigma_1)^2(1 - l^2 - m^2)l^2 \quad (2\text{-}39)$$

Differentiating with respect to l, we find that

$$2\tau \frac{d\tau}{dl} = 2lm^2(\sigma_1 - \sigma_2)^2 - 2lm^2(\sigma_2 - \sigma_3)^2 + 2l(\sigma_3 - \sigma_1)^2(1 - l^2 - m^2)$$

or
$$\tau \frac{d\tau}{dl} = l(\sigma_1 - \sigma_3)[2m^2(\sigma_3 - \sigma_2) + (1 - 2l^2)(\sigma_1 - \sigma_3)] \quad (2\text{-}40)$$

Similarly, differentiating with respect to the other variable m,

$$\tau \frac{d\tau}{dm} = m(\sigma_2 - \sigma_3)[2l^2(\sigma_3 - \sigma_1) + (1 - 2m^2)(\sigma_2 - \sigma_3)] \quad (2\text{-}41)$$

The stationary values are found where the conditions of (2-40) and (2-41) are simultaneously zero. From (2-40) it is seen that $l = 0$ would satisfy a zero value of $d\tau/dm$. Putting this value in (2-41), we obtain the condition

$$m(1 - 2m^2)(\sigma_2 - \sigma_3)^2 = 0$$

Because $(\sigma_2 - \sigma_3)$ is not generally equal to 0 we have the condition that $m = 0$, or $m = \pm 1/\sqrt{2}$. Using these values in $l^2 + m^2 + n^2 = 1$, we obtain the direction cosines of the three stationary values:

(1) $l = 0$ $m = 0$ $n = 1$

(2) $l = 0$ $m = \dfrac{\pm 1}{\sqrt{2}}$ $n = \dfrac{\pm 1}{\sqrt{2}}$

(3) $l = 0$ $m = \dfrac{\pm 1}{\sqrt{2}}$ $n = \dfrac{\mp 1}{\sqrt{2}}$

By using the same reasoning on the pair of equations similar to those of (2-40) and (2-41) that can be derived from (2-39) by eliminating m, and on another pair by eliminating l, we obtain six other stationary values:

(4) $l = 1$ $m = 0$ $n = 0$

(5) $l = \dfrac{\pm 1}{\sqrt{2}}$ $m = 0$ $n = \dfrac{\pm 1}{\sqrt{2}}$

(6) $l = \dfrac{\pm 1}{\sqrt{2}}$ $m = 0$ $n = \dfrac{\mp 1}{\sqrt{2}}$

(7) $l = 0$ $m = 1$ $n = 0$

(8) $l = \dfrac{\pm 1}{\sqrt{2}}$ $m = \dfrac{\pm 1}{\sqrt{2}}$ $n = 0$

(9) $l = \dfrac{\pm 1}{\sqrt{2}}$ $m = \dfrac{\mp 1}{\sqrt{2}}$ $n = 0$

Three of these stationary values, (1), (4), and (7), refer to principal planes and these must represent minimum values of the shearing stresses. The others all refer to stationary values which are maximum values of shearing stress, and these planes are all oriented so that they contain one of the principal axes of stress and make an angle of 45° with the other two axes (Fig. 2-9).

The values of the shearing stresses on these planes are found by substituting

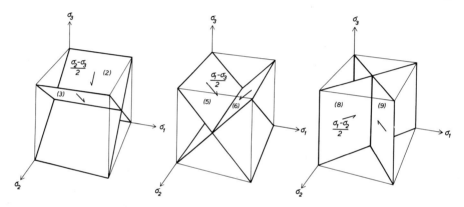

Figure 2-9

The planes of maximum shearing stress.

the known direction cosines in (2-38).

$$\text{(2) and (3) give maximum } \tau = \pm \frac{\sigma_2 - \sigma_3}{2}$$

$$\text{(5) and (6) give maximum } \tau = \pm \frac{\sigma_1 - \sigma_3}{2}$$

$$\text{(8) and (9) give maximum } \tau = \pm \frac{\sigma_1 - \sigma_2}{2}$$

Therefore the greatest shearing stress always occurs on the planes which contain the σ_2 axis and make an angle of 45° to the principal stresses σ_1 and σ_3 *irrespective of the signs or values of the principal stresses.*

2-12   HYDROSTATIC STRESS

Where the principal stresses are equal it will be seen from (2-34) that the normal stresses that act on any plane surface are equal, and also from (2-38) that no shearing stresses exist in the material. This state of stress is known as *hydrostatic stress*, and its effect is to cause only a change in volume, a change which may be elastic (that is recoverable on removal of the stress system) or it may be permanent dilation.

2-13   DEVIATORIC STRESS

In any stress system with principal stresses σ_1, σ_2, and σ_3, we can define a *mean stress* $\bar{\sigma}$

$$\bar{\sigma} = \frac{\sigma_1 + \sigma_2 + \sigma_3}{3} = \frac{I_1}{3} \tag{2-42}$$

This can be thought of as the hydrostatic part of the stress system and as such causes only volume changes in the material. The normal stresses σ which act across any plane can be divided into (1) a hydrostatic part, and (2) another part σ' which departs or deviates from the first part.

$$\sigma' = \sigma - \bar{\sigma} = \sigma - \left(\frac{\sigma_1 + \sigma_2 + \sigma_3}{3}\right) \tag{2-43}$$

This second component is known as a *stress deviator* or a *deviatoric stress*. Along the three principal axes we have *three principal deviatoric stresses* σ'_1, σ'_2, and σ'_3.

$$\sigma'_1 = \sigma_1 - \bar{\sigma} = \frac{2\sigma_1 - \sigma_2 - \sigma_3}{3}$$

$$\sigma'_2 = \sigma_2 - \bar{\sigma} = \frac{2\sigma_2 - \sigma_1 - \sigma_3}{3} \tag{2-44}$$

$$\sigma'_3 = \sigma_3 - \bar{\sigma} = \frac{2\sigma_3 - \sigma_1 - \sigma_2}{3}$$

The effect of these deviatoric stresses is to produce distortion (as distinct from the dilation produced by the hydrostatic part of the stress system), a strain which may be elastic and recoverable or plastic and permanent.

Any stress system can therefore be divided into two parts in the manner shown diagrammatically in Fig. 2-10. Likewise, the effects of the stress system on a material substance can be divided into two parts: one, dilation and the other, distortions which can be separated from the volume change.

It is found that the hydrostatic part of a stress system seems to exert no influence on the actual initiation of yield and it may be because we can add

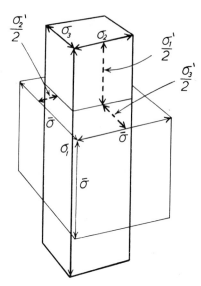

Figure 2-10

Diagrammatic representation of the deviatoric stresses σ'_1, σ'_2, and σ'_3 and hydrostatic stress $\bar{\sigma}$ in terms of the principal stresses σ_1, σ_2, and σ_3.

or subtract a hydrostatic part to any stress system without altering the values of shearing stress. This can be proved very easily by replacing σ_1, σ_2, and σ_3 in Eq. (2-38) by $\sigma_1 + \bar{\sigma}$, $\sigma_2 + \bar{\sigma}$, and $\sigma_3 + \bar{\sigma}$. The actual initiation of plastic yield is related to two factors: first, some material constant depending on the nature of the substance, and second, some function of an invariant part of the stress tensor, yet not the invariant I_3 as this is connected with the hydrostatic part of the stress.

The principal axes of deviatoric stress coincide with those for total stress. The cubic equation (2-22) which gives the three principal stresses in terms of six components of stress can be modified using (2-43) and (2-44) so as to express the three principal deviatoric stresses in these components:

$$\sigma'^3 - \frac{1}{6}\left[(\sigma_x - \sigma_y)^2 + (\sigma_y - \sigma_z)^2 + (\sigma_z - \sigma_x)^2 + 6\tau_{xy}^2 + 6\tau_{yz}^2 + 6\tau_{zx}^2\right]\sigma'$$

$$- \frac{2}{27}(\sigma_x^3 + \sigma_y^3 + \sigma_z^3)$$

$$+ \frac{1}{9}(\sigma_x\sigma_y^2 + \sigma_x\sigma_z^2 + \sigma_y\sigma_x^2 + \sigma_y\sigma_z^2 + \sigma_z\sigma_x^2 + \sigma_z\sigma_y^2 - 4\sigma_x\sigma_y\sigma_z)$$

$$- \frac{\tau_{xy}^2}{3}(\sigma_x + \sigma_y - 2\sigma_z) - \frac{\tau_{yz}^2}{3}(\sigma_y + \sigma_z - 2\sigma_x)$$

$$- \frac{\tau_{zx}^2}{3}(\sigma_z + \sigma_x - 2\sigma_y) - 2\tau_{xy}\tau_{yz}\tau_{zx} = 0 \quad (2\text{-}45)$$

By rearranging the total stress components into components of deviatoric stress using (2-43), we obtain a simpler equation for the principal deviatoric stresses.

$$\sigma'^3 + (\sigma'_x\sigma'_y + \sigma'_y\sigma'_z + \sigma'_z\sigma'_x - \tau_{xy}{}^2 - \tau_{yz}{}^2 - \tau_{zx}{}^2)\sigma'$$
$$- (\sigma'_x\sigma'_y\sigma'_z + 2\tau_{xy}\tau_{yz}\tau_{zx} - \sigma'_x\tau_{yz}{}^2 - \sigma'_y\tau_{zx}{}^2 - \sigma'_z\tau_{xy}{}^2) = 0 \quad (2\text{-}46)$$

Equations (2-45) and (2-46) are in the form

$$\sigma'^3 + I'_2\sigma' - I'_3 = 0 \qquad\qquad (2\text{-}47)$$

where I'_2 and I'_3 are the *deviatoric stress invariants*. As the coefficient of σ'^2 is zero, $I'_1 = 0$. By comparing Eqs. (2-45) and (2-22) it will be seen that these invariants are not completely independent but are related to the invariants I_1, I_2, and I_3 of the total stress system.

$$I'_2 = 3\bar\sigma^2 - I_2 \quad \text{or} \quad I'_2 = \frac{I_1^2}{3} - I_2 \qquad\qquad (2\text{-}48)$$

and
$$I'_3 = I_3 - I_2\bar\sigma + 2\bar\sigma^3 \quad \text{or} \quad I'_3 = I_3 - \frac{I_1 I_2}{3} - \frac{I_1^3}{27} \qquad (2\text{-}49)$$

2-14   VARIATION IN STRESS THROUGH A BODY

In Sec. 2-2 it was shown that the state of stress at any point (x,y,z) may be described in terms of nine components of which only six are independent if the body is in equilibrium.

$$\sigma_x \; \tau_{yx} \; \tau_{zx} \qquad \text{acting in the } x \text{ direction}$$
$$\tau_{xy} \; \sigma_y \; \tau_{zy} \qquad \text{acting in the } y \text{ direction}$$
$$\tau_{xz} \; \tau_{yz} \; \sigma_z \qquad \text{acting in the } z \text{ direction}$$

In general the state of stress varies through a body. These components will vary from point to point (Fig. 2-11), and at another position in the body, with coordinates $(x + \delta x, y + \delta y, z + \delta z)$, the stress components will be given by

$$(\sigma_x + \delta\sigma_x) \; (\tau_{yx} + \delta\tau_{yx}) \; (\tau_{zx} + \delta\tau_{zx}) \qquad \text{acting in the } x \text{ direction}$$
$$(\tau_{xy} + \delta\tau_{xy}) \; (\sigma_y + \delta\sigma_y) \; (\tau_{zy} + \delta\tau_{zy}) \qquad \text{acting in the } y \text{ direction}$$
$$(\tau_{xz} + \delta\tau_{xz}) \; (\tau_{yz} + \delta\tau_{yz}) \; (\sigma_z + \delta\sigma_z) \qquad \text{acting in the } z \text{ direction}$$

If δx, δy, and δz are small, then these changes in stress $\delta\sigma_x$, $\delta\tau_{yx}$, . . ., etc., may be considered as linear variations which depend on the rates of change of the stress, that is,

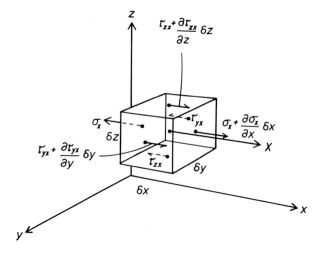

Figure 2-11

Variations in stress state from the point (x,y,z) to the point (x + δx, y + δy, z + δz).

$$\delta\sigma_x = \frac{\partial\sigma_x}{\partial x}\, \partial x \tag{2-50}$$

$$\delta\tau_{zx} = \frac{\partial\tau_{zx}}{\partial z}\, \delta z \tag{2-51}$$

Let us now determine the resultant force in the body which tends to cause it to accelerate in the x direction. If the body forces are X, Y, and Z in the x, y, and z directions, then the accelerating forces in the x direction are given by

$$\left(\sigma_x + \frac{\partial\sigma_x}{\partial x}\delta x\right)\delta y\,\delta z - \sigma_x\,\delta y\,\delta z + \left(\tau_{yx} + \frac{\partial\tau_{yx}}{\partial y}\delta y\right)\delta x\,\delta z - \tau_{yx}\delta x\,\delta z$$

$$+ \left(\tau_{zx} + \frac{\partial\tau_{zx}}{\partial z}\delta z\right)\delta x\,\delta y - \tau_{zx}\,\delta x\,\delta y + X\,\delta x\,\delta y\,\delta z = \text{accelerating force}$$

$$= (\text{acceleration} \times \text{mass})$$

$$= \frac{\partial^2 u}{\partial t^2}\rho\,\delta x\,\delta y\,\delta z$$

where ρ is the density of the material, u is the displacement in the x direction, and t is time. Dividing by the volume of the body, $\delta x\,\delta y\,\delta z$, we obtain

$$\frac{\partial\sigma_x}{\partial x} + \frac{\partial\tau_{yx}}{\partial y} + \frac{\partial\tau_{zx}}{\partial z} + X = \frac{\partial^2 u}{\partial t^2}\rho \tag{2-52a}$$

By using similar reasoning and investigating the accelerating forces in the

y and z directions, the following further equations are obtained, where v and w are the displacements in the y and z directions, respectively:

$$\frac{\partial \tau_{xy}}{\partial x} + \frac{\partial \sigma_y}{\partial y} + \frac{\partial \tau_{zy}}{\partial z} + Y = \frac{\partial^2 v}{\partial t^2} \rho \qquad (2\text{-}52b)$$

$$\frac{\partial \tau_{xz}}{\partial x} + \frac{\partial \tau_{yz}}{\partial y} + \frac{\partial \sigma_z}{\partial z} + Z = \frac{\partial^2 w}{\partial t^2} \rho \qquad (2\text{-}52c)$$

These three equations are known as the *stress equations of small motion.* If the element is in static equilibrium, then $\partial^2 u/\partial t^2 = \partial^2 v/\partial t^2 = \partial^2 w/\partial t^2 = 0$, and we obtain the *equations of stress equilibrium* which are very important in the solution of problems of stress distribution in static bodies:

$$\frac{\partial \sigma_x}{\partial x} + \frac{\partial \tau_{yx}}{\partial y} + \frac{\partial \tau_{zx}}{\partial z} + X = 0$$

$$\frac{\partial \tau_{xy}}{\partial x} + \frac{\partial \sigma_y}{\partial y} + \frac{\partial \tau_{zy}}{\partial z} + Y = 0 \qquad (2\text{-}53)$$

$$\frac{\partial \tau_{xz}}{\partial x} + \frac{\partial \tau_{yz}}{\partial y} + \frac{\partial \sigma_z}{\partial z} + Z = 0$$

The integration of these differential equations leads to the appearance of certain arbitrary constants. Before a solution to a specific problem can be found, it is essential to have states of stress at certain parts of the body specified. These essential conditions are known as *boundary conditions*; in general, three conditions are necessary on each coordinate boundary for a complete solution to be accomplished.

2-15 STRESS TRAJECTORIES

The three-dimensional variations in stress state that exist in a body are most easily envisaged by considering the way the principal axes of stress vary in orientation and in value. The orthogonal lines representing the directions of the principal stresses are known as *stress trajectories.*

In two-dimensional problems the two sets of orthogonal lines representing maximum and minimum stress directions can be very simply represented (Figs. 2-12, 2-13, 2-15). If the curvature of these lines varies so that the adjacent trajectories come closer together, then this indicates a stress concentration at this position. All points where the principal stresses have equal values are known as *isotropic points.* If these values are both zero, then the point is called a *singular point* (Figs. 2-12, 2-13). Isotropic points can be classified into two types: those with interlocking stress trajectories ($+ve$ isotropic point, Fig. 2-12) or those with noninterlocking stress trajectories ($-ve$ isotropic point, Fig. 2-13).

From the stress trajectories it is possible to construct another network of orthogonal curves representing the position of lines (planes in three dimensions) of maximum shearing stress. These two sets of curves represent lines of right-handed and left-handed maximum shearing stress. Both sets always converge toward isotropic points ($+ve$ and $-ve$) with each other and with adjacent members of the same set.

It is possible to illustrate the variations in intensity of the two principal stresses in two sets of curves, but it is often more useful to record the intensity of the stress difference since this records the value of the maximum shearing stress $(\sigma_1 - \sigma_2)/2$.

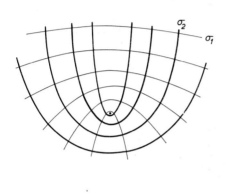

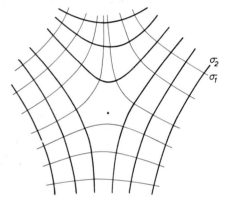

Figure 2-12

Stress trajectories with a positive isotropic point.

Figure 2-13

Stress trajectories with a negative isotropic point.

2-16   SUPERPOSITION OF STRESS SYSTEMS

The stress system that exists in the crust at any one time is generally the result of the interaction of stresses produced by the influence of gravity and supplementary tectonic stresses. Sometimes this pattern is further complicated by the superposition of two or more supplementary systems. The effect of superposing two stress fields is to change both the direction and values of the principal stresses at any point. These changes may be simply computed by determining the changes in the stress components. For example, if two stress systems have components acting at one point with reference to the same coordinate directions x, y, and z, given by

$$\begin{matrix} \sigma_{x1} & \tau_{xy1} & \tau_{xz1} \\ \tau_{yx1} & \sigma_{y1} & \tau_{yz1} \\ \tau_{zx1} & \tau_{zy1} & \sigma_{z1} \end{matrix} \quad \text{and} \quad \begin{matrix} \sigma_{x2} & \tau_{xy2} & \tau_{xz2} \\ \tau_{yx2} & \sigma_{y2} & \tau_{yz2} \\ \tau_{zx2} & \tau_{zy2} & \sigma_{z2} \end{matrix}$$

then the stress components of the combined systems are given by

$$(\sigma_{x1} + \sigma_{x2}) \quad (\tau_{xy1} + \tau_{xy2}) \quad (\tau_{xz1} + \tau_{xz2})$$

$$(\tau_{yx1} + \tau_{yx2}) \quad (\sigma_{y1} + \sigma_{y2}) \quad (\tau_{yz1} + \tau_{yz2})$$

$$(\tau_{zx1} + \tau_{zx2}) \quad (\tau_{zy1} + \tau_{zy2}) \quad (\sigma_{z1} + \sigma_{z2})$$

From these quantities the resultant principal stresses and their orientations may be calculated using Eqs. (2-22) and (2-27).

One example of the way calculated compound stress fields have been used to explain the distribution of structures results from the work of Odé (1957). This study was made on the complex pattern of dikes around a central volcanic complex in the Spanish Peaks area, Colorado (Fig. 2-14). The dikes radiate from the volcanic center. Those on the east side tend to be longer

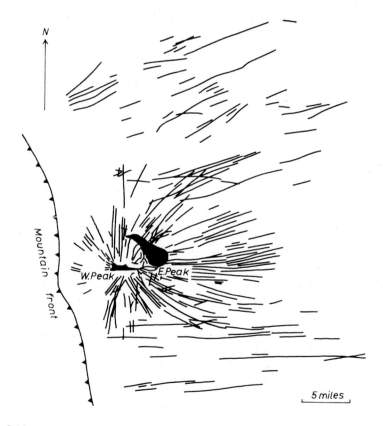

Figure 2-14

Dike pattern of the Spanish Peaks, Colorado area. (After Knopf, 1936.)

than those on the west side, and they also tend to curve toward the east. At distances of more than about 8 miles from the center, the dikes have a more regular and linear plan. Odé suggested that this dike pattern probably had some causal connection with the stress distribution that existed at the time of their emplacement, and that they probably developed along the trajectories of least principal tensile stress (σ_2). He computed the total stress pattern that would be produced by the superposition of a radial stress distribution around the volcanic center on a regional field related to the mountain front situated to the west of Spanish Peaks. The stress trajectories of this combined stress field (Fig. 2-15) are remarkably close to the stress distribution pattern determined from the dikes.

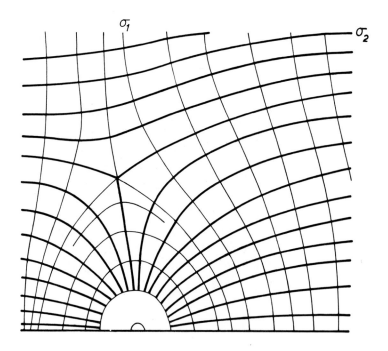

Figure 2-15

Stress trajectories resulting from the superposition of a stress system around a volcanic center on a regional field (cf. upper part of Figure 2-14). (After Odé, 1957.)

A number of other dyke distribution plans in other volcanic regions of the world show somewhat similar patterns and may also be the result of the superposition of local stress fields of variable intensity around the volcanic centers on more uniform fields of regional stress.

2-17   PHOTOELASTIC TECHNIQUES FOR DETERMINING STRESS DISTRIBUTION

The mathematical analysis of stress distribution in anisotropic substances, such as are found in the earth's crust, are often very difficult and time consuming, and three-dimensional problems may be practically intractable. The techniques offered by photoelastic analysis enables problems of stress distribution to be solved very rapidly and with a high degree of accuracy. The technique consists of making models of the structure from certain transparent celluloid, plastic, or gelatin materials, subjecting them to stress, and observing the models by transmitted polarized light. The model materials which are normally isotropic become optically active and doubly refracting. When viewed through a polariscope a series of dark lines (isoclinic lines) are seen in the model (Fig. 2-16) from which the orientation of the principal stresses (to $\pm 1°$) and their values may be determined. (For details of this technique see Coker and Filon, 1931; Filon, 1936; Frocht, 1948, 1963.)

Figure 2-16

Photoelastic study of a buckled rubber strip in gelatin. (After an experiment by Currie, Patnode, and Trump, 1962.)

For many years this method has been employed by engineers for determining the distribution of the stress in beams and frames of complex shape, and around holes and notches. It has been used for discovering the stresses in some geological structures (e.g., Currie, Patnode, and Trump, 1962), and it may well become an important tool for future research.

For the investigation of stress distribution in three dimensions the normal photoelastic technique outlined above is modified. Normally, when the load is removed from the model, the optical activity disappears as the elastic strains are recovered. If, however, the model is loaded and then heated and allowed to cool while under load, the stress patterns disappear as the material internally adjusts itself to the load. If now the model is cooled and the load removed, the model becomes optically active. The patterns of the isoclinic lines in this *frozen stress pattern* are identical to those produced normally by loading. The model can be sliced up, and the stress distribution in the slices is unaltered. This technique would appear to have a considerable potential for investigating the stress in complex three-dimensional structures. For further details the reader is referred to the work of Hetényi (1950).

REFERENCES AND SUGGESTED FURTHER READING
Stress

Arges, K. P., and A. E. Palmer: "Mechanics of Materials," McGraw-Hill Book Company, New York, 1962.

Conway, H. D.: "Mechanics of Materials," Prentice-Hall, Inc., Englewood Cliffs, N.J., 1950.

Crandall, S. H., and N. C. Dahl: "An Introduction to the Mechanics of Solids," McGraw-Hill Book Company, New York, 1959.

Deyarmond, A., and A. Arslan: "Fundamentals of Stress Analysis," Aero Publication, Glendale, Calif., 1942.

Durelli, A. J., E. A. Phillips, and C. H. Tsao: "Introduction to the Theoretical and Experimental Analysis of Stress and Strain," McGraw-Hill Book Company, New York, 1958.

Ford, H.: "Advanced Mechanics of Materials," Longmans, Green & Co., Ltd., London, 1963.

Jaeger, J. C.: "Elasticity, Fracture and Flow," Methuen & Co., Ltd., London, and John Wiley & Sons, Inc., New York, 1962.

Jeffreys, H.: "Cartesian Tensors," Cambridge University Press, London, 1931.

Knopf, A.: Igneous Geology of the Spanish Peaks Region, Colorado, *Geol. Soc. Am. Bull.*, **47**: 1727-1784 (1936).

Lee, G. H.: "An Introduction to Experimental Stress Analysis," John Wiley & Sons, Inc., New York, 1950.

Long, R. R.: "Mechanics of Solids and Fluids," Prentice-Hall, Inc., Englewood Cliffs, N.J., 1961.

Nadai, A.: "Theory of Flow and Fracture of Solids," McGraw-Hill Book Company, New York, 1950.

Odé, H.: Mechanical Analysis of the Dike Pattern of the Spanish Peaks Area, Colorado, *Geol. Soc. Am. Bull.*, **68**: 567-578 (1957).

Roark, R. J.: "Formulae for Stress and Strain," McGraw-Hill Book Company, New York, 1938.

Synge, J. L., and B. A. Griffith: "Principles of Mechanics," McGraw-Hill Book Company, New York, 1959.

Timoshenko, S. P., and J. M. Gere: "Theory of Elastic Stability," 2d ed., McGraw-Hill Book Company, New York, 1961.

Stress analysis, photoelastic techniques

Coker, E. G., and L. N. G. Filon: "A Treatise on Photelasticity," Cambridge University Press, London, 1931.

Currie, J. B., H. W. Patnode, and R. P. Trump: Development of Folds in Sedimentary Strata, *Geol. Soc. Am. Bull.*, **73**: 655-674 (1962).

Filon, L. N. G.: "A Manual of Photoelasticity for Engineers," Cambridge University Press, London, 1936.

Fried, B., and R. Weller: Photoelastic Analysis of Two- and Three-dimensional Stress Systems, *Ohio State Univ. Bull.*, 106, 1940.

Frocht, M. M.: "Photoelasticity," John Wiley & Sons, Inc., New York, 1948.

Frocht, M. M. (Ed.): "Proc. Intern. Symposium on Photoelasticity," Pergamon Press, New York, London, 1963.

Hetényi, M.: "Handbook of Experimental Stress Analysis," John Wiley & Sons, Inc., New York, 1950.

Jessop, H. T., and F. C. Harris: "Photoelasticity," Cleaver-Hume, London, 1949.

3 ‖ Strain in two dimensions

STRESS conditions within the earth's crust change during the progress of time, and these changes often lead to the permanent deformation of the crustal rocks. One of the prime aims of the structural geologist is to determine the nature and amount of these displacements. It is generally possible to determine the sequence and amounts of deformation that any one part has undergone and it is theoretically possible to relate these deformations to the changes in state of stress that produced them. Thus it should be possible to understand the kinematics of the processes which went on beneath the surface layers and which were the cause of the crustal instability. However, the relationships between stress and strain in such anisotropic and inhomogeneous substances as the rock materials of the crustal layers are extremely complex. In practice, the problem of determining the changing stress history from an end product which reflects the sum total of all the distortion produced by the stresses is generally insoluble.

The study of strain is primarily a geometrical problem and the aim of this chapter is to discuss in some detail the geometrical effects of displacement and internal distortion and to describe how measurements made on deformed material can be used to determine the nature and amount of the strain.

For the purposes of geometrical analysis, the process of the rock displacement can be described under two headings. If a unit of rock $ABCD$ is displaced to a position $A'B'C'D'$ (Fig. 3-1), there is

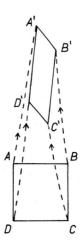

Figure 3-1

The original unit ABCD is deformed and takes up a new position A'B'C'D'. The dashed lines represent the movement paths of the four points. The displacements result in a bulk transport or translation, and a change in relative position or strain.

    1. A bulk transport of the unit or a *translation* of the mass as a whole, and

    2. A distortion, deformation or change in the position of the particles relative to each other, known as *strain.*

The amounts of these two displacements may vary greatly and, although they generally take place simultaneously, one is often more important than the other. For example, the Prealpine nappes of the Swiss and French Alps are great sheets of Mesozoic and Tertiary sediments which have been moved long distances from the site of their original deposition and now lie in an exotic position on top of other sedimentary materials. Although it is certain that these sheets have traveled horizontally for a distance of at least 50 km, it is remarkable how slight is the internal deformation in many of the sedimentary strata. Here translation was a much more important factor than strain. On the other hand, in parts of some slate belts it seems likely that internal deformation may be a more important phenomenon than mass translation of the rock.

    An investigation of the geometry of deformed objects (e.g., ooids, fossils) within the rock may make it possible to determine exactly the state of strain, but the amount of bulk transport must be found from other evidence (e.g., relation of root zone to transported nappe sheets). In practice it may be impossible to determine exactly the amount of translation.

3-1   THE MEASUREMENT OF STRAIN

As a result of deformation there are changes in both the lengths of lines and in the angles between intersecting lines. By measuring these changes it is possible to compute the state of strain.

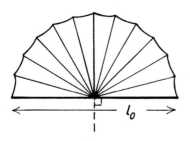 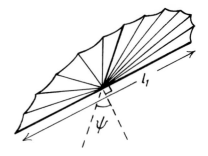

Figure 3-2

The parameters used for the measurement of strain. The original undeformed brachiopod shell with bilateral symmetry and with a hinge line of initial length l_o is distorted so that the hinge line is now of length l_1 and the angle between the median and hinge line (originally 90°) is deflected through an angle ψ.

Longitudinal strain   There are several ways of recording changes of length that are commonly used in an analysis of strain. Some of these parameters are more useful than others in the numerical solution of certain problems. The most useful for recording small strains describes the linear extension, and this particular parameter is used extensively in engineering problems involving small elastic deformations which do not greatly change the internal configuration of the body. In many geological problems, however, the internal deformation is large, and other strain parameters simplify the strain equations. No one parameter is more natural or fundamental than another; their use is entirely a matter of convenience with respect to the problem in hand.

The *extension e* is defined as the change in unit length. For example, in Fig. 3-2 if the initial length of the hinge line of the brachiopod is l_0, and the length in the distorted fossil is l_1, then the extension (either positive or negative in value depending on whether it has been expanded or contracted) for this particular direction is

$$e = \frac{l_1 - l_0}{l_0} = \frac{\delta l}{l_0}$$

In geological computation one of the most convenient measurements of change of length is the *quadratic elongation* λ, defined as the square of the length of a line of originally unit dimensions.

$$\lambda = \left(\frac{l_1}{l_0}\right)^2 = (1 + e)^2$$

Another useful measure of large longitudinal strain is known as *logarithmic, natural, or true strain* (ε). A large strain is envisaged as the sum of a series of progressive small increments of extension and the measure is given by

$$\varepsilon = \sum_{l_0}^{l_1} \frac{\delta l}{l}$$

If δl is an infinitely small increment, then the strain is

$$\varepsilon = \int_{l_0}^{l_1} \frac{dl}{l} = \log_e \frac{l_1}{l_0} = \log_e (1 + e)$$

In many ways this measure expresses more realistically the changes in length. For example, in a comparison of different parameters of longitudinal strain of two lines, one of which is contracted to half its original length and the other expanded to twice its original length, the extensions of the two lines are $e = -0.5$ and $e = 1.0$, respectively, whereas the natural strains are $\varepsilon = -\log_e 2$ and $\varepsilon = +\log_e 2$. With very great contractions e approaches -1, whereas ε approaches $-\infty$. The measure of natural strain is a particularly useful measure of strain if stress-strain relations are discussed because in homogeneous isotropic substances the increment of logarithmic strain has a direct linear relationship to the state of stress.

Shear strain   During deformation a change generally takes place in the angle between intersecting lines. If the angle between two lines was initially $90°$, any deflection from the right angle is defined as the *angular shear* ψ (Fig. 3-2) and the *shear strain* γ as

$$\gamma = \tan \psi$$

If the angular shear is small, then $\tan \psi = \psi$ in radians and $\gamma = \psi$. If the material above the shear plane is deflected to the right, the strain is considered positive; if to the left, the sign is negative.

3-2   HOMOGENEOUS AND INHOMOGENEOUS STRAIN

The type of strain shown by a distorted body can be classified as homogeneous or inhomogeneous, based on the following geometrical criteria:

HOMOGENEOUS DEFORMATION (Fig. 3-3*B*)

1. Straight lines remain straight after deformation.
2. Parallel lines remain parallel after deformation.
3. All lines in the same direction in the strained body have constant values of e, λ, ψ, and γ.

INHOMOGENEOUS DEFORMATION (Fig. 3-3*C*)

1. Straight lines become curved after deformation.

2. Parallel lines lose their parallelism after deformation.

3. For any one given direction in the body after deformation, the values of e, λ, ψ, and γ are variable.

The general mathematical theory of the geometry of inhomogeneous strain is extremely complex and for practical purposes it is generally considered to be almost useless. Although some of the geometrical features of inhomogeneous strain are of great importance in certain types of geological phenomena, the section which follows will deal for the most part with an analysis of homogeneous strain in two and three dimensions. This theory can be applied to geological problems because even in rocks which have suffered inhomogeneous strain, the state of strain in small units can be considered homogeneous. In geological problems the sizes of the elements that can be considered homogeneous vary greatly, and often homogeneity only extends over a field of a few cubic millimeters. On a very small scale the concept of homogeneity breaks down, for the amount and nature of the deformation of the individual crystals making up the rock varies through the aggregate. The properties and behaviors of aggregates under stress is a subject that is still in its infancy, and the understanding of strain in terms of aggregate structure will mark a fundamental development in the science of solid state physics.

Some of the geometrical problems of inhomogeneous strain will be discussed after the description of homogeneous strain by considering the effect of the variation in strain in small adjacent homogeneously strained units.

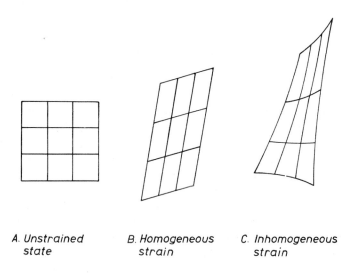

A. *Unstrained state* B. *Homogeneous strain* C. *Inhomogeneous strain*

Figure 3-3

The geometrical properties of homogeneous and inhomogeneous strain.

3-3 FINITE AND INFINITESIMAL STRAIN

Any deformed substance has an involved history, for it passes from its initial condition through a whole series of deformed states before it eventually arrives at its final state of strain. This process is known as *progressive deformation*. The geologist investigating rocks deformed by natural tectonic processes sees only the end product of the deformation processes known as the *finite state of strain* of the material. The structures that are developed in rocks depend to a certain extent on this finite strained state, but many probably depend on the actual progress of the deformation. The final strained state can be evaluated by the geologist by observing the nature of the deformation of objects of known original shape, and a large part of the discussion which follows is aimed at a thorough understanding of finite strain and the techniques that can be employed to determine the state of strain. The actual progress of the deformation process, however, has very important geometrical consequences which depend on the way the state of finite strain at any one time is modified by further increments of distortion. Progressive deformation can be envisaged as the modifications of a particular strained state by small incremental distortions or *infinitesimal strains*. The changes of the finite strain are related to the nature of the infinitesimal strains and also to the actual stress states that exist at any time. Although the geologist studying tectonic structures generally finds it impossible to determine the complete strain history and the nature of the successive incremental strains, the effects of these processes will be discussed in theory and related to some of the structures that are seen in naturally deformed rocks.

3-4 FINITE STRAIN IN TWO DIMENSIONS

In view of the importance of geological problems involving distortion in two dimensions (e.g., deformation of fossil impressions on bedding surfaces), the two-dimensional aspects of the strain theory will be discussed in some detail. Many problems of strain in three dimensions are solved by using two-dimensional computations, and there seems little doubt that a thorough understanding of the geometry of deformation in two dimensions greatly assists the understanding of the more involved theory of three-dimensional strain.

To begin this analysis of two-dimensional strain we set up a pair of rectangular coordinate axes Ox, Oy. We shall now consider what happens if we keep one point on the body fixed, the point $(0,0)$, and let the other points be displaced according to some linear law. In this way we can investigate the change in the orientation of the points relative to one another, that is, by definition the strain. Some simple and rather special types of displacement should give the reader a clear picture of the process, and these will lead to the most general displacements which describe the strained state.

1. Simple extension parallel to one axis (Fig. 3-4A) The coordinates x, y of any point before deformation are displaced to positions x_1, y_1 such that

$$x_1 = (e + 1)x \qquad y_1 = y \tag{3-1}$$

2. Extension parallel to both axes (Fig. 3-4B)

$$x_1 = (e_x + 1)x \qquad y_1 = (e_y + 1)y \tag{3-2}$$

The change in area after straining is $(e_x + 1)(e_y + 1) - 1$. Where $(e_x + 1) = 1/(e_y + 1)$ there is no change in area and the strain is known as *pure shear*. Taking logarithms of both sides we have $\log(e_x + 1) = -\log(e_y + 1)$; or using the parameters of natural strain, $\varepsilon_x = -\varepsilon_y$.

3. Simple shear (Fig. 3-4C)   Particles are displaced in a direction parallel to the x axis so that the ordinate line of any point is sheared through an angle ψ_1.

$$x_1 = x + \tan \psi_1 y \qquad y_1 = y \tag{3-3}$$

There is no area change.

4. Superimposed simple shears (Fig. 3-4D, E) If the strained body of Fig. 3-4C is subjected to another shear parallel to the y axis so that the lines parallel to the x axis are deflected through an angle ψ_2 (Fig. 3-4D), then

$$y_2 = y_1 + \tan \psi_2 x_1 \qquad x_2 = x_1$$

or, substituting (3-3),

$$x_2 = x + \tan \psi_1 y \qquad y_2 = y(1 + \tan \psi_1 \tan \psi_2) + x \tan \psi_2 \tag{3-4}$$

If the order but not the amount of successive shear deformation is changed, the final transformation is not the same as that of Eq. (3-4) (see Fig. 3-4E).

First deformation:

$$x_1 = x \qquad y_1 = y + \tan \psi_2 x$$

Second deformation:

$$x_2 = x_1 + \tan \psi_1 y_1 \qquad y_2 = y_1$$

or

$$x_2 = x(1 + \tan \psi_1 \tan \psi_2) + \tan \psi_1 y$$
$$y_2 = y + \tan \psi_2 x \tag{3-5}$$

It is generally true that when two strains are superimposed the nature of the final product depends upon the order of the deformations. The superposition of linear transformations is a *noncommutative* process.

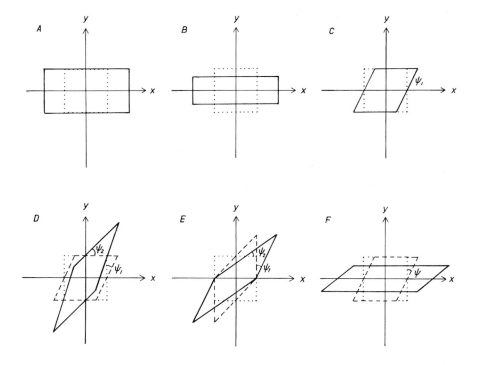

Figure 3-4

Types of special translation which give rise to a state of strain. The dotted square represents the undeformed state.

5. Simple shear and extension parallel to both axes (Fig. 3-4*F*) If the deformation by simple shear (Fig. 3-4*C*) is followed by that of Fig. 3-4*B*, the coordinates of the points after the two deformations become

$$x_2 = (e_x + 1)(x + \tan \psi_1 \, y) \qquad y_2 = (e_y + 1) \, y \qquad (3\text{-}6)$$

Again, if the order of the deformation is reversed, the final transposition is different from that of Eq. (3-6).

$$x_2 = (e_x + 1)x + \tan \psi_1 \, (e_y + 1)y \qquad y_2 = (e_y + 1)y \qquad (3\text{-}7)$$

6. General transformation The most general strain displacement is given by

$$x_1 = ax + by \qquad y_1 = cx + dy \qquad (3\text{-}8)$$

(where $a > 0$ and $d > 0$), or

$$x = \frac{dx_1 - by_1}{ad - bc} \qquad y = \frac{-cx_1 + ay_1}{ad - bc}$$

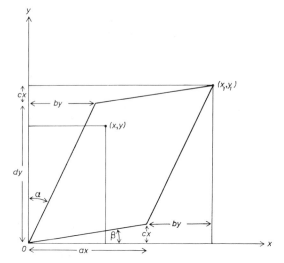

Figure 3-5

The general strain translation.

This transformation is illustrated in Fig. 3-5. The original rectangle formed by joining the four points $(0,0)$, $(x,0)$, (x,y) and $(0,y)$ is deformed into the parallelogram $(0,0)$, (ax,cx), $(ax + by, cx + dy)$, (by,dy). The physical significance of each of the constants a, b, c, and d is as follows: a and d can be considered as components of logitudinal strain parallel to the x and y coordinate axes, respectively; whereas b and c are in part shear components that reflect the angular displacements of the originally perpendicular sides of the rectangle, that is, $b = d \tan \alpha$ and $c = a \tan \beta$.

By substituting the values of x and y of (3.8) in the general equation of a straight line

$$y = mx + k \tag{3-9}$$

it is found that the line is transformed into another straight line

$$y_1 = \frac{c + dm}{a + bm} x_1 + \frac{ad - bc}{a + bm} k \tag{3-10}$$

and therefore it is concluded that this displacement fits the criteria for homogeneous strain given above.

The effect of the general displacement (3-8) on points which lie on the circle $x^2 + y^2 = 1$ is a transformation into the equation

$$(c^2 + d^2)x_1^2 - 2(ac + bd)x_1 y_1 + (a^2 + b^2)y_1^2 = (ad - bc)^2 \tag{3-11}$$

This is an ellipse, the *strain ellipse*, and the major and minor axes of this ellipse represent the positions of *maximum and minimum longitudinal strain* of lengths $1 + e_1 = \sqrt{\lambda_1}$, and $1 + e_2 = \sqrt{\lambda_2}$.

If the general displacement (3-8) is applied to the general equation of an ellipse centered at the origin ($lx^2 - 2mxy + ny^2 = 1$), this ellipse is transformed into another ellipse

$$px_1^2 - 2qx_1y_1 + ry_1^2 = 1 \tag{3-12}$$

where
$$p = \frac{ld^2 + 2mcd + nc^2}{(ad - bc)^2}$$

$$q = \frac{-m(ad + bc) + lbd + nac}{(ad - bc)^2}$$

$$r = \frac{lb^2 + 2mab + na^2}{(ad - bc)^2}$$

This is important because it means that the geometrical effects of several superimposed finite homogeneous two-dimensional strains can always be described in a single strain ellipse. There is, however, one ellipse that is transformed to the circle $x_1{}^2 + y_1{}^2 = 1$, given by the equation

$$(a^2 + c^2)x^2 + 2(ab + cd)xy + (b^2 + d^2)y^2 = 1 \tag{3-13}$$

This is known as the *reciprocal strain ellipse*.

In general, the deformation (3-8) leads to a change in the angle between any two originally perpendicular lines. There are, however, two lines that are perpendicular before deformation that remain so after straining. The equation of a line through the origin initially perpendicular to $y = mx$ is

$$y = -\frac{x}{m} \tag{3-14}$$

and after strain this will have a slope [from (3-10)]

$$\frac{mc - d}{ma - b} \tag{3-15}$$

If these two perpendicular lines of initial slopes m and $-1/m$ are to remain perpendicular, the product of their slopes after deformation must be -1.

$$\frac{(md + c)(mc - d)}{(mb + a)(ma - b)} = -1$$

$$m^2 + m\frac{a^2 - b^2 + c^2 - d^2}{ab + cd} - 1 = 0 \tag{3-16}$$

Providing that $ab + cd$ is not zero, the roots of this equation are always real, and therefore there are always two initially perpendicular lines which remain perpendicular after deformation. The *original orientations* of the lines of slope

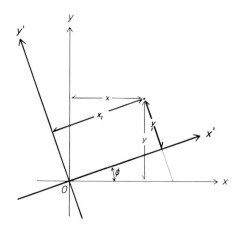

Figure 3-6

The special translation of body rotation without internal deformation.

m and $-1/m$ are defined as the *principal axes of strain*. In their deformed condition, these coincide with the directions of maximum elongation and minimum elongation, that is, the axes of the strain ellipse. In general, the original orientation of these perpendicular lines does not coincide with the position after strain, and the deformation is called *rotational strain*. If, however, they do coincide, the deformation is known as *irrotational strain*. The condition for irrotational strain is that the slopes before and after deformation are the same.

$$\frac{md + c}{mb + a} = m$$

or
$$m^2b + m(a - d) - c = 0 \qquad (3\text{-}17a)$$

and
$$\frac{mc - d}{ma - b} = -\frac{1}{m}$$

or
$$m^2c + m(a - d) - b = 0 \qquad (3\text{-}17b)$$

When these conditions hold simultaneously, we obtain the necessary condition for irrotational strain: $b = c$. *Simple rotation* is a transformation that may go on without strain. The displacement of the point (x,y) can be determined by rotating the x and y coordinate axes through an angle ϕ to take up new positions x' and y'. The relationships between (x,y) and (x_1,y_1) are, from Fig. 3-6,

$$x_1 = x \cos \phi + y \sin \phi \qquad (3\text{-}18a)$$

$$y_1 = -x \sin \phi + y \cos \phi \qquad (3\text{-}18b)$$

Rotational deformation of this type generally goes on together with strain, and any general strain can be envisaged as the product of two types of displacement, an irrotational strain followed by a finite rotation. For example, the point (x,y) may be displaced to (x_1,y_1) by a finite irrotational strain [remembering the condition to be applied to (3-8) is that $b = c$] and the general displacement becomes

$$x_1 = ax + by \qquad (3\text{-}19a)$$

$$y_1 = bx + dy \qquad (3\text{-}19b)$$

If this now undergoes a finite rotation through an angle ϕ, then, using the transformation (3-18), we obtain the final position of any point (x_2,y_2).

$$x_2 = Ax + By \qquad y_2 = Cx + Dy$$

where

$$A = a \cos \phi + b \sin \phi \qquad (3\text{-}20a)$$

$$B = b \cos \phi + d \sin \phi \qquad (3\text{-}20b)$$

$$C = -a \sin \phi + b \cos \phi \qquad (3\text{-}20c)$$

$$D = -b \sin \phi + d \cos \phi \qquad (3\text{-}20d)$$

As $B - C = (a + d) \sin \phi$, and $A + D = (a + d) \cos \phi$, we can obtain the rotational part ϕ of any strain if we know the values of A, B, C, and D for

$$\tan \phi = \frac{B - C}{A + D} \qquad (3\text{-}21a)$$

Similarly it is possible to obtain values of a, b, and d of the irrotational part of the strain in terms of A, B, C, D, and ϕ. Multiplying (3-20a) by $\cos \phi$, (3-20c) by $\sin \phi$, and substracting, we obtain

$$a = A \cos \phi - C \sin \phi \qquad (3\text{-}21b)$$

Multiplying (3-20a) by $\sin \phi$, (3-20c) by $\cos \phi$, and adding,

$$b = A \sin \phi + C \cos \phi \qquad (3\text{-}21c)$$

Multiplying (3-20b) by $\sin \phi$, (3-20d) by $\cos \phi$, and adding,

$$d = B \sin \phi + D \cos \phi \qquad (3\text{-}21d)$$

As we have values of a, b, and d of the irrotational transformation, we can now obtain the principal strains of the irrotational part as follows. In Fig. 3-7 let the point $P(x,y)$ be positioned at unit distance from the origin, and let OP be one of the principal axes of strain making an angle θ with the x axis. Let the deformed position of P be $P_1(x_1,y_1)$; then, because the strain is irrotational and P lies on a principal axis of strain, angle P_1Ox = angle POx = θ. Because P_1O will coincide with one of the principal axes of the strain

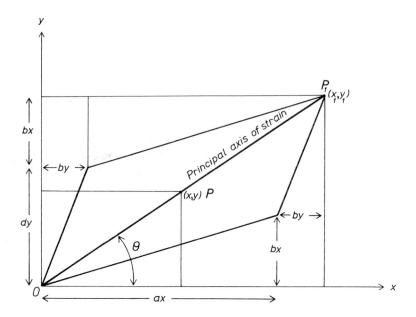

Figure 3-7

Determination of the two principal strains. P(x,y) has been chosen so that it lies on a principal axis of strain; it is displaced to $P_1(x_1,y_1)$

ellipse, the length of $P_1O = (1 + e)$ will be a principal strain. Then

$$\cos \theta = x = \frac{ax + by}{1 + e} \quad \text{and} \quad \sin \theta = y = \frac{bx + dy}{1 + e}$$

$$0 = x[a - (1 + e)] + by \tag{3-22a}$$

$$0 = bx + y[d - (1 + e)] \tag{3-22b}$$

Eliminating x and y from these equations

$$[a - (1 + e)][d - (1 + e)] = b^2$$

$$(1 + e)^2 - (1 + e)(a + d) + ad - b^2 = 0 \tag{3-23}$$

and the two roots of this equation are the two principal strains $1 + e_1$ and $1 + e_2$. To determine the value of θ we replace x by $\cos \theta$ and y by $\sin \theta$ in Eqs. (3-22).

$$0 = [a - (1 + e)] + b \tan \theta \tag{3-24a}$$

$$0 = \frac{b}{\tan \theta} + [d - (1 + e)] \tag{3-24b}$$

Combining these equations we obtain a quadratic equation in tan θ defining the position of the two mutually perpendicular strain-ellipse axes.

$$\tan^2 \theta + \frac{a - d}{b} \tan \theta - 1 = 0 \qquad (3\text{-}25)$$

Summary   At this stage it is perhaps advisable to summarize the main conclusions of Sec. 3-4 with a view to deciding which features of the strain are likely to be calculated from observations made on naturally deformed rocks.

First, it is important to understand the significance of the fact that we made no special choice of the coordinate axes that we used as a framework to describe the displacements of points in space. If we had chosen other axes, we could equally well have described the same displacements; the final equation would then be slightly different but the displacements themselves would be unchanged. The actual distortions or strains that result from the displacements are also the same, no matter what axes are used to describe them. Mathematical entities that are completely independent of their reference frames are known as *tensors*. Some physical quantities such as length and volume are specified by magnitude alone and are called *scalars*. They need no reference frame to describe them, and such quantities are known further as *tensors of the zero order*. Other quantities, termed *vectors*, require further definition: for example, force and velocity need magnitude, direction, and a sense of direction to define them completely. This can be done with any arbitrary reference frame, and these quantities are known as *tensors of the first order*. The features that we are so concerned with in the geological processes of deformation, namely stress and strain, are three-dimensional entities that require either six or nine quantities for their complete specification, and these are known as *tensors of the second order* or more usually simply as *tensors*. If they have six compenents to define them (e.g., stress), the tensor is known as a *symmetric tensor;* if they require nine components (e.g., finite homogeneous strain in three dimensions), the tensor is known as an *asymmetric tensor*. Because the actual state of stress or strain is independent of the reference axes, there are certain quantities which can be measured in terms of these axes which are constant no matter what the orientation of the axes might be. These quantities are known as *invariants* and much of the rather specialized mathematics known as *tensor calculus* used to define concisely the properties of tensors is built up on a study of the invariant relations involved in changing from one coordinate system to another.

Having chosen the set of reference axes, we have seen that finite displacements can generally be described (in two dimensions) by the use of four terms A, B, C, and D. These can be transformed into other measurements which are of more assistance in understanding the deformation. The first two are the principal elongations $1 + e_1$, $1 + e_2$ which represent the lengths of the axes of the strain ellipse derived from the deformation of an initial unit circle; the third is the angle θ between these ellipse axes and one of the axes of the coordinate frame; and the fourth is the angle of rotation ϕ made by the lines that eventually become the ellipse axes. These effects of any displacement can be grouped into two parts:

1. The factors $1 + e_1$, $1 + e_2$, and θ defining the strain ellipse.
2. The angle ϕ defining the rotation of the principal strain axes.

The strain tensor is an asymmetric tensor because when the complete deformation is taken into account the geometry has a low symmetry. In the two-dimensional case that we have discussed, because no variations exist in the third dimension, the complete three-dimensional symmetry is monoclinic. In the general three-dimensional case it is triclinic. The symmetry of an asymmetric tensor can be divided into two parts, one symmetric and the other rotational. The first part of the deformation (1 above) is the symmetric part. Because the strain ellipse is symmetrical about its two axes, distortions, when related to the strain ellipse, are symmetrical on either side of the ellipse, since the strain conditions on one side of the ellipse are the mirror image of those on the other. If we ignore (or do not know) the initial orientation of the deformed substance and observe only the strain ellipse, then all distortions with reference to the axes are symmetrical. For example, a line making an angle of 20° with one of the principal axes of the ellipse has the same extension and shear strain as one making an angle of $-20°$ with the same axis.

The second component of the deformation is the rotational part (2 above) defined by the angle ϕ. If the deformation is irrotational, then this angle is zero and the whole displacement and deformation is a symmetric one; however, most deformations have a rotational component. Before we can define the rotational part of the strain tensor it is imperative to know the undeformed state, and in naturally deformed rocks we do not know this. We may have some knowledge which enables us to make a guess at it (for example, the bedding planes before deformation may have been horizontal), but we must always recognize that it is a guess even though the assumption may be a very reasonable one.

Providing that the structural geologist has suitable material, he can define exactly the symmetric part of the strain tensor and determine the strain ellipse or strain ellipsoid. The rotational part of the tensor can never be fixed precisely, although it may be possible to approximate very closely to it.

For further mathematical analysis of the finite-strain properties of the strain ellipse, it is convenient to rearrange the coordinate axes so that they coincide with the principal axes of the strain ellipse and to consider the features of the symmetrical part of the strain tensor. The equation of this ellipse with principal axes of lengths $1 + e_1 = \sqrt{\lambda_1}$ and $1 + e_2 = \sqrt{\lambda_2}$ is

$$\frac{x^2}{\lambda_1} + \frac{y^2}{\lambda_2} = 1 \tag{3-26}$$

The surface area of the ellipse is $\pi(\lambda_1\lambda_2)^{\frac{1}{2}}$. Since this ellipse is derived from a circle of unit radius and area π, any area change after deformation is recorded by

$$\Delta = \frac{\pi(\lambda_1\lambda_2)^{\frac{1}{2}} - \pi}{\pi} = (\lambda_1\lambda_2)^{\frac{1}{2}} - 1 \tag{3-27}$$

3-5 CHANGE IN LENGTH OF LINES

After a given finite strain the lengths of most lines are altered by an amount $(1 + e)$ or $\lambda^{\frac{1}{2}}$ (see Fig. 3-8). If θ is the angle made by any line OP and the x axis before deformation, and θ' the angle after deformation, then

$$(1 + e)^2 = \lambda = x_1{}^2 + y_1{}^2$$

But as

$$x_1 = x(1 + e_1) = \cos\theta\,\lambda_1{}^{\frac{1}{2}} \qquad y_1 = \sin\theta\,\lambda_2{}^{\frac{1}{2}} \tag{3-28}$$

then

$$\lambda = \lambda_1 \cos^2\theta + \lambda_2 \sin^2\theta \tag{3-29}$$

This equation gives the quadratic elongation of lines with reference to the angle θ that they made with the x coordinate axis *in the unstrained state*. For the practical purposes of geological computation it is often more useful to

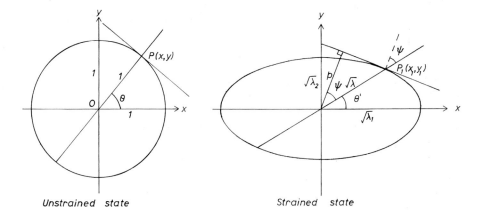

Unstrained state   Strained state

Figure 3-8

The point $P(x,y)$ lies on a unit circle and, as a result of a homogeneous strain, takes up a new position $P_1(x_1,y_1)$ on the strain ellipse $x^2/\lambda_1 + y^2/\lambda_2 = 1$.

have the lengths in terms of θ'; thus

$$\cos \theta = x = \frac{x_1}{\lambda_1^{\frac{1}{2}}} = \frac{\lambda^{\frac{1}{2}} \cos \theta'}{\lambda_1^{\frac{1}{2}}}$$

and

$$\sin \theta = \frac{\lambda^{\frac{1}{2}} \sin \theta'}{\lambda_2^{\frac{1}{2}}} \tag{3-30}$$

Putting these values in $\cos^2 \theta + \sin^2 \theta = 1$

$$\frac{\lambda}{\lambda_1} \cos^2 \theta' + \frac{\lambda}{\lambda_2} \sin^2 \theta' = 1$$

or

$$\frac{\cos^2 \theta'}{\lambda_1} + \frac{\sin^2 \theta'}{\lambda_2} = \frac{1}{\lambda}$$

By replacing $1/\lambda_1$ by λ_1' and $1/\lambda_2$ by λ_2' (λ_1' and λ_2' being the quadratic extensions of the principal axes of the reciprocal strain ellipse), and defining the reciprocal value of the quadratic extension as λ', a simple equation is obtained for lengths of lines with reference to the angle they make with the x coordinate axis *in the strained state*:

$$\lambda' = \lambda_1' \cos^2 \theta' + \lambda_2' \sin^2 \theta' \tag{3-31}$$

Lines of no finite elongation   After a finite strain, provided that $\lambda_1 > 1 > \lambda_2$, lines in some directions will have been extended and in others they will have been contracted, but there will always be two lines within the strain ellipse which have suffered *no finite longitudinal strain*, that is, lines which have the same length as in the unstrained state ($\lambda = 1$). The positions of such lines can be found where the circle $x^2 + y^2 = 1$ cuts the ellipse $x^2/\lambda_1 + y^2/\lambda_2 = 1$, or more simply by putting $\lambda = 1$ in (3-29):

$$1 = \lambda_1 \cos^2 \theta + \lambda_2 \sin^2 \theta$$

$$\cos^2 \theta = \frac{1 - \lambda_2}{\lambda_1 - \lambda_2} \tag{3-32a}$$

$$\sin^2 \theta = \frac{\lambda_1 - 1}{\lambda_1 - \lambda_2} \tag{3-32b}$$

$$\tan^2 \theta = \frac{\lambda_1 - 1}{1 - \lambda_2} \tag{3-32c}$$

Similarly, from (3-31), the positions of lines of no finite elongation can be referred to angles measured in the strained state.

$$\cos^2 \theta' = \frac{\lambda_2' - 1}{\lambda_2' - \lambda_1'} \tag{3-33a}$$

$$\sin^2 \theta' = \frac{1 - \lambda_1'}{\lambda_2' - \lambda_1'} \tag{3-33b}$$

$$\tan^2 \theta' = \frac{1 - \lambda_1'}{\lambda_2' - 1} \tag{3-33c}$$

Where $\lambda_1 > \lambda_2 > 1$ or $1 > \lambda_1 > \lambda_2$ there are no real solutions to these equations, for lines within the strain ellipse are either all longer or all shorter than their original lengths.

3-6 CHANGES IN ANGLES

After deformation the angle any line makes with the x direction is changed.

$$\tan \theta' = \frac{y_1}{x_1} = \frac{y(\lambda_2)^{\frac{1}{2}}}{x(\lambda_1)^{\frac{1}{2}}} = \tan \theta \left(\frac{\lambda_2}{\lambda_1} \right)^{\frac{1}{2}} \tag{3-34}$$

This formula, giving the angle after deformation if the angle before deformation and the shape of the strain ellipse are known, is dependent only on the *ratios* of the lengths of the principal axes of the strain ellipse. It is independent of the absolute values of λ_1 and λ_2 and is therefore independent of area change.

3-7 SHEAR STRAIN

The aim of this section is to determine the value of shear strain γ at any point on the strain ellipse. If the point $P(x,y)$ (Fig. 3-8) on the circle $x^2 + y^2 = 1$ is displaced to $P_1(x_1,y_1)$ on the strain ellipse $x_1^2/\lambda_1 + y_1^2/\lambda_2 = 1$, the equation of the tangent to the ellipse at this point (x_1,y_1) is

$$\frac{xx_1}{\lambda_1} + \frac{yy_1}{\lambda_2} = 1$$

or using (3-28)

$$\frac{x \cos \theta}{(\lambda_1)^{\frac{1}{2}}} + \frac{y \sin \theta}{(\lambda_2)^{\frac{1}{2}}} = 1 \tag{3-35}$$

The length of the perpendicular p from the origin to this line is given by

$$p = \left(\frac{1}{\cos^2 \theta/\lambda_1 + \sin^2 \theta/\lambda_2} \right)^{\frac{1}{2}} \tag{3-36}$$

$$\sec \psi = \frac{(\lambda)^{\frac{1}{2}}}{p} = \left[\lambda \left(\frac{\cos^2 \theta}{\lambda_1} + \frac{\sin^2 \theta}{\lambda_2} \right) \right]^{\frac{1}{2}} \tag{3-37}$$

$$\gamma^2 = \tan^2 \psi = \sec^2 \psi - 1 = \lambda \left(\frac{\cos^2 \theta}{\lambda_1} + \frac{\sin^2 \theta}{\lambda_2} \right) - 1 \tag{3-38}$$

Substituting for λ (3-29) and solving for γ using $(\cos^2 \theta + \sin^2 \theta)^2 = 1$,

$$\gamma = \frac{\lambda_1 - \lambda_2}{(\lambda_1 \lambda_2)^{\frac{1}{2}}} \cos \theta \sin \theta \qquad (3\text{-}39)$$

This gives the shear strain in terms of the angle θ with reference to the *unstrained state*. Equation (3-39) can be rewritten

$$\gamma = \left(\frac{\lambda_1}{\lambda_2} + \frac{\lambda_2}{\lambda_1} - 2\right)^{\frac{1}{2}} \cos \theta \sin \theta \qquad (3\text{-}40)$$

and therefore the values of γ are dependent only on the *ratios* of the principal axes of the ellipse.

If the substitutions (3-30) involving angles measured in the strained body are made in (3-39),

$$\gamma = \frac{\lambda_1 - \lambda_2}{(\lambda_1 \lambda_2)^{\frac{1}{2}}} \frac{(\lambda)^{\frac{1}{2}} \cos \theta'}{(\lambda_1)^{\frac{1}{2}}} \frac{(\lambda)^{\frac{1}{2}} \sin \theta'}{(\lambda_2)^{\frac{1}{2}}}$$

or

$$\frac{\gamma}{\lambda} = \left(\frac{1}{\lambda_2} - \frac{1}{\lambda_1}\right) \sin \theta' \cos \theta' \qquad (3\text{-}41)$$

The term γ' is now defined such that $\gamma' = \gamma/\lambda$. This strain parameter is chosen entirely for the convenience of simplifying the equation. One of its important properties is that it has the same value but opposite sign along any two perpendicular directions in a strained material. By using γ' and the values for the reciprocal quadratic extensions, i.e., $\lambda_1' = 1/\lambda_1$ and $\lambda_2' = 1/\lambda_2$, Eq. (3-41) simplifies to give a standard formula for shear strain with reference to the *strained state*

$$\gamma' = (\lambda_2' - \lambda_1') \sin \theta' \cos \theta' \qquad (3\text{-}42)$$

Maximum value of the finite shear strain   Differentiating (3-39) with respect to θ,

$$\frac{d\gamma}{d\theta} = \frac{\lambda_1 - \lambda_2}{(\lambda_1 \lambda_2)^{\frac{1}{2}}} \cos 2\theta \qquad (3\text{-}43)$$

Maximum values of this expression occur where $d\gamma/d\theta = 0$

where $\qquad\qquad \cos 2\theta = 0 \qquad$ or $\qquad \theta = 45°$ or $135°$ $\qquad (3\text{-}44)$

Using (3-34) the position of lines of maximum shearing strain after deformation are found

$$\tan \theta' = \left(\frac{\lambda_2}{\lambda_1}\right)^{\frac{1}{2}} \qquad (3\text{-}45)$$

Lines which initially make an angle of 45° with the final position of the strain ellipse axes and which make an angle $\tan^{-1}(\lambda_2/\lambda_1)^{\frac{1}{2}}$ after deformation show a maximum shear strain. The value of this strain is found by putting $\theta = 45°$ or 135° in (3-39).

$$\gamma = \pm \frac{\lambda_1 - \lambda_2}{2(\lambda_1\lambda_2)^{\frac{1}{2}}} \tag{3-46}$$

3-8 THE MOHR DIAGRAM

An extremely convenient method of graphically representing the variations in values of the components of tensor quantities such as stress or strain was developed by the German engineer, Otto Mohr, in 1882. To make this construction Eqs. (3-29), (3-39), (3-31), and (3-42) are first rearranged by making standard substitutions for double angles [that is, $\cos^2 \theta = (1 + \cos 2\theta)/2$; $\sin^2 \theta = (1 - \cos 2\theta)/2$; $\sin \theta \cos \theta = (\sin 2\theta)/2$].

$$\lambda = \frac{\lambda_1 + \lambda_2}{2} + \frac{\lambda_1 - \lambda_2}{2} \cos 2\theta \tag{3-47}$$

$$\gamma = \frac{1}{(\lambda_1\lambda_2)^{\frac{1}{2}}} \frac{\lambda_1 - \lambda_2}{2} \sin 2\theta \tag{3-48}$$

$$\lambda' = \frac{\lambda'_1 + \lambda'_2}{2} - \frac{\lambda'_2 - \lambda'_1}{2} \cos 2\theta' \tag{3-49}$$

$$\gamma' = \frac{\lambda'_2 - \lambda'_1}{2} \sin 2\theta' \tag{3-50}$$

The equation of any circle (Fig. 3-9) of center $(c,0)$ and radius r can be written

$$x = c - r \cos \alpha \tag{3-51a}$$

$$y = r \sin \alpha \tag{3-51b}$$

and the equation of any ellipse (Fig. 3-10) of center $(c,0)$ and principal axes $2r$ and $2r'$ is

$$x = c + r \cos \alpha \tag{3-52a}$$

$$y = r' \sin \alpha \tag{3-52b}$$

It will be seen that the equations which define the state of strain with reference to angles measured in the unstrained condition (3-47) and (3-48) are in the form of (3-52), where

$$x = \lambda \qquad y = \gamma \qquad r = \frac{\lambda_1 - \lambda_2}{2} \qquad r' = \frac{\lambda_1 - \lambda_2}{2(\lambda_1\lambda_2)^{\frac{1}{2}}} \qquad c = \frac{\lambda_1 + \lambda_2}{2} \qquad \alpha = 2\theta$$

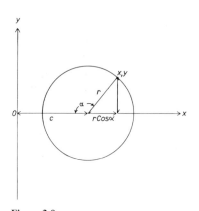

Figure 3-9

The circle representing the equation
x = c − r cos α and y = r sin α.

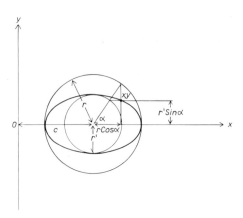

Figure 3-10

The ellipse representing the equation
x = c + r cos α and y = r′ sin α.

Longitudinal strain and shear strain along a line in any direction θ from the principal axis of longitudinal strain can therefore be represented by points on an ellipse. If $\lambda_1\lambda_2 = 1$ (i.e., from (3-27) the deformation has resulted in no change of area), $r = r'$, and the states of strain can be represented by points on a circle.

The equations which give the state of strain with reference to angles measured in the strained condition (3-49) and (3-50) are in the form of (3-51), where

$$x = \lambda' \qquad y = \gamma' \qquad c = \frac{\lambda_1' + \lambda_2'}{2} \qquad r = \frac{\lambda_2' - \lambda_1'}{2} \qquad \alpha = 2\theta'$$

and therefore the states of strain for any strain ellipse, irrespective of area change during deformation can, with reference to measurements in the strained state, be represented by points on a circle.

These two graphical constructions (Mohr diagrams) enable the quadratic elongation λ and shear strain γ to be evaluated simply for any line of known angle to the principal axes of strain before or after deformation without recourse to numerical solution of the equations. Many of the geometrical properties of the strain ellipse are particularly clear in these diagrams. The Mohr constructions can also be used to compute absolute values of λ_1 and λ_2 or the ratio λ_1/λ_2 using measurements made on distorted objects oriented randomly in the strain ellipse. Used in this way they afford an especially valuable method for manipulating the type of geological data obtained from measurements of strained objects in deformed rocks (Brace, 1961).

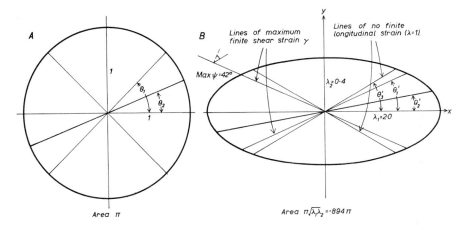

Figure 3-11

The relationships between lines in the undeformed initial unit circle and in a finite strain ellipse with principal strains $\lambda_1 = 2.0$ and $\lambda_2 = 0.4$.

Mohr's constructions with reference to strained and unstrained conditions For any state of strain it is possible to construct two Mohr diagrams. The diagram which gives the strain with reference to angles measured in the deformed condition is always a circle; it is simple to construct and has more immediate application to practical geological problems than the diagram for the undeformed condition.

A numerical example is given below to illustrate how these diagrams are constructed and how they can be used to discover the finite geometrical properties of a particular strain ellipse.

MOHR DIAGRAM WITH REFERENCE TO THE UNSTRAINED STATE   The circle illustrated in Fig. 3-11A is deformed to an ellipse with principal strains $\lambda_1 = 2.0$ and $\lambda_2 = 0.4$ (Fig. 3-11B). To construct the Mohr diagram (Fig. 3-12)

1. Draw two perpendicular coordinate axes; the abscissa axis will record λ, the ordinate axis γ.

2. Describe a circle center $(\lambda_1 + \lambda_2)/2$, 0, that is, (1.2, 0) and radius $(\lambda_1 - \lambda_2)/2$, that is, (0.8). This will intersect the λ axis at position $\lambda_1 = 2.0$ and $\lambda_2 = 0.4$.

3. As $(\lambda_1\lambda_2)^{\frac{1}{2}} = 0.894$, there has been a decrease in area caused by the deformation, and the variation in strain will therefore be represented as an ellipse. Modify all ordinates by an amount $1/(\lambda_1\lambda_2)^{\frac{1}{2}}$, that is, by a factor 1.12, so that they fall on this ellipse. The γ and λ values of all points on this ellipse satisfy the conditions of Eqs. (3-47) and (3-48).

PROPERTIES OF THE STRAIN ELLIPSE DEDUCED FROM THE MOHR DIAGRAM   If θ is the angle a line made with that line which became the major axis (x axis)

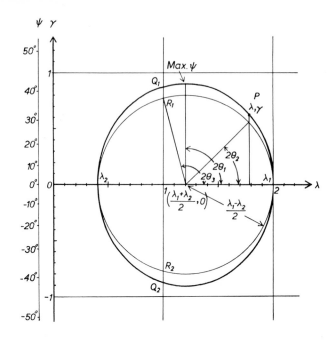

Figure 3-12

Mohr diagram representation of the strain ellipse of Fig. 3-11 with reference to angles measured in the unstrained state.

of the strain ellipse (Fig. 3-11A), then the maximum value of γ will occur where $2\theta_1$ is 90 or 270°, where θ_1 = 45 or 135° [this is the graphical solution of Eq. (3-44)], and at this point $\gamma = \pm 0.89°$ and $\psi = \pm 42°$. These represent the graphical solutions of Eq. (3-46), the negative sign referring to shear in the opposite sense to that indicated by the positive sign. It is also clear from the diagrams that the shear strains are zero where $2\theta = 0$ or 180°, that is, in directions parallel to the principal axes of the strain ellipse.

To determine the conditions of strain in any direction making an angle θ_2 (for example, 22°) with the x axis before deformation, set off from the center of the circle a radius line making an angle of $2\theta_2 = 44°$ from λ_1. At the point where this line cuts the circle, construct a line parallel to the γ axis to meet the Mohr ellipse at P. The abscissa and ordinate of P give the values of $\lambda = 1.78$ and $\gamma = 0.63$, respectively, for the θ_2 direction.

The positions of lines of no finite longitudinal strain before deformation are found at the two points Q_1 and Q_2 on the ellipse which satisfy the condition $\lambda = 1$. At each of these points, construct the line $\lambda = 1$ to meet the circle at R_1 and R_2. The angles measured from the center of the circle between λ_1 and R_1 and λ_1 and R_2 give the angles $2\theta_3$. The angles before deformation

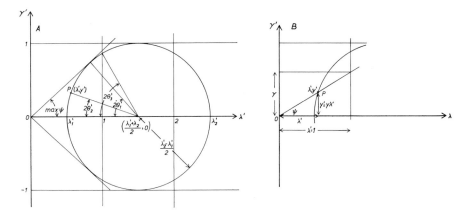

Figure 3-13

Mohr diagram representation of the strain ellipse of Fig. 3-11 with reference to angles measured in the strained state. Part B illustrates the method of determining the angular shear strain.

made by the lines of no finite longitudinal strain with the principal strain axis are therefore 52 and 128°.

MOHR DIAGRAM WITH REFERENCE TO THE STRAINED STATE   The Mohr circle giving states of strain with respect to angles measured in the deformed condition is constructed as follows (Fig. 3-13A):

1. Calculate the values of λ'_1 ($1/\lambda_1 = 0.5$) and λ'_2 ($1/\lambda_2 = 2.5$).
2. Construct two perpendicular coordinate axes, abscissa λ' and ordinate γ'.
3. Describe a circle of center $(\lambda'_1 + \lambda'_2)/2$, 0, that is, (1.5, 0), and radius $(\lambda'_2 - \lambda'_1)/2 = 1.0$.

PROPERTIES OF THE STRAIN ELLIPSE DEDUCED FROM THE MOHR DIAGRAM   The abscissa and ordinate of any point P on this circle satisfy the conditions of λ' and γ' in Eqs. (3-49) and (3-50) for any length of line making an angle θ' with the principal axis (x axis) of longitudinal strain. Because the position of (λ',γ') relative to the λ' axis is different from that in the Mohr diagram for the unstrained state, the angle $2\theta'$ is set off in a slightly different way (Fig. 3-13A).

Because of the definition $\gamma' = \gamma/\lambda$,

$$\gamma = \frac{\gamma'}{\lambda'} \tag{3-53}$$

the tangent of any angle measured between a line from the origin to the point $P(\lambda',\gamma')$ and the λ' axis is γ'/λ', and therefore this angle is ψ, the angular shear strain for the point P (Fig. 3-13B). A rapid construction for determining the value of γ is to find where the line joining the origin to point P cuts the line $\lambda' = 1$ (Fig. 3-13B). The ordinate of this intersection (measured in the same units as γ') gives the value of γ.

The position of lines in the ellipse where γ reaches a maximum value can be found by drawing two tangents to the Mohr circle from the origin O and measuring the angle $2\theta'$ ($2\theta' = \pm 48°$). The angles between the lines of maximum shear strain and the x axis of the ellipse are therefore $\theta'_1 = \pm 24°$, and along these directions $\psi = \pm 42°$.

To determine the state of strain in any direction making an angle θ'_2 (for example, 9.5°) with the axis of principal longitudinal strain, construct a line from the center of the Mohr circle making an angle $2\theta'_2 = 19°$ with the λ' axis. The abscissa and ordinate of the point P where this line cuts the circle give the λ' and γ' values, respectively. The shear strain γ can be evaluated using the constructions described above.

The positions of lines of no finite longitudinal strain can be found where the line $\lambda' = 1$ cuts the Mohr circle. The angle $2\theta'_3$ records the direction of these lines from the principal x axis of the ellipse, $\theta'_3 = \pm 30°$.

Computation of λ_1 and λ_2 from measurements of randomly oriented objects of known original shape   In certain geological problems it may be possible to compute the shape of the strain ellipse from measurements of noncircular objects contained in the surface. The general theory and principles behind the solution of such problems will now be discussed. The particular techniques to be employed in special geological problems with deformed fossils and with boudinaged and buckled strata will be described in Chap. 5 after the theory of three-dimensional finite strain has been discussed.

In problems concerning two-dimensional strain the Mohr construction often offers a particularly convenient method of solving the strain equations and of determining the orientations and lengths of the principal axes of the strain ellipse. The computations that are feasible depend entirely on the number and nature of the strain measurements that can be made (Table 3-1).

ONE ANGLE ALONE   A knowledge of the orientation of distorted lines which were initially perpendicular may be useful for the purpose of calculating the principal strain ratio because the angular shear strain ψ can be directly measured. With only one value of ψ it is impossible to compute anything about the shape of the strain ellipse; but if the directions of the principal axes of strain are known, it is possible to calculate the ratio λ_1/λ_2. The value of ψ is first determined, together with the angle θ' between one of the lines and the principal axis of longitudinal strain. The Mohr construction now proceeds as follows:

CONSTRUCTION 1

1. Draw perpendicular axes $c\gamma'$ and $c\lambda'$, where c is a constant of unknown value.

Table 3-1 *Calculation of principal strains from randomly oriented objects in two dimensions*

Data available	Computations possible	Mohr construction
Deformed angles only:		
1. One angle between two line elements before and after deformation	None	
2. One angle initially 90° between two line elements	Angular shear strain for the direction	
3. One angle initially 90° between two line elements and the directions of the principal strain axes	λ_1/λ_2	1
4. Two angles between three line elements	λ_1/λ_2, orientation of the principal strain axes	2A and 2B
Deformed lines only:		
5. Lengths of one or two lines before and after deformation	None	
6. Lengths of two lines (λ_i and λ_j) and the orientation of the principal strain axes	λ_1, λ_2	3
7. Lengths of three line elements ($\lambda_i, \lambda_j, \lambda_k$)	λ_1, λ_2, orientations of the principal strain axes	4A and 4B
Deformed angles and lines:		
8. Length of one line and one angle between two line elements before and after deformation	None	
9. Length of one line and one angle initially 90° and the directions of the principal axes of strain	λ_1, λ_2	
10. Lengths of two lines and the angle between them	λ_1, λ_2, orientation of the principal strain axes	
11. Length of one line and two angles between three line directions	λ_1, λ_2, orientation of the principal strain axes	

2. At the origin O construct a line OP making an angle ψ with the λ' axis (Fig. 3-14*A*).

3. At any point P construct a line PC making an angle $2\theta'$ from a line drawn through P parallel to the λ' axis (Fig. 3-14*A*).

4. Construct the Mohr circle of center C, radius CP. The two points where this circle cuts the λ' axis represent the two principal axes of the strain ellipse (Fig. 3-14*B*). Determine the values of $c\lambda'_1$ and $c\lambda'_2$; hence

$$\frac{\lambda_1}{\lambda_2} = \frac{c\lambda'_2}{c\lambda'_1}$$

If the angle between the two lines was not originally 90°, no strain components can be determined from measurements of its deformed state.

TWO OR MORE ANGLES   If the dimensions of two angles before (α and β) and after (α' and β') deformation are known, it is possible to calculate the angular shear strain for three independent directions in the surface using a construction detailed below. This construction also enables the direction of the principal axes, together with the ratios of the principal longitudinal strains λ_1/λ_2, to be determined.

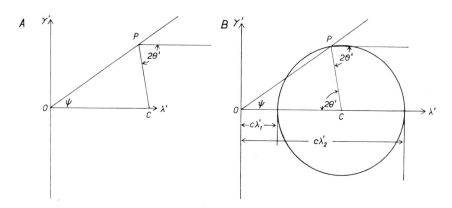

Figure 3-14

Mohr construction for calculating the ratios of the principal strains when one value of and the directions of the strain axes are known (see construction 1, p. 74).

CONSTRUCTION 2A

1. Draw any line AB and a triangle ABC such that $BAC = \alpha$ and $ABC = \beta$ (Fig. 3-15*A*).

2. Drop a perpendicular from C onto AB, meeting AB at D.

3. Construct a triangle ABC' such that $BAC' = \alpha'$ and $ABC' = \beta'$; draw $C'D$.

4. The angle CDC' gives ψ_1 for the direction of line AB, making an unknown angle θ' with the principal axis of longitudinal strain.

This construction is valid because the ratio AD/DB is unaltered by homogeneous strain and because the angular relationships of the points $ABCD$ before deformation and $ABC'D$ after deformation are dependent only on the ratios of the principal strain, not on their absolute values. Because it is possible to calculate the angles ACB and $AC'B$, a similar construction can be made to calculate the angular shear strain ψ_2 for the direction AC (Fig. 3-15B).

The primary data we have so far obtained are

1. Shear strain ψ_1 for direction AB
2. Shear strain ψ_2 for direction AC'
3. Angle α' between AB and AC'

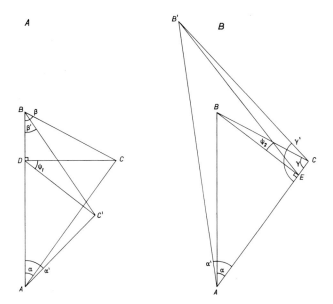

Figure 3-15

Mohr construction for determining the values of the shear strain when two or more angles before and after deformation are known.

A completed Mohr diagram is drawn in Fig. 3-16. The following relationships between two points P_1 and P_2 representing the strained states of AB and AC, respectively, on the circle are apparent:

$$\frac{p}{r} = \frac{\sin\left[180 - (\psi_1 + 2\theta')\right]}{\sin\psi_1} \tag{3-54a}$$

$$\frac{p}{r} = \frac{\sin\left[180 - (\psi_2 + 2\theta' + 2\alpha')\right]}{\sin\psi_2} \tag{3-54b}$$

Eliminating p and r and solving for $2\theta'$ (noting especially the $+ve$ or $-ve$ sign of the angular shear strain)

$$\tan 2\theta' = \frac{\sin\psi_1 \sin\psi_2 - \sin\psi_1 \sin(\psi_2 + 2\alpha')}{\cos(\psi_2 + 2\alpha')\sin\psi_1 - \cos\psi_1 \sin\psi_2} \tag{3-55}$$

where θ' is the angle between AB and the principal elongation. From this equation it is possible to compute $2\theta'$ from the data already obtained. The construction of the Mohr circle proceeds as follows:

CONSTRUCTION 2B

1. Draw the perpendicular coordinate axes γ' and λ'.
2. Construct two lines from the origin making angles ψ_1 and ψ_2 with the λ' axis (Fig. 3-16).
3. Take any point P_1 on the line making an angle ψ_1 with the λ' axis and draw P_1C at an angle $2\theta'$ with a line drawn through P_1 parallel to the λ' axis.
4. Draw the circle of center C and radius CP_1. Where this circle cuts the line of slope $\tan^{-1}\psi_2$ drawn through the origin, is P_2; check that $P_1CP_2 = 2\alpha'$.
5. The ratio λ_1/λ_2 is determined from $c\lambda'_2/c\lambda'_1$, as in construction 1.

Another graphical method of solving this problem involving less computation but more construction is presented in Sec. 5-12.

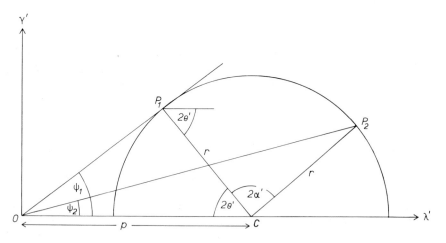

Figure 3-16

Mohr construction for determining the ratios of the principal strains when the values of two angles before and after deformation are known.

MEASUREMENTS OF LONGITUDINAL STRAIN ONLY   If only one measurement of the state of elongation can be determined, no further computation of absolute values or ratios of the principal strain can be determined even when the directions of the principal axes of strain are known.

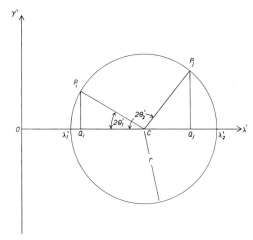

Figure 3-17

Mohr construction for determining the principal strains when the extensions along two lines and the orientation of the strain axis are known.

Where the elongations of two nonparallel lines (λ_i and λ_j) and the orientations of the principal axes of strain are known, then values of the principal longitudinal strains can be found as follows:

CONSTRUCTION 3

1. Draw any line to represent the λ' axis, mark any point C on it, and draw a circle of any convenient radius r (Fig. 3-17).
2. If θ'_1 is the angle between one of the lines and the principal axis of the strain ellipse, construct a line through C at an angle $2\theta'_1$ to the λ' axis. Where this cuts the circle at P_1, drop a perpendicular P_iQ_i on to the λ' axis.
3. Repeat step 2 above for the second line and determine the position of P_jQ_j perpendicular to the λ' axis.
4. As the quadratic elongations λ_i and λ_j are known, calculate their reciprocals λ'_i and λ'_j. The position of the origin O on the λ' axis can now be determined for $OQ_i = \lambda'_i$ and $OQ_j = \lambda'_j$.
5. Determine the values of λ'_1 and λ'_2 where the circle cuts the λ' axis.

Where the quadratic elongations of three lines are known (λ_i, λ_j, λ_k) and the angles between the lines in the strained condition (α' between lines i and j, β' between j and k), both the orientation of the principal strains and values for the principal extensions can be found. The first part of the construction entails calculating the angular shear strain for two of the line elements.

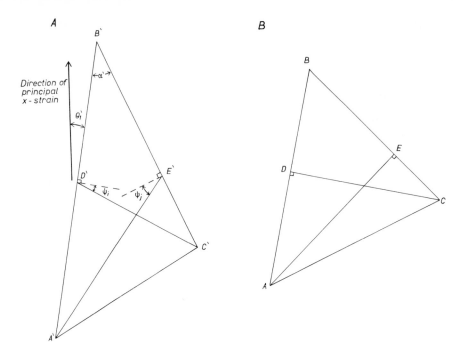

Figure 3-18

Mohr construction for determining the values of shear strain when the extensions along three lines are known.

CONSTRUCTION 4*A*

1. Draw three lines $A'B'$, $B'C'$, and $C'A'$ of any length but having the correct angular relationships of the lines i, j, k in the deformed state (Fig. 3-18A).

2. Calculate the undeformed lengths of lines $A'B'$, $B'C'$, and $C'A'$ ($AB = A'B'/(\lambda_i)^{\frac{1}{2}}$, etc.) and construct the triangle ABC representing $A'B'C'$ in an undeformed state (Fig. 3-18B).

3. Drop a perpendicular CD onto AB, and a perpendicular AE onto BC.

4. Locate the positions of D and E in the deformed state (D' and E', respectively) knowing that $A'D'/D'B' = AD/DB$, and $B'E'/E'C' = BE/EC$. Draw $C'D'$ and $A'E'$. The deflections of angles $A'D'C'$ and $B'E'A'$, from the perpendicular give the angular shear strain ψ_i and ψ_j for the directions of $A'B'$ and $B'C'$, respectively. For the lines i and j we now know both the quadratic elongation (λ_i, λ_j) and shear strain ($\gamma_i = \tan \psi_i$, $\gamma_j = \tan \psi_j$). It is now possible to construct a Mohr diagram and determine the orientations and amounts of the principal strains.

CONSTRUCTION 4*B* (Fig. 3-19)

1. Draw two perpendicular coordinate axes λ' and γ'.

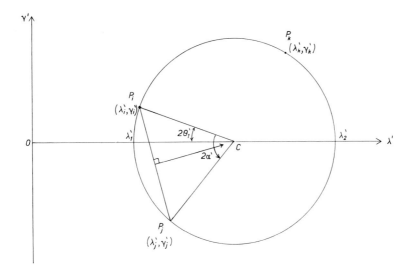

Figure 3-19

Mohr construction for determining the principal strains when the extensions along three lines are known.

2. Calculate values of λ'_i, λ'_j, $\gamma'_i = \gamma_i/\lambda_i$, and γ'_j. Plot the points $P_i(\lambda'_i, \gamma'_i)$ and $P_j(\lambda'_j, \gamma'_j)$; these must both lie on the Mohr strain circle, and therefore the right bisector of the line joining them will intersect the λ' axis at C, the center of this circle. Draw the circle of radius CP_i. Check that angle $P_iCP_j = 2\alpha'$.

3. The angle P_iCO is $2\theta'_i$, where θ'_i is the angle between the principal axis of longitudinal strain and the line i; the values of principal strains λ'_1 and λ'_2 are found where the Mohr circle cuts the λ' axis.

3-9 STRAIN INVARIANTS

There are certain properties of the strain ellipse that are obviously independent of the orientation of the coordinate axes which are used to describe it mathematically. Consider any strain ellipse with principal axes having reciprocal quadratic extensions λ'_1 and λ'_2 and with the major axis λ'_1 making an angle θ with the x axis. The reciprocal quadratic extension in directions parallel to the coordinate axes (λ'_x and λ'_y) are given by (3-31)

$$\lambda'_x = \lambda'_1 \cos^2 \theta + \lambda'_2 \sin^2 \theta$$

$$\lambda'_y = \lambda'_1 \cos^2 (90 - \theta) + \lambda'_2 \sin^2 (90 - \theta)$$

$$= \lambda'_1 \sin^2 \theta + \lambda'_2 \cos^2 \theta$$

Therefore
$$\lambda'_x + \lambda'_y = \lambda'_1 + \lambda'_2 = J_1 \tag{3-56}$$

This important relationship describes what is known as a *strain invariant*. In words it means that whenever we sum the reciprocal quadratic extensions along two mutually perpendicular directions in the ellipse, this sum is always a constant quality (J_1) and equal to the sum of the principal reciprocal quadratic extensions.

The product of the reciprocal quadratic extensions $\lambda'_x \lambda'_y$ can similarly be obtained.

$$\lambda'_x \lambda'_y = (\lambda'_1 \cos^2 \theta + \lambda'_2 \sin^2 \theta)(\lambda'_1 \sin^2 \theta + \lambda'_2 \cos^2 \theta)$$

$$= \lambda'^2_1 \sin^2 \theta \cos^2 \theta + \lambda'_1 \lambda'_2(\sin^4 \theta + \cos^4 \theta) + \lambda'^2_2 \sin^2 \theta \cos^2 \theta$$

$$= \lambda'_1 \lambda'_2(\sin^2 \theta + \cos^2 \theta)^2 + (\lambda'_1 - \lambda'_2)^2 \sin^2 \theta \cos^2 \theta$$

From Eq. (3-42) this simplifies to give the following relationship:

$$\lambda'_1 \lambda'_2 = \lambda'_x \lambda'_v - \gamma'^2_{xy} = J_2 \tag{3-57}$$

This equation expresses the *second strain invariant* of the strain ellipse.

Principal strains determined from the invariants   If the two reciprocal principal strains are $\lambda' = \lambda'_1$ and $\lambda' = \lambda'_2$, they can be thought of as the roots of the quadratic equation

$$(\lambda' - \lambda'_1)(\lambda' - \lambda'_2) = 0$$

or $$\lambda'^2 - (\lambda'_1 + \lambda'_2)\,\lambda' + \lambda'_1 \lambda'_2 = 0 \tag{3-58}$$

Because the two invariants express the sum and product of the principal strains, this equation can be rewritten using the invariant quantities J_1 and J_2.

$$\lambda'^2 - J_1 \lambda' + J_2 = 0 \tag{3-59}$$

The principal strains expressed in terms of the invariants are

$$\lambda'_1 = \frac{J_1 - (J_1^2 - 4J_2)^{\frac{1}{2}}}{2} \tag{3-60a}$$

$$\lambda'_2 = \frac{J_1 + (J_1^2 - 4J_2)^{\frac{1}{2}}}{2} \tag{3-60b}$$

The value of θ is determined from the relationship (3-31)

$$\lambda'_x = \lambda'_1 \cos^2 \theta + \lambda'_2 \sin^2 \theta$$

that is, $$\sin \theta = [(\lambda'_x - \lambda'_1)/(\lambda'_2 - \lambda'_1)]^{\frac{1}{2}} \tag{3-61}$$

The method for determining the values and orientations of the principal strains using the calculation of invariants is best illustrated by the following numerical example:

Line AB, initially of unit length, is now 0.89 unit, and the angular shear along AB is 14°. Line CD is perpendicular to AB; it was initially of unit length but is now 1.15 units long.

From these data we calculate the reciprocal quadratic extensions for each line and the value of γ' for AB.

$$\lambda'_{AB} = \left(\frac{1}{0.89}\right)^2 = 1.25$$

$$\lambda'_{CD} = \left(\frac{1}{1.15}\right)^2 = 0.75$$

$$\gamma' = \lambda'_{AB} \tan \psi = 1.25 \tan 14 = 0.31$$

From (3-56) and (3-57)

$$J_1 = 1.25 + 0.75 = 2.0$$

$$J_2 = 1.25 \times 0.75 - (0.31)^2 = 0.84$$

The principal strains are therefore

$$\lambda'_1 = \frac{2 - (4 - 3.36)^{\frac{1}{2}}}{2} \qquad \lambda'_2 = \frac{2 + (4 - 3.36)^{\frac{1}{2}}}{2}$$

$$= \underline{0.61} \qquad\qquad = \underline{1.39}$$

The angle between the λ'_1 axis and the line AB is calculated thus

$$\sin \theta = \left(\frac{1.25 - 0.61}{1.39 - 0.61}\right)^{\frac{1}{2}} \qquad \theta = 65.1°$$

Although the determination of the principal strains in two dimensions is generally more easily carried out using techniques which employ the Mohr diagram, a thorough understanding of this method is important in determining the principal strains in three-dimensional problems.

3-10 SIMPLE SHEAR

One special type of rotational deformation produced by the shearing of a substance along a series of discrete (although often closely spaced) shear planes is known as *simple shear* and is a rather common feature of certain geological processes. The orientations of the principal axes of the finite-strain ellipse depend on the amount of shear.

If we take a point (x,y) on a circle of unit radius $x^2 + y^2 = 1$, this point, after distortion by a simple shear strain γ, is displaced to the position (x_1, y_1) (Fig. 3-20) such that

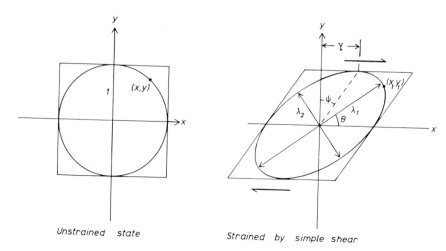

Figure 3-20

Distortion of a unit circle by simple shear.

$$y_1 = y \tag{3-62a}$$

$$x_1 = x + \gamma y \quad \text{or} \quad x = x_1 - \gamma y_1 \tag{3-62b}$$

We can find the equation of the strain ellipse by replacing (x,y) by these values in $x^2 + y^2 = 1$, obtaining

$$(x_1 - \gamma y_1)^2 + y_1^2 = 1$$

or
$$x_1^2 - 2\gamma x_1 y_1 + (1 + \gamma^2) y_1^2 = 1 \tag{3-63}$$

To find the lengths of the principal axes of this ellipse (that is, the principal strains), we proceed by the normal technique of coordinate geometry and find where the circle $x^2 + y^2 = r^2$ of radius r just touches the ellipse. With these conditions the radius will equal either one or the other of the principal axes of the ellipse, that is, $r^2 = \lambda_1$ or λ_2. The circle $x^2 + y^2 = r^2$ cuts the ellipse in a pair. of straight lines (real or imaginary depending on the value of r) which are given by subtracting $x^2/r^2 + y^2/r^2 = 1$ from (3-63),

$$x^2 \left(1 - \frac{1}{r^2}\right) - 2\gamma xy + y^2 \left(1 + \gamma^2 - \frac{1}{r^2}\right) = 0 \tag{3-64}$$

If the circle just touches the ellipse, the pair of straight lines will coincide. If this equation represents the coincident lines

$$(ax - by)(ax - by) = 0$$

that is,
$$a^2 x^2 - 2abxy + b^2 y^2 = 0$$

then the product of the coefficients of the terms in x^2 and y^2 must equal the

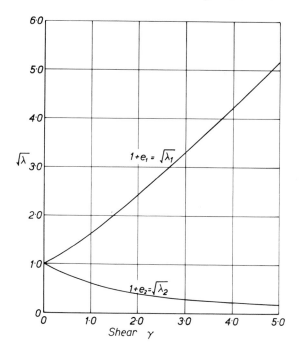

Figure 3-21

Values of the principal strains in simple shear.

square of half the coefficient of the term in xy. Applying this condition to (3-64),

$$\left(1 - \frac{1}{r^2}\right)\left(1 + \gamma^2 - \frac{1}{r^2}\right) = \gamma^2 \qquad (3\text{-}65)$$

Solving this for r

$$r^4 - r^2(\gamma^2 + 2) + 1 = 0 \qquad (3\text{-}66)$$

or
$$r^2 = \lambda_1 \text{ or } \lambda_2 = \frac{\gamma^2 + 2 \pm \gamma(\gamma^2 + 4)^{\frac{1}{2}}}{2} \qquad (3\text{-}67)$$

This equation gives the two principal strains and has been solved for some values of γ in Fig. 3-21.

To calculate the angle θ between the major axis and the shear plane, we determine the slope of the coincident straight lines.

Multiplying (3-64) by $(1 + \gamma^2 - 1/\lambda_1)$ and replacing r^2 by λ_1,

$$x^2\left(1 - \frac{1}{\lambda_1}\right)\left(1 + \gamma^2 - \frac{1}{\lambda_1}\right) - 2\gamma\left(1 + \gamma^2 - \frac{1}{\lambda_1}\right)xy + y^2\left(1 + \gamma^2 - \frac{1}{\lambda_1}\right)^2 = 0 \qquad (3\text{-}68)$$

and using condition (3-65),

$$x^2\gamma^2 - 2\gamma\left(1 + \gamma^2 - \frac{1}{\lambda_1}\right)xy + y^2\left(1 + \gamma^2 - \frac{1}{\lambda_1}\right)^2 = 0$$

which is a perfect square

$$\left[\gamma x - \left(1 + \gamma^2 - \frac{1}{\lambda_1}\right)y\right]^2 = 0 \tag{3-69}$$

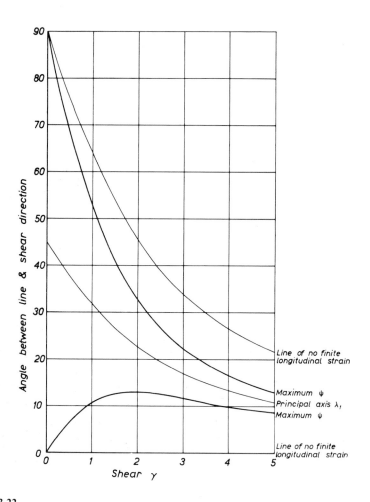

Figure 3-22

Orientations of the principal extensions, lines of maximum finite shear strain, and lines of no finite longitudinal strain in simple shear.

The slope of these coincident lines gives θ

$$\tan \theta = \frac{\gamma}{1 + \gamma^2 - 1/\lambda_1} \qquad (3\text{-}70)$$

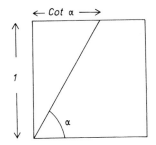
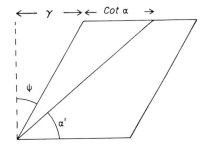

Figure 3-23

Displacement of a line making an angle α with the planes of simple shear.

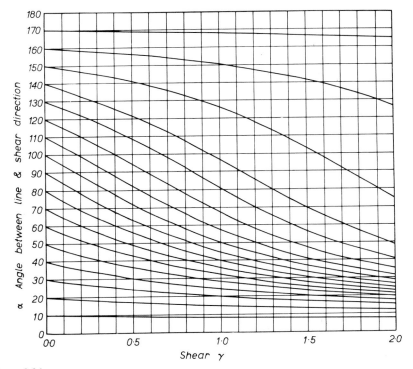

Figure 3-24

Graphical solution to Eq. (3-71).

In Fig. 3-22 this is solved for values of γ ranging from 0 to 5. Using Eqs. (3-33) and (3-45), the angles made by lines of no finite elongation and maximum shearing strain to the plane of simple shear have been calculated. The directions of the principal axes do not coincide with the positions of the diagonals of the parallelogram made by deforming an original square (Fig. 3-20), and neither of the lines of maximum finite shearing strain coincide with the actual surface of shear.

Another feature of simple shear deformation is that lines originally making an angle α with the shear plane are sheared to make a new angle α' with the plane. From Fig. 3-23 the relationships before and after deformation lead to

$$\cot \alpha' = \cot \alpha + \gamma \tag{3-71}$$

which is graphically recorded in Fig. 3-24.

Structural features in rocks which are believed to have been deformed by simple shear often accord well with this analysis. At the start of simple shear deformation the principal axes of strain make an angle of 45° with the shear plane. Providing that the rock does not contain fractures already, tension joints and fissures may develop perpendicular to the maximum elongation at an

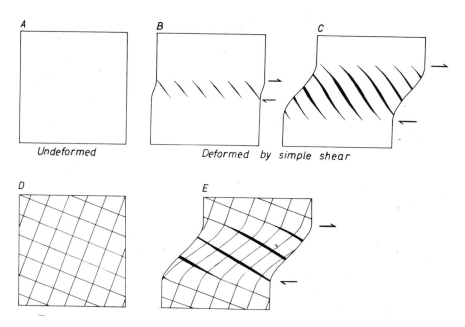

Figure 3-25

The development of zones of en echelon tension fissures by simple shear deformation. Parts A, B, and C illustrate the structures that are likely to develop as a result of progressive simple shear in an isotropic rock. Parts D and E illustrate the structures which might form in a previously jointed rock.

angle of 45° to the shearing surface. These joints are most frequently arranged in "en echelon" fashion (Fig. 3-25B, 3-26). As the tension fissures open up, it is common for material to crystalize in the cavity; especially common are quartz and carbonates. Often these crystals show a fibrous structure with their longest dimensions oriented parallel to the maximum elongation or dilation across the vein (Fig. 3-27). Occasionally a slaty cleavage is developed perpendicular to the maximum finite compressive strain and perpendicular to the tension fissures (Fig. 3-26). As the simple shear deformation proceeds, the already-formed en echelon cracks are themselves distorted according to the relationships of Eq. (3-71) and Fig. 3-24. The amount of shear displacement in the shear zone varies across the zone, and this variation gives rise to sigmoidally distorted tension gashes with the axis of folding perpendicular to the shear displacement (Fig. 3-28). The nature of the sigmoidal form can be used to confirm the sense of shear motion; and also, because the tension cracks were initially formed at an angle of 45° (135° in the sense shown in Fig. 3-24), deflections from this angle can be used to calculate the amount of shear displacement along this zone.

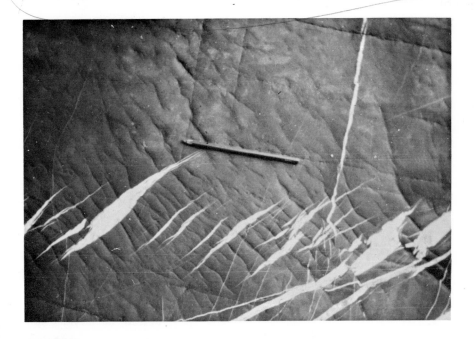

Figure 3-26

A zone of en echelon quartz-filled tension gashes produced by simple shear. A crude slaty cleavage is also developed perpendicular to the maximum compressive strain and to the tension cracks. Carboniferous graywacke, near Bude, Devon, England.

Figure 3-27

Fibrous intergrowth of quartz and calcite in tension gashes in deformed sediments, Valais, Switzerland. The curving form of the fibers results from distortion during progressive deformation.

Figure 3-28

Sigmoidally folded en echelon tension gashes, Tertiary sediments, Klausen Pass, Switzerland.

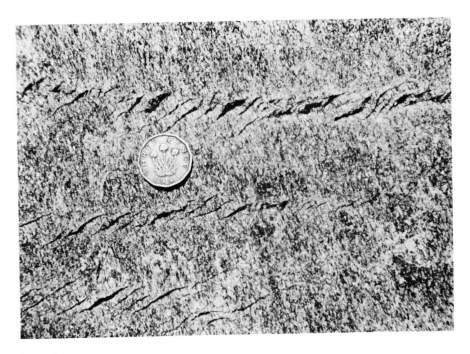

Figure 3-29

Several zones of en echelon gashes produced by the opening of preexisting joint planes in the surrounding granitic gneiss (cf. Fig. 3-25E), Lower Pennine nappes, Ticino, Switzerland.

It must again be stressed that before such calculations can be made it is imperative to verify that no fractures existed in the rock before the simple shear process. These may be opened during the deformation and the geometry of dilated preexisting fractures is different from that of fractures produced during the process of strain. For example, the actual dilation movements will not be perpendicular to the walls of the fissure, and frequently zones of en echelon cracks which are initiated at angles different from 45° are formed (Fig. 3-25D, E; Fig. 3-29). Sometimes the actual dilation in such veins is indicated by the "stretching fiber" in the crystals that have grown in the fissure.

3-11   THE SUPERPOSITION OF TWO FINITE STRAINS

It has been shown in Eq. (3-12) that where two homogeneous strains are superimposed one on the other the result is a total homogeneous strain which can be analyzed in terms of a single strain ellipse. The overprinting of one deformation on another is fairly common in geological deformations, and so the geometry of this process will be examined in further detail.

The equation of any strain ellipse with principal elongation λ_1 situated at an angle θ from the x axis can be determined by rotating the ellipse $x^2/\lambda_1 + y^2/\lambda_2 = 1$ through θ using the rotation transformation (3-18) for the new positions of x and y.

$$\frac{(x \cos \theta + y \sin \theta)^2}{\lambda_1} + \frac{(-x \sin \theta + y \cos \theta)^2}{\lambda_2} = 1$$

$$(\cos^2 \theta\, \lambda_1' + \sin^2 \theta\, \lambda_2')x^2 - 2(\lambda_2' - \lambda_1')\sin \theta \cos \theta\, xy$$

$$+ (\sin^2 \theta\, \lambda_1' + \cos^2 \theta\, \lambda_2')y^2 = 1 \quad (3\text{-}72)$$

It will be seen from (3-31) and (3-42), however, that the coefficients of x^2, y^2, and xy are, respectively, λ_x', λ_y', and $2\lambda_{xy}'$ for the x and y axis directions. This ellipse can be expressed in a very simple form

$$\lambda_x' x^2 - 2\gamma_{xy}' xy + \lambda_y' y^2 = 1 \quad (3\text{-}73)$$

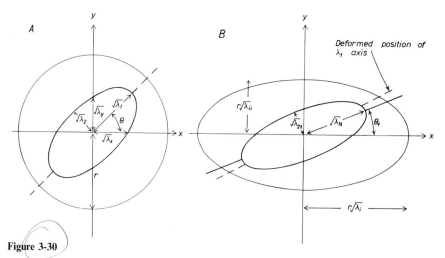

Figure 3-30

The homogeneous deformation of an ellipse; A, before deformation; B, after deformation.

We shall now superimpose on this ellipse (Fig. 3-30A) a second strain, a homogeneous irrotational deformation with principal axes parallel to the x and y axes and with principal reciprocal quadratic extensions of λ_i' and λ_{ii}' (Fig. 3-30B). The points lying on the ellipse (3-73) are transformed according to

$$x' = x(\lambda_i')^{\frac{1}{2}} \quad \text{or} \quad x = x'(\lambda_i')^{\frac{1}{2}}$$

$$y' = y(\lambda_{ii}')^{\frac{1}{2}} \quad \text{or} \quad y = y'(\lambda_{ii}')^{\frac{1}{2}}$$

and the finite ellipse representing the total strain becomes

$$\lambda_x' \lambda_i' x^2 - 2\gamma_{xy}'(\lambda_i' \lambda_{ii}')^{\frac{1}{2}} xy + \lambda_y' \lambda_{ii}' y^2 = 1 \quad (3\text{-}74)$$

From the form of (3-73) we see that this ellipse has coefficients of terms in x^2, y^2, and xy that can be interpreted as follows:

$\lambda'_x \lambda'_i = \lambda'_{xt}$ the reciprocal quadratic extension in the x direction

$\lambda'_y \lambda'_{ii} = \lambda'_{yt}$ the reciprocal quadratic extension in the y direction

$\gamma'_{xy}(\lambda'_i \lambda'_{ii})^{\frac{1}{2}} = \gamma'_t$ the value of γ' measured along the x and y directions

With these coefficients we can easily determine the orientation and length of the principal axes of the total strain ellipse. This can be done most conveniently by applying the strain invariant theory (3-60), (3-61). The two invariants are

$$J_1 = \lambda'_x \lambda'_i + \lambda'_y \lambda'_{ii}$$

$$J_2 = \lambda'_x \lambda'_i \lambda'_y \lambda'_{ii} - \gamma'_{xy}{}^2 \lambda'_i \lambda'_{ii}$$

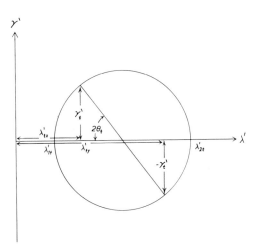

Figure 3-31

Mohr diagram solution to the problem of the strain of initially elliptical particles.

The problem can also be solved graphically using the Mohr circle construction (Fig. 3-31). The two points $(\lambda'_{tx}, -\gamma'_t)$ and $(\lambda'_{ty}, -\gamma'_t)$ are plotted, and where the line which joins them cuts the λ' axis, lies the center of the Mohr circle which gives the principal total strains λ'_{1t} and λ'_{2t}. The angle θ_t between the λ'_{1t} axis and the x axis is determined by measuring the angle $2\theta_t$ in the Mohr diagram.

The positions of the lines of maximum and minimum elongation are of special interest because in deformed rocks these lines may mark the position of some special structural feature. For example, the structure known as *slaty cleavage* is known to develop perpendicular to the maximum shortening direction; if it had developed as a result of the first deformation shown in Fig. 3-30A, it would be located in a plane normal to the surface of the diagram through line λ_1. If this slaty cleavage is now deformed by a second strain,

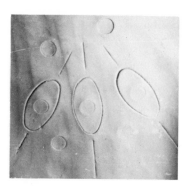

Figure 3-32

Ellipses before straining, with the lines marking the major axes of each ellipse.

Figure 3-33

Ellipses of Fig. 3-32 after a homogeneous strain (measured by the small uniform strain ellipses).

where are we likely to find the slaty cleavage? Any cleavage that developed entirely as a result of the second strain would be found in a plane containing the x axis, that is, normal to the maximum shortening along the y axis, λ_{ii}. It is also possible, however, that cleavage might reflect the total strain and be developed along a plane containing λ_{1t}. There is also the possibility that a preexisting structure formed along λ_1 would be passively sheared by the second rotation according to Eq. (3-34) and take up the position indicated in Fig. 3-30B (see also Figs. 3-32 and 3-33).

3-12   METHODS OF GRAPHICALLY RECORDING THE SHAPE OF THE FINITE-STRAIN ELLIPSE

It is often useful to be able to illustrate the shapes of strain ellipses in some diagrammatic way. If the data available lends itself only to the calculation of the ratios of the principal strains, then it is not possible to represent the linear variations of different ellipses very sensibly using graphical methods. If, however, we know the absolute values of the principal strains, then we can plot any strain ellipse on a graph in which the abscissa records one strain and the ordinate the other. Figure 3-34 is an example of this type of diagram, recording λ_1 as the abscissa and λ_2 as the ordinate. Because, by definition, $\lambda_1 \geqslant \lambda_2$, the plots of points representing strain ellipses all fall on or below a line of unit slope drawn through the origin. The point (1, 1) represents an undeformed circle, and all dilations where $\lambda_1 = \lambda_2$ fall along the line of unit slope through the origin. All other strain ellipses fall into one of three fields: (1) those which fall above the ordinate $\lambda_2 = 1$ where both principal extensions are positive, (2) those which fall to the left of the abscissa $\lambda_1 = 1$ where both principal extensions are negative, and (3) those between two fields where one

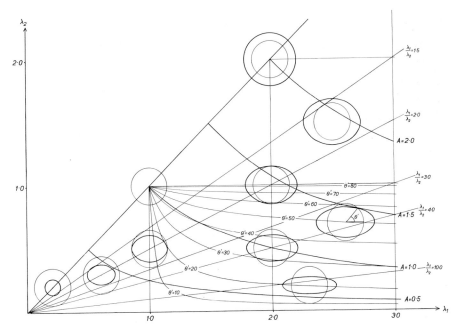

Figure 3-34

Graphical method of representing the shape of any strain ellipse.

is positive and the other negative. Only in the third field do the ellipses have lines of no finite longitudinal strain making angles of θ' with the axis λ_1. Lines of equal θ' have been plotted in this diagram using Eqs. (3-32c) and (3-34) in the form

$$\lambda_2 = \frac{\tan^2 \theta'}{\tan^2 \theta' + 1 - 1/\lambda_1} \qquad (3\text{-}75)$$

Ellipses with the same ratios of principal axes λ_1/λ_2 lie along lines of different slope through the origin. Only along one hyperbola in this diagram (the line $A = 1$ where $\lambda_1 = 1/\lambda_2$) is the area of the ellipse equal to the area of the undeformed circle from which it was derived. All other ellipses show an area change; lines of equal change in area are shown by the various hyperbolas $A = 1.5$, $A = 0.5$, and $A = 2.0$.

Although a two-dimensional diagram such as Fig. 3-34 is very useful in helping one to visualize different types of strain ellipse, it does not completely represent the type of ellipse because it does not indicate whether the deformation was a rotational or irrotational strain. This means, for example, that ellipses formed by pure shear are not separated from those formed by simple shear. Every finite two-dimensional strain can be described completely in terms

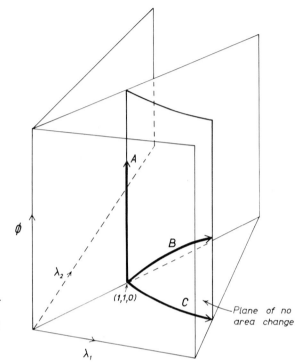

Figure 3-35

Three-dimensional graph for completely representing the two-dimensional finite-strain state. Irrotational strains are recorded on the basal plane of the diagram ($\phi = 0$).

of the strain-ellipse axes and the angle of rotation ϕ made by the principal axes. If, therefore, we make a three-dimensional diagram (Fig. 3-35) and plot the rotation ϕ of the ellipse in a direction perpendicular to the $\lambda_1\lambda_2$ plane, we can completely represent the strains and rotations within the two-dimensional surface. The point (1,1,0) represents the undeformed circle, and all the finite ellipses of no area change lie on a surface containing this point, the line $A = 1.0$ on Fig. 3-34 and the vertical direction. All ellipses (in fact, circles) which lie on line A in Fig. 3-35 show only a rigid-body rotation and no internal strain, those along line B are formed by simple shear [their λ_1, λ_2, and ϕ values satisfy (3-67) and (3-70) simultaneously for any given value of shear], and those which lie along line C represent the ellipses formed by pure shear. It is clear from this diagram that there are many other equal-area ellipses which have not been formed by any of these special processes.

3-13   INFINITESIMAL STRAIN IN TWO DIMENSIONS

The theory of infinitesimal strain is an investigation into the nature of the states of strain that are produced by very small displacements. It is developed by assuming that certain expressions which normally appear in the equations

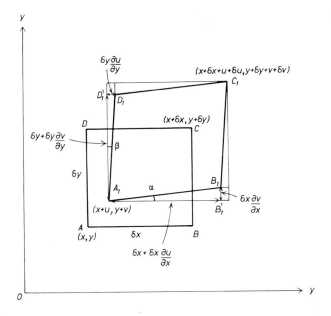

Figure 3-36

The general distortion of infinitesimal strain. The original rectangle ABCD is transformed to the new position $A_1B_1C_1D_1$.

for finite strain and which depend on products, squares, and higher powers of these very small quantities can be neglected without leading to any significant inaccuracy. These simplifying assumptions are true when the strains are less than about 1 percent, but begin to break down as soon as the strain reaches 2 percent.

Infinitesimal-strain theory is the standard strain analysis used by the engineer who deals with small elastic distortion, but obviously for the large finite strains commonly developed in deformed rocks its assumptions are invalid. However, if infinitesimal strains are thought of in a slightly different way, the theory does have considerable application to geological deformation. Every large strain can be thought of as the product of a great number of very small or infinitesimal strains which are superimposed on each other; thus at any one time during the history of deformation, although the finite strain may be large, the changes that are actually going on depend on the infinitesimal-strain conditions at that instant. The infinitesimal strains reflect the rate of change of the finite strain in different directions in the substance. It will be shown later how lines oriented in a certain direction that have been contracted by finite strain may, at a certain instant, be in the act of expanding. Structures which suggest a compression followed by an extension may not always imply the superposition of two separate and distinct deformations; they can be the result of a single progressive deformation.

To develop the theory of infinitesimal strain in two dimensions we consider two points which in the undeformed state have coordinates (x,y) and $(x + \delta x, y + \delta y)$ located at the opposite corners A and C of a rectangle $ABCD$ (Fig. 3-36). These are infinitesimally deformed and take up new positions $A_1(x + u, y + v)$ and $C_1(x + \delta x + u + \delta u, y + \delta y + v + \delta v)$ at the opposite corners of the parallelogram $A_1B_1C_1D_1$.

The rates of change of the displacement u in the x and y directions are $\partial u/\partial x$ and $\partial u/\partial y$, and the rates of change of the displacement v in the x and y directions are $\partial v/\partial x$ and $\partial v/\partial y$, respectively. Because the changes in length of AB are very small and the length of A_1B_1 does not differ very much from that of its projection onto the x axis A_1B_1', we may conclude that the deformed length is very nearly equal to δx plus δx multiplied by the rate of change of displacement of u in the x direction, that is,

$$A_1B_1 = A_1B_1' = \delta x \left(1 + \frac{\partial u}{\partial x}\right)$$

Using similar reasoning,

$$A_1D_1 = A_1D_1' = \delta y \left(1 + \frac{\partial v}{\partial y}\right)$$

$$B_1B_1' = \delta x \frac{\partial v}{\partial x}$$

$$D_1D_1' = \delta y \frac{\partial v}{\partial y}$$

The infinitesimal strains in the x direction are

$$e_x = \frac{A_1B_1 - AB}{AB} = \frac{\delta x(1 + \partial u/\partial x) - \delta x}{\delta x} = \frac{\partial u}{\partial x} \tag{3-76}$$

$$e_y = \frac{A_1D_1 - AD}{AD} = \frac{\partial v}{\partial y} \tag{3-77}$$

Because the strains are very small and $\delta x(\partial u/\partial x)$ is very small compared with δx,

$$\tan \alpha = \alpha = \frac{\partial v}{\partial x} \quad \text{and} \quad \tan \beta = \beta = \frac{\partial u}{\partial y}$$

As angle DAB was originally 90°, the infinitesimal shear strain $\gamma = \psi = \alpha + \beta$

$$\gamma_{xy} = \frac{\partial v}{\partial x} + \frac{\partial u}{\partial y} \tag{3-78}$$

We have now determined the three fundamental parameters of infinitesimal strain; but as we have expressed them in terms of displacements with reference only to the two directions x and y, they are not all independent, for

$$e_x = \frac{\partial u}{\partial x} \qquad \frac{\partial e_x}{\partial y} = \frac{\partial^2 u}{\partial x \partial y} \qquad \frac{\partial^2 e_x}{\partial y^2} = \frac{\partial^3 u}{\partial x \partial y^2}$$

$$e_y = \frac{\partial v}{\partial y} \qquad \frac{\partial e_y}{\partial x} = \frac{\partial^2 v}{\partial x \partial y} \qquad \frac{\partial^2 e_y}{\partial x^2} = \frac{\partial^3 v}{\partial x^2 \partial y}$$

$$\gamma_{xy} = \frac{\partial v}{\partial x} + \frac{\partial u}{\partial y} \qquad \frac{\partial \gamma_{xy}}{\partial x} = \frac{\partial^2 v}{\partial x^2} + \frac{\partial^2 u}{\partial x \partial y} \qquad \frac{\partial^2 \gamma_{xy}}{\partial x \partial y} = \frac{\partial^3 v}{\partial x^2 \partial y} + \frac{\partial^3 u}{\partial x \partial y^2}$$

$$\frac{\partial^2 \gamma_{xy}}{\partial x \partial y} = \frac{\partial^2 e_x}{\partial y^2} + \frac{\partial^2 e_y}{\partial x^2} \tag{3-79}$$

This equation expresses what is known as the *compatibility condition* for the strains.

The displacements given by the relationships derived from Fig. 3-36,

$$\delta u = \frac{\partial u}{\partial x} \delta x + \frac{\partial u}{\partial y} \delta y \tag{3-80a}$$

$$\delta v = \frac{\partial v}{dx} \delta x + \frac{\partial v}{\partial y} \delta y \tag{3-80b}$$

are perfectly linear transformations which obey the geometrical principles for homogeneous strain; that is, straight lines are transformed to straight lines, and a circle is transformed into an ellipse known as the *infinitesimal-strain ellipse*.

3-14 CHANGES IN LENGTH OF LINES

We shall now determine the infinitesimal elongation of a line which made an angle θ with the x axis before deformation. Let the line AC, of length l, be inclined to the x axis at an angle θ. Then the length of the sides of the rectangle $ABCD$ are given by

$$\delta x = l \cos \theta \qquad \delta y = l \sin \theta$$

After deformation the line AC is extended to $A_1 C_1$ of length l_1; then by Pythagoras

$$l_1^2 = \left[l \cos \theta \left(1 + \frac{\partial u}{\partial x} \right) + l \sin \theta \frac{\partial u}{\partial y} \right]^2 + \left[l \sin \theta \left(1 + \frac{\partial v}{\partial y} \right) + l \cos \theta \frac{\partial v}{dx} \right]^2$$

$$\tag{3-81}$$

Multiplying out, ignoring squares and products of small quantities like $\partial u/\partial x$, and putting the differentials in the forms of strain components,

$$l_1^2 = l^2(1 + 2e_x \cos^2 \theta + 2e_y \sin^2 \theta + 2\gamma_{xy} \sin \theta \cos \theta)$$

Taking the square roots, remembering that squares and products of e_x, e_y, and γ are very small indeed,

$$l_1 = l(1 + e_x \cos^2 \theta + e_y \sin^2 \theta + \gamma_{xy} \sin \theta \cos \theta) \tag{3-82}$$

The extension along the line AC is given by $e = (l_1 - l)/l$

$$e = e_x \cos^2 \theta + e_y \sin^2 \theta + \gamma_{xy} \sin \theta \cos \theta \tag{3-83}$$

The lines of maximum and minimum extension can be found by determining where $de/d\theta = 0$, that is, where

$$2 \sin \theta \cos \theta \, (e_y - e_x) + \gamma_{xy}(\cos^2 \theta - \sin^2 \theta) = 0$$

Dividing through by $\cos^2 \theta$

$$2(e_y - e_x) \tan \theta + \gamma_{xy}(1 - \tan^2 \theta) = 0$$

$$\tan^2 \theta + \frac{2(e_x - e_y)}{\gamma_{xy}} \tan \theta - 1 = 0 \tag{3-84}$$

The two roots of this equation give the original orientation of the two perpendicular lines which become the principal axes of the infinitesimal-strain ellipse, and these directions are known as the *principal axes of infinitesimal strain*. The values of the *principal infinitesimal strains* along these lines are $e_1 \geqslant e_2$, and they can be determined by substituting the two calculated values of θ from (3-84) in Eq. (3-83).

The two perpendicular lines which, before deformation, are the principal axes of strain may or may not coincide exactly with the principal extensions of the infinitesimal-strain ellipse. If they coincide, the deformation is *irrotational*; if they do not, the deformation is a *rotational strain*. As was found with finite strain, rotational infinitesimal strain can be divided into two components: first, the small angular rotation ω of a rigid body, and second, an irrotational part producing the strain ellipse. The *rigid-body rotation* is defined as the difference in angle between the bisectors of the angles BAD (or $B_1'A_1D_1'$) and $B_1A_1D_1$ (Figs. 3-36 and 3-37).

$$\omega = \frac{\pi}{4} - \left(\frac{\pi}{4} - \frac{\alpha + \beta}{2}\right) - \beta = \frac{\alpha}{2} - \frac{\beta}{2} = \frac{1}{2}\left(\frac{\partial v}{\partial x} - \frac{\partial u}{\partial y}\right) \tag{3-85}$$

Lines of no infinitesimal longitudinal strain The positions of these lines in terms of the strain components referred to the axes x and y can be found

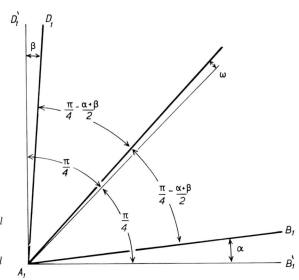

Figure 3-37

The significance of infinitesimal rotational strain ω in terms of the angular distortion made by the sides of the block A_1B_1 and A_1D_1 in Fig. 3-36.

by simply putting $e = 0$ in (3-83) and dividing by $\cos^2 \theta$,

$$\tan^2 \theta + \frac{\gamma_{xy}}{e_y} \tan \theta + \frac{e_x}{e_y} = 0 \tag{3-86}$$

which gives two values of θ such that

$$\tan \theta = \frac{-\gamma_{xy} \pm (\gamma_{xy}^2 - 4e_x e_y)^{\frac{1}{2}}}{2e_y}$$

However, the positions of these lines are much more simply referred to the principal strains of the infinitesimal-strain ellipsoid e_1 and e_2 by adapting (3-29) to infinitesimal-strain conditions.

$$(1 + e)^2 = (1 + e_1)^2 \cos^2 \theta + (1 + e_2)^2 \sin^2 \theta$$

With the usual approximations, this becomes

$$e = e_1 \cos^2 \theta + e_2 \sin^2 \theta \tag{3-87}$$

where θ refers to the angle measured from the principal strain axis e_1. Where there is no longitudinal strain $e = 0$, and (3-87) gives the condition

$$\tan^2 \theta = \frac{-e_1}{e_2} \tag{3-88}$$

This equation, which shows that the positions of lines of no infinitesimal longitudinal strains are dependent only on the *ratios* of the principal strains, will only have roots where the signs of the principal strains are different. It should be compared with the more complicated relationships for finite strain; the positions of lines of no longitudinal strain given by Eq. (3-32) depend on absolute values of the principal finite strains.

3-15 INFINITESIMAL SHEAR STRAIN

The variations in shear strain are also most simply described in terms of the angle θ that the direction makes with the principal axis e_1 of the infinitesimal-strain ellipse. Converting the equation for finite shear strain (3-39) to infinitesimal-strain conditions with the usual approximations

$$\gamma = \frac{(1 + e_1)^2 - (1 + e_2)^2}{(1 + e_1)(1 + e_2)} \sin \theta \cos \theta$$

$$= \frac{2(e_1 - e_2)}{1 + e_1 + e_2} \sin \theta \cos \theta \qquad (3\text{-}89)$$

Where the strains are very small, $(1 + e_1 + e_2)$ does not differ significantly from 1, and (3-89) approximates very closely to

$$\gamma = 2(e_1 - e_2) \sin \theta \cos \theta \qquad (3\text{-}90)$$

If we make the usual substitution for double angles in (3-87) and (3-90), we get

$$e = \frac{e_1 + e_2}{2} + \frac{e_1 - e_2}{2} \cos 2\theta \qquad (3\text{-}91)$$

$$\frac{\gamma}{2} = \frac{e_1 - e_2}{2} \sin 2\theta \qquad (3\text{-}92)$$

These two equations are in the form of (3-51a) (3-51b), respectively, and therefore a complete representation of the state of infinitesimal strains can be made with a Mohr circle construction with e as abscissa and $\gamma/2$ as ordinate.

Maximum infinitesimal shear strain Maximum values of γ will occur where $d\gamma/d\theta = 0$ in (3-90)

$$\frac{d\gamma}{d\theta} = 2(e_1 - e_2) \cos 2\theta = 0 \qquad (3\text{-}93)$$

and this will occur where $\theta = \pm 45°$. No matter what the values of the principal infinitesimal strains are, maximum shearing strain always occurs along directions at 45° to the principal axis and the value of this strain is $\pm(e_1 - e_2)$. These directions should be compared with the finite-strain conditions where maximum shearing strain occurs along lines making an angle of $\tan^{-1} (\lambda_2/\lambda_1)^{\frac{1}{4}}$ to the principal elongation.

3-16 INFINITESIMAL AREA CHANGE

Any change in area during strain is given by the difference in area between that of the infinitesimal-strain ellipse and the unit circle from which it was derived.

$$\text{Area change} = \Delta = \frac{\pi(1 + e_1)(1 + e_2) - \pi}{\pi} = e_1 + e_2 \qquad (3\text{-}94)$$

3-17 GEOLOGICAL IMPLICATIONS OF DEFORMATION IN TWO DIMENSIONS

The analysis of two-dimensional finite strain has shown that deformation results in systematic changes in length, and that in any homogeneously deformed plane surface there are always perpendicular directions of maximum and minimum elongation. These changes in length often have very considerable consequences in controlling the types of minor structures that are developed. If the rock substances that are being deformed are perfectly homogeneous throughout, then the material undergoes only a homogeneous shape change. However, most rocks are heterogeneous, the commonest type of heterogeneity being a layering. If some layers in the rock flow less easily than others—a property known as *competence difference*—then they will behave differently from their surrounding and more ductile material. The various structures that develop in layered inhomogeneous materials depend primarily on whether the deformation has led to extension or contraction.

Structures which result from extension   If there is extension in the plane of banding, the more competent bands often break apart. The more competent band will not extend as easily as the ductile matter enclosing it; large tensile

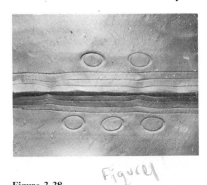

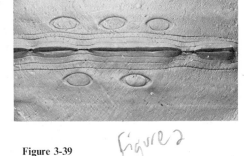

Figure 3-38   Figure 3-39

Experiments showing the effect of progressive elongation of a competent layer enclosed in a more ductile matrix, the development of pinch-and-swell structure and boudinage.

stresses build up around flaws in the structure and at these points the less ductile bands become thinned (Fig. 3-38). Metallurgists have made detailed studies of a very similar type of structure which is developed in metal rods and sheets that are subjected to extensive strain, a structure termed *necking*. In competent bands of rock the development of these regions of thinned material is often more or less evenly spaced and gives rise to what is called *pinch-and-swell structure* (Figs. 3-40, 3-41). If the material suffers further straining, the thinning at the neck continues until rupture occurs (Figs. 3-39, 3-42), and the competent band becomes separated into a number of long cylindrical pieces lying side by side. These structures were originally compared by Lohest, Stanier, and Fourmarier (1909) to sausages lying side by side and called *boudins.* The process by which they are formed, i.e., extension perpendicular to their length, is known as *boudinage.*

The appearance of the cross section of boudins varies greatly and seems to depend on two main factors: first, the competence difference between the band and its enclosing material, and second, the values of the principal extensions of the two-dimensional strain ellipse on a cross section of the boudins. If the difference of competence is great, the initial failure of the most competent bands often takes place with very small strain and often on well-defined cross joints without any development of necks (Fig. 3-44A, band 1). Further extensive strain now leads to separation of the rectangular cross sections (Fig. 3-44B, band 1). If the competence difference is less strongly marked, however, the competent bands are often considerably thinned and necked before the rupture and separation of the boudins occur (Fig. 3-44A, B, and C, band 2). If the competence difference is very slight, the more competent bands may only develop pinch-and-swell structures without the separation into boudins (Fig. 3-44, band 3). The ductile material which surrounds separated boudins often flows into the gaps between the boudins (Figs. 3-43, 3-44A, B, and C, bands 1 and 2); in so doing, it sometimes modifies the edges of the competent bands by shear and gives the cross sections of the

Figure 3-40

Pinch-and-swell structure in pegmatite veins contained in a semipelitic schist, Khan Gorge, Southwest Africa.

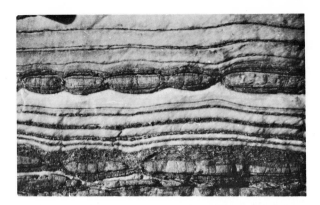

Figure 3-41

Figure 3-42

Figure 3-43

Examples of the progressive development of boudinage structure in naturally deformed rocks (cf. Figs. 3-38, 3-39, 3-44). Calc-silicate bands in marble, Khan Gorge, Southwest Africa.

boudins a barrel-like shape. Sometimes this modification can completely alter the rectangular cross section into a lenticular form (Fig. 3-44C, band 2). It is occasionally found, however, that these features produced by ductile flow into the neck are not shown to any marked extent; instead there are spaces filled with crystalline material of pegmatitic origin, quartz, or carbonates. This phenomenon seems to occur where the strain ellipse of the cross section has one main direction of extension and where the other principal extension has a value of λ_2 close to 1, or even greater than 1, that is, where there is an increase in area as a result of deformation. The results of two experiments which clearly illustrate how the length of the minor axis of the strain ellipse influences the structures around boudins are illustrated in Figs. 3-45, 3-46, and 3-47. A sheet of wood was cut into a series of strips with rectangular cross sections, and these were placed side by side in a ductile material (Fig. 3-45). The ductile material and the wood blocks were then deformed together and the blocks became separated into boudins. Where the strain-ellipse axes were $\lambda_1 = 4.9$, $\lambda_2 = 0.2$ the surrounding ductile material flowed into the necks between the separated wood block boudins (Fig. 3-46); but on changing the deformation in a second experiment so that $\lambda_1 = 4.4$, $\lambda_2 = 0.7$, although the blocks were separated by about the same amount, open voids were developed between the boudins (Fig. 3-47). These are the openings where in naturally deformed rocks solutions would be drawn in and any material carried in solution would crystallize.

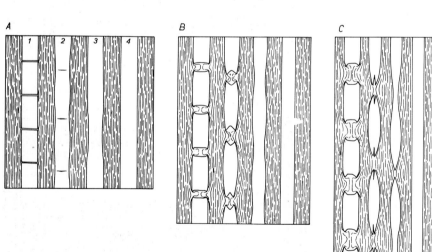

Figure 3-44

The progressive development of boundiage. The competent bands 1, 2, 3, and 4 are arranged in decreasing order of competence, and band 4 has the same properties as the matrix.

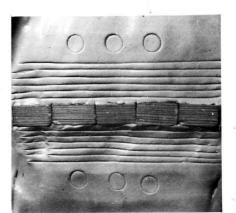

Figure 3-45

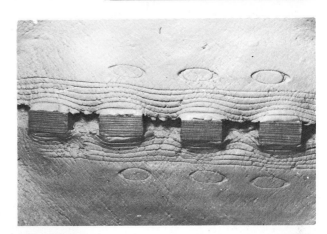

Figure 3-46

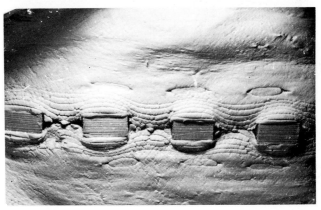

Figure 3-47

A series of experiments to show the type of structure which develops at boudin necks; Fig. 3-45, undeformed; Fig. 3-46, deformed with little contraction perpendicular to the banding; Fig. 3-47, deformed with considerable contraction perpendicular to the banding.

Figure 3-48

Apparent rotation of a boudin relative to the banding in the surrounding strata, Khan Gorge, Southwest Africa.

Figure 3-49

Sigmoidal boudinage of competent calc-silicate layer contained in marble, Sokomfjell, Norway.

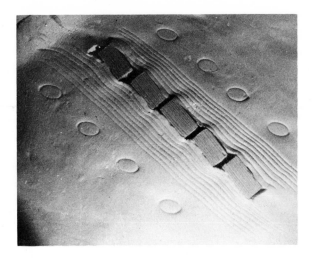

Figure 3-50

Experimental production of "rotated boudinage." The competent layer was initially arranged obliquely to the principal axes of strain, and the layering in the boudinaged layer rotated less than that in the surrounding ductile material.

It is a common feature of rocks which have been boudinaged for the boudins to appear to show a rotation relative to the banding of the material which encloses them (Fig. 3-48) and sometimes individual boudins are connected by a thin strip of competent material with a sigmoidal shape (Fig. 3-49). This phenomenon is sometimes explained as being the result of shear along the layering of the rocks which enclose the boudins. However, experiments suggest that "rotated" boudins are simply the result of an asymmetric arrangement of the competent band within the strain ellipse. If the wood blocks of a previous experiment (Fig. 3-45) are aligned so that they are not parallel to a principal extension direction of the strain ellipse and then the whole is deformed (Fig. 3-50), it is found that although the banding in the ductile material changes orientation according to Eq. (3-34), the banding in the boudins lags behind. The angle of rotation of the boudin is always less than that of the matrix but varies according to the ratio of thickness to length of the cross section of the boudins. If the boudins are equidimensional in cross section and the strain is irrotational, it is found that they do not rotate at all even after considerable strain of their matrix.

Structures which result from contraction   Where there is shortening within a competent layer, the stresses which build up generally lead to an instability which gives rise to buckling of the layer, and to the production of folds.

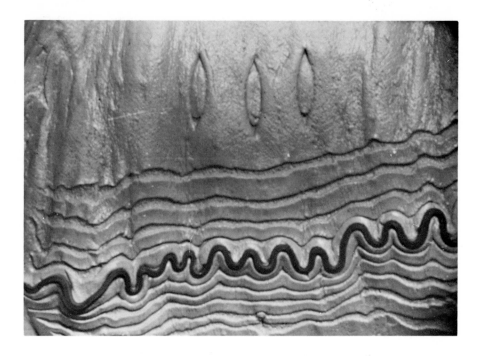

Figure 3-51

Ptygmatic structure produced experimentally by the contraction of a competent band contained in a more ductile material.

The folds which are formed have a very characteristic shape (Figs. 3-51, 3-52, and 3-53) and are sometimes known as *ptygmatic veins*. Usually the thickness of the competent band is approximately constant through the fold, and many folds formed in this way can be classified as parallel or concentric flexural folds. It is found that the initial length of the buckled band may be closely approximated by measuring the length of a central line through the folded competent layer. This is a useful technique because, since we know the length of the layer after deformation, we can evaluate the extension in certain directions in a deformed rock.

The wavelength of the ptygmatic folds depends on two factors: the thickness t of the competent band and the ratio of the competencies of the band and the matrix around it (viscosity ratio μ_1/μ_2). There is a simple relationship which links the initial wavelength W_i of the folds with these parameters (see Sec. 7-6),

$$W_i = 2\pi t \left(\frac{\mu_1}{6\mu_2}\right)^{\frac{1}{3}}$$

Figure 3-52

Ptygmatic structure in a thin pegmatite vein cutting through semipelitic schists, Ross of Mull, Scotland.

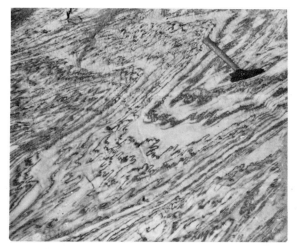

Figure 3-53

Ptygmatic structures in calc-silicate layers in marble, Khan Gorge, Southwest Africa.

Geological significance of the three types of strain ellipse It has been shown, when discussing the methods used to represent strain in two dimensions graphically, that all strain ellipses can be classified as one of three main

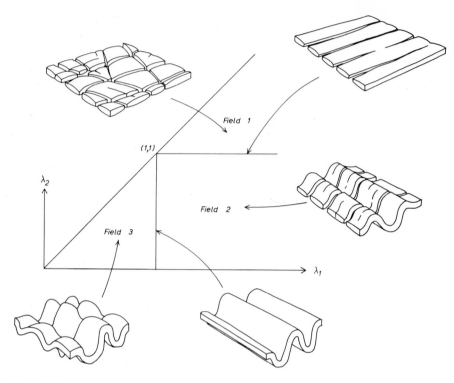

Figure 3-54

The geological structure that may be developed in the three fields of strain ellipse.

types according to the values of their principal extensions, and that these types fall into three main fields when plotted on a graph (Figs. 3-34 and 3-54):

Field 1 $\lambda_1 > \lambda_2 > 1$

Field 2 $\lambda_1 > 1 > \lambda_2$

Field 3 $1 > \lambda_1 > \lambda_2$

If strain occurs in the plane of a competent layer embedded in a more ductile material, different types of minor structures will develop in the layer according to the deformation field in which the strain ellipse is situated (Fig. 3-54). If it falls in field 1, all directions within the layer have suffered extension and the structures that develop will be entirely of the boudinage type. The boudinage "necks" will have no uniform linear orientation, although probably most will be subperpendicular to the principal extension. The intersection of boudinage in various directions within the competent layer often leads to the phenomenon termed by Wegmann *chocolate tablet structure* (Fig. 3-54). If the strains are large, the slabs of competent rock bounded by fractures become separated and isolated from each other (Fig. 3-55).

If the strain ellipse lies between fields 1 and 2 on the line $\lambda_2 = 1$, one set of boudins is developed on the surface, and the lengths of the boudins are perpendicular to the principal axis λ_1 (Fig. 3-54). If any joints or fractures existed in the rock before the development of the strain, these will be preferentially used as separation lines within the competent band, and the boudins formed need not be perpendicular to the principal elongations. Similarly, in jointed rocks there may be a complicated crossing arrangement of boudin separation lines inherited from the preexisting joint pattern even though λ_2 is not greater than 1.

In field 2, strain ellipses have one principal extension and one principal contraction. This arrangement produces folding in one direction and boudinage or cross-jointing perpendicular to the lines of the fold axes (Fig. 3-54).

Between fields 2 and 3 along the line $\lambda_1 = 1$ there is no extension; only a single principal contraction and a single set of folds is produced.

If the ellipse lies in field 3, all directions are contracted. Competent bands undergo a very complex crumpling. The folds have little regularity, although probably more axes are subperpendicular to the principal strain λ_2 than in other directions (Ramberg, 1959).

Figure 3-55

"Chocolate tablet" boudinage produced by extension in all three directions within a competent layer. Calc-silicate band in marble, Sokomfjell, Norway.

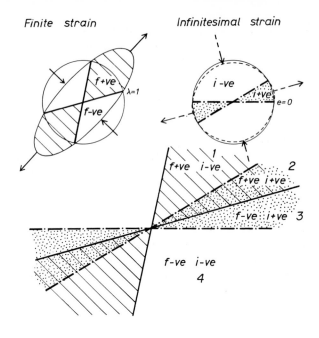

Figure 3-56

The relationship of finite and infinitesimal strain at any stage during the deformation, and the four zones that may result from an asymmetric combination of the two phenomena.

Progressive deformation   In the study of infinitesimal strain it was pointed out that progressive deformation could be thought of as a process whereby infinitesimal strains were integrated and that at any one instant in the deformation history we could analyze the strain properties into a finite part (what has happened) and an incremental part (what is happening). Figure 3-56 illustrates the finite- and infinitesimal-strain ellipses at one time during the deformation history. The finite-strain ellipse can be divided into sectors according to whether lines have been elongated ($f + ve$) or contracted ($f - ve$), these being separated by the lines of no finite longitudinal strain ($\lambda = 1$). The infinitesimal-strain ellipse can also be subdivided into sectors according to whether lines are expanding ($i + ve$) or contracting ($i - ve$). The principal axes of the infinitesimal-strain ellipse only coincide with those of the finite-strain ellipse if the deformation is irrotational. In general there is a rotational component in deformation and the combination of the two ellipses and strain features associated with them is markedly asymmetric. In Fig. 3-56 it will be seen that the superposition of the various sectors of the two ellipses leads to four zones in which lines have had markedly different histories:

Zone 1 lines have expanded but are now contracting.
Zone 2 lines have expanded and are still expanding.
Zone 3 lines have contracted but are now expanding.
Zone 4 lines have contracted and are still contracting.

The structures that are formed in competent bands lying in these zones are as follows:

Zone 1 boudins which are being pressed together or folded (Fig. 3-57)
Zone 2 boudins with the separation between them still increasing
Zone 3 folds being unfolded or being disrupted by boudinage (Figs. 3-58, 3-59, 3-60)
Zone 4 folds decreasing in wavelength and increasing in amplitude

The processes of folding and boudinage in a progressive deformation usually follow a complex sequence. Figure 3-61 shows the state of finite strain at intervals during an irrotational progressive deformation (pure shear) and illustrates the history of the changes in length of four lines of different orientation. Lines *a* and *b* have a fairly simple history of continuous contraction and continuous expansion respectively, but lines *c* and *d* have a much more

Figure 3-57

Banded hornblendic gneisses which have suffered elongation along the banding and then subsequent contracting. The boudins formed during one stage of the deformation were folded at a later date. Lower Pennine nappes, Ticino, Switzerland.

Figure 3-58

A thin calc-silicate layer enclosed in marble which was contracted along its length and developed ptygmatic structures, and later extended along its length. Khan Gorge, Southwest Africa.

Figure 3-59

Strongly boudinaged folds which indicate an initial contraction along the banding followed by an extension. Khan Gorge, Southwest Africa.

Figure 3-60

Folds which have been completely disrupted as a result of very strong elongation after their formation. Lower Pennine nappes, Ticino, Switzerland.

complex history. Line c begins by shortening, but the rate decreases until at one position it reaches a static condition where it becomes parallel to the lines of no infinitesimal longitudinal strain. With further strain, line c rotates toward the principal axis of finite extension and increases in length, but in ellipse 4 it still has not expanded to its original length. Line d begins by contracting; it passes through the line of no infinitesimal longitudinal strain and then expands, passing through the position of lines of no finite longitudinal strain. At any one time during the deformation the distortions can be analyzed into three zones as shown in Fig. 3-62:

ZONE 1   Lines in this zone have been elongated and are still being elongated, and boudinage is the dominant process that has gone on in competent bands. This zone may be subdivided into 1a and 1b by the lines making an angle θ from the principal finite extension given by $\tan \theta = (\lambda_2/\lambda_1)^{\frac{1}{2}}$. This line initially made an angle of 45° to the principal axis of the first infinitesimal-strain ellipse and was the line of no infinitesimal longitudinal strain in that ellipse. Although all lines in zone 1 have a finite elongation, those in 1b had at an early stage in their history a period of contraction like that of line d in Fig. 3-61. Competent bands with orientations which place them in zone 1b sometimes show small remnants of disrupted folds and isolated fold hinges.

ZONE 2   This is the zone in which lines have been contracted and are now

Change in strain during progressive irrotational
deformation.

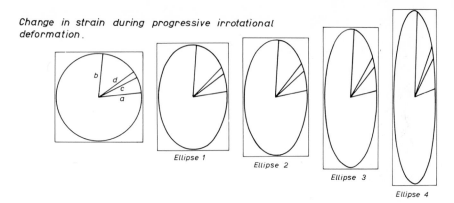

Ellipse 1

Ellipse 2

Ellipse 3

Ellipse 4

Changes in the longitudinal strains of lines a, b, c & d, and their effects in competent bands.

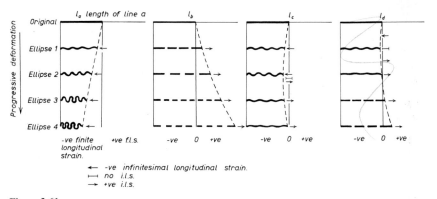

← -ve infinitesimal longitudinal strain.
⊢—⊣ no i.l.s.
→ +ve i.l.s.

Figure 3-61

Changes in lengths of lines as a result of progressive pure shear

expanding, and here one finds folds which are becoming unfolded or boudin-aged.

ZONE 3   Lines with this orientation have had a history of continuous contraction, and this is the region of ptygmatic folds.

It should be pointed out that the structures which may develop in competent layers depend entirely on the orientation of the layer; if there are several crossing competent bands in a rock (frequently one finds crossing veins in migmatite complexes), then it is possible to get boudinage of one and folding of the other in the same rock at the same time depending on their orientation.

In folded rocks one finds a great variation in orientation of the layering and it is in this environment that one can apply most significantly the results of the studies of progressive deformation. Although at this stage no detailed discussion on fold formation will be given, we shall assume that many folds

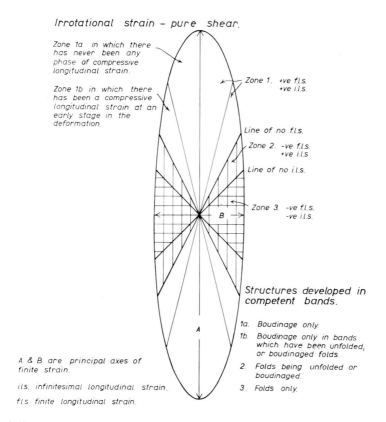

Irrotational   strain – pure   shear.

Zone 1a in which there
has never been any
phase of compressive
longitudinal strain.

Zone 1b in which there
has been a compressive
longitudinal strain at an
early stage in the
deformation.

Zone 1. +ve f.l.s.
         +ve i.l.s.

Line of no f.l.s.

Zone 2. -ve f.l.s.
         +ve i.l.s.

Line of no i.l.s.

B

Zone 3. -ve f.l.s.
         -ve i.l.s.

Structures developed in
competent bands.

1a. Boudinage only.

1b. Boudinage only in bands
     which have been unfolded,
     or boudinaged folds.

2.   Folds being unfolded or
     boudinaged.

3.   Folds only.

A

A & B are principal axes of
finite strain.

i.l.s. infinitesimal longitudinal strain.

f.l.s. finite longitudinal strain.

Figure 3-62

Zonal arrangement within a strain ellipse developed by progressive pure shear.

express a regional compressive strain in a direction which is subperpendicular to their axial planes. The minor structures that are likely to be developed in thin competent bands within a major fold structure are illustrated in Fig. 3-63; ptygmatic folds will be formed at the crest of the structure, boudinaged folds on either side of the crest, and boudinage only on the fold limbs.

We have seen that it is possible to develop any finite-strain ellipse in a number of ways, and that the rotational component is generally an important one in natural deformation. It is interesting to compare the finite-strain ellipse produced by a progressive simple shear (Fig. 3-64) with that produced by pure shear (Fig. 3-62). Although the finite-strain properties of these two ellipses are identical, the infinitesimal-strain phenomena are quite different. In Fig. 3-64 one of the lines of no infinitesimal longitudinal strain coincides with one of the lines of no finite longitudinal strain, and both are parallel to the shear line. The combined properties of finite and infinitesimal strain show

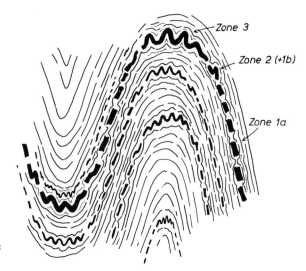

Figure 3-63

Zonal development of structures within a simple fold.

a marked asymmetry with respect to the axes of the finite-strain ellipse and likewise so will the minor structures which reflect the different strain histories of the different zones. The recognition of these asymmetric distributions of structures is one way in which irrotational and rotational strains can be distinguished from each other.

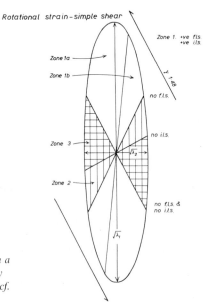

Figure 3-64

Zonal arrangement within a strain ellipse developed by progressive simple shear (cf. Fig. 3-62).

4 | Strain in three dimensions

A discussion about the three-dimensional aspects of strain forms a most vital part of the understanding of the processes of rock deformation and the structures that develop. Although the formulas that result are often long and tedious to manipulate, the mathematical methods are, in themselves, not at all complex and form a natural extension to those used previously in the study of two-dimensional deformation. Much of the mathematics can be made more concise using the subject known as *tensor analysis*, but since this is rarely covered in the training of a geologist it has been thought best to concentrate primarily on the more elementary mathematical techniques, especially those which can be used to manipulate the data which the geologist has collected.

4-1 FINITE STRAIN IN THREE DIMENSIONS

To begin the study of strain in three dimensions we must establish a reference frame to describe the geometry in space; these are three mutually perpendicular cartesian axes designated x, y, and z. The most general linear displacement of the point (x,y,z) in its undeformed position to the point (x_1, y_1, z_1) in its deformed position follows logically from the general transformation for strain in two dimensions previously established [Eq. (3-8)]:

$$x_1 = ax + by + cz$$
$$y_1 = dx + ey + fz \qquad (4\text{-}1)$$
$$z_1 = gx + hy + iz$$

Three of the nine components (a,e,i) can be thought of as longitudinal strains parallel to the reference axes x, y, and z, respectively, while the other six (b,c,d,f,g,h) have shear components parallel to one axis and perpendicular to another axis. Their physical significance is shown in Fig. 4-1. If this transformation is applied to any straight line or plane, the result is another straight line or plane surface. The transformations are therefore in accord with the conditions of homogeneous strain. When the transformation is applied to points situated on a unit sphere, after laborious but not difficult manipulation, the result is the equation of an ellipsoid, the *strain ellipsoid*. The three principal axes of this ellipsoid have lengths which are the *principal longitudinal strains*, $(1 + e_1) \geqslant (1 + e_2) \geqslant (1 + e_3)$, with *principal quadratic extensions* $\lambda_1 \geqslant \lambda_2 \geqslant \lambda_3$. Only three mutually perpendicular directions before deformation remain perpendicular after strain, and their *initial* orientations are known as the *principal axes of strain*; their deformed position coincides with the principal axes of the strain ellipsoid. If the orientation of these lines before and after

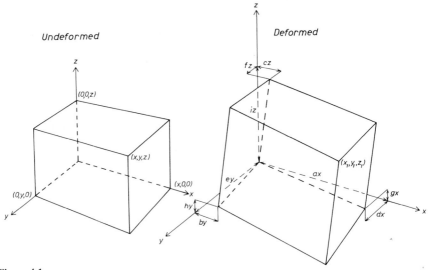

Figure 4-1

The general transformation for three-dimensional finite strain. The point (x,y,z) in the undeformed state is displaced to new position (x_1, y_1, z_1).

deformation is the same, the strain is called *irrotational strain*. The condition for this is that $b = d$, $c = g$, and $f = h$. In all other deformations the principal axes change position during deformation, and the strain is known as *rotational strain*. As with strain in two dimensions, any finite strain can be described in terms of an irrotational-strain ellipsoid and a rigid-body rotation of the axes of this ellipsoid. The three mutually perpendicular planes which contain the principal strain axes of the ellipsoid are known as the *principal planes*.

The volume of the strain ellipsoid is $4(\lambda_1\lambda_2\lambda_3)^{\frac{1}{2}}/3\pi$ derived from a unit sphere of volume $4/3\pi$, and *unit volume change* or *dilation* (Δ) is recorded by $\Delta = (\lambda_1\lambda_2\lambda_3)^{\frac{1}{2}} - 1$.

4-2 THE FINITE-STRAIN TENSOR

The equations which describe the general transformation for homogeneous strain in three dimensions define what is known as an asymmetric second-order tensor. It has been shown previously how, in two dimensions, the transformation equations can be separated into two parts, one connected with the irrotational part and the other with the rotational component of the deformation. All transformations in three dimensions can likewise be separated into these two parts. This is done by rearranging the terms in the displacement equation (4-1) as follows:

$$x_1 = ax + \left[\tfrac{1}{2}(b + d) + \tfrac{1}{2}(b - d)\right]y + \left[\tfrac{1}{2}(c + g) + \tfrac{1}{2}(c - g)\right]z$$
$$y_1 = \left[\tfrac{1}{2}(d + b) + \tfrac{1}{2}(d - b)\right]x + ey + \left[\tfrac{1}{2}(f + h) + \tfrac{1}{2}(f - h)\right]z$$
$$z_1 = \left[\tfrac{1}{2}(g + c) + \tfrac{1}{2}(g - c)\right]x + \left[\tfrac{1}{2}(h + f) + \tfrac{1}{2}(h - f)\right]y + iz$$

and separating the equations into two parts

$$x_1 = ax + \tfrac{1}{2}(b + d)y + \tfrac{1}{2}(c + g)z + Ox + \tfrac{1}{2}(b - d)y + \tfrac{1}{2}(c - g)z$$
$$y_1 = \tfrac{1}{2}(b + d)x + ey + \tfrac{1}{2}(f + h)z + \tfrac{1}{2}(d - b)x + Oy + \tfrac{1}{2}(f - h)z$$
$$z_1 = \tfrac{1}{2}(g + c)x + \tfrac{1}{2}(h + f)y + iz + \tfrac{1}{2}(g - c)x + \tfrac{1}{2}(h - f) + Oz$$

In terms of matrix notation the original asymmetric tensor representing the complete equation is given by the matrix

$$\begin{vmatrix} a & b & c \\ d & e & f \\ g & h & i \end{vmatrix}$$

and this tensor can be decomposed into the sum of two parts. The first part of the equation can be written as a matrix with six independent terms which show a symmetrical relationship across the main diagonal of the matrix; hence this is known as the *symmetric* part of the tensor.

$$
\begin{vmatrix}
a & \dfrac{b+d}{2} & \dfrac{c+g}{2} \\[2ex]
\dfrac{b+d}{2} & e & \dfrac{f+h}{2} \\[2ex]
\dfrac{c+g}{2} & \dfrac{f+h}{2} & i
\end{vmatrix}
$$

This defines the irrotational part of the strain and describes the strain ellipsoid. The second part has three independent terms which are arranged so that on one side of the zero diagonal lie the terms which are a mirror image of those on the other side, but the signs of the terms are changed.

$$
\begin{vmatrix}
0 & \dfrac{b-d}{2} & -\dfrac{g-c}{2} \\[2ex]
-\dfrac{b-d}{2} & 0 & \dfrac{f-h}{2} \\[2ex]
\dfrac{g-c}{2} & -\dfrac{f-h}{2} & 0
\end{vmatrix}
$$

This is known as a *skew-symmetric matrix* and relates to the rotational part of the strain.

Further discussion of the properties of the strain ellipsoid are best considered with reference to coordinate axes x, y, and z which have been rotated into a new position so that they coincide with the directions of the three principal strains of the ellipsoid $1 + e_1$, $1 + e_2$, $1 + e_3$, respectively. With reference to these new axes a unit sphere is distorted into the ellipsoid.

$$
\frac{x^2}{\lambda_1} + \frac{y^2}{\lambda_2} + \frac{z^2}{\lambda_3} = 1
$$

4-3 CHANGES IN LENGTH OF LINES

As a result of deformation the lengths of lines within the original sphere are changed, and the first study is to investigate the amount of this distortion. Consider a point $P(x,y,z)$ on the surface of a unit sphere $x^2 + y^2 + z^2 = 1$ (Fig. 4-2A) where the coordinate axes are chosen to coincide with the principal strain axes of the ellipsoid produced from the sphere after deformation (Fig. 4-3A).

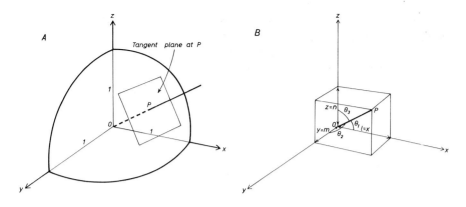

Figure 4-2

P is a point on the surface of a sector of an undeformed unit sphere $x^2 + y^2 + z^2 = 1$. The direction of OP is defined by the angles θ_1, θ_2, and θ_3 and direction cosines l, m, and n.

It is necessary to describe the orientation of the line OP in terms of the axes x, y, and z. This is most conveniently done using what are known as *direction cosines l, m,* and n (see Fig. 4-2B) defined as follows:

$$\cos \text{ angle } POx = \cos \theta_1 = l \qquad (4\text{-}2a)$$

$$\cos \text{ angle } POy = \cos \theta_2 = m \qquad (4\text{-}2b)$$

$$\cos \text{ angle } POz = \cos \theta_3 = n \qquad (4\text{-}2c)$$

As OP is of unit length, being a radius of the sphere, then

$$x = l \qquad y = m \qquad z = n$$

also

$$x^2 + y^2 + z^2 = 1$$

or

$$l^2 + m^2 + n^2 = 1 \qquad (4\text{-}3)$$

From (4-3) it will be seen that not all direction cosines are independent, and it is only necessary to give two to fix completely the direction of any line.

The sphere is strained and the point P is distorted to $P'(x',y',z')$ with direction cosines $l' = \cos \theta'_1$, $m' = \cos \theta'_2$, $n' = \cos \theta'_3$ (see Fig. 4-3A, B) situated on the surface of the ellipsoid $x^2/\lambda_1 + y^2/\lambda_2 + z^2/\lambda_3 = 1$ where λ_1, λ_2 and λ_3 are the three principal quadratic extensions parallel to x, y, and z, respectively. The relationships between the coordinates of P and P' are

$$x' = x\lambda_1^{\frac{1}{2}} = l\lambda_1^{\frac{1}{2}} \qquad (4\text{-}4a)$$

$$y' = y\lambda_2^{\frac{1}{2}} = m\lambda_2^{\frac{1}{2}} \qquad (4\text{-}4b)$$

$$z' = z\lambda_3^{\frac{1}{2}} = n\lambda_3^{\frac{1}{2}} \qquad (4\text{-}4c)$$

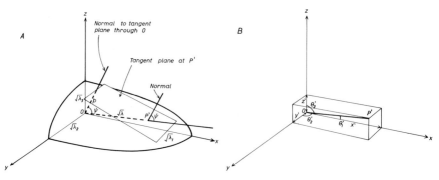

Figure 4-3

P' represents the deformed position of P shown in Fig. 4-2, and lies on the surface of the strain ellipsoid $x^2/\lambda_1 + y^2/\lambda_2 + z^2/\lambda_3 = 1$. The angle normal to the tangent plane at P' makes an angle with the line OP'. The perpendicular distance from the tangent plane to the origin is p. The direction of OP is defined by the angles θ'_1, θ'_2, and θ'_3.

From Pythagoras, $\lambda = x'^2 + y'^2 + z'^2$; therefore

$$\lambda = l^2\lambda_1 + m^2\lambda_2 + n^2\lambda_3 \tag{4-5}$$

an equation which gives the length of a line with direction cosines l, m, and n *before deformation* in terms of the principal strains. In a similar way to that of the two-dimensional analysis, it is possible to arrive at another relationship which refers to angles measured in the *strained state* (θ'_1, θ'_2, and θ'_3). From Fig. 4-3B:

$$\cos\theta'_1 = l' = \frac{x'}{OP'} = \frac{l\lambda_1^{\frac{1}{2}}}{\lambda^{\frac{1}{2}}}$$

therefore

$$l = \frac{l'\lambda^{\frac{1}{2}}}{\lambda_1^{\frac{1}{2}}} \qquad m = \frac{m'\lambda^{\frac{1}{2}}}{\lambda_2^{\frac{1}{2}}} \qquad n = \frac{n'\lambda^{\frac{1}{2}}}{\lambda_3^{\frac{1}{2}}} \tag{4-6}$$

Substituting these values in (4-3)

$$\frac{l'^2\lambda}{\lambda_1} + \frac{m'^2\lambda}{\lambda_2} + \frac{n'^2\lambda}{\lambda_3} = 1$$

$$\frac{l'^2}{\lambda_1} + \frac{m'^2}{\lambda_2} + \frac{n'^2}{\lambda_3} = \frac{1}{\lambda}$$

which can be put in terms of the reciprocal quadratic extension ($1/\lambda_1 = \lambda'_1$, etc.)

$$\lambda' = l'^2\lambda'_1 + m'^2\lambda'_2 + n'^2\lambda'_3 \tag{4-7}$$

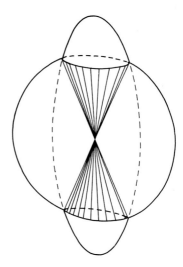

Figure 4-4

Lines of no finite longitudinal strain ($\lambda = 1$) in the strain ellipsoid positioned at the intersection of the strain ellipsoid and the unit sphere.

Lines of no finite longitudinal strain Within the strain ellipsoid, lines whose deformed length is equal to their undeformed length are found by substituting $\lambda' = 1$ in (4-7), which together with (4-3) leads to

$$1 = l'^2 \lambda'_1 + m'^2 \lambda'_2 + (1 - l'^2 - m'^2) \lambda'_3$$

or

$$m' = \left[\frac{\lambda'_3 - 1}{\lambda'_3 - \lambda'_2} - \frac{l'^2(\lambda'_3 - \lambda'_1)}{\lambda'_3 - \lambda'_2} \right]^{\frac{1}{2}} \tag{4-8}$$

Choosing a value for l', we find that there will in general be two values of m' equal but opposite in sign. By varying the values of l', the orientations of successive lines of no finite longitudinal strain will generally define a surface. This surface is the cone formed by the intersection of the original unit sphere with the strain ellipsoid (Fig. 4-4). Because $0 \leqslant l'^2 \leqslant 1$, Eq. (4-8) has no roots if $\lambda'_3 - 1 < 0$, that is, if $\lambda'_3 < 1$ or if $(\lambda'_3 - \lambda'_1) < (\lambda'_3 - 1)$; that is, $\lambda'_1 > 1$. The two limits occur when the original sphere lies either completely outside or completely within the strain ellipsoid. If there is no alteration in the length of the intermediate axis of the strain ellipsoid ($\lambda'_2 = 1$), the lines of no finite longitudinal strain lie on two planes which intersect on the y axis and which make an angle of $\pm\cos^{-1} \left[(\lambda'_3 - 1)/(\lambda'_3 - \lambda'_1) \right]^{\frac{1}{2}}$ with the x axis. These two planes are two *circular sections* of the ellipsoid. Every ellipsoid has a pair of circular sections passing through the y axis; but only where there is no strain along this intermediate axis is there no longitudinal strain within these sections. The circular sections can be found by discovering the locus of lines which have a quadratic extension λ_2, where

$$\lambda'_2 = \lambda'_1 l'^2 + \lambda'_2 m'^2 + \lambda'_3 n'^2$$

Rearranging this, making use of $l'^2 + m'^2 + n'^2 = 1$, we find this surface is given by

$$[l(\lambda_2' - \lambda_1')^{\frac{1}{2}} + n(\lambda_3' - \lambda_2')^{\frac{1}{2}}][l(\lambda_2' - \lambda_1')^{\frac{1}{2}} - n(\lambda_3' - \lambda_2')^{\frac{1}{2}}] = 0 \qquad (4\text{-}9)$$

which is the equation of a pair of planes passing through the intermediate strain axis and making angles of

$$\pm \cos^{-1} \left(\frac{\lambda_3' - \lambda_2'}{\lambda_3' - \lambda_1'} \right)^{\frac{1}{2}}$$

with the x axis. All lines within these sections have the same longitudinal strain $(1 + e_2)$. In geological literature there is sometimes confusion about the significance of the circular sections and their relationship to the lines of no finite longitudinal strain and to positions within the ellipsoid of maximum shearing strain. In general these three features do not coincide.

4-4   SHEAR STRAIN

Shear strain in three dimensions is defined in terms of the shear distortion of a normal to a plane surface. In Fig. 4-2A the tangent to the sphere at the point P is illustrated and OP is normal to this tangent plane. As a result of deformation (Fig. 4-3A), the orientation of this tangent plane is changed; it becomes a tangent plane to the ellipsoid surface at the point P'. The angle between the normal to this reoriented tangent plane and OP' is defined as the *angular shear strain* ψ; and $\tan \psi = \gamma$, the *shear strain*. To determine the amount of shear strain at any point we proceed in a way similar to that of the two-dimensional analysis (Sec. 3-7).

The equation of the tangent plane to the ellipsoid at the point $P'(x',y',z')$ is

$$\frac{xx'}{\lambda_1} + \frac{yy'}{\lambda_2} + \frac{zz'}{\lambda_3} = 1 \qquad (4\text{-}10)$$

which from (4-4) is

$$\frac{lx}{\lambda_1^{\frac{1}{2}}} + \frac{my}{\lambda_2^{\frac{1}{2}}} + \frac{nz}{\lambda_3^{\frac{1}{2}}} = 1 \qquad (4\text{-}11)$$

The length of the perpendicular (Fig. 4-3A) from the origin to this tangent plane is

$$p = \frac{1}{(l^2/\lambda_1 + m^2/\lambda_2 + n^2/\lambda_3)^{\frac{1}{2}}} \qquad (4\text{-}12)$$

Also

$$\sec \psi = \frac{OP'}{p} = \frac{\lambda^{\frac{1}{2}}}{p} = \lambda^{\frac{1}{2}} \left(\frac{l^2}{\lambda_1} + \frac{m^2}{\lambda_2} + \frac{n^2}{\lambda_3} \right)^{\frac{1}{2}} \qquad (4\text{-}13)$$

$$\gamma^2 = \tan^2\psi = \sec^2\psi - 1 = \lambda\left(\frac{l^2}{\lambda_1} + \frac{m^2}{\lambda_2} + \frac{n^2}{\lambda_3}\right) - 1 \qquad (4\text{-}14)$$

Substituting for λ [Eq. (4-5)], this becomes

$$\gamma^2 = (l^2\lambda_1 + m^2\lambda_2 + n^2\lambda_3)\left(\frac{l^2}{\lambda_1} + \frac{m^2}{\lambda_2} + \frac{n^2}{\lambda_3}\right) - 1$$

and using the condition $(l^2 + m^2 + n^2)^2 = 1$, this simplifies to

$$\gamma^2 = \frac{(\lambda_1 - \lambda_2)^2}{\lambda_1\lambda_2}l^2m^2 + \frac{(\lambda_2 - \lambda_3)^2}{\lambda_2\lambda_3}m^2n^2 + \frac{(\lambda_3 - \lambda_1)^2}{\lambda_3\lambda_1}n^2l^2 \qquad (4\text{-}15)$$

This is the important standard formula for shear strain in terms of the principal strains and angles measured in the *undeformed state*. It is an ordered equation that is easy to remember when it is seen how the terms repeat themselves in a cyclic manner. In most geological problems it is useful to have the relations referred to angles measured in the *deformed state*, and using (4-6) in (4-15) it is found that

$$\frac{\gamma^2}{\lambda^2} = \left(\frac{1}{\lambda_1} - \frac{1}{\lambda_2}\right)^2 l'^2m'^2 + \left(\frac{1}{\lambda_2} - \frac{1}{\lambda_3}\right)^2 m'^2n'^2 + \left(\frac{1}{\lambda_3} - \frac{1}{\lambda_1}\right)^2 n'^2l'^2 \qquad (4\text{-}16)$$

which is rewritten using the usual substitution $\gamma' = \gamma/\lambda$, $\lambda' = 1/\lambda$, etc.

$$\gamma'^2 = (\lambda_1' - \lambda_2')^2 l'^2m'^2 + (\lambda_2' - \lambda_3')^2 m'^2n'^2 + (\lambda_3' - \lambda_1')^2 n'^2l'^2 \qquad (4\text{-}17)$$

Equations (4-17) and (4-7) form the basis for much of the following discussion of three-dimensional strain and from them we can build a relationship of γ' and λ' which will enable us to apply Mohr's construction to three-dimensional problems (Sec. 4-8).

4-5   CHANGES IN ANGLE DURING DEFORMATION

As a result of deformation, lines with direction cosines l, m, and n become distorted to a new position and have direction cosines l', m', and n'. From (4-5) and (4-6) the relationships between these cosines can be established. (Two similar equations can be derived for m'^2 and n'^2.)

$$l'^2 = \frac{\lambda_1 l^2}{\lambda} = \frac{\lambda_1 l^2}{\lambda_1 l^2 + \lambda_2 m^2 + \lambda_3 n^2} \qquad (4\text{-}18)$$

Or, using (4-3) and expressing the relationships in terms of the independent original direction cosines l and m,

$$l'^2 = \frac{\lambda_1 l^2}{l^2(\lambda_1 - \lambda_3) + m^2(\lambda_2 - \lambda_3) + \lambda_3} \qquad (4\text{-}19a)$$

$$m'^2 = \frac{\lambda_2 m^2}{l^2(\lambda_1 - \lambda_3) + m^2(\lambda_2 - \lambda_3) + \lambda_3} \qquad (4\text{-}19b)$$

As with Eq. (3-34), these changes are independent of absolute values on the principal strains; they depend only on the *ratios* of the principal quadratic extensions. For example,

$$l'^2 = \frac{(\lambda_1/\lambda_3)l^2}{l^2(\lambda_1/\lambda_3 - 1) + m^2(\lambda_2/\lambda_3 - 1) + 1} \qquad (4\text{-}20)$$

Although Eqs. (4-19) can be used to determine the changes in orientation of lines, a much more convenient practical construction—which can be used to determine the change in orientation of both planes and lines—uses the angles that intersections of planes make with the principal planes of the ellipsoid. Figure 4-5 shows the orientation of a plane and its intersections with the principal planes of the ellipsoid before and after deformation. The following angular relationships hold from Eq. (3-34):

$$\tan \theta'_1 = \tan \theta_1 \left(\frac{\lambda_3}{\lambda_1}\right)^{\frac{1}{2}} \qquad (4\text{-}21a)$$

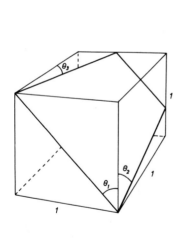
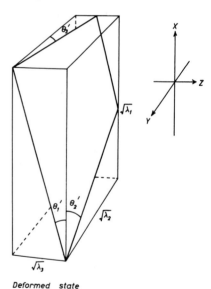

Undeformed state   Deformed state

Figure 4-5

The relationships of the angles θ_1, θ_2, and θ_3 made by the intersections of a plane surface with the principal strain plane sections XZ, XY, and YZ and the X and Y strain directions before deformation, and θ'_1, θ'_2, and θ'_3 after deformation.

$$\tan \theta'_2 = \tan \theta_2 \left(\frac{\lambda_2}{\lambda_1}\right)^{\frac{1}{2}}$$ (4-21b)

$$\tan \theta'_3 = \tan \theta_3 \left(\frac{\lambda_3}{\lambda_2}\right)^{\frac{1}{2}}$$ (4-21c)

A numerical example will illustrate the method of determining the orientation of a plane before and after deformation. From the following data we can find the orientation of the bedding surface before straining:

Principal x axis of ellipsoid plunges 70° toward 138°.
Principal y axis of ellipsoid plunges 10° toward 21°.
Principal axes of the ellipsoid have the ratio $\lambda_1 : \lambda_2 : \lambda_3 = 9 : 4 : 1$.
Deformed bedding surface strikes 174° and dips 60° to the east.

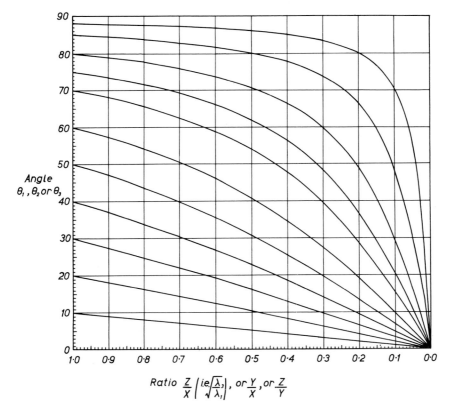

Figure 4-6

Graphical solution of the relationships of θ_1 and θ'_1, θ_2 and θ'_2, and θ_3 and θ'_3 of Fig. 4-5.

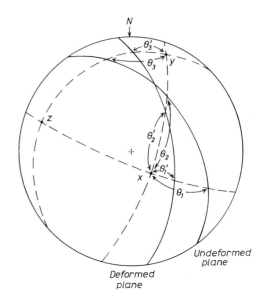

Figure 4-7

*Equal-area net construction of
the spatial orientation of a plane
before and after deformation.*

Undeformed
plane

Deformed
plane

1. On an equal-area net plot the points representing the principal axes of the strain ellipsoid and draw the principal planes (Fig. 4-7).

2. Construct the great circle which represents the bedding plane and the positions of the intersections on the three principal planes. Measure the angles $\theta'_1 = 19°$, $\theta'_2 = 44°$, $\theta'_3 = 19°$. Although only two of these values are necessary for a solution of the problem, use of the third is a good check for accuracy.

3. The strain-ellipse ratios on the three principal sections are $Z/X = 0.33$, $Y/X = 0.66$, $Z/Y = 0.5$. Use the graph (Fig. 4-6) and read back the undeformed angles $\theta_1 = 46°$, $\theta_2 = 56°$, and $\theta_3 = 35°$. Plot the three points on the principal sections.

4. Find the great circle which passes through all three points; this represents the unstrained position of the bedding planes (strike 146°, dip 44° NE).

Using a similar construction, the change in orientation of any line in space as the result of deformation can also be easily determined. The line is envisaged as forming at the intersection of two planes: one containing the line and the principal Y axis, the other containing the line and the principal X axis. The relationships of θ and θ' before and after deformation that are shown in Fig. 4-8 are again those of Eq. (4-21). Consider a line which plunges at 25° to 77° after deformation in the block shown in Fig. 4-8. To find its unstrained position the following construction is employed:

1. On an equal-area net plot the points representing the principal axes X, Y, and Z of the ellipsoid and draw the principal planes (Fig. 4-9).

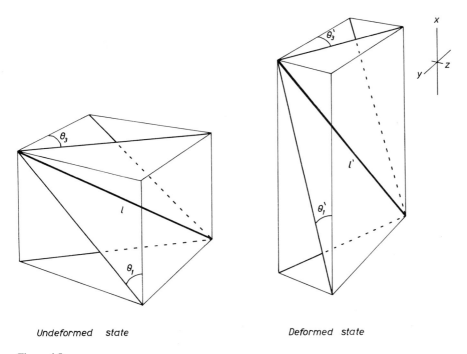

Undeformed   state   Deformed   state

Figure 4-8

The deformation of a linear structure l as the result of homogeneous strain to a new orientation l'.

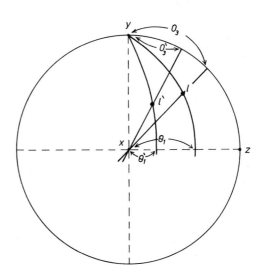

Figure 4-9

Equal-area net construction of the spatial orientation of a lineation before and after deformation.

2. Plot the point P' representing the line, and draw two great circles passing through P' and Y, and through P' and X. Measure the angles θ'_1 and θ'_3.

3. Knowing the ratios of the principal strain ellipses Z/X and Z/Y, use the graph in Fig. 4-6 to compute the angles θ_1 and θ_3 for their unstrained state and plot these points on the principal sections.

4. Draw the great circle through one of these points and the Y axis, and through the other point and the X axis; the point where these great circles intersect represents the unstrained position of P.

4-6 METHODS OF GRAPHICALLY RECORDING THE COMPONENTS OF FINITE STRAIN

There are many ways of illustrating states of finite strain by graphs. Some of these methods are particularly valuable in enabling one to visualize clearly the properties of certain types of ellipsoids and to describe the changes that go on during a progressive deformation. The state of finite strain is described mathematically by an asymmetric second-order tensor with nine independent components, and if we wish to record exactly the state of strain we must have some method of recording all nine components. We cannot get all this information clearly onto one graph, but when several diagrams are combined, we can define all these components comprehensively. The nine components of the strain tensor are derived from the nine components of displacement, p. 122), but these displacements can be put into a form more pertinent to our problem. The most useful nine terms which completely describe the finite strained state are (1), (2), and (3), the principal extensions e_1, e_2, and e_3; (4) and (5), the direction cosines l_1 and m_1 of the initial position of one of the principal axes; and (6), the cosine l_2 of another principal axis measured from any cartesian coordinate axes x, y, and z. Only one direction cosine is necessary to fix the second principal axis, because when one axis is known, the second axis has only one degree of freedom in the plane perpendicular to the first axis. When two axes are fixed, the third is completely defined. The direction cosines l_3 and m_3, (7) and (8), of one of the principal strain axes after deformation, and the direction cosine l_4, (9), of one other, complete the specification. These last three components allow us to define the orientation of the strain ellipsoid in space with reference to the arbitrarily chosen coordinate framework x, y, and z. Of these nine components the first three and the last three describe the strain ellipsoid, that is, the symmetric part of the tensor; the other three describe how the axes of the strain ellipsoid have been rotated from their initial position, that is, the skew-symmetric part of the tensor.

Strain components (1), (2), and (3) The shape of the strain ellipsoid can be illustrated in several ways depending on whether absolute values of the principal strains are known or whether ratios only have been determined.

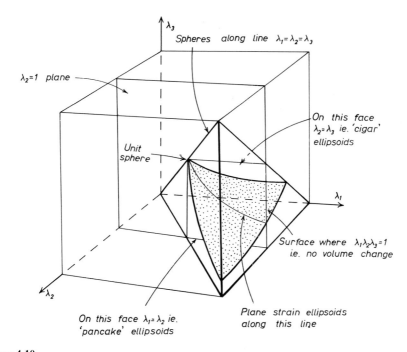

Figure 4-10

Three-dimensional graph for representing the shape of the finite-strain ellipsoid.

METHOD 1   The first graph (Fig. 4-10) is a three-dimensional one following the general methods used to represent the shape of strain ellipses (Fig. 3-34). The values of λ_1, λ_2, and λ_3 are recorded in three mutually perpendicular directions. Because of the definition that all the values are positive and $\lambda_1 \geqslant \lambda_2 \geqslant \lambda_3$, not all the space contained between these axes is occupied by points representing the ellipsoids. The plots will only fall in the space between the planes $\lambda_3 = 0$, $\lambda_1 = \lambda_2$, and $\lambda_2 = \lambda_3$, and this space extends to infinity on the right-hand side of Fig. 4-10. The undeformed unit sphere is (1,1,1); all other spheres plot along the diagonal line $\lambda_1 = \lambda_2 = \lambda_3$. All *oblate uniaxial ellipsoids* (pancake shapes) lie along the plane $\lambda_1 = \lambda_2$; and all *prolate uniaxial ellipsoids* (cigar shapes), along the plane $\lambda_2 = \lambda_3$. Ellipsoids which have had no change in their intermediate axis lie along the vertical plane $\lambda_2 = 1$. Where there has been no volume change during deformation the plots fall on the hyperboloid $\lambda_1\lambda_2\lambda_3 = 1$ shown as a dotted surface in Fig. 4-10. All points above this curved surface are ellipsoids which have increased in volume; those below it have decreased in volume. The special ellipsoids in which there is no volume change and where the length of the intermediate axis is unchanged, known as the *plane strain ellipsoids*, are found along the line where the equal-area hyperboloid is cut by the plane $\lambda = 1$.

METHOD 2   A second method, which is basically similar to the first, plots the logarithms of the principal lengths ε_1, ε_2, and ε_3 along three mutually perpendicular directions (Fig. 4-14A). Points representing all possible ellipsoids are found in the space between the planes $\varepsilon_1 = \varepsilon_2$ and $\varepsilon_2 = \varepsilon_3$. This method gives a more equal distribution of the ellipsoids of volume expansion and volume contraction on either side of the curved surface of no volume change given by $\varepsilon_1 + \varepsilon_2 + \varepsilon_3 = 0$.

METHOD 3   Data from naturally deformed rocks do not always enable a determination of the absolute values of the principal extensions, but it is sometimes possible by measuring the distortions of known original angles between intersecting lines to determine the ratios of the principal extensions.

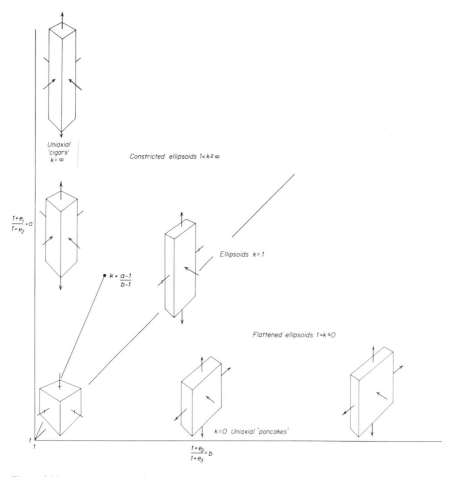

Figure 4-11

Graphical plot to represent ellipsoids in terms of the ratios of the principal strain.

If this can be done, then it is possible to represent the strain-ellipsoid shape on a two-dimensional graph by plotting the ratios $(1 + e_1)/(1 + e_2) = a$ as ordinate and $(1 + e_2)/(1 + e_3) = b$ as abscissa (Fig. 4-11). This is a method originally employed by Zingg for recording the shape of pebbles, and one which has since been considerably developed by Flinn (1962). Ellipsoids occur where $a \geq 1$ and $b \geq 1$, and the origin (1,1) represents a sphere. All oblate uniaxial ellipsoids occur along the line $a = 1$, and all prolate uniaxial ellipsoids along $b = 1$. All ellipsoids having the property $(1 + e_2)^2 = (1 + e_1)(1 + e_3)$ are found along the line $a = b$, the plane-strain equal-volume ellipsoids satisfy this condition and occur along this line. All other ellipsoids fall either between the lines $a = 1$ and $a = b$, or between $a = b$ and $b = 1$. The former are known as the *constriction-type* ellipsoids (in terms of optical mineralogy they are positive) where $(1 + e_2)^2 < (1 + e_1)(1 + e_3)$; the latter where $(1 + e_2)^2 > (1 + e_1)(1 + e_3)$ are the *flattening-type* ellipsoids (negative). It should be pointed out that because this graph does not record volume change the line $a = b$ does not divide ellipsoids which have their intermediate axes expanded, $(1 + e_2) > 1$, from those where they have been contracted, $(1 + e_2) < 1$. Flinn has suggested the parameter k for describing some factors of the shapes of ellipsoids on this graph; k is the slope of the line joining the plot of the ellipsoid to the point (1,1) (see Fig. 4-11).

$$k = \frac{a - 1}{b - 1} = \frac{(1 + e_3)(e_1 - e_2)}{(1 + e_2)(e_2 - e_3)} \tag{4-22}$$

Although this is a very useful parameter to describe the ellipsoid generally, it is a parameter of convenience more than one of fundamental significance with respect to the actual process of deformation. Flinn has used this parameter to divide the constant-volume ellipsoids into five types:

$k = 0$	uniaxial oblate types
$1 > k > 0$	flattening types
$k = 1$	plane-strain types
$\infty > k > 1$	constriction types
$k = \infty$	uniaxial prolate types

This method of plotting data has great application to geological problems. Its main disadvantages are, first, that it may be restrictive when it comes to realizing the effects of volume changes during deformation, and second, that if a wide range of ellipsoid shapes is plotted the points representing low-deformation ellipsoids tend to have an unnaturally close grouping. The differences in shape of ellipsoids in slightly deformed material are best investigated by plotting $(1 + e_2)/(1 + e_1)$ as ordinate $(1/a)$ and $(1 + e_3)/(1 + e_2)$ as abscissa $(1/b)$ (Fig. 4-12). This is a finite two-dimensional plot: (1,1) represents a sphere and all ellipsoids lie between the lines $1/a = 1$, $1/a = 0$, $1/b = 1$, $1/b = 0$.

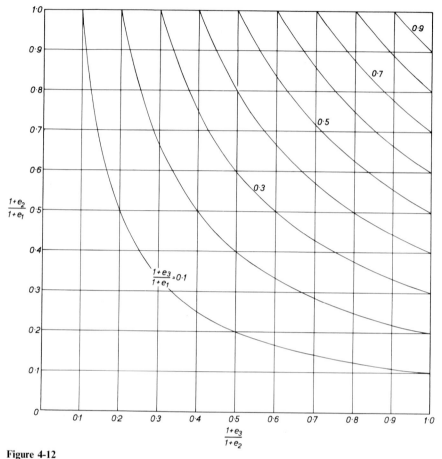

Figure 4-12

The best graphical plot for illustrating strain ratios in slightly deformed material.

METHOD 4   The natural logarithms of ratios of the principal strain $\log_e [(1 + e_1)/(1 + e_2)] = \varepsilon_1 - \varepsilon_2$ and $\log_e [(1 + e_2)/(1 + e_3)] = \varepsilon_2 - \varepsilon_3$ are plotted as ordinate and abscissa, respectively. This has advantages over method 3 in that deformation paths are more easily interpreted in terms of their incremental strain components (see Sec. 6-15). For example, it will be shown that with certain types of progressive irrotational deformation the slope of the deformation path is directly related to the shape of the incremental strain ellipsoid.

METHOD 5   This method is based on a record of the ratios of the three principal extensions and these parameters are expressed in a triangular diagram. The technique is an adaptation of one originally suggested by

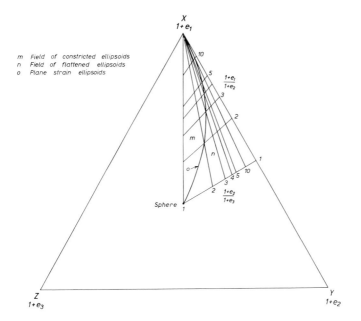

Figure 4-13

Triangular diagram for illustrating strain ratios. Constriction ellipsoids are found in field m, flattened ellipsoids in field n, and equivolume plane-strain ellipsoids lie along line o.

Harland and Bayley (1958) for illustrating the values of principal stresses. The three principal extensions $1 + e_1$, $1 + e_2$, and $1 + e_3$ are placed at the apexes of an equilateral triangle (Fig. 4-13) and straight lines of equal ratios $(1 + e_1)/(1 + e_2)$ and $(1 + e_2)/(1 + e_3)$ are drawn through the triangle. Because of the definition of the order of size of the three strains, only one-sixth of the area of the triangle is occupied by possible ellipsoids. The point at the center of the triangle represents a sphere; prolate uniaxial ellipsoids are found along the straight line which joins the sphere to the apex $(1 + e_1)$, oblate ellipsoids occur along the line $(1 + e_1)/(1 + e_2) = 1$ and equal-volume plane-strain ellipsoids lie along the line o. Although this may be found useful, it has drawbacks when deformation paths are to be expressed in mathematical terms.

Strain components (4) through (9)

METHOD 1 These components can all be expressed on one graph. The angles are plotted as ordinates against a principal extension (e.g., Fig. 4-15, ε) as abscissa. For any one deformation there are six single points. For a progressive deformation, the variations in the final and initial orientations of the principal axes of the strain-ellipsoid trace out continuous curves (Fig. 4-15); these curves, together with the deformation paths showing the variations in

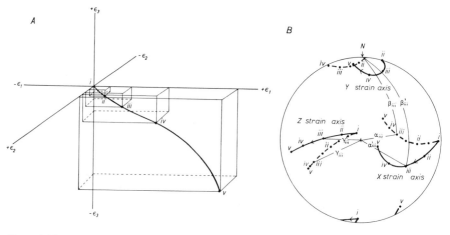

Figure 4-14

The three-dimensional graph in part A enables the changes in the amount of the principal strains to be recorded. The line joining i, ii, iii, iv, and v represents a deformation path. Part B shows the changes in orientation of the axes of the strain ellipsoid (solid lines) and of the principal axes of strain (broken lines) during the progress of deformation. The two graphical methods define completely the change in the finite-strain tensor during progressive deformation.

amounts of principal strains, describe the complete history of the finite-strain tensor.

METHOD 2   Another method which can be used to represent the orientations of lines and which is easier to interpret employs an equal-area net (Fig. 4-14B). For any given finite strain (e.g., position *iii*), the six components can be plotted on a stereogram: the angles α_{iii} and β_{iii} define the initial position of the principal X axis, and γ_{iii} fixes the position of the principal Y axis. Similarly we can locate the three axes of the strain ellipsoid by using the angles α'_{iii}, β'_{iii}, and γ'_{iii}. The complete representation of these six angles during a progressive deformation is shown by the six movement paths of the three principal axes and the three principal elongations of the successive ellipsoids. If this diagram is combined with Fig. 4-14A, all nine components of the strain tensor are precisely fixed at any point during the deformation history and Fig. 4-14 therefore completely describes the progress of the deformation.

Limitations of applications to data from deformed rocks   The applications of the complete representation of the state of strain in naturally deformed rocks are limited and it is unlikely that all nine components of finite tensor strain can be established. Thus the initial orientations of the principal strain axes which define the rotational part of the tensor can never be determined exactly when we only have deformed materials to investigate. However, it has been

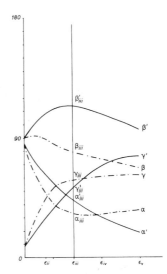

Figure 4-15

Graphical methods for showing changes in orientation of the axes of the strain ellipsoid and the principal axes of strain during progressive deformation. This forms a method of illustrating strain components (4) through (9) alternative to that of Fig. 4-14B.

suggested that it is often possible to make plausible guesses at these components.

Given suitable material, it should always be possible to fix the other six components of the finite-strain tensor, that is, those components which describe the orientation and principal extensions of the strain ellipsoid.

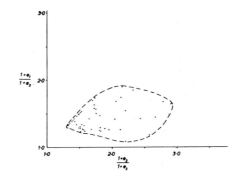

Figure 4-16

The deformation field occupied by three components of the strain tensor. The data are from the study of oolite deformation in Maryland made by Cloos (1947).

From one finite state of strain it is also impossible to fix the deformation path and trace the variations in all nine components during the deformation history. It seems likely, however, that what will become important in the future is the interpretation of what are best termed *deformation fields* (Fig. 4-16). These are the regions occupied by plots of six strain components on the variation graphs described above and which represent variations in components of the strain tensor. So far, little has been done to record these variations and even less to interpret them.

4-7   DETERMINING THE PRINCIPAL STRAINS OF THE STRAIN ELLIPSOID FROM TWO-DIMENSIONAL DATA

Much of the exact information that a geologist can obtain about the deformed condition of a rock in nature is of a two-dimensional nature. For example, fossils are often only preserved as impressions on the bedding surfaces, and measurements made on the distortion suffered by these impressions will enable the principal axes of the strain ellipse on that surface to be determined. Other sections through the rock may enable further strain ellipses to be determined. The discussions which follow will describe the most practical ways of evaluating the principal elongations from the two-dimensional strain ellipses.

Measurements of strain made on any three mutually perpendicular planes
The technique to be described is of very great importance and forms the basic method for determining the orientations of the three principal axes of the strain ellipsoid and the values of the three principal extensions assuming that no information is available about the orientation of the principal planes and axes of the ellipsoid. If the geologist has some knowledge of this orientation, as is sometimes the case, then a somewhat simpler mathematical method which is to be described later will give the unknown quantities. The importance of this technique is that it does not rely on any assumptions whatsoever. Any section through an ellipsoid is an ellipse (the two circular sections are special ellipses with equal principal extensions), and a section through a strain ellipsoid that is derived from distortion of a unit sphere is a two-dimensional strain ellipse. Consider three mutually perpendicular sections through a strain ellipsoid where the intersections of the sections are the mutually perpendicular axes x, y, and z, respectively. The general equation for each of the strain-ellipse sections in terms of the quadratic elongations and the angle θ that the principal axis of elongation in the ellipse makes with the x, y, or z coordinate axes is given by combining Eq. (3-26) with (3-18) (for the xy principal section); thus

$$\frac{(x \cos \theta + y \sin \theta)^2}{\lambda_{xy1}} + \frac{(-x \sin \theta + y \cos \theta)^2}{\lambda_{xy2}} = 1$$

$$\left(\frac{\cos^2 \theta}{\lambda_{xy1}} + \frac{\sin^2 \theta}{\lambda_{xy2}}\right)x^2 - 2 \sin \theta \cos \theta \left(\frac{1}{\lambda_{xy2}} - \frac{1}{\lambda_{xy1}}\right)xy + \left(\frac{\sin^2 \theta}{\lambda_{xy1}} + \frac{\cos^2 \theta}{\lambda_{xy2}}\right)y^2 = 1$$

$$(4-23)$$

where λ_{xy1} and λ_{xy2} are the principal axes of the ellipse.
   This equation can be put more simply in terms of the reciprocal quadratic extension along x and y and the γ' value for these directions. Using Eqs. (3-31) and (3-42) the three strain-ellipse sections on xy, yz, and zx, respectively, become

$$\lambda'_x x^2 - 2\gamma'_{xy} xy + \lambda'_y y^2 = 1 \qquad (4\text{-}24a)$$

$$\lambda'_y y^2 - 2\gamma'_{yz} yz + \lambda'_z z^2 = 1 \qquad (4\text{-}24b)$$

$$\lambda'_z z^2 - 2\gamma'_{zx} zx + \lambda'_x x^2 = 1 \qquad (4\text{-}24c)$$

The general equation for the unknown ellipsoid in terms of the coordinate axes x, y, and z is

$$ax^2 + by^2 + cz^2 - 2dxy - 2eyz - 2fzx = 1 \qquad (4\text{-}25)$$

By replacing $z = 0$ in this equation we obtain the equation for the strain ellipse on the xy section, and with $x = 0$ and $y = 0$ we obtain the ellipses on the yz and zx sections, respectively.

$$ax^2 - 2dxy + by^2 = 1 \qquad (4\text{-}26a)$$

$$by^2 - 2eyz + cz^2 = 1 \qquad (4\text{-}26b)$$

$$cz^2 - 2fzx + ax^2 = 1 \qquad (4\text{-}26c)$$

These ellipses are identical to those of Eqs. (4-24) and therefore by comparing coefficients we obtain the equation of the unknown ellipsoid in terms of six strain parameters measured on the three sections.

$$\lambda'_x x^2 + \lambda'_y y^2 + \lambda'_z z^2 - 2\gamma'_{xy} xy - 2\gamma'_{yz} yz - 2\gamma'_{zx} zx = 1 \qquad (4\text{-}27)$$

The next problem is to determine the values of the principal axes of this ellipsoid. This is done by a method analogous to that employed previously for finding the principal axes of the strain ellipse produced by simple shear (Chapter 3, p. 84). In that analysis we found the two values of r where a circle of radius r just touched the ellipse; these were the lengths of the principal axes of the strain ellipse. In three dimensions we find where the sphere of radius r, given by

$$x^2 + y^2 + z^2 = r^2 \qquad (4\text{-}28)$$

just touches the ellipsoid. We know from the general properties of the ellipsoid that this will occur at three positions, and that the radius r that appears in the three solutions will represent the length of the principal axes of the ellipsoid.

Differentiating (4-27) and (4-28),

$$2\lambda'_x x \, dx + 2\lambda'_y y \, dy + 2\lambda'_z z \, dz - 2\gamma'_{xy} x \, dy - 2\gamma'_{xy} y \, dx - 2\gamma'_{yz} y \, dz$$

$$- 2\gamma'_{yz} z \, dy - 2\gamma'_{zx} z \, dx - 2\gamma'_{zx} x \, dz = 0$$

or

$$(\lambda'_x x - \gamma'_{xy} y - \gamma'_{zx} z) \, dx + (-\gamma'_{xy} x + \lambda'_y y - \gamma'_{yz} z) \, dy$$

$$+ (-\gamma'_{zx} x - \gamma'_{yz} y + \lambda'_z z) \, dz = 0 \qquad (4\text{-}29)$$

and

$$x \, dx + y \, dy + z \, dz = 0 \qquad (4\text{-}30)$$

These two differential equations are solved by Lagrange's method; we first multiply (4-30) by a factor q of, as yet, unknown significance

$$qx\,dx + qy\,dy + qz\,dz = 0 \qquad (4\text{-}31)$$

Now comparing the coefficients of (4-29) and (4-31) and rearranging the terms, we obtain three linear equations in x, y, and z:

$$(\lambda'_x - q)x - \gamma'_{xy}y - \gamma'_{zx}z = 0 \qquad (4\text{-}32a)$$

$$-\gamma'_{xy}x + (\lambda'_y - q)y - \gamma'_{yz}z = 0 \qquad (4\text{-}32b)$$

$$-\gamma'_{zx}x - \gamma'_{yz}y + (\lambda'_z - q)z = 0 \qquad (4\text{-}32c)$$

The meaning of the term q is found by multiplying (4-32a) by x, (4-32b) by y, and (4-32c) by z and adding, for we obtain the equation

$$\lambda'_x x^2 + \lambda'_y y^2 + \lambda'_z z^2 - 2\gamma'_{xy}xy - 2\gamma'_{yz}yz - 2\gamma'_{zx}zx - q(x^2 + y^2 + z^2) = 0$$

$$(4\text{-}33)$$

which, from (4-27) and (4-28), is

$$1 - qr^2 = 0$$

or

$$q = \frac{1}{r^2} \qquad (4\text{-}34)$$

which is, by definition, the reciprocal quadratic extension of a line of initial unit length and strained length r, that is, the reciprocal quadratic extension of the axes of the ellipsoid $1/r^2 = \lambda' = \lambda'_1, \lambda'_2,$ or λ'_3. We shall therefore replace the factor q by λ', and solve Eqs. (32) by eliminating x, y, and z.

First multiply (4-32a) by γ'_{yz} and (4-32b) by γ'_{zx}, and subtract

$$x(\lambda'_x\gamma'_{yz} - \lambda'\gamma'_{yz} + \gamma'_{xy}\gamma'_{zx}) + y(\lambda'\gamma'_{zx} - \lambda'_y\gamma'_{zx} - \gamma'_{xy}\gamma'_{yz}) = 0 \qquad (4\text{-}35)$$

Then multiplying (4-32b) by $(\lambda'_z - \lambda')$, (4-32c) by γ'_{yz}, and add

$$x(-\gamma'_{xy}\lambda'_z + \gamma'_{xy}\lambda' - \gamma'_{zx}\gamma'_{yz}) + y(\lambda'_y\lambda'_z - \lambda'\lambda'_y - \lambda'\lambda'_z + \lambda'^2 - \gamma'_{yz}{}^2) = 0 \qquad (4\text{-}36)$$

$$-\frac{x}{y} = \frac{\lambda'\gamma'_{zx} - \lambda'_y\gamma'_{zx} - \gamma'_{xy}\gamma'_{yz}}{\lambda'_x\gamma'_{yz} - \lambda'\gamma'_{yz} + \gamma'_{xy}\gamma'_{zx}} = \frac{\lambda'_y\lambda'_z - \lambda'\lambda'_y - \lambda'\lambda'_z + \lambda'^2 - \gamma'_{yz}{}^2}{-\gamma'_{xy}\lambda'_z + \gamma'_{xy}\lambda' - \gamma'_{zx}\gamma'_{yz}}$$

Cross multiplying and collecting terms, we obtain a cubic equation in the variable λ',

$$\lambda'^3 - (\lambda'_x + \lambda'_y + \lambda'_z)\lambda'^2 + (\lambda'_x\lambda'_y + \lambda'_y\lambda'_z + \lambda'_z\lambda'_x - \gamma'_{xy}{}^2 - \gamma'_{yz}{}^2 - \gamma'_{zx}{}^2)\lambda'$$

$$- (\lambda'_x\lambda'_y\lambda'_z - 2\gamma'_{xy}\gamma'_{yz}\gamma'_{zx} - \lambda'_x\gamma'_{yz}{}^2 - \lambda'_y\gamma'_{zx}{}^2 - \lambda'_z\gamma'_{xy}{}^2) = 0 \qquad (4\text{-}37)$$

This equation could, of course, have been obtained more elegantly by finding the nontrivial solution where the determinant of the coefficient of Eqs. (4-32) vanishes, where

$$\begin{vmatrix} \lambda'_x - \lambda & -\gamma'_{xy} & -\gamma'_{zx} \\ -\gamma'_{xy} & \lambda'_y - \lambda' & -\gamma'_{yz} \\ -\gamma'_{zx} & -\gamma'_{yz} & \lambda'_z - \lambda' \end{vmatrix} = 0$$

that is,

$$(\lambda' - \lambda'_x)(\lambda' - \lambda'_y)(\lambda' - \lambda'_z) - \gamma'^2_{yz}(\lambda' - \lambda'_x)$$
$$- \gamma'^2_{zx}(\lambda' - \lambda'_y) - \gamma'^2_{xy}(\lambda' - \lambda'_z) + 2\gamma'_{xy}\gamma'_{yz}\gamma'_{zx} = 0$$

We assumed that we knew the states of strain given by the three strain ellipses on sections xy, yz, and zx and it is therefore a simple matter to calculate the six strain parameters necessary to define Eq. (4-37). By finding the three roots we can discover the reciprocal quadratic elongation of the three principal strain ellipsoid axes (λ'_1, λ'_2, and λ'_3).

The importance of the constant functions known as invariants has been mentioned before (Secs. 2-9, 3-9). Equation (4-37) must be equivalent to $(\lambda' - \lambda'_1)(\lambda' - \lambda'_2)(\lambda' - \lambda'_3) = 0$, and by comparing coefficients we obtain the *three finite-strain invariants* J_1, J_2, and J_3.

$$\lambda'_1 + \lambda'_2 + \lambda'_3 = \lambda'_x + \lambda'_y + \lambda'_z = J_1 \tag{4-38}$$

$$\lambda'_1\lambda'_2 + \lambda'_2\lambda'_3 + \lambda'_3\lambda'_1 = \lambda'_x\lambda'_y + \lambda'_y\lambda'_z + \lambda'_z\lambda'_x - \gamma'^2_{xy} - \gamma'^2_{yz} - \gamma'^2_{zx} = J_2 \tag{4-39}$$

$$\lambda'_1\lambda'_2\lambda'_3 = \lambda'_x\lambda'_y\lambda'_z - 2\gamma'_{xy}\gamma'_{yz}\gamma'_{zx} - \lambda'_x\gamma'^2_{yz} - \lambda'_y\gamma'^2_{zx} - \lambda'_z\gamma'^2_{xy} = J_3 \tag{4-40}$$

The solution of the three roots of (4-37) necessitates first the calculation of these three invariant quantities from the six strain parameters. The roots of the cubic equation are then most readily obtained by determining one (λ'_2) by trial and error, using a desk calculating machine, knowing that it lies between the turning points of the cubic given by $3\lambda'^2 - 2J_1\lambda' + J_2 = 0$, that is, between $\lambda' = (J_1 \pm (J_1^2 - 3J_2)^{\frac{1}{2}})/3$. When this has been obtained with sufficient accuracy, λ'_1 and λ'_3 are found by dividing $\lambda'^3 - J_1\lambda'^2 + J_2\lambda' - J_3 = 0$ by $(\lambda' - \lambda'_2)$; they are the roots of the quadratic

$$\lambda'^2 + (\lambda'_2 - J_1)\lambda' + (\lambda'^2_2 - J_1\lambda'_2 + J_2) = 0 \tag{4-41}$$

Generally this technique leads to a fairly rapid solution. However, if an exact solution is required we can solve the cubic by Cardon's method. The equation

$$\lambda'^3 - J_1\lambda'^2 + J_2\lambda' - J_3 = 0$$

is rearranged into the new cubic equation in $(\lambda' - J_1/3)$

$$\left[\lambda'^3 - J_1\lambda'^2 + 3\left(\frac{J_1}{3}\right)^2\lambda' - \left(\frac{J_1}{3}\right)^3 \right]$$

$$+ \left[-3\left(\frac{J_1}{3}\right)^2\lambda' + 3\left(\frac{J_1}{3}\right)^3 + J_2\lambda' - J_2\frac{J_1}{3} \right]$$

$$+ \left[-J_3 + J_2\left(\frac{J_1}{3}\right) - 2\left(\frac{J_1}{3}\right)^3 \right] = 0$$

or

$$\left(\lambda' - \frac{J_1}{3}\right)^3 - \left[3\left(\frac{J_1}{3}\right)^2 - J_2\right]\left(\lambda' - \frac{J_1}{3}\right) - \left[J_3 - J_2\frac{J_1}{3} + 2\left(\frac{J_1}{3}\right)^3\right] = 0$$

(4-42)

This cubic is in the form

$$x^3 - K_1 x - K_2 = 0 \tag{4-43}$$

where $x = (\lambda' - J_1/3)$ and K_1 and K_2 are invariant quantities dependent only on the strain invariants J_1, J_2, and J_3.

This new cubic is solved as follows. Let

$$a = \left(\frac{2K_1}{3}\right)^{\frac{1}{2}} \qquad K_1 = \frac{3a^2}{2} \tag{4-44}$$

Then if one of the roots of (4-43) is x_1 such that

$$x_1 = a(2)^{\frac{1}{2}}\cos\alpha \tag{4-45}$$

substitute this in (4-43) and multiply throughout by $(2)^{\frac{1}{2}}$.

$$a^3(4\cos^3\alpha - 3\cos\alpha) - K_2\sqrt{2} = 0 \tag{4-46}$$

As $4\cos^3\alpha - 3\cos\alpha = \cos 3\alpha$, one value of α will be given by

$$\cos 3\alpha = \frac{K_2\sqrt{2}}{a^3} \tag{4-47}$$

and will lie between 0 and $\pi/3$.

The three roots of (4-43) are therefore

$$x_1 = a\sqrt{2}\cos\alpha \tag{4-48a}$$

$$x_2 = a\sqrt{2}\cos\left(\alpha + \frac{2\pi}{3}\right) \tag{4-48b}$$

$$x_3 = a\sqrt{2}\cos\left(\alpha - \frac{2\pi}{3}\right) \tag{4-48c}$$

and $\lambda_1' = x_1 + J_1/3$, etc.

The orientations of the principal axes are found by discovering the co-ordinates (x, y, and z) of their end points. Applying similar modifications to Eqs. (4-32*b* and *c*) such as produced Eq. (4-36) from Eq. (4-32*a* and *b*), we find that

$$-\frac{x}{z} = \frac{\lambda'_y \lambda'_z - \lambda' \lambda'_y - \lambda' \lambda'_z + \lambda'^2 - \gamma'^2_{yz}}{\gamma'_{xy}\gamma'_{yz} + \lambda'_y \gamma'_{zx} - \lambda' \gamma'_{zx}} \tag{4-49}$$

and therefore from (4-36) and (4-49)

$$\frac{x}{\lambda' \lambda' - \lambda' \lambda' - \lambda' \lambda' + \lambda'^2 - \gamma'^2} = \frac{-y}{-\gamma'_{xy}\lambda'_z + \gamma'_{xy}\lambda' - \gamma'_{zx}\gamma'_{yz}}$$

$$= \frac{-z}{\gamma'_{xy}\gamma'_{yz} + \lambda'_y \gamma'_{zx} - \lambda' \gamma'_{zx}} \tag{4-50}$$

Inserting values of the strain components and values of one of the principal strains (λ'_1), we obtain the relations of the coordinates x_1, y_1, and z_1, and these can be established completely by using the additional relationship

$$x_1^2 + y_1^2 + z_1^2 = \lambda_1 \tag{4-51}$$

To establish the direction cosines (l_1, m_1, and n_1) of the axes of this ellipsoid we use the relationships

$$l_1 = \frac{x_1}{\lambda_1^{\frac{1}{2}}} \qquad m_1 = \frac{y_1}{\lambda_1^{\frac{1}{2}}} \qquad n_1 = \frac{z_1}{\lambda_1^{\frac{1}{2}}}$$

Similarly the direction cosines (l_2, m_2, and n_2) of the intermediate strain axis may be obtained.

Measurements of strain made on any three nonparallel plane surfaces The solution of this problem depends on first finding the values of six basic strain parameters on any three mutually perpendicular surfaces, and then proceeding by the techniques of the previous problem.

This is done by first plotting three planes A, B, and C as great circles on a projection (Fig. 4-17A). Any three mutually perpendicular axes x, y, and z are now chosen, the great circles representing the planes xy, yz, and zx are drawn, and the nine intersections of the planes A, B, and C with xy, yz and zx (A_{xy}, A_{yz}, A_{zx}, B_{xy}, . . . , C_{zx}) are determined. For an accurate final result it is most important to select the axes x, y, and z so that on any of the planes containing them (e.g., plane xy) the angles $A_{xy}B_{xy}$, $B_{xy}C_{xy}$ and $C_{xy}A_{xy}$ are as nearly equal as possible.

Taking plane A, we know the state of strain for this surface; and using Eqs. (3-31) and (3-42) or a Mohr diagram, we compute the reciprocal quadratic elongation for the lines A_{xy}, A_{yz}, and A_{zx}. We repeat this for the surfaces B and C and calculate the elongations for B_{xy}, B_{yz}, . . . , C_{zx}. Thus for each of the planes xy, yz, and zx, we calculate the elongations of three lines and

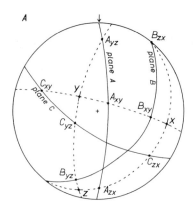

Figure 4-17A

Conversion of strain data from any three planes onto three mutually perpendicular planes

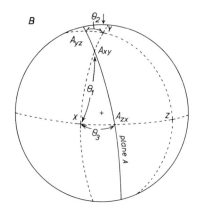

Figure 4-17B

Calculation of three principal strains knowing the state of strain on surface A and the orientation of the strains.

their angular relationships in the strained state. Using a technique previously described (Chap. 3, p. 79), it is possible to calculate the state of strain for each of these planes and therefore determine λ'_x, λ'_y, λ'_z, γ'_{xy}, γ'_{yz}, γ'_{zx}.

From a knowledge of the state of strain on a single plane surface when the orientation of the principal axes of the ellipsoid is known   In strongly deformed rocks it is sometimes possible to determine the orientation of the principal axes of the strain ellipsoid. For example, a strong linear fabric or "stretching" direction may mark the position of the maximum elongation in the strain ellipsoid. A structure known as slaty cleavage may be formed parallel to the xy plane of the finite-strain ellipsoid and perpendicular to the direction of maximum shortening. Where such information can be used to establish the principal axes of the strain ellipsoid, a complete analysis of the state of strain in three dimensions is possible using only the two-dimensional strain state on a single plane surface providing that this surface does not include a principal axis of the ellipsoid. The method which is described is very useful and particularly simple in application.

The principal axes of the strain ellipsoid X, Y, and Z are plotted as a stereographic projection and the principal planes constructed (Fig. 4-17B). Plane A is plotted as a great circle, and the angles between its intersections with the principal planes (A_{xy}, A_{yz}, A_{zx}) and the principal ellipsoid axes are determined, $\theta_1 = xA_{xy}$, $\theta_2 = yA_{yz}$, and $\theta_3 = xA_{zx}$. The reciprocal quadratic extensions of the line $a_{xy} = \lambda'_{xy}$, $A_{yz} = \lambda'_{yz}$, and $A_{zx} = \lambda'_{zx}$ are calculated from Eq. (3-31) or by means of a Mohr diagram for the deformed state of plane A. It follows from (3-31) that

$$\lambda'_{xy} = \lambda'_1 \cos^2 \theta_1 + \lambda'_2 \sin^2 \theta_1 \qquad (4\text{-}52a)$$

$$\lambda'_{yz} = \lambda'_2 \cos^2 \theta_2 + \lambda'_3 \sin^2 \theta_2 \qquad (4\text{-}52b)$$

$$\lambda'_{zx} = \lambda'_1 \cos^2 \theta_3 + \lambda'_3 \sin^2 \theta_3 \qquad (4\text{-}52c)$$

Solving these for λ'_1, λ'_2, and λ'_3, we have

$$\lambda'_1 = \frac{\lambda'_{xy} \sin^2 \theta_3 \cos^2 \theta_2 - \lambda'_{yz} \sin^2 \theta_1 \sin^2 \theta_3 + \lambda'_{zx} \sin^2 \theta_1 \sin^2 \theta_2}{\cos^2 \theta_1 \cos^2 \theta_2 \sin^2 \theta_3 + \sin^2 \theta_1 \sin^2 \theta_2 \cos^2 \theta_3} \qquad (4\text{-}53a)$$

$$\lambda'_2 = \frac{\lambda'_{xy} \sin^2 \theta_2 \cos^2 \theta_3 + \lambda'_{yz} \cos^2 \theta_1 \sin^2 \theta_3 - \lambda'_{zx} \cos^2 \theta_1 \sin^2 \theta_2}{\cos^2 \theta_1 \cos^2 \theta_2 \sin^2 \theta_3 + \sin^2 \theta_1 \sin^2 \theta_2 \cos^2 \theta_3} \qquad (4\text{-}53b)$$

$$\lambda'_3 = \frac{-\lambda'_{xy} \sin^2 \theta_2 \cos^2 \theta_3 + \lambda'_{yz} \sin^2 \theta_1 \cos^2 \theta_3 + \lambda'_{zx} \cos^2 \theta_1 \sin^2 \theta_2}{\cos^2 \theta_1 \sin^2 \theta_2 \sin^2 \theta_3 + \sin^2 \theta_1 \cos^2 \theta_2 \cos^2 \theta_3} \qquad (4\text{-}53c)$$

If plane A contains the y axis of the strain ellipsoid but intersects the xz plane at some position between x and z, then Eqs. (4-52) become

$$\lambda'_{xy} = \lambda'_2 \qquad (4\text{-}54a)$$

$$\lambda'_{yz} = \lambda'_2 \qquad (4\text{-}54b)$$

$$\lambda'_{zx} = \lambda'_1 \cos^2 \theta_3 + \lambda'_3 \sin^2 \theta_3 \qquad (4\text{-}54c)$$

and it follows that a complete solution cannot be found without additional information. This complete solution is best discovered by determining the shear strain along line A_{zx} and solving the two-dimensional strain problem on the zx plane by the technique described on p. 74.

If plane A is a principal plane of the ellipsoid, it is obvious that the principal axes of the strain ellipse on plane A represent two of the principal planes in three dimensions, but that no information about the value of the third principal strain can be derived.

4-8   MOHR'S CONSTRUCTION FOR REPRESENTING STATES OF STRAIN IN THREE DIMENSIONS

The state of elongation and shear in any direction described by the direction cosines l', m', and n' in a strain ellipsoid can be described in two equations:

$$\lambda' = l'^2 \lambda'_1 + m'^2 \lambda'_2 + n'^2 \lambda'_3 \qquad (4\text{-}7)$$

$$\gamma'^2 = (\lambda'_1 - \lambda'_2)^2 l'^2 m'^2 + (\lambda'_2 - \lambda'_3)^2 m'^2 n'^2 + (\lambda'_3 - \lambda'_1)^2 n'^2 l'^2 \qquad (4\text{-}17)$$

We shall now examine these equations to see whether it is possible to represent the states of strain in three dimensions by some graphical method.

First we can represent the strain states on the three principal planes of the ellipsoid because on these planes Eqs. (4-7) and (4-17) degenerate to simple two-dimensional forms which can be represented by Mohr circles.

On the YZ plane $l = 0$,

$$\lambda' = m'^2\lambda'_2 + n'^2\lambda'_3 = \lambda'_2 \cos^2 \theta_2 + \lambda'_3 \sin^2 \theta_2$$

$$= \frac{\lambda'_2 + \lambda'_3}{2} - \frac{(\lambda'_3 - \lambda'_2) \cos 2\theta_2}{2} \tag{4-55a}$$

$$\gamma' = (\lambda'_3 - \lambda'_2)m'n' = (\lambda'_3 - \lambda'_2) \cos \theta_2 \sin \theta_2 = \frac{(\lambda'_3 - \lambda'_2) \sin 2\theta_2}{2} \tag{4-55b}$$

and two similar sets of two equations for the XZ plane where $m = 0$, and the XY plane where $n = 0$, can be derived. The states of strain on the principal planes can therefore be represented as three circles of center $(\lambda'_1 + \lambda'_2)/2$, $(\lambda'_2 + \lambda'_3)/2$, $(\lambda'_3 + \lambda'_1)/2$ with radii $(\lambda'_2 - \lambda'_1)/2$, $(\lambda'_3 - \lambda'_2)/2$, and $(\lambda'_3 - \lambda'_1)/2$, respectively (Fig. 4-18). The problem now is to see if it is possible to represent the λ' and γ' values of any line as a point in this same diagram.

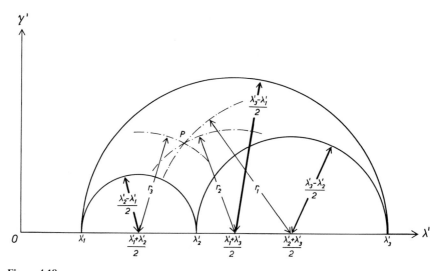

Figure 4-18

Mohr circle construction for three-dimensional strain.

If we take Eqs. (4-7) and (4-17), combine them with $l'^2 + m'^2 + n'^2 = 1$, and eliminate either m' and n' or n' and l' or l' and m', we obtain three equations:

$$\gamma'^2 + \left(\lambda' - \frac{\lambda'_2 + \lambda'_3}{2}\right)^2 = \left(\frac{\lambda'_3 - \lambda'_2}{2}\right)^2 + l'^2(\lambda'_3 - \lambda'_1)(\lambda'_2 - \lambda'_1) \tag{4-56a}$$

$$\gamma'^2 + \left(\lambda' - \frac{\lambda_3' + \lambda_1'}{2} \right)^2 = \left(\frac{\lambda_3' - \lambda_1'}{2} \right)^2 + m'^2 (\lambda_1' - \lambda_2')(\lambda_3' - \lambda_2') \quad (4\text{-}56b)$$

$$\gamma'^2 + \left(\lambda' - \frac{\lambda_1' + \lambda_2'}{2} \right)^2 = \left(\frac{\lambda_2' - \lambda_1'}{2} \right)^2 + n'^2 (\lambda_1' - \lambda_3')(\lambda_2' - \lambda_3') \quad (4\text{-}56c)$$

Now the equation of any circle of radius r and center at the point $(k,0)$ is

$$y^2 + (x - k)^2 = r^2 \quad (4\text{-}57)$$

and it will be seen that Eqs. (4-2) are exactly in this form where the variables are $y = \gamma'$ and $x = \lambda'$. It therefore follows that the state of strain of all lines with a fixed value of l', that is, at a constant angle to the principal axis X (λ_1), can be represented as points on a circle of center $[(\lambda_2' + \lambda_3')/2, 0)]$ and radius

$$\left[\left(\frac{\lambda_3' - \lambda_2'}{2} \right)^2 + l'^2 (\lambda_3' - \lambda_1')(\lambda_2' - \lambda_1') \right]^{\frac{1}{2}}$$

and that state of strain with different values of l' can be represented as a series of concentric circles. On the YZ plane where $l' = 0$, the radius of the circle is $(\lambda_3' - \lambda_2')/2$, but where l' lies between 0 and 1 the radius of the circle (Fig. 4-18, r_1) always *exceeds* $(\lambda_3' - \lambda_2')/2$ because by definition $\lambda_1' \leqslant \lambda_2' \leqslant \lambda_3'$ and therefore both the terms $(\lambda_3' - \lambda_1')$ and $(\lambda_2' - \lambda_1')$ take a positive value.

Similarly the states of strain along any line in the ellipsoid at a fixed angle to the Y axis can be represented as points on a circle of center $(\lambda_1' + \lambda_3')/2$, and where $m = 0$ (the XZ principal plane) the radius is $(\lambda_3' - \lambda_2')/2$. The radii of all other circles where $m > 0$ (Fig. 4-18, r_2) are *less than* $(\lambda_3' - \lambda_1')/2$ because $(\lambda_1' - \lambda_2')$ is always negative and $(\lambda_3' - \lambda_1')$ always positive and therefore their product is always negative.

Using the same reasoning, the strains in any direction with fixed values of n lie on concentric circles of center $(\lambda_1' + \lambda_2')/2$, and radii (r_3) *equal to* or *greater than* $(\lambda_2' - \lambda_1')/2$.

Because of the limitations we have established in the radii of the sets of concentric circles, it will be apparent that the λ' and γ' values for any direction in an ellipsoid can be represented in the type of diagram shown in Fig. 4-18 as a point P which plots in the area contained between the three circles representing strain states on the three principal planes.

It should also be pointed out here that the equations which relate to states of stress in three dimensions (2-34) and (2-38) are also in the form of (4-7) and (4-17) and therefore can be represented in a similar diagram where normal stress σ is plotted as abscissa and shearing stress τ as ordinate.

To determine the elongation and shear along any line in the strain ellipsoid all we need to know are the radii and centers of two circles satisfying any

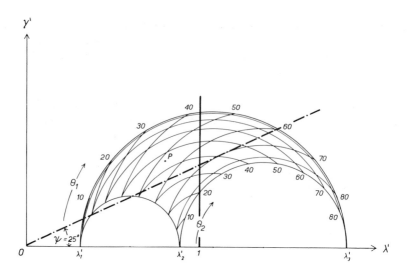

Figure 4-19

Completed Mohr diagram for three-dimensional strain. The vertical line $\lambda' = 1$ gives the orientation of all lines with no finite longitudinal strain; the line at 25° to the λ' axis gives the orientation of all directions having an angular shear ψ of 25°.

two of the three equations (4-56a, b, c). There is a method, however, by which we can draw these circles very rapidly with no arithmetic computation. This method consists of locating a series of points on the principal sections with known l, m, and n values. Consider first the principal plane XZ where $m = 0$. We know how to determine the strain characteristics of any line in this plane located at a known angle θ_1 to the X axis (λ_1) by using the standard two-dimensional Mohr construction. We set up an angle 2θ from the λ' axis at the center of the Mohr circle $(\lambda_1' + \lambda_3')/2$. The points on this circle at various angles of θ_1 (see Fig. 4-19) are values of l, for $\cos^{-1} \theta_1 = l$. Therefore to draw any circle representing states of strain of lines oriented at, say, 30° from the X axis ($l = \cos^{-1} 30°$) we draw a circle of center $(\lambda_2' + \lambda_3')/2$ passing through the point $\theta_1 = 30°$. Similarly it is possible to draw a whole series of concentric circles with values of $\cos^{-1} l$ of 10, 20, 30° . . . , etc. Using identical reasoning and considering the YZ plane where $l = 0$, it is possible to establish points representing strain states of lines at angles θ_2 from the Y axis, and these will all be values of $\cos^{-1} m$. Therefore a series of concentric circles of centers $(\lambda_1 + \lambda_3)/2$ passing through these points will satisfy (4-56b) and represent the state of strain on any line oriented at θ_2 ($\cos^{-1} m$) from the Y axis. These two sets of circles intersect and build up a coordinate framework of lines of equal values of l and m (Fig. 4-19). Because the direction of every line in the ellipsoid can be specified in terms of l and

m, we can directly plot its position in this Mohr diagram and define the elongation and shear. For example, lines oriented at an angle of 53° from the X axis and 45° from the Y axis have strain characteristics which are given by the abscissa and ordinate of the point P in Fig. 4-19.

Lines of no finite longitudinal strain It has been proved earlier (4-8) that if $\lambda_1 > 1 > \lambda_3$ lines of no finite elongation are arranged on the surface of a cone passing through the center of the ellipsoid. It is possible to utilize the three-dimensional Mohr diagram to find the orientation of these lines and define the cone very rapidly. The lines of no finite longitudinal strain are found along the line $\lambda' = 1$ in Fig. 4-19. The grid of concentric circles enables us to find the values of θ_1 and θ_2 for all points on this line. It is then possible to transfer these values onto a projection (Fig. 4-20) and to trace the orientation of this cone surface.

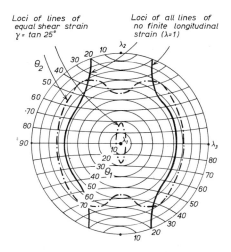

Figure 4-20

Equal-area projection of the surface of lines of no finite longitudinal strain and surface of lines of angular shear 25° derived from the Mohr diagram of Fig. 4-19.

Lines of equal shear strain It has been noted previously (p. 73, Fig. 3-13A) that if a line is drawn through the origin making an angle ψ with the λ' axis then this line intersects the Mohr circles at lines of equal angular shear. Using the grid of lines for equal θ_1 and θ_2, we can therefore determine the orientations of all lines of equal finite shear, and these can be transferred to a projection where the relationships to the axes of the ellipsoid can be seen more easily. The *maximum shear strain* occurs where the line of greatest value of ψ is tangent to the circle of center $(\lambda_1' + \lambda_3')/2$, radius $(\lambda_3' - \lambda_1')/2$. This maximum value is always found on the XZ principal plane along lines which make an angle given by (3-45), that is, $\tan \theta_1 = (\lambda_3/\lambda_1)^{\frac{1}{2}}$, and the value of this strain is given by (3-46), that is, $\gamma = \pm(\lambda_1 - \lambda_3)/2(\lambda_1\lambda_3)^{\frac{1}{2}}$.

4-9   FINITE-STRAIN PROPERTIES OF THE FIVE TYPES OF CONSTANT-VOLUME ELLIPSOIDS

Two of the most important contributions to the understanding of three-dimensional strain have been made by Ramberg (1959) and Flinn (1962). Ramberg's study of folding and boudinage led him to consider the effects of progressive deformation in three dimensions, and he examined three special types of strain ellipsoid: those formed by pure shear ($\lambda_2 = 1$), those where there is one principal contraction and uniform extension in the plane perpendicular to it ($\lambda_1 = \lambda_2 > 1 > \lambda_3$), and those with equal compressions in two directions and extension in the third ($\lambda_1 > 1 > \lambda_2 = \lambda_3$). Flinn

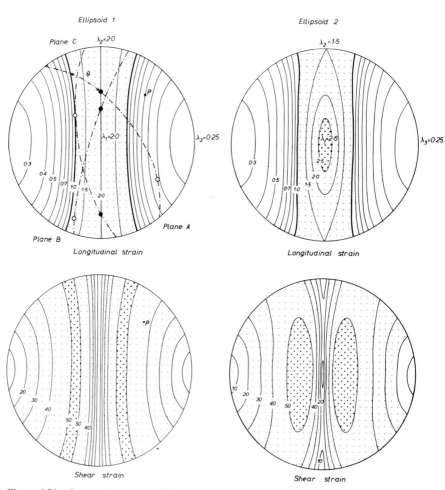

Figure 4-21   *See caption on page 156.*

considerably developed Ramberg's work and investigated the strain properties of other types of ellipsoids, recognizing five main types of constant-volume ellipsoids and describing them using the parameter k [see Eq. (4-22)].

The properties of these five types are illustrated in Figs. 4-21 and 4-22. For each of the ellipsoids 1 to 5 shown in the graph (Fig. 4-21), two equal-area projections have been constructed to give (1) the state of elongation and (2) the angular shear for each ellipsoid. These were constructed using the technique described above employing the Mohr diagram to determine the three-dimensional strain state. The ellipsoids in Fig. 4-21 are all oriented with their principal extension (X) vertical, the intermediate axis (Y) N-S horizontal,

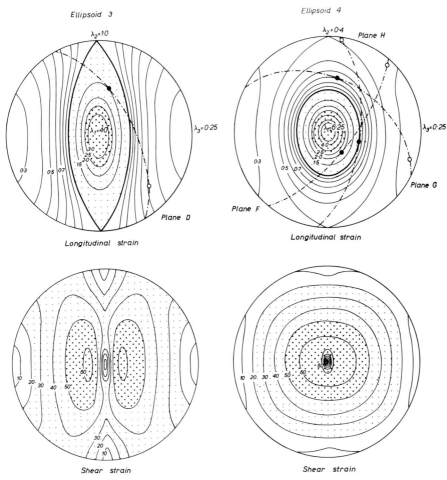

Figure 4-21 *See caption on page 156.*

and the (Z) axis E-W horizontal. If any line within the strain ellipsoid of known angular relationship to the principal axes is plotted in these diagrams, its elongation and angular shear can be determined immediately. For example in ellipsoid 1, a line making an angle 60° with X and 50° with Y and represented by the point P has been decreased in length ($\lambda = 0.59$) and has an angular shear of 44°.

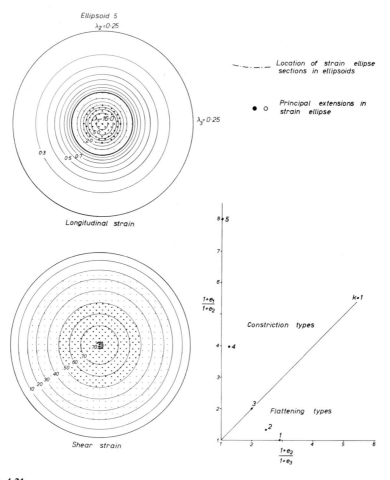

Figure 4-21

The properties of five types of equal-volume ellipsoids each with a compressive strain of 50 percent ($\lambda_3 = 0.25$). For each ellipsoid two equal-area projections have been plotted to show the variations in λ and ψ. The surfaces of no finite longitudinal strain are indicated with heavy lines. The dashed lines represent strain-ellipse cross sections through the ellipsoids discussed on pages 158-161.

All the five ellipsoids illustrated in Fig. 4-21 have the same compressive strain $[50$ percent or $(1 + e_3) = 0.5)]$ along the Z axis, and the strains along the other two principal axes have been accommodated in different ways so that the resulting ellipsoids have different axial ratios and fall in different parts of the graph of Fig. 4-21. The ellipsoids have the following k values: (1) $k = 0$, (2) $k = 0.22$, (3) $k = 1.0$, (4) $k = 11.35$, (5) $k = \infty$.

The strain properties of the two uniaxial ellipsoids 1 and 5, where $k = 0$ and ∞ respectively, show an axial symmetry distribution; all the other ellipsoids have orthorhombic symmetry. In these diagrams only the symmetrical part of the finite-strain tensor is represented; the geometry of the rotational part will be discussed later (Sec. 4-17).

The positions of lines of no finite longitudinal strain are shown in thick lines in the stereogram; in general, they form cones. In ellipsoids 2 and 4 these cones have elliptical cross sections, while in ellipsoids 1 and 5 they have circular cross sections. The lines of no finite longitudinal strain in ellipsoid 3

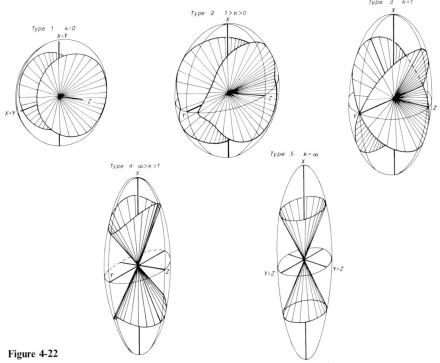

Figure 4-22

Diagrammatic representation of the five types of equal-volume strain ellipsoids and the position of the surfaces of no finite longitudinal strain.

lie on two planes intersecting along the intermediate (Y) strain axis—the two circular sections of this ellipsoid (Fig. 4-22).

The finite shear strain is always zero along the principal axes X, Y, and Z, and has maximum values along the principal planes and an absolute maximum which is always situated on the XZ plane at an angle of $\tan^{-1}(1 + e_3)/(1 + e_1)$ to the X axis.

Strain-ellipse sections of the five types of ellipsoids   Any plane surface section through the strain ellipsoids has strain properties which can be analyzed in a strain ellipse, and the various types of ellipsoids show a variety of types of these strain ellipses. If any great circle representing a section through the ellipsoid is drawn in the projection in Fig. 4-21, then the finite strains in this surface can be read off directly from the diagram (see ellipsoid 1,

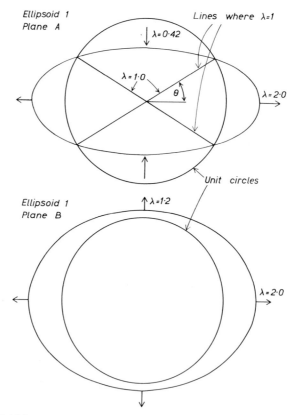

Figures 4-23 and 4-24

The shapes of two strain-ellipse sections through the ellipsoid 1 of Fig. 4-21 to show the relationships of two- and three-dimensional strain states.

plane A). Unless the plane contains the intermediate axis Y and is a circular section, it will be found that there is always one direction where the finite longitudinal strain reaches a maximum value, and another at right angles to it where it has a minimum value. These are the principal strains of the strain ellipse, and from them the ellipse can be constructed (Fig. 4-23). The two lines where the section cuts the cone surface $\lambda = 1$ will represent the two lines of no finite longitudinal strain in the strain ellipse oriented at an angle θ to the principal extension (Fig. 4-23). At some position the great circle representing the plane will not intersect the cone $\lambda = 1$. For example, in ellipsoid 1 all directions in plane B are extended and the strain ellipse lies completely outside the initial unit circle (Fig. 4-24), and in ellipsoid 4 all lines in plane G are contracted and the strain ellipse lies completely within a unit circle. It will be apparent that the features of deformation in two dimensions that were discussed in Chap. 3 are dependent on deformation in three dimensions, and that the variation in the shapes of strain ellipses is essentially a consequence of the properties of the ellipsoid in which they are situated, and also the orientation of the section within the ellipsoid.

It will be seen that with ellipsoids 1 and 2 (and all other equal-volume ellipsoids with the property $1 > k \geqslant 0$) there are only three possible types of strain-ellipse section. These are illustrated by the three planes A, B, and C with strain ellipses which have the properties

Plane A	$\lambda_1 > 1 > \lambda_2$
Plane B	$\lambda_1 > \lambda_2 > 1$
Plane C	$\lambda_1 > \lambda_2 = 1$

Using the nomenclature of the strain-ellipse fields developed in Chap. 3, we can say that all the ellipses lie in fields 1 and 2 or on the line which divides these fields from each other. For any specific ellipsoid we can locate precisely the extent of these fields; this is done for ellipsoids 1 and 2 in Fig. 4-25. If the pole of any plane is plotted on the equal-area projections in this diagram, it will fall into either the stippled zone or the vertically ruled zone and the strain ellipse is either in field 1 or 2, respectively. As a consequence of this, competent layers in equivolume ellipsoids may show the complex crossing boudinage structures described previously (Chap. 3, p. 112) if their poles plot in field 1.

The strain ellipses in practically every plane section within ellipsoid 3 have one principal axis extended and the other contracted and fall into ellipse field 2. Plane D is typical of most sections through the ellipsoid, and the strain ellipse for this section is shown in Fig. 4-26. In this diagram the traces of the principal planes of the ellipsoid on the section have been constructed; and these show a complete lack of symmetry to the principal extensions of the

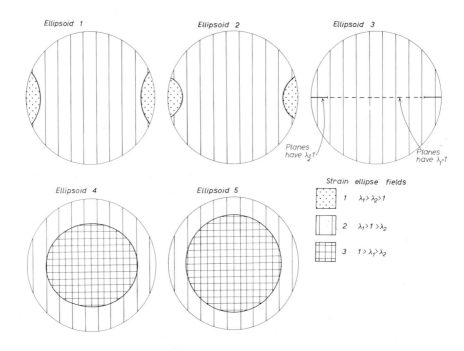

Figure 4-25

The strain-ellipse fields of the five types of equal-volume ellipsoids. Planes whose poles fall in the various zones have different types of finite-strain ellipses.

ellipse, and the principal elongation within plane D does not lie on any principal plane of the ellipsoid. This means that it should never be assumed that the intersection of slaty cleavage (parallel to XY) on any random section will mark the longest axis of the strain ellipse in that section. There are some special sections through ellipsoid 3, namely, those containing the intermediate axis Y whose strain ellipses do not lie in field 2. These sections lie to either side of the circular sections and have the strain characteristics $\lambda_1 = 1 > \lambda_2$ and $\lambda_1 > 1 = \lambda_2$, and they lie on the line separating strain ellipse fields 1 and 2 or on the lines between fields 2 and 3. The poles of the planes which have these properties lie along the XZ plane of the ellipsoid and their positions are shown by Fig. 4-25 (ellipsoid 3).

The strain properties of sections through equivolume ellipsoids with $k > 1$ may be any of three types. Some sections have a principal extension and a principal contraction ($\lambda_1 > 1 > \lambda_2$) (Fig. 4-21, ellipsoid 4, plane F). The other type of ellipse shows contraction in all directions ($1 > \lambda_1 > \lambda_2$) (plane G) or contraction in all except one direction ($\lambda_1 = 1 > \lambda_2$) where the section just touches the cone of lines with no finite longitudinal strain (plane H). Thus where $k > 1$, all the strain ellipse sections fall into ellipse fields 2 or 3 or on

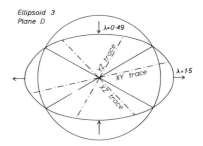

Figure 4-26

Section of ellipsoid 3 (Fig. 4-21 for location of section) to show the relationships of the principal two-dimensional strains and the traces of the principal planes of the ellipsoid on the section.

the line separating them. The locations of the poles of planes having these various properties are shown in Fig. 4-25 (ellipsoids 4 and 5). It will be clear from the preceding discussion on strain in two dimensions that competent layers whose poles fall in field 3 may show crumpling and complex folding with fold axes oriented in several directions.

The finite strain properties of the equivolume ellipsoids can be summarized as follows:

ELLIPSOIDS WITH $\infty \geqslant k > 1$ Ellipse sections fall in fields 1 or 2; therefore, in competent layers the structures that may develop are either complex crossing boudinage or other extension structures, or alternatively, ptygmatic folding in one direction with boudinage at right angles to the fold axes. The axes of folding will be perpendicular to the principal contraction in the competent layer, and therefore need have no simple relationship to the principal planes and axes of the strain ellipsoid.

ELLIPSOIDS WITH $k = 1$ Ellipse sections all fall into field 2 or on the boundary lines of this field. The dominant structures are folds with structures indicating extension subperpendicular to the fold axes. In some special sections folds only ($\lambda_1 = 1$) or boudinage only ($\lambda_2 = 1$) will develop; these sections contain the principal intermediate axis Y of the strain ellipsoid.

ELLIPSOIDS WITH $1 > k \geqslant 0$ Ellipse sections are situated in fields 2 or 3. Competent layers will show folding and boudinage if in field 2, or complex folding in several directions if in field 3. Both Ramberg and Flinn have pointed out that the development of fold axes which run in several directions and which cross one another are not necessarily indicative of the superposition of several deformations; such structures can form as a result of a single constriction-type deformation.

Changes in volume during deformation   If the strain ellipsoid does not have the same volume as the unit sphere from which it was derived, then the previous conclusions must be somewhat modified. The line $k = 1$ will not separate ellipse field 1 from field 3 because the intermediate axis Y of an ellipsoid with $k = 1$ will have had its length changed, either increased if the volume has increased, or decreased if the volume has decreased. The line which will separate the ellipse fields 1 and 3 will contain all ellipsoids where $(1 + e_2) = 1$.

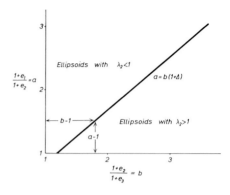

Figure 4-27

Position of true constriction and flattening fields with volume change Δ. Compare with the deformation field of Fig. 4-16. This field might be explained as the result of true flattening (e₂ positive) (if Δ = 0) or it could also result from a variable volume reduction with e₂ = 0.

If the unit volume change is Δ, then because $1 + e_2 = 1$,

$$(1 + e_1)(1 + e_3) = 1 + \Delta \qquad (4\text{-}58)$$

The locus of all points on the graph shown in Fig. 4-27 where $(1 + e_2) = 1$ is given by

$$a = 1 + e_1 \qquad (4\text{-}59a)$$

$$b = \frac{1}{1 + e_3} \qquad (4\text{-}59b)$$

Using condition (4-58) in (4-59)

$$a = b(1 + \Delta) \qquad (4\text{-}60)$$

a straight line which only passes through the point (1,1) where $\Delta = 0$. If, for example, there is a 15-percent volume reduction, the line dividing ellipsoids with ellipse fields 1 from those with fields 3 will be positioned as shown on Fig. 4-27 on the line $a = 0.85b$. Ellipsoids which lie on this line have variable k values depending on their ratios $(1 + e_2)/(1 + e_3) = b$ which are given by

$$k = \frac{(1 + \Delta)b - 1}{b - 1} \qquad (4\text{-}61)$$

4-10   DISPERSION OF LINES AND PLANES AS A RESULT OF STRAIN

The orientation of lines and planar surfaces within a strain ellipsoid is different from that in the undeformed condition. The change in orientation depends entirely on the ratios of the principal extensions, and it is a simple matter to use the techniques described in Sec. 4-5 to compute these changes. Quite different rearrangements of lines and planes occur in the various types of finite strain. Figure 4-28A is an equal-area projection of a number of lines dis-

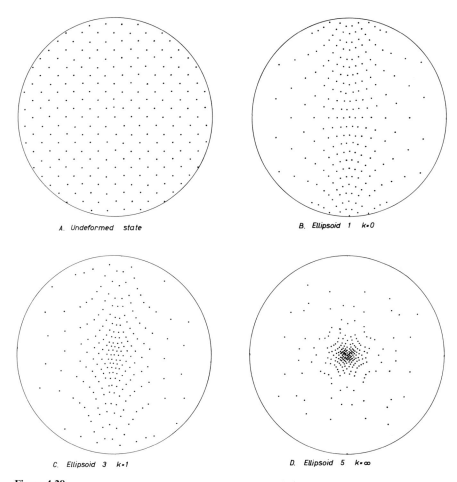

A. *Undeformed   state*

B. *Ellipsoid   1   k=0*

C. *Ellipsoid   3   k=1*

D. *Ellipsoid   5   k=∞*

Figure 4-28

The dispersion of lines as a result of various types of finite strain.

persed at approximately equal angular distances from each other in the undeformed state, and Fig. 4-28*B*, *C*, and *D* shows the final positions of these lines after finite deformation in the strain ellipsoids 1, 3, and 5 of Fig. 4-21, respectively. With flattening strain (ellipsoid 1) the lines approach the XY principal plane, whereas with constricting strain (ellipsoid 5) lines approach the X strain axis. In ellipsoid 3 both of these features are combined; there is a movement toward the XY plane but a concentration toward the X direction within this plane.

   In a similar way the rearrangements of planar surfaces which were uniformly oriented before deformation (Fig. 4-29*A*) can be calculated in the various types of strain ellipsoids 1, 3, and 5 (Fig. 4-29*B*, *C*, and *D*, respectively). In ellipsoid 1

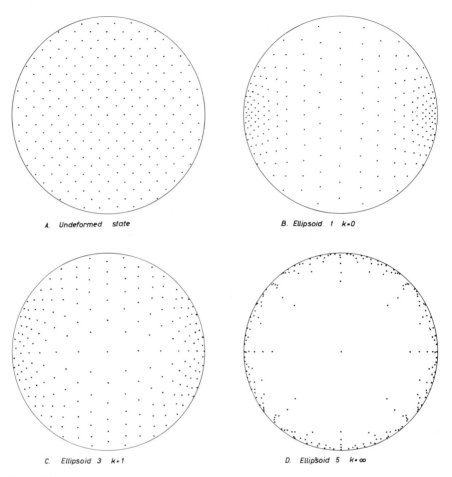

A. *Undeformed state*   B. *Ellipsoid 1 k=0*

C. *Ellipsoid 3 k=1*   D. *Ellipsoid 5 k=∞*

Figure 4-29

The dispersion of poles to surfaces as a result of various types of finite strain.

the poles of the planes become concentrated around the principal contraction axis Z of the ellipsoid, whereas in ellipsoid 5 they approach the YZ principal plane. In the plane-strain ellipsoid 3 these features are combined and the poles approach the YZ plane but concentrate around the Z axis in this plane.

These rearrangements lead to the production of different types of final fabrics, and Sander (1950) and Flinn (1962) have suggested that the crystal particles in rocks would undergo reorientation in the ways described above depending first on their shape (if prismatic or platy) and second on the shape of the strain ellipsoid. Using this hypothesis, rocks where the strain ellipsoids have $k > 1$ would tend to show a linear fabric parallel to the main axis of extension X of the ellipsoid and be classed as *linear tectonites,* whereas those where the strain ellipsoids have $k < 1$ would tend to show a planar fabric parallel to the XY principal plane and be classed as *planar tectonites*.

There is no doubt that in some deformation environments the rock fabrics are probably to be interpreted in this manner; however it is most unlikely that all planar and linear fabrics have come about in this way. This analysis does not take into account the rotational part of the finite-strain tensor. Rotation during deformation is often a very important factor and if there are particles in the rock which do not change shape in the same way as their matrix, then the resultant rearrangement is quite different from that expected from a simple irrotational deformation. There is an important field for future investigation and research along these lines to determine exactly how rigid or slightly deformable particles behave when they and their enclosing, more ductile, matrix are deformed in different ways. We shall return to this in Chap. 5 where we will discuss the problem of the determination of finite strain of rock from the shapes and orientations of more competent pieces (e.g., conglomerate pebbles) contained in it.

4-11 SUPERPOSITION OF TWO FINITE STRAINS

The problem of superimposed strains has already been investigated in two dimensions, and the three-dimensional analysis that follows will use essentially the same methods as those described in Sec. 3-11. The results of this analysis are particularly applicable to the methods used to determine the principal strains in a rock from measurements of initially ellipsoidal particles (e.g., ellipsoidal oolites) and to find how it is possible to isolate the original and tectonic parts of their ellipsoids.

The equation of any ellipsoid whose axes are inclined obliquely to the coordinate axes x, y, and z can be expressed in terms of the strain parameters measured on the xy, yz, and zx planes; thus

$$\lambda'_x x^2 + \lambda'_y y^2 + \lambda'_z z^2 - 2\gamma'_{xy}xy - 2\gamma'_{yz}yz - 2\gamma'_{zx}zx = 1 \qquad (4\text{-}27)$$

This ellipsoid will now be deformed by a homogeneous irrotational strain with principal axes arranged so that they coincide with the x, y, and z axes and with the principal strains $\lambda_i^{\frac{1}{2}}$, $\lambda_{ii}^{\frac{1}{2}}$ and $\lambda_{iii}^{\frac{1}{2}}$ along these directions. Any point (x,y,z) on ellipsoid (4-27) will be transposed to a new position (x_1,y_1,z_1) such that

$$x = x_1(\lambda_i')^{\frac{1}{2}} \qquad y = y_1(\lambda_{ii}')^{\frac{1}{2}} \qquad z = z_1(\lambda_{iii}')^{\frac{1}{2}}$$

and (4-27) is transformed into a new ellipsoid given by

$$\lambda'_x\lambda'_i x^2 + \lambda'_y\lambda'_{ii} y^2 + \lambda'_z\lambda'_{iii} z^2 - 2\gamma'_{xy}(\lambda'_i\lambda'_{ii})^{\frac{1}{2}}xy$$

$$- 2\gamma'_{yz}(\lambda'_{ii}\lambda'_{iii})^{\frac{1}{2}}yz - 2\gamma'_{zx}(\lambda'_{iii}\lambda'_i)^{\frac{1}{2}}zx = 1 \qquad (4\text{-}62)$$

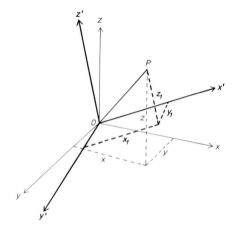

Figure 4-30

The change in coordinates of the point P
(x,y,z) as a result of rotation of the coordinate
axes into new positions x′, y′, z′.

If this new ellipsoid is compared with that of (4-27), it will be seen that its strain components measured on the xy, yz, and zx planes must be given by

$$\lambda'_{xt} = \lambda'_x \lambda'_i \qquad \gamma'_{xyt} = \gamma'_{xy}(\lambda'_i \lambda'_{ii})^{\frac{1}{2}}$$

$$\lambda'_{yt} = \lambda'_y \lambda'_{ii} \qquad \gamma'_{yzt} = \gamma'_{yz}(\lambda'_{ii} \lambda'_{iii})^{\frac{1}{2}}$$

$$\lambda'_{zt} = \lambda'_z \lambda'_{iii} \qquad \gamma'_{zxt} = \gamma'_{zx}(\lambda'_{iii} \lambda'_i)^{\frac{1}{2}}$$

where the subscript t refers to the component of total strain.

The values of the total extension and orientation of the axes of this new ellipsoid are determined by using these six new strain components to determine the cubic equation (4-37) the roots of which are the three principal extensions for the combined deformation.

If the second strain has a rotational component, the equation of the ellipsoid (4-62) will be modified, although obviously the principal strains will remain the same; only the axes will change orientation.

The equation formed by a rigid-body rotation of ellipsoid (4-62) can be found by rotating the coordinate axes to a new position. The relationships between the coordinates of any point (x,y,z) referred to another set of cartesian co-ordinates are found as follows.

Let the two sets of cartesian axes x, y, and z and $x′$, $y′$, and $z′$ have a common origin O and let the direction cosines of $x′$ with respect to x, y, and z be l_1, m_1, n_1; those of $y′$ be l_2, m_2, n_2; and those of $z′$ be l_3, m_3, n_3. Then if the point $P(x,y,z)$ has coordinates (x_1,y_1,z_1) referred to the axes $x′$, $y′$, and $z′$ (Fig. 4-30), the projection of OP on $Ox′$ is equal to the sum of the projections of lengths x, y, and z on OP; that is,

$$x_1 = xl_1 + ym_1 + zn_1 \tag{4-63a}$$

$$y_1 = xl_2 + ym_2 + zn_2 \tag{4-63b}$$

$$z_1 = xl_3 + ym_3 + zn_3 \qquad (4\text{-}63c)$$

Using the same method of projection, the old coordinates can be expressed in terms of the new coordinates

$$x = x_1 l_1 + y_1 l_2 + z_1 l_3 \qquad (4\text{-}64a)$$

$$y = x_1 m_1 + y_1 m_2 + z_1 m_3 \qquad (4\text{-}64b)$$

$$z = x_1 n_1 + y_1 n_2 + z_1 n_3 \qquad (4\text{-}64c)$$

By using these substitutions in (4-62) the equation of the new ellipsoid can be determined.

4-12   INFINITESIMAL STRAIN IN THREE DIMENSIONS

In Chap. 3 it was shown how the study of infinitesimal strains and changes in finite strain were particularly important in understanding what modifications occur during the progress of deformation. We shall now examine the geometry of infinitesimal strain in three dimensions using the same simplifying conditions as were employed in the two-dimensional analysis, i.e., that the displacements and strains are so small that their squares and products can be neglected without leading to any significant error.

To develop the notation we first consider a rectangular block element $ABCDEFGH$ with its edges oriented parallel to three rectangular coordinate axes x, y, and z (Fig. 4-31). The point A has coordinates (x,y,z) and the lengths of the sides of the block parallel to x, y, and z are δx, δy, and δz, respectively.

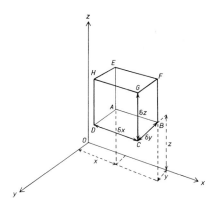

Figure 4-31

A small rectangular block of material with sides of length δx, δy, and δz in the undeformed state.

The block is now infinitesimally deformed and takes up the new position shown in Fig. 4-32. A is displaced to $A_1(x + u, y + v, z + w)$ and the sides of the block are deformed so that the lengths of their projections on x, y, and z are $\delta x + \delta u_x$, $\delta y + \delta v_y$, and $\delta z + \delta w_z$, respectively. The nine rates of change

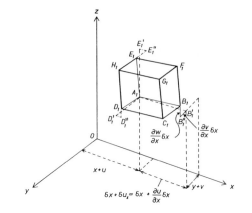

Figure 4-32

The infinitesimally strained condition of the block of Fig. 4-31.

of the displacements u, v, and w in the directions x, y, and z are $\partial u/\partial x$, $\partial u/\partial y$, $\partial u/\partial z$, $\partial v/\partial x$, ..., $\partial w/\partial z$.

Because the strains are small, the length of A_1B_1 does not differ significantly from that of A_1B_1', and this is equal to the original length δx together with the extension, which is δx multiplied by the rate of change of u in the x direction, that is,

$$A_1B_1 = A_1B_1' = \delta x + \frac{\partial u}{\partial x}\delta x$$

and therefore the extension in the x direction is given by

$$e_x = \frac{A_1B_1 - AB}{AB} = \frac{\delta x(1 + \partial u/\partial x) - \delta x}{\delta x} = \frac{\partial u}{\partial x} \qquad (4\text{-}65a)$$

Similarly

$$e_y = \frac{\partial v}{\partial y} \qquad (4\text{-}65b)$$

and

$$e_z = \frac{\partial w}{\partial z} \qquad (4\text{-}65c)$$

To determine the shear strain on the xy plane we compare the rectangular shape of $ABCD$ with the shape of the projection $A_1B_1C_1D_1$ on the xy section.

As the extensions along A_1B_1 are very small, A_1B_1' and A_1B_1 do not differ very much from δx and therefore

$$\tan B_1'A_1B_1'' = \frac{(\partial v/\partial x)\,\delta x}{\delta x} = \frac{\partial v}{\partial x}$$

and as angle $B_1'A_1B_1''$ is very small, the tangent is equal to the angle measured in radians. Using similar reasoning,

$$\tan D_1' A_1 D_1'' = \frac{\partial u}{\partial y}$$

Since the angle DAB was originally 90°, the small angular shear in plane xy is given by

$$\gamma_{xy} = \psi_{xy} = D_1' A_1 B_1' - D_1'' A_1 B_1'' = B_1' A_1 B_1'' + D_1' A_1 D_1''$$

$$\gamma_{xy} = \frac{\partial v}{\partial x} + \frac{\partial u}{\partial y} \qquad (4\text{-}66a)$$

Similarly

$$\gamma_{yz} = \frac{\partial v}{\partial z} + \frac{\partial w}{\partial y} \qquad (4\text{-}66b)$$

and

$$\gamma_{zx} = \frac{\partial w}{\partial x} + \frac{\partial u}{\partial z} \qquad (4\text{-}66c)$$

Using the same notation and reasoning as was developed in the two-dimensional investigation (Chap. 3, p. 100), we can define three small rigid-body rotations about the x, y, and z axes, respectively, as

$$\omega_1 = \frac{1}{2} \left(\frac{\partial w}{\partial y} - \frac{\partial v}{\partial z} \right) \qquad (4\text{-}67a)$$

$$\omega_2 = \frac{1}{2} \left(\frac{\partial u}{\partial z} - \frac{\partial w}{\partial x} \right) \qquad (4\text{-}67b)$$

$$\omega_3 = \frac{1}{2} \left(\frac{\partial v}{\partial x} - \frac{\partial u}{\partial y} \right) \qquad (4\text{-}67c)$$

If $\partial w/\partial y = \partial v/\partial z$, $\partial u/\partial z = \partial w/\partial x$, and $\partial v/\partial x = \partial u/\partial y$, the deformation is an *irrotational strain*; if these conditions do not hold, it is a *rotational strain*.

We can express the displacements of the point G_1 relative to that of the point A_1. If these displacements are δu, δv, and δw measured along the x, y, and z axes, respectively, then

$$\delta x + \delta u = A_1' B_1' - D_1' D_1'' - E_1' E_1''$$

or

$$\delta u = \frac{\partial u}{\partial x} \delta x + \frac{\partial u}{\partial y} \delta y + \frac{\partial u}{\partial z} \delta z \qquad (4\text{-}68a)$$

and similarly

$$\delta v = \frac{\partial v}{\partial x} \delta x + \frac{\partial v}{\partial y} \delta y + \frac{\partial v}{\partial z} \delta z \qquad (4\text{-}68b)$$

$$\delta w = \frac{\partial w}{\partial x} \delta x + \frac{\partial w}{\partial y} \delta y + \frac{\partial w}{\partial z} dz \qquad (4\text{-}68c)$$

These displacements form a linear transformation identical to that established as the general transformation equation for homogeneous strain (Chap. 4, p. 122). These transformation equations therefore define the tensor quantity given by

$$
\begin{vmatrix}
\dfrac{\partial u}{\partial x} & \dfrac{\partial u}{\partial y} & \dfrac{\partial u}{\partial z} \\[2mm]
\dfrac{\partial v}{\partial x} & \dfrac{\partial v}{\partial y} & \dfrac{\partial v}{\partial z} \\[2mm]
\dfrac{\partial w}{\partial x} & \dfrac{\partial w}{\partial y} & \dfrac{\partial w}{\partial z}
\end{vmatrix}
\tag{4-69}
$$

This is an asymmetric tensor and can therefore be decomposed into two parts. The first is a symmetric tensor which gives the irrotational part of the deformation,

$$
\begin{vmatrix}
\dfrac{\partial u}{\partial x} & \dfrac{1}{2}\left(\dfrac{\partial u}{\partial y}+\dfrac{\partial v}{\partial x}\right) & \dfrac{1}{2}\left(\dfrac{\partial u}{\partial z}+\dfrac{\partial w}{\partial x}\right) \\[3mm]
\dfrac{1}{2}\left(\dfrac{\partial v}{\partial x}+\dfrac{\partial u}{\partial y}\right) & \dfrac{\partial v}{\partial y} & \dfrac{1}{2}\left(\dfrac{\partial v}{\partial z}+\dfrac{\partial w}{\partial y}\right) \\[3mm]
\dfrac{1}{2}\left(\dfrac{\partial w}{\partial x}+\dfrac{\partial u}{\partial z}\right) & \dfrac{1}{2}\left(\dfrac{\partial w}{\partial y}+\dfrac{\partial v}{\partial z}\right) & \dfrac{\partial w}{\partial z}
\end{vmatrix}
\tag{4-70}
$$

and which can be written more simply in terms of the strain components Eqs. (4-65) and (4-66)

$$
\begin{vmatrix}
e_x & \tfrac{1}{2}\gamma_{xy} & \tfrac{1}{2}\gamma_{zx} \\[2mm]
\tfrac{1}{2}\gamma_{xy} & e_y & \tfrac{1}{2}\gamma_{zy} \\[2mm]
\tfrac{1}{2}\gamma_{zx} & \tfrac{1}{2}\gamma_{zy} & e_z
\end{vmatrix}
\tag{4-71}
$$

The second part of the decomposed asymmetric tensor is the skew-symmetric part and it defines the rigid-body rotation

$$
\begin{vmatrix}
0 & \dfrac{1}{2}\left(\dfrac{\partial u}{\partial y}-\dfrac{\partial v}{\partial x}\right) & \dfrac{1}{2}\left(\dfrac{\partial u}{\partial z}-\dfrac{\partial w}{\partial x}\right) \\[3mm]
\dfrac{1}{2}\left(\dfrac{\partial v}{\partial x}-\dfrac{\partial u}{\partial y}\right) & 0 & \dfrac{1}{2}\left(\dfrac{\partial v}{\partial z}-\dfrac{\partial w}{\partial y}\right) \\[3mm]
\dfrac{1}{2}\left(\dfrac{\partial w}{\partial x}-\dfrac{\partial u}{\partial z}\right) & \dfrac{1}{2}\left(\dfrac{\partial w}{\partial y}-\dfrac{\partial v}{\partial z}\right) & 0
\end{vmatrix}
\tag{4-72}
$$

which can be rewritten in terms of the rotations [Eqs. (4-67)]

$$\begin{vmatrix} 0 & -\omega_3 & \omega_2 \\ \omega_3 & 0 & -\omega_1 \\ -\omega_2 & \omega_1 & 0 \end{vmatrix} \tag{4-73}$$

If the points $(x,\ y,\ z)$ situated on a unit sphere $x^2 + y^2 + z^2 = 1$ are displaced according to Eqs. (4-68), they come to lie on an ellipsoid surface known as the *infinitesimal-strain ellipsoid*, the principal axes of which are the principal extensions e_1, e_2, and e_3. These can be determined from the matrix of the irrotational part of the tensor (4-71). The ellipsoid has the equation

$$(1 + 2e_x)x^2 + (1 + 2e_y)y^2 + (1 + 2e_z)z^2$$

$$- 2\gamma_{xy}xy - 2\gamma_{yz}yz - 2\gamma_{zx}zx = 1 \tag{4-74}$$

and the solution to the problem lies in finding the three principal axes of this ellipsoid. This can be done by finding the values of r, where a sphere of radius r given by

$$x^2 + y^2 + z^2 = r^2 \tag{4-75}$$

just touches the ellipsoid, using the methods previously described to determine the principal extensions of the finite-strain ellipse.

Differentiating (4-74) and (4-75) we obtain

$$\left[(1 + 2e_x)x - \gamma_{xy}y - \gamma_{zx}z\right] dx + \left[-\gamma_{xy}x + (1 + 2e_y)y - \gamma_{yz}z\right] dy$$

$$+ \left[-\gamma_{zx} - \gamma_{yz} + (1 + 2e_z)z\right] dz = 0 \tag{4-76}$$

$$x\ dx + y\ dy + z\ dz = 0 \tag{4-77}$$

Multiplying (4-77) by $(1 + 2e)$

$$(1 + 2e)x\ dx + (1 + 2e)y\ dy + (1 + 2e)z\ dz = 0 \tag{4-78}$$

Comparing coefficients of (4-78) and (4-76)

$$2(e_x - e) - \gamma_{xy} - \gamma_{zx} = 0 \tag{4-79a}$$

$$-\gamma_{xy} + 2(e_y - e) - \gamma_{yz} = 0 \tag{4-79b}$$

$$-\gamma_{zx} - \gamma_{yz} + 2(e_z - e) = 0 \tag{4-79c}$$

Solving a, b, c, for e we obtain the cubic

$$e^3 - (e_x + e_y + e_z)e^2 + \left(e_x e_y + e_y e_z + e_z e_x - \frac{\gamma_{xy}^2}{4} - \frac{\gamma_{yz}^2}{4} - \frac{\gamma_{zx}^2}{4}\right)e$$

$$- \left(e_x e_y e_z - \frac{1}{4}\gamma_{xy}\gamma_{yz}\gamma_{zx} - \frac{e_x\gamma_{yz}^2}{4} - \frac{e_y\gamma_{zx}^2}{4} - \frac{e_z\gamma_{xy}^2}{4}\right) = 0 \tag{4-80}$$

The three roots of this equation are the three principal extensions e_1, e_2, and e_3. The coefficients of e^2, e, together with the constant term, are known as the *three infinitesimal-strain invariants* K_1, K_2, and K_3, respectively.

The solution can be obtained more readily using matrix algebra by finding the eigenvalues of the characteristic matrix of (4-71), i.e., the roots of

$$0 = \begin{vmatrix} e_x & \frac{1}{2}\gamma_{xy} & \frac{1}{2}\gamma_{zx} \\ \frac{1}{2}\gamma_{xy} & e_y & \frac{1}{2}\gamma_{zy} \\ \frac{1}{2}\gamma_{zx} & \frac{1}{2}\gamma_{zy} & e_z \end{vmatrix} - e\begin{vmatrix} 1 & 0 & 0 \\ 0 & 1 & 0 \\ 0 & 0 & 1 \end{vmatrix}$$

or

$$0 = \begin{vmatrix} e_x - e & \frac{1}{2}\gamma_{xy} & \frac{1}{2}\gamma_{zx} \\ \frac{1}{2}\gamma_{xy} & e_y - e & \frac{1}{2}\gamma_{zy} \\ \frac{1}{2}\gamma_{zx} & \frac{1}{2}\gamma_{zy} & e_z - e \end{vmatrix} \qquad (4\text{-}81)$$

Expanding this determinant we find that the three eigenvalues are the roots of

$$(e - e_x)(e - e_y)(e - e_z) - \frac{\gamma_{yz}^2}{4}(e - e_x) - \frac{\gamma_{zx}^2}{4}(e - e_y)$$

$$- \frac{\gamma_{xy}^2}{4}(e - e_z) + \frac{\gamma_{xy}\gamma_{yz}\gamma_{zx}}{4} = 0 \quad (4\text{-}82)$$

Now that the axes of the infinitesimal-strain ellipsoid have been fixed, the other features of the strain are most simply described by rotating the x, y, and z coordinate axes so that they coincide with the directions of the ellipsoid axes.

4-13   INFINITESIMAL LONGITUDINAL STRAIN

The changes in length of lines of known direction can be determined from first principles by evaluating the changes in the length of the diagonal AG of the rectangular block in Fig. 4-31. These changes are, however, more readily determined by adapting the equations already established for finite strain to conditions of infinitesimal strain. Equation (4-5) for finite strain can be used to determine the lengths of lines which have direction cosines l, m, and n.

$$\lambda = l^2\lambda_1 + m^2\lambda_2 + n^2\lambda_3 \qquad (4\text{-}5)$$

$$(1 + e)^2 = l^2(1 + e_1)^2 + m^2(1 + e_2)^2 + n^2(1 + e_3)^2$$

With the usual approximations for small strains this becomes

$$e = l^2 e_1 + m^2 e_1 + n^2 e_3 \qquad (4\text{-}83)$$

Lines of no infinitesimal longitudinal strain are found where $e = 0$ and lie on the cone given by

$$0 = l^2 e_1 + m^2 e_2 + n^2 e_3 \tag{4-84}$$

Because (4-84) can be put in the form

$$0 = l^2 \frac{e_1}{e_3} + m^2 \frac{e_2}{e_3} + n^2 \tag{4-85}$$

the orientation of these lines depends on the *ratios* of the principal strains.

4-14 INFINITESIMAL SHEAR STRAIN

The shear strain in any direction can be found by adapting (4-15) to infinitesimal strain conditions.

$$\gamma^2 = \frac{(\lambda_1 - \lambda_2)^2}{\lambda_1 \lambda_2} l^2 m^2 + \frac{(\lambda_2 - \lambda_3)^3}{\lambda_2 \lambda_3} m^2 n^2 + \frac{(\lambda_3 - \lambda_1)^2}{\lambda_3 \lambda_1} n^2 l^2 \tag{4-15}$$

$$\gamma^2 = \frac{[(1 + e_1)^2 - (1 + e_2)^2]^2}{(1 + e_1)^2 (1 + e_2)^2} l^2 m^2 + \frac{[(1 + e_2)^2 - (1 + e_3)^2]^2}{(1 + e_2)^2 (1 + e_3)^2} m^2 n^2$$

$$+ \frac{[(1 + e_3)^2 - (1 + e_1)^2]^2}{(1 + e_3)^2 (1 + e_1)^2} n^2 l^2$$

$$\left(\frac{\gamma}{2}\right)^2 = \frac{(e_1 - e_2)^2 l^2 m^2}{1 + 2e_1 + 2e_2} + \frac{(e_2 - e_3)^2 m^2 n^2}{1 + 2e_2 + 2e_3} + \frac{(e_3 - e_1)^2 n^2 l^2}{1 + 2e_3 + 2e_1} \tag{4-86}$$

Because the strains are small, terms like $(e_1 - e_2)^2/(1 + 2e_1 + 2e_2)$ approximate closely to $(e_1 - e_2)^2$; therefore

$$\left(\frac{\gamma}{2}\right)^2 = (e_1 - e_2)^2 l^2 m^2 + (e_2 - e_3)^2 m^2 n^2 + (e_3 - e_1)^2 n^2 l^2 \tag{4-87}$$

This is a very similar function to Eq. (2-38) for shearing stress, and therefore the discussion of the position of maximum shearing stress in Sec. 2-11 can be applied directly to the position of maximum infinitesimal shearing strain. The absolute maximum shearing strain occurs along lines which lie in the principal plane containing e_1 and e_3, and these lines always make an angle of $45°$ to e_1 and e_3 no matter what values are taken by the principal strains.

4-15 MOHR DIAGRAM FOR REPRESENTING INFINITESIMAL STRAIN IN THREE DIMENSIONS

The equations that have been established for infinitesimal elongation (4-84) and shear strain (4-87) are in the form of those which describe finite elongation

(4-7) and shear strain (4-17), and it follows that all the discussion for the representation of finite strain in three dimensions made in Sec. 4-8 may be applied to infinitesimal strains. The Mohr circles are drawn with e as abscissa and $\gamma/2$ as ordinate.

4-16   INFINITESIMAL VOLUME CHANGE

The volume of an infinitesimal-strain ellipsoid derived from a unit sphere is $4\pi(1 + e_1)(1 + e_2)(1 + e_3)/3 = 4\pi(1 + e_1 + e_2 + e_3)/3$. The volume change or dilation is given by

$$\Delta = e_1 + e_2 + e_3 \tag{4-88}$$

4-17   PROGRESSIVE DEFORMATION IN THREE DIMENSIONS

The finite-strain ellipsoid is developed by the superposition of successive small or infinitesimal strains, and we shall now study some of the geometrical features of this process. In Sec. 4-3 it was shown how lines within the finite-strain ellipsoid may be either expanded or contracted, or they may be of unchanged length depending on their orientation. With any particular state of finite strain in naturally deformed rocks, it must be remembered that it is most likely that the lines of no finite longitudinal strain have been through a complex history of both expansion and contraction in a compensatory manner so that the resultant state is one of no finite elongation. The zones of positive and negative finite elongation within a finite-strain ellipsoid with principal strain directions, X_f, Y_f, and Z_f are indicated in Fig. 4-33A. Consider now the changes that go on during the next increment of straining. In general the axes of the infinitesimal incremental strain ellipsoid X_i, Y_i, and Z_i will not coincide with those of the principal finite strain ellipsoid (Fig. 4-33A). During the superposition of the strain increment, lines in some directions are expanding, some are contracting, and some are not undergoing any change in length. The zones of positive and negative infinitesimal longitudinal strain are symmetrically disposed about the axes of the infinitesimal strain ellipsoid; their positions are located in Fig. 4-33A. Because the axes of the finite- and infinitesimal-strain ellipsoids are not generally coincident, the changes in length that are going on during the progress of deformation do not coincide with those finite changes already established. Within the deforming material at any time during deformation it is possible to delimit various zones depending on the nature of the changes in longitudinal strain.

Zone 1 lines elongated and undergoing elongation (f +ve, i +ve)
Zone 2 lines elongated and undergoing contraction (f +ve, i −ve)
Zone 3 lines contracted and undergoing contraction (f −ve, i −ve)
Zone 4 lines contracted and undergoing elongation (f −ve, i +ve)

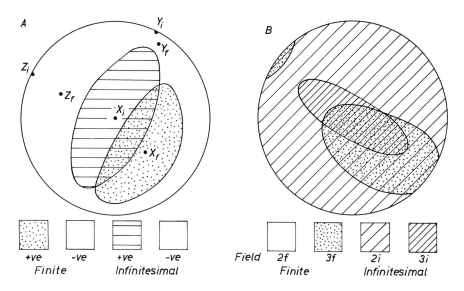

Figure 4-33

An equal-area plot showing the relationships of the fields of contraction and expansion in a finite-strain ellipsoid (axes X_f, Y_f, and Z_f) with an infinitesimal-strain increment (axes X_i, Y_i, and Z_i). Part B shows the pole positions of the various types of two-dimensional ellipse fields.

Lines having directions such that they lie on the boundaries of any two of the zones have one of the following properties:

Lines of unchanged length undergoing elongation
Lines of unchanged length undergoing contraction
Lines of unchanged length undergoing no change in length
Lines elongated undergoing no change in length
Lines contracted undergoing no change in length

The structural potentialities of this general arrangement of the zones is very complex. The triclinic symmetry of the general type of progressive deformation means that *it is extremely unlikely that the minor structures formed during the progress of deformation will ever be arranged symmetrically with respect to the axes of the finite-strain ellipsoid.* Structures formed during one part of the deformation history will be deformed as successive new strain increments are superimposed.

4-18 RELATIONSHIP BETWEEN PROGRESSIVE DEFORMATION IN TWO AND THREE DIMENSIONS

Any section through the three-dimensional strain ellipsoid represents a two-dimensional strain ellipse, and in Sec. 4-9 it was shown how the properties of these various types of finite-strain ellipse are related to those of the type of ellipsoid by which they are contained. Figure 4-33*B* illustrates the properties of the various sections through the deforming ellipsoid of Fig. 4-33*A*. All planes which have normals falling into the region 3*f* have $-ve$ extensions in all directions and they fall into the field 3 finite-strain ellipses illustrated in Fig. 3-54. Those in the region 2*f* are field 2 finite-strain ellipses. In this diagram it is also possible to represent the positions of various types of infinitesimal-strain ellipses. Regions 2*i* and 3*i* show the positions of poles of planes having infinitesimal-strain ellipses of fields 2 and 3, respectively. Surfaces of different orientations within the deforming material will therefore have different finite-strain states and will be undergoing different strains at any one time during the deformation. To illustrate this point consider a type of irrotational progressive deformation shown in Fig. 4-34. The original cube (*i*) is progressively strained and (*ii*) and (*iii*) represent its appearance at certain stages during the deformation history. The two-dimensional strains on the principal sections and on surface *A* obliquely inclined to the principal strain direction have been computed and their deformation paths determined. If surface *A* were occupied by a competent layer embedded in a more ductile material, then progressive deformation would lead to the development of folds and possibly to the

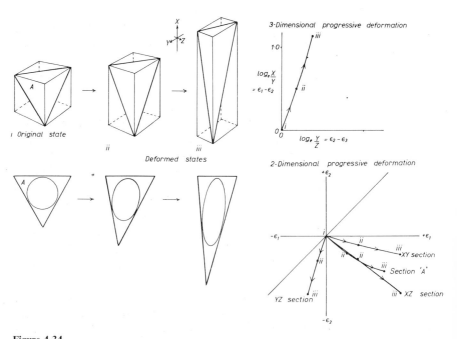

Figure 4-34

The relationship between progressive deformation in two dimensions and in three dimensions.

development of extension structures (boudinage, tension fissures). The axes of the folds would be initiated perpendicular to the original maximum shortening direction of the first infinitesimal strains and the tension fissures would be developed at right angles to these axes. As deformation proceeds, the axes of the infinitesimal-strain ellipses are altered and the maximum incremental shortening of the layer at any time will not be perpendicular to the axes of the previously initiated folds. The fold might continue its development by taking up this oblique shortening, or secondary en echelon folds might develop with axes perpendicular to the principal contraction of the infinitesimal-strain ellipse. In a similar way the original tension fissures will also be obliquely distorted and no longer be perpendicular to the lines of the original fold hinges. Thus several directions of folding could be developed in this surface as the result of a single deformation, even though the finite-strain ellipse at any time during the deformation was of a field 2 type (cf. Fig. 3-54 and text on p. 113). The axes of the various folds would be obliquely inclined to all three of the principal three-dimensional finite strains at all stages during the deformation.

4-19   THE SIGNIFICANCE OF CLEAVAGE AND SCHISTOSITY

The constituents of a rock generally undergo certain rearrangements as a result of strain. These changes may be physical displacements and rotation of relatively rigid rock or crystal particles or internal deformation within individual crystals by the process of lattice dislocations. Many rocks undergo structural changes as a result of solution, recrystallization of the constituents, and the growth of new minerals. These modifications lead to the formation of certain types of anisotropy and to the formation of a fabric related to the symmetry of the strain. The relative importance of the various mechanisms of fabric formation and modification depend on the physical and chemical environment of the rock during deformation and on the rate of strain.

One of the commonest types of planar fabric produced during tectonic deformation is known as *cleavage*. A number of different types of cleavage have been described, but the nomenclature is still rather confused and there exists no general agreement on the terminology or on the mechanical significance of the various types. An excellent summary of these problems of nomenclature has been given by Wilson (1961). *Slaty cleavage* (Fig. 4-35) and *schistosity* (Fig. 4-36) are terms best applied to a planar fabric that is uniformly developed (or penetrative) throughout all the rock material. For example, roofing slates are made from rocks possessing true slaty cleavage, and these slabs can be made of almost any thickness because the structural weakness pervades the whole rock. In contrast to slaty cleavage, the structures variously known as *fracture cleavage*, *strain-slip cleavage*, or *crenulation cleavage* do not penetrate on all scales through the rock but from discrete planar discontinuities which give rise to localized structural weaknesses in the rock (Figs. 4-37, 4-38). In view of the uncertain mechanical significance of all the various types of cleavage, it would appear advisable to use a descriptive, nongenetic nomenclature for these structures such as *s* surface (see discussion in Turner and Weiss, 1963, pp. 22-36), avoiding genetic terms like flow-, fracture- and shear-cleavage, etc.

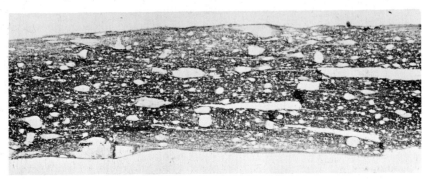

Figure 4-35

Thin section of a rock with slaty cleavage showing the deformation of clastic grains in the sediment. Lac du Chambon, near Bourg d'Oisans, French Alps.

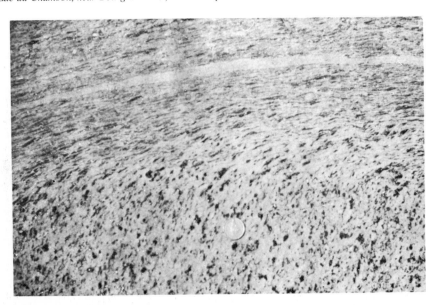

Figure 4-36

The development of schistosity from an almost homogeneous granite rock as a result of deformation. Cristallina region, Pennine nappes, Swiss Alps.

Figure 4-37

Thin section of crenulation cleavage in a clastic sediment. Brianconnais region, French Alps.

Figure 4-38

Thin section of fine crenulation cleavage in Jurassic black slate. l'Argentière, French Alps.

From a study of strained objects in slates it is clear that *slaty cleavage* forms perpendicular to the direction of maximum finite shortening in the rocks. This was originally established by a series of studies made in the nineteenth century by Sharpe (1847, 1849), Sorby (1853, 1856), Haughton (1856), and Phillips (1857). All the more detailed studies made since these early discoveries have confirmed this view (e.g., Heim, 1919; Cloos, 1947); and from the outstanding studies of Cloos it appears that the presence of slaty cleavage indicates that the compressive strain exceeds about 30 percent. All transitions between slaty cleavage and schistosity may be found in low- and medium-grade metamorphic rocks, and the correspondence of these structures leads to the suggestion that schistosity probably has the same mechanical significance as slaty cleavage. Because there can only be one direction of maximum finite compressive strain in a deformed material, there should only be one *true* slaty cleavage even if the rocks have been subjected to more than one phase of tectonic activity. If rocks have been repeatedly deformed to produce several sets of cleavage structures, then the later cleavages are invariably of the crenulation-cleavage type.

Slaty cleavage in low-grade metamorphic rocks is generally produced by the mechanical reorientation of the crystal particles with some new growth of micaceous minerals. Schistosity in more highly metamorphosed rocks is produced by more or less complete recrystallization of the components

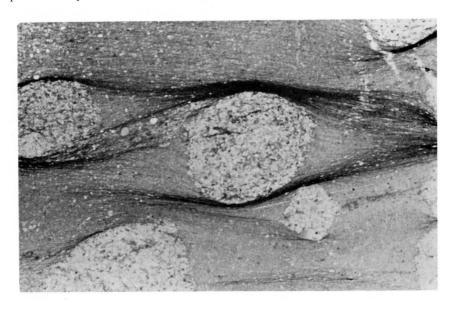

Figure 4-39

Slaty cleavage curving around worm tubes filled with coarse detritus. North Devon, England.

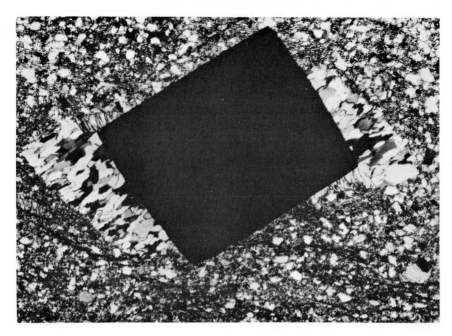

Figure 4-40

Pyrite cube with quartz-filled sigmoidal pressure-shadow zones. Rio Tinto, Spain.

(Turner, 1948), and it is not always certain how much of their preferred orientation is inherited from previously established strain anisotropy and how much of the growth is controlled by particular stress states that exist at different times during the metamorphic history.

Curved slaty-cleavage surfaces indicate that the XY principal plane of the strain varies in orientation from place to place, and also implies that the values of the principle strains are also variable (Fig. 4-39). Slaty-cleavage planes converge toward the region of highest strain. A linear fabric often accompanies slaty cleavage and takes the orientation of the maximum elongation (X direction) within the XY plane (Fig. 5-2 on p. 189). Deflection of the cleavage or schistosity surfaces often occurs where competent particles were present in the rock before deformation (e.g., conglomerate pebbles) or grew during deformation (e.g., porphyroblasts). Carbonate minerals, quartz, or feldspar may sometimes crystallize along the sides of these porphyroblasts facing the direction of maximum extension in the so-called "*pressure shadow*" regions (Figs. 4-40, 4-41). The particles may be extended in this direction as a series of broken, disconnected fragments. If the axes of incremental strain change their orientation during the process of deformation, these recrystallized zones may take up a sigmoidal form (Fig. 4-40).

Several of the general references given at the end of Chap. 2 describe strain; but the geologist should note that these discussions often center on problems of elasticity and of small strains and the conclusions and mathematical results may not be generally applicable to strongly deformed material.

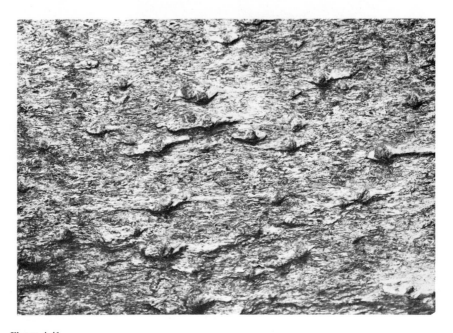

Figure 4-41

Broken garnet porphyroblasts with pressure-shadow zones elongated in the direction of maximum finite extension. Hornblende-schist, North Scotland.

REFERENCES AND SUGGESTED READING

Becker, G. F.: Finite Homogeneous Strain, Flow and Rupture of Rocks, *Geol. Soc. Am. Bull.*, **4**: 13-90 (1893).

Griggs, D. T.: The Strain Ellipsoid As a Theory of Rupture, *Am. J. Sci.*, **30**: 121-137 (1936).

Jaeger, J. C.: "Elasticity, Fracture and Flow," Methuen & Co., Ltd., London and John Wiley & Sons, Inc., New York, 1962.

Nadai, A.: "Theory of Flow and Fracture of Solids," McGraw-Hill Book Company, New York, 1950.

O'Driscoll, E. S.: Rheid and Rigid Rotations, *Nature*, **203**: 832-835 (1964*a*).

O'Driscoll, E. S.: Simple and Pure shear, *Nature*, **201**: 672-674 (1964*b*).

Application of theory to deformed rocks

1. General application

Brace, W. F.: Analysis of Large Two-dimensional Strain in Deformed Rocks, *21st Intern. Geol. Congr., Copenhagen,* **18**: 261-269 (1960).

Brace, W. F.: Mohr Construction in the Analysis of Large Geologic Strain, *Geol. Soc. Am. Bull.,* **72**: 1059-1080 (1961).

Cloos, E.: Lineation, a Critical Review, *Geol. Soc. Am., Mem.,* **18**, 1946, with supplementary review of the literature added in 1953.

Cloos, E.: Oolite Deformation in South Mountain Fold Maryland, *Geol. Soc. Am. Bull.,* **58**: 843-917 (1947).

Fairbairn, H. W.: "Structural Petrology of Deformed Rocks," Addison-Wesley Publishing Company, Inc., Reading, Mass., 1949.

Harland, E. B., and M. B. Bayley: Tectonic Regimes, *Geol. Mag.,* **95**: 89-104 (1958).

Knopf, E. B., and E. Ingerson: Structural Petrology, *Geol. Soc. Am., Mem.,* **6** (1938).

Turner, F. J., and L. E. Weiss: "Structural Analysis of Metamorphic Tectonites," McGraw-Hill Book Company, New York, 1963.

Wilson, G.: The Tectonic Significance of Small Scale Structures and Their Importance to the Geologist in the Field, *Ann. Soc. Geol. Belg.,* **84**: 423-548 (1961).

Zingg, T.: Beitrag zur Schotteranalyse, *Schweiz. Mineral. Petrog. Mitt.,* **15**: 39-140 (1935).

2. Cleavage and lineation

Anderson, E. M.: On Lineation and Petrofabric Structure and the Shearing Movement by Which They Have Been Produced, *Quart. J. Geol. Soc.,* **104**: 99-132 (1948).

Balk, R.: Structural and Petrologic Studies in Dutchess County, New York, *Geol. Soc. Am. Bull.,* **47**: 685-774 (1936).

Bates, T. F.: "Investigation of the Micaceous Minerals in Slate," *Am. Mineralogist,* **32**: 625-636 (1947).

Becker, G. F.: Experiments on Schistosity and Slaty Cleavage, *U.S. Geol. Surv. Bull.* 241, 1904.

Becker, G. F.: Current Theories of Slaty Cleavage, *Am. J. Sci.,* **24**: 1-17 (1907).

Broughton, J. G.: An Example of the Development of Cleavage, *J. Geol.,* **54**: 1-18 (1947).

Buessen, W. R., and B. Nagy: The Mechanism of the Deformation of Clay, *Nat. Acad. Sci. Res. Council Pub.,* **327**: 480-491 (1954).

Charlsworth, H. A. K., and C. R. Evans: Cleavage — Boudinage in Pre-Cambrian Rocks at Jasper, Alberta, *Geol. Mijnbouw,* **41**: 356-362 (1962).

Colette, B. J.: On the Origin of Schistosity, *Proc. Koninkl. Ned. Akad. Wetenschap.,* **61**: 121-139 (1958).

Colette, B. J.: On Helicitic Structure and the Occurrence of Elongate Crystals in the Direction of the Axis of a Fold, *Proc. Koninkl. Ned. Akad. Wetenschap.,* **62**: 161-171 (1959).

Crook, K. A. W.: Cleavage in Weakly Deformed Mudstones, *Am. J. Sci.,* **262**: 523-531 (1964).

Dale, T. N.: On Plicated Cleavage-Foliation, *Am. J. Sci.,* **43**: 317-319 (1892).

Dale, T. N.: The Slate Belt of Eastern New York and West Vermont, *U.S. Geol. Surv., 19th Ann. Report,* 153-307 (1899).

Fairbairn, H. W.: Elongation in Deformed Rocks, *J. Geol.,* **44**: 670-680 (1936).

Furtak, H.: Die "Brechung" der Schiefrigkeit, *Geol. Mitt. Aachen,* **2**: 177-196 (1962).

Goguel, J.: Sur l'origine mécanique de la schistosite, *Bull. Soc. Geol. France,* **15**: 509-522 (1945).

Harker, A.: On Slaty Cleavage and Allied Rock Structures, *Brit. Assoc. Advan. Sci.,* **813** (1885).

Haughton, S.: On Slaty Cleavage and the Distortion of Fossils, *Phil. Mag.,* **12**: 409-421 (1856).

Heim, A.: "Geologie der Schweiz," Tauchnitz, Leipzig, 1919.

Hoeppener, R.: Zum Problem der Bruchbildung, Schieferung und Faltung. *Geol. Rundschau.,* **45**: 247-283 (1956).

Knill, J. L.: A Classification of Cleavages, *21st Int. Geol. Congr., Copenhagen,* **18**: 317-325 (1960).

Leith, C. K.: Rock Cleavage, *U.S. Geol. Surv. Bull.* 239, 1905.

Maxwell, J. C.: Origin of Slaty and Fracture Cleavage in the Delaware Water Gap Area, New Jersey and Pennsylvania, in "Petrologic Studies," *Geol. Soc. Am., Buddington vol.:* 281-311 (1962).

Morris, T. O., and W. G. Fearnsides: The Stratigraphy and Structure of the Cambrian Slate-belt of Nantlle (Caernarvonshire), *Quart. J. Geol. Soc.,* **82**: 250-303 (1926).

Phillips, J.: Report on Cleavage and Foliation in Rocks and on the Theoretical Explanation of These Phenomena, *Brit. Assoc. Advan. Sci.,* **269**: 60-61 (1857).

Sander, B.: Einfuhrung in die Gefugekunde der geologischen Korper, **1**, **2**, Springer-Verlag OHG, Vienna, 1948 and 1950.

Rickard, M. J.: A Note on Cleavages and Crenulated Rocks, *Geol. Mag.,* **98**: 324-332 (1961).

Sharpe, D.: On Slaty Cleavage, *Quart. J. Geol. Soc.,* **3**: 74-104 (1847).

Sharpe, D.: On Slaty Cleavage, *Quart. J. Geol. Soc.,* **5**: 111-115 (1849).

Sorby, H. C.: On the Origin of Slaty Cleavage, *New Phil. J. (Edinburgh),* **55**: 137-148 (1853).

Sorby, H. C.: On the Theory of the Origin of Slaty Cleavage, *Phil. Mag.,* **12**: 127-129 (1856).

Swanson, C. O.: Flow Cleavage in Folded Beds, *Geol. Soc. Am. Bull.,* **52**: 1245-1263 (1941).

Turner, F. J.: Review of Current Hypotheses and Tectonic Significance of Schistosity in Metamorphic Rocks, *Am. Geophys. Union Trans.,* **29**: 558-564 (1948).

White, W. S.: Cleavage in East-Central Vermont, *Am. Geophys. Union Trans.,* **30**: 587-594 (1949).

Wilson, G.: The Relationship of Slaty Cleavage and Kindred Structures to Tectonics, *Proc. Geol. Assoc.,* **57**: 263-302 (1946).

Wilson, G.: The Tectonics of the Tintagel Area, North Cornwall, *Quart. J. Geol. Soc.,* **106**: 393-432 (1951).

3. Folding with reference to finite strain

Flinn, D.: On Folding during Three-dimensional Progressive Deformation, *Quart. J. Geol. Soc.,* **118**: 385-433 (1962).

Ramberg, H.: Evolution of Ptygmatic Folding, *Norsk. Geol. Tidsskr.,* **39**: 99-151 (1959).

4. Boudinage

Coe, K.: Boudinage Structure in West Cork, Ireland, *Geol. Mag.,* **96**: 191-200 (1959).

Cloos, E.: Boudinage, *Am. Geophys. Union Trans.,* **28**: 626-632 (1947).

Lohest, M., X. Stanier, and P. Fourmarier: C. R. de la session extraordinaire de la Soc. Geol. de Belgique, *Ann. Soc. Belg.,* **35**: 351-414 (1909).

Quirke, T. T.: Boudinage, an Unusual Structural Phenomenon, *Geol. Soc. Am. Bull.,* **34**: 649-660 (1923).

Ramberg, H.: Natural and Experimental Boudinage and Pinch and Swell Structures, *J. Geol.,* **63**: 512-526 (1955).

Rast, N.: The Origin and Significance of Boudinage, *Geol. Mag.,* **93**: 401-408 (1956).

Wegmann, E.: Note sur le Boudinage, *Bull. Soc. Geol. France,* **2**: 477-489 (1932).

4. Simple shear

Blyth, F. G. H.: The sheared porphry dykes of South Galloway, *Quart. J. Geol. Soc.,* **105**: 393-421 (1950),

Riedel, W.: Zur Mechanik geologischer Brucherscheinungen, *Centralblatt Min. Geol. Pal., Abt. B,* 354-369 (1929).

Shainin, V. E.: Conjugate Sets of En Echelon Tension Fractures in the Athens Limestone at Riverton, Virginia, *Geol. Soc. Am. Bull.,* **61**: 509-517 (1950).

Wilson, G.: The Tectonics of the "Great Ice Chasm", Filchner Ice Shelf, Antarctica, *Proc. Geol. Assoc. (Engl.),* **71**: 130-138 (1960).

5 ‖ Determination of finite strain in rocks

MANY factors hinder the accurate determination of the distortion suffered by naturally deformed rocks. It is, however, one of the most rewarding lines of research and one that seems likely to be much developed in the future. The analysis of the variation in amount of finite strain in a deformed zone is of the utmost importance in helping to understand the structural geometry and hence the structural history of the rocks. Although we cannot derive all the mechanical history of formation from a single end product, it seems likely that, by making detailed studies of the state of strain in deformed zones which show a variation in intensity of deformation, we shall be able to see these structures arrested at different stages of their development. This method of understanding the mechanisms of formation of structures in deformed rocks has limitations in that it depends on the correct recognition and correlation of types of structures from place to place and it also assumes that the early stages of formation of a strongly deformed structure were the same as those we now see in less-deformed rocks. Although these assumptions may not always be correct they are probably the most reasonable for the geologist to adopt if he wishes to use naturally deformed rocks to determine a deformation history.

It is unfortunate that only a few rocks contain the sort of features that are necessary for the exact determination of the state of strain. Even so, much valuable information can be obtained from measurements of strain, information that it is generally impossible to establish in any other way, and a careful search for the right material is amply rewarded. Basically we have to find rocks which contain one or more of these features: lines of known original length, lines of known original relative length, or the initial angle made between intersecting lines. Rarely do we have any knowledge of the exact original length of lines in deformed rocks; because of this, we can rarely compute the actual extensions along the axes of the strain ellipsoid—only the ratios of these extensions. Also, without the actual extensions we can never compute the volume change. Many of the studies that have been made on the state of strain in deformed rocks have assumed certain values for the volume change (generally no volume change) and in order to solve the three-dimensional problems from inadequate two-dimensional data, some researchers have assumed that the ellipsoid is of some special type (e.g., uniaxial, $k = 0$). This is unfortunate, for though the data that are presented can be recalculated and put into some form which does not employ these assumptions, it seems more useful to present the results without any assumptions. Inadequate data for the solution of the problems of three-dimensional strain are generally more productive and stimulating to future research than superficially adequate results that rely on dubious assumptions. The amount of volume change during deformation is one of the most interesting unknown parameters and may have great relevance in explaining some of the structures found in deformed rocks. For example, the prevalence of "flattening type" ($1 > k > 0$) strain ellipsoids in slates may be explained by loss of volume during deformation, and not necessarily by actual expansion along the intermediate axis of the strain ellipsoid (page 162).

The significance of many of the structural features seen in deformed rocks was perceived in the middle of the nineteenth century by the careful measurement of objects in the rocks. Studies of deformed fossils were made by Phillips as early as 1843, and Sharpe (1846) was the first to use deformed fossils to determine the significance of slaty cleavage. He saw that the maximum shortening in the rock was perpendicular to the cleavage and that the maximum elongation was in the direction of dip of the cleavage planes. Haughton (1856) arrived at the same conclusion on the mechanical significance of slaty cleavage and was one of the first to apply the strain ellipsoid to rock deformation. Outstanding contributions to the study of deformation were made by Sorby (1855) who measured the shapes of the green reduction spots in the purple Cambrian slates of North Wales (Fig. 5-3). He pointed out that in undeformed rocks spots of this type are generally spherical or are uniaxial ellipsoids ($k = 0$) with the longest dimensions in the plane of the bedding. In these Cambrian slates the reduction spots are now strongly elongated in the cleavage planes and compressed along a line perpendicular to the cleavage.

The results of these studies had a great influence on work on deformed rocks both during and since that time. Controversies that developed as a result of this work on deformed rocks and fossils stimulated others to take an active interest in these topics. Renevier (1856), Daubrée (1876), and Harker (1884) brought these techniques to a wider geological audience in Europe and developed them further. The studies by Heim (1878, 1921) carried out in the deformed Mesozoic and Tertiary sediments in the Alps and by Wettstein (1886) on deformed oolites and fossils were masterly contributions to the understanding of rock deformation; they made geologists aware of the intensity and variations in intensity of strains suffered by rocks deformed in orogenic processes. The most valuable work yet published is probably that of Cloos (1947). Here are presented the results of a systematic and detailed survey of the strained state of oolitic limestones determined at several hundred localities in Cambro-Ordovician rocks on the western side of the Appalachian fold belt.

Breddin (1956, 1957) has published what are probably the most useful modern reference works on the actual techniques of measuring deformed fossil impressions. By making these measurements over wide areas, he has been able to prepare deformation maps showing the variation in intensity of strain within parts of the Rheinische Schiefergebirge and in the Carboniferous strata around Aachen, Germany. Although some of the calculations made to prepare these maps employ assumptions that are open to question, there is no doubt that this work has greatly stimulated studies of deformed fossils.

5-1 DEFORMATION OF INITIALLY SPHERICAL OBJECTS

One of the simplest methods for determining the state of finite strain in naturally deformed rocks entails measuring the shape of objects that initially had a spherical form. The objects that might be useful for this purpose are found mostly in sedimentary rocks.

1. OOIDS These are usually subspherical bodies generally about a millimeter or less in diameter (Fig. 5-1) and they are found in limestones formed in a shallow water environment. Ooids generally have a conspicuously laminated structure and are concentrically banded. Although they are most commonly made up of carbonates or aragonite, some consist almost entirely of concentric layers of chamosite. Chamositic ooids only rarely have an originally spherical form; most are ellipsoidal or rather irregular in shape.

2. SPHERULITES These are spherical particles generally about the same size as ooids and commonly found in the same environments. Their characteristic feature is a radiating crystalline structure.

3. PISOLITES AND SPHERICAL MUD PELLETS Some carbonate deposits are made up almost entirely of aggregates of small pellets of mud sometimes with, sometimes without, a concentric structure. The pellets may be spherical in form, but many have an original shape which approximates to an ellipsoid.

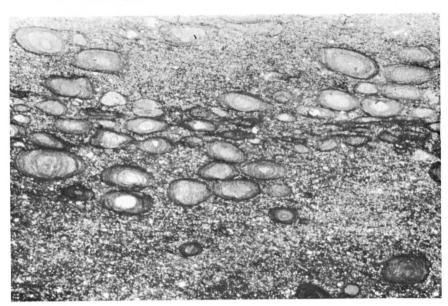

Figure 5-1

Thin section of concentrically banded ooids in limestone. Fernigen, Switzerland.

Some are formed in the same way as oolites, but other pellet limestones are formed by invertebrate organisms.

In some pyroclastic rocks pisolites are common. The concentrically layered pellet tuffs known as *chalazoidites* or "bird's-eye tuffs" (Fig. 5-2) appear to be formed in the same manner as hailstones by accretion of volcanic material in the clouds of dust above eruption centers. Studies of geologically recent chalazoidites show that even when they have fallen from a considerable height they generally retain a spherical shape.

4. REDUCTION SPOTS In environments where red and green shales, marls, and sandstones are deposited, reduction of the ferric components sometimes takes place around certain points in the rocks and produces spherical or ellipsoidal green spots with a surrounding purple or red matrix. If the original shape of this environment is known, these spots are especially useful for strain computation because they generally have almost exactly the same physical properties as the surrounding unaltered sediment (Fig. 5-3).

5. CONCRETIONS, BALLS, AND NODULES In some sediments the soluble calcium phosphate or carbonate may be redistributed around certain nuclei. The resulting concretion can have a subspherical or ellipsoidal form but often has a more irregular shape. Deformed nodules are difficult to use for strain computation partly because this initial shape factor is not always known, but also because, on account of their composition, most nodules deform rather differently from their surrounding sediment.

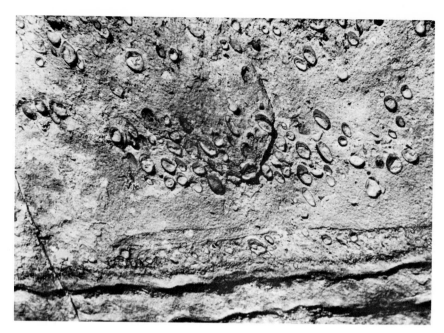

Figure 5-2

Pisolitic tuff of volcanic origin. Capel Curig, North Wales. The face of the rock viewed here is a cleavage surface. Traces of the bedding planes can be seen in the lower part of the diagram. These traces are parallel to the axes of major folding.

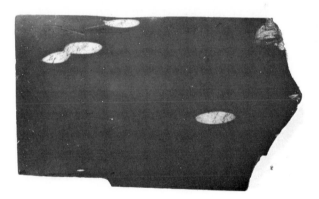

Figure 5-3

Reduction spots in slate. The section is perpendicular to the slaty cleavage. Bethesda, North Wales.

6. ORGANIC MATERIAL Some small organisms such as algae, certain Foraminifera and Radiolaria have a shape which closely approximates to

that of a sphere, and these can be very useful for determination of strain if they are preserved in deformed rocks.

7. AMYGDULES AND VESICLES Extrusive and intrusive volcanic rocks often have originally spherical gas or liquid "bubble" inclusions (Fig. 5-4), and some chilled acid and igneous rocks often show a spherulitic structure as a result of devitrification of the glassy material (Fig. 5-5).

8. HORNFELS OR SPOTTED ROCKS Thermally altered pelitic rocks around igneous intrusions often show small subspherical spots formed by the segregation of carbonaceous material, andalusite, mica, and quartz.

The two principal difficulties that arise when one attempts to determine the finite-strain state of the rock from distorted objects of the types listed above concern the original shape and the amount of deformation. Only rarely are these objects perfectly spherical, and if only deformed material is available for study it may be difficult (but not necessarily impossible) to evaluate the primary shape factor. The other problem is that the object may deform differently from the surrounding rock and therefore measurements made of its shape will indicate the strain of the particle and not the matrix. Both these problems will be discussed in some detail later; the sections which now follow will describe the techniques that are applicable to the measurement of originally spherical objects which are deformed homogeneously with their matrix.

Figure 5-4

Deformed chlorite and calcite filled amygdales in a vesicular basic dike. Tryfan, North Wales.

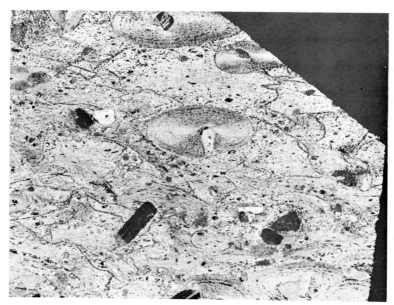

Figure 5-5

Thin section of deformed spherulites from an ignimbritic flow. Tryfan, North Wales.

Collection and preparation of material   It is imperative that the material that is collected in the field be very carefully oriented. The specimens are broken from the rock surface, fitted back into their original position, and held in place with adhesive tape. A horizontal line is then drawn on any convenient flat surface on the specimen (joint surface and bedding surface) with an indelible marker; the direction of this line is determined with a compass and then an arrow is drawn at the end of the line to indicate the strike direction. The angle and direction of dip of the surface are then marked by a tick on the side of the horizontal strike line. All the data are carefully recorded in a notebook, together with as much other information about the macroscopic structures (lineations, fold directions, cleavage planes, joints, etc.) as can be gathered from the outcrop.

If the ellipsoidal objects are small (oolites, pisolites), it is essential to prepare thin sections for microscope study; if the objects are more than a centimeter across, polished sections must be prepared. It is frequently thought that it is essential to make these sections parallel to the principal planes of the ellipsoid, but this is not essential using the technique described below and, in fact, it invites assumptions that can lead to considerable inaccuracy. Cloos (1947) has pointed out that it is often difficult to locate the principal planes in slightly deformed rocks. In strongly deformed rocks, slight deviations of the sections from the principal planes introduce large errors in the values of the principal

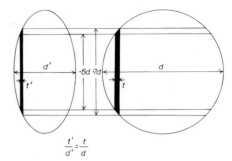

Figure 5-6

The frequencies of sections through the circle and ellipse having a size range 0.6 to 0.7d are equal.

$$\frac{t'}{d'} = \frac{t}{d}$$

strains if it has been assumed that the sections contain these axes. Any three sections through the rock can be used to determine the principal strains and orientations of the ellipsoid axes, but less computation is required if the three sections are mutually perpendicular.

If the original spheres were of approximately equal volume, it will be found that most of the elliptical sections have almost the same size. The frequency of ellipse sections of a given size in the ellipse in Fig. 5-6 is the same as that in the circle with diameter d. A histogram showing the frequency of the sections of known diameter through the circle diameter d is shown in Fig. 5-7; 8.6 percent of the sections through this circle have a diameter between 0.6 and 0.7d, and 86.6 percent of the sections have a diameter greater than 0.5d.

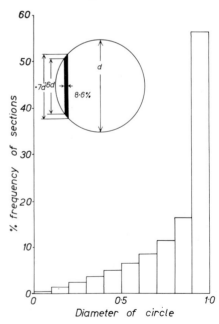

Figure 5-7

Histogram of the frequencies of sections of certain size made through a circle of diameter d. That is, 8.6 percent of the sections have a value between 0.6 and 0.7d.

5-2   DETERMINATION OF THE STRAIN ELLIPSE

There are several methods which can be used to evaluate the principal extensions and their ratios in the sections.

Method 1   The most obvious, but not always the most accurate, method is by direct measurement of the principal axes of each ellipse. If the particles are small, measurements are made in thin sections with an occular grid, systematically traversing the slide with a mechanical stage. It is often the more convenient to prepare enlarged photographs of the whole section and from these measure the lengths of the axes of each ellipse together with the orientation of the major axis. Generally about 50 measurements are sufficient to define the strain ellipse with accuracy, and it is usual to take the arithmetic mean of the individual observations. Another way of recording the data is to represent each ellipse on a graph by plotting the length of the short axis as abscissa and that of the long axis as ordinate (Fig. 5-8). The points should fall on or about a straight line passing through the origin, and the slope of this line gives the mean ratio of the principal strains in the section. This method generally produces results very close to those of the simple arithmetic mean. Sometimes, however, all the ellipses do not lie on a straight line. This is an interesting phenomenon that may be explained in several ways. For example, in contrast to small particles, the larger particles might not have deformed homogeneously with their matrix.

Cloos (1947) discussed some of the difficulties of measuring strongly deformed oolites by this method of directly recording the lengths of the principal axes. He found that often the ends of the oolites have a "frayed" appearance that makes it difficult to find the lengths of the major axes accurately, and with very high deformation he found it impossible to accurately determine the principal strains. Sometimes he was able to make measurements on the inner shells of the oolites and thus avoid their irregular margins. Another problem with the direct-measurement technique is that it is not always easy to locate the direction of the principal extension where the deformation is low, particularly if the oolites initially deviated from a perfectly spherical shape.

Method 2   The second method is particularly useful with both slightly deformed and strongly deformed material, and does not rely on direct measurement of the principal axes of the ellipses. It is most convenient to use enlarged photographs of the thin sections. First the centers of each ellipse are located and any three directions are selected, and then for each ooid the lengths of the three chords passing through their centers and parallel to the three chosen directions are recorded (Fig. 5-9, lengths *a*, *b*, and *c*). The sums of the lengths of the chords along each direction are now calculated. Assuming

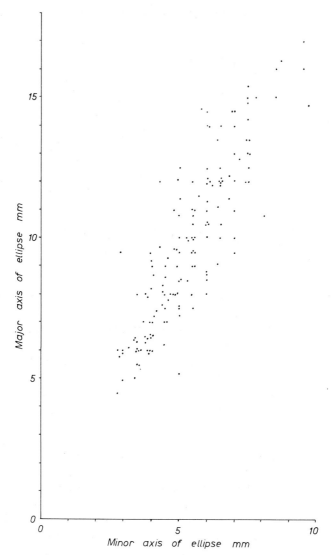

Figure 5-8

Plot of the major and minor axes of a series of elliptical cross sections of deformed pisolites. Capel Curig, North Wales.

that these three lengths were originally equal because all the ellipses were originally circles, we can calculate the ratios of elongation along the three directions. Since the angle between the lines is known, we can use a Mohr construction to determine the principal strains and the orientations of the strain axes (Sec. 3-8, constructions 4*A* and *B*).

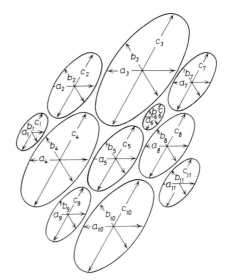

Figure 5-9

Determination of the values of principal strain ratio and orientations of the axes of the strain ellipse by making measurements in any three directions.

Method 3   This is a technique developed for rocks which have suffered pressure solution, but it is generally applicable to rocks containing particles which were initially distributed equally through the rocks or which had a known unequal distribution.

In Fig. 5-10 the individual ooids have not been deformed as much as the whole rock. They show elliptical cross sections, yet much of the change of shape of the rock has been accomplished by solution along certain zones and by the redeposition of the carbonate material in other places. This is the phenomenon generally known as *pressure solution*, and there is little doubt that it is a very important process by which rocks can change their shape. The process is generally explained by Riecke's principle—material goes into solution on the sides of the objects which face the principal compressive stress and is redeposited on the side facing the principal tensile stress. In Fig. 5-10 it is evident that the ooids have been pressed into one another; the carbonate has been dissolved away along their mutual contacts and deposited in the region which separates the pointed ends of the ooids. Along the zones where carbonate solution has taken place, there is a concentration of the dark insoluble matter of the oolite.

To determine the strain in rocks like that in Fig. 5-10 we obviously cannot rely on the elliptical shape of the ooid being an indication of the shape of the strain ellipse. However, we can determine the way in which the centers of the ooids have been displaced relative to one another. Consider first a section through an undeformed oolitic rock which initially had an approximately uniform distribution of the spheres as illustrated in Fig. 5-11. A line is drawn joining the centers of any two adjacent circular sections and the distance d

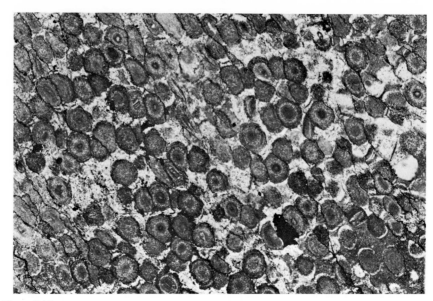

Figure 5-10

Deformed oolitic limestone from Ilfracombe, N. Devon, with pressure solution zones between adjacent ooids.

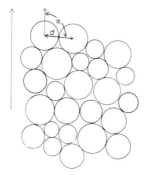

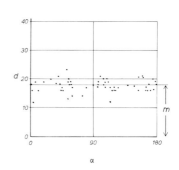

Figure 5-11

Section through a rock made up of an aggregate of undeformed spheres.

Figure 5-12

Plot of d against α for the material of Fig. 5-11.

is measured. This is now plotted against the orientation of the line measured from some known azimuth in the section (α). In undeformed material the points on the graph would scatter along some mean distance m depending on the frequency distribution of the diameters of the original spheres and the frequency of sections of certain sizes through these spheres as seen in Fig.

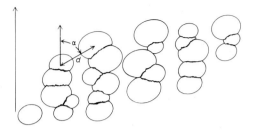

Figure 5-13

Section through a deformed aggregate of partially deformed spheres which have undergone partial pressure solution along their mutual contacts.

5-12. If this material is now deformed (Fig. 5-13), the distances between adjacent ooid centers is altered and plots of d against α show a distribution which is proportional to the amount of longitudinal strain in the direction α. From the graph (Fig. 5-14) we can easily determine the direction of maximum and minimum extension (α_1 and α_2), and by measuring the distances mX and mY we obtain the ratios of the principal strains $mX/mY = X/Y$. If the original size and distribution of the ooids is known it is possible to determine the value of m and therefore calculate the two principal quadratic extensions. It may sometimes be possible to restore the deformed shapes to their original form geometrically and calculate the value of m.

This technique is essential for measuring strain in rock which has suffered pressure solution, but it is generally applicable to any rock containing originally uniformly distributed particles, and it often affords a valuable check on the assumption that the included objects are deforming homogeneously with their matrix.

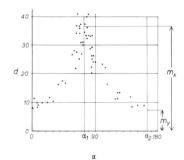

Figure 5-14

Plot of d against α for the material of Fig. 5-13.

Method 4   This method uses measurements of distorted angles of radial and tangential lines in the ellipse sections and again is best carried out on enlarged photographs of the section. It is very useful for those cases where the original spheres contained a radial spherulitic texture.

A line is first chosen as a reference direction in the section. At points on the periphery of the ellipse the angles between the tangent to the ellipse surface and the radial lines of the spherulitic structure are measured and the deflection

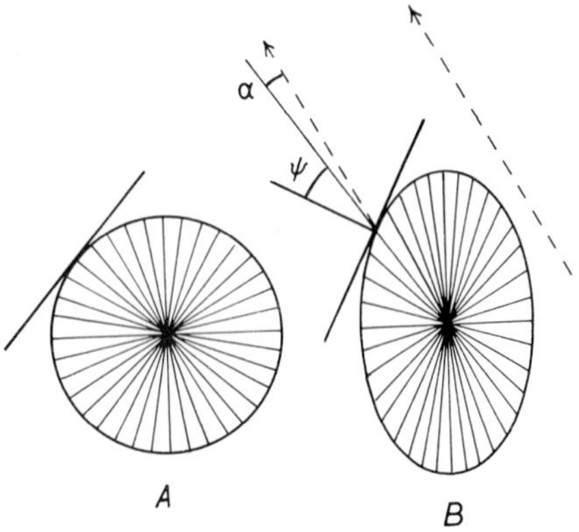

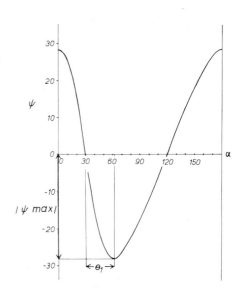

Figure 5-15

Method of measuring shearing strain ψ for different directions in a deformed spherulite (part B) derived from the undeformed spherulite A.

from 90° (ψ) is recorded (Fig. 5-15B). A graph is now prepared showing the variation in ψ along the various radial lines making an angle α with the reference direction (Fig. 5-16).

Figure 5-16

Variation in shearing strain ψ in a deformed spherulite of Fig. 5-15.

From this graph we find the two positions where $\psi = 0$, and these give the orientations of the principal axes of the strain ellipse. Next the angle θ_1 (Fig. 5-16) where lines have a maximum ψ value is determined and the ratios of the principal strains can be calculated from (3-45) for $\tan^2 \theta_1 = \lambda_2/\lambda_1$. Another independent method of calculating these ratios is from the maximum value of the shearing strain (Fig. 5-16, ψ_{max}). From (3-46) this value is dependent on the ratios of the principal quadratic extensions.

$$\tan \psi_{max} = \frac{\lambda_1 - \lambda_2}{2(\lambda_1\lambda_2)^{\frac{1}{2}}} \tag{3-46}$$

$$4 \tan^2 \psi_{max} = \frac{\lambda_1}{\lambda_2} + \frac{\lambda_2}{\lambda_1} - 2 \tag{5-1}$$

Replacing λ_1/λ_2 by R we find

$$0 = R^2 - 2R(2 \tan^2 \psi_{max} + 1) + 1 \tag{5-2}$$

There are two real roots to this equation and because the constant term is $+1$ it follows that if one is R, the other will be $1/R$. Solving (5-2) for R we obtain

$$R = 1 + 2 \tan^2 \psi_{max} + 2 \tan \psi_{max} \sec \psi_{max} \tag{5-3}$$

This equation enables us to find the ratio of the principal quadratic extensions in the section knowing ψ_{max}.

5-3   CALCULATIONS OF THE STRAIN COMPONENTS FROM THE STRAIN ELLIPSE

When the orientation of the strain ellipse and its principal extensions or extension ratios have been determined, we must compute the strain components with reference to two other perpendicular sections. Let us denote these three sections xy, yz, and zx with lines of intersection x, y, and z.

On the xy section we determine the angle θ_x between the major axis of the strain ellipse and the x axis. The strain components λ'_x, λ'_y, and γ'_{xy} are calculated from

$$\lambda'_x = \lambda'_{xy1} \cos^2 \theta_x + \lambda'_{xy2} \sin^2 \theta_x \tag{5-4}$$

$$\lambda'_y = \lambda'_{xy1} \sin^2 \theta_x + \lambda'_{xy2} \cos^2 \theta_x \tag{5-5}$$

$$\gamma'_{xy} = (\lambda'_{xy2} - \lambda'_{xy1}) \sin \theta_x \cos \theta_x \tag{5-6}$$

where λ'_{xy1} and λ'_{xy2} are the reciprocal principal quadratic extensions of the strain ellipse on the xy section. If the quadratic extension for the strain ellipse in the yz section is known, we can calculate λ'_y, λ'_z, and γ'_{yz}; and likewise on the zx section, λ'_z, λ'_x, and γ'_{zx}. Thus we calculate all six of the independent strain components. For each of three components (λ'_x, λ'_y, λ'_z) we shall have two independent calculations which offer a check on the accuracy of the previous work. The basic technique for determining the principal strains and orientations of the axes of strain ellipsoids from these strain components has been described previously (Sec. 4-7).

If only the ratios of the axes of the strain ellipse in each section have been found—and generally this is all that can be determined from distorted spheres of unknown original diameter—we proceed in a slightly different way. Let the ratios be $R_{xy} = \lambda_{xy1}/\lambda_{xy2}$, $R_{yz} = \lambda_{yz1}/\lambda_{yz2}$, and $R_{zx} = \lambda_{zx1}/\lambda_{zx2}$. Then the strain components in the xy section are given by

$$\lambda'_x = \lambda'_{xy1}(\cos^2 \theta_x + R_{xy} \sin^2 \theta_x) \tag{5-7}$$

$$\lambda'_y = \lambda'_{xy1}(\sin^2 \theta_x + R_{xy} \cos^2 \theta_x) \tag{5-8}$$

$$\gamma'_{xy} = \lambda'_{xy1}(R_{xy} - 1) \sin \theta_x \cos \theta_x \tag{5-9}$$

On the yz section the strain components are given by

$$\lambda'_y = \lambda'_{yz1}(\cos^2 \theta_y + R_{yz} \sin^2 \theta_y) \tag{5-10}$$

$$\lambda'_z = \lambda'_{yz1}(\sin^2 \theta_y + R_{yz} \cos^2 \theta_y) \tag{5-11}$$

$$\gamma'_{yz} = \lambda'_{yz1}(R_{yz} - 1) \sin \theta_y \cos \theta_y \tag{5-12}$$

Because both (5-8) and (5-10) give values of λ'_y we can find λ'_{yz1} in terms of λ'_{xy1} for

$$\lambda'_{yz1} = \frac{\lambda'_{xy1}(\cos^2 \theta_x + R_{xy} \cos^2 \theta_x)}{\cos^2 \theta_y + R_{yz} \sin^2 \theta_y} \tag{5-13}$$

It therefore follows that we can express λ'_z and γ'_{yz} in the same way as λ'_x, λ'_y, and λ'_{xy}, that is, with the same unknown multiplying factor $\lambda'_{xy1} = k$. The strain components on the zx section are

$$\lambda'_z = \lambda'_{zx1}(\cos^2 \theta_z + R_{zx} \sin^2 \theta_z) \tag{5-14}$$

$$\lambda'_x = \lambda'_{zx1}(\sin^2 \theta_z + R_{zx} \cos^2 \theta_z) \tag{5-15}$$

$$\gamma'_{zx} = \lambda'_{zx1}(R_{zx} - 1) \sin \theta_z \cos \theta_z \tag{5-16}$$

and because

$$\lambda'_{zx1} = \frac{\lambda'_{xy1}(\cos^2 \theta_x + R_{xy} \sin^2 \theta_x)}{\sin^2 \theta_z + R_{zx} \cos^2 \theta_z} \tag{5-17}$$

all can be given in terms of the same unknown multiplying factor k. Again there are checks on the general accuracy of measurements and computing because there are two independent ways of determining λ'_z from (5-11) and from (5-14).

As a result of these calculations we determine the six strain components in terms of the same unknown constant multiplier $k = \lambda'_{xy1}$. These components can now be used to find the principal strains $k\lambda'_1$, $k\lambda'_2$, and $k\lambda'_3$ and their orientation in the way described in Sec. 4-7.

5-4 MEASUREMENT OF THE INDIVIDUAL ELLIPSOIDS

The usual method of determining the shape and orientation of the strain ellipsoid is from measurement of three perpendicular sections as described above; but with some material it is possible to obtain results more rapidly by making measurements of the three axes of each ellipsoid. For example, it is sometimes possible to disintegrate the matrix of oolitic limestones with hydrogen peroxide and measure the individual ooids thus obtained with a micrometer. Another technique entails reconstructing each ellipsoid from a number of parallel cross sections. The rock containing the ellipsoidal particles is cut and polished carefully on two sides so that the two surfaces are absolutely parallel. One of these surfaces is now progressively ground away keeping the ground face parallel to the other face. A record is kept of the changing shapes and positions of the elliptical sections of the particles (photographs and camera lucida drawings) using a micrometer to measure the thickness of the material that has been removed by grinding. From the succession of ellipses the shape and orientation of each ellipsoid can be completely reconstructed. This technique is particularly useful where the ellipsoidal strain markers are sparsely distributed through the rock and where individual sections do not pass through every ellipsoid to help one to determine the ratios of the principal strains on that section. The main disadvantage of this method is that the material is completely destroyed during the grinding process.

Figure 5-17

Deformed pebbles in a conglomerate. Barberton, South Africa. The pebbles are all flattened in the plane of the slaty cleavage and have their longest axes subparallel to the maximum finite elongation.

Figure 5-18

Strongly deformed rodlike pebbles in a schist. Jerrisbaken, Sweden.

5-5 DEFORMATION OF NONSPHERICAL OBJECTS

It is rare that the objects listed in Sec. 5-1 are exactly spherical in their unstrained state and some may deviate greatly from original spheres. Conglomerate pebbles come into this category. Deformed conglomerates are of fairly widespread development (Figs. 5-17, 5-18) and are potentially good material for investigation of strain in metamorphic schists and in pre-Cambrian rocks. Although many studies have been made of the shapes of pebbles in deformed conglomerates, few have considered the effects of variable primary shape on the deformed pebble shapes. This is one of the several complex problems that must be solved before the strain can be computed.

The observations made on deformed oolites in the South Mountain fold of Maryland by Cloos (1947) make the best starting point for a discussion on the effects of an original shape factor.

Fluctuation The investigations made by Cloos of the shapes of deformed oolites showed clearly that the major axes of the elliptical sections of the ooids were rarely parallel. Cloos termed this variation *fluctuation in orientation* and he recorded graphically the maximum fluctuations in 144 thin sections and plotted them against the ratios of the major and minor axes of the strain

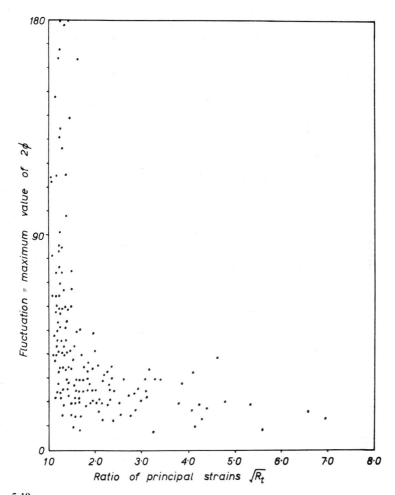

Figure 5-19

Plots of maximum variation (fluctuation) in the orientation of major axes of deformed ooids from South Mountain fold, Maryland. (After Cloos, 1947.)

ellipse (Fig. 5-19). The fluctuation was often very large at low deformation, but it decreased markedly in highly deformed rocks. Cloos suggested that this fluctuation was the result of original eccentricity and that the ooids initially deviated from a perfect spherical form. He noted that these original variations could introduce errors in orientation of up to 10 percent, but that they carried little weight in deformation that exceeded 100 percent. Although the errors in orientation of the principal axes of strain are not likely to be large, it is not always realized that the primary-shape factor has a great influence on the lengths of the major and minor axes of the elliptical sections of the deformed ooids and that it can never be overprinted no matter how great the deformation. For example, let us suppose the original ellipse had a ratio of major axis to minor of 2; then, if a tectonic deformation with a principal strain ratio of 10 was superimposed coaxially so that its longest dimension coincided with the major axis of the original ellipse, the resulting ellipse would have an axial ratio of 20. If they were superimposed so that the principal tectonic elongation coincided with the original minor axis, the final ellipse would have an axial ratio of 5. Where the two sets of axes coincide, the shape of the resulting ellipse is always a product not a sum of the two ratios.

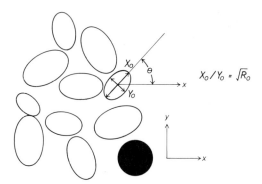

Figure 5-20

A group of elliptical objects each with the same axial ratio $X_o/Y_o = (R_o)^{\frac{1}{2}}$ but variably oriented in the material.

$X_o/Y_o = \sqrt{R_o}$

Let us now consider in more detail, and in two dimensions, the effect of an originally noncircular shape on the resulting form after deformation. Suppose that the original shape is that of an ellipse. The shape and orientation of the final ellipse depends on three factors: the ratios of the principal axes of the original ellipse, the ratios of the principal tectonic strains, and the orientations of the axes of the original ellipse with reference to the principal strain directions. Figure 5-20 shows a series of ellipses of different sizes, all with the same axial ratios (X_o/Y_o), and with their long axes oriented in several directions. If $1/X_o^2 = \lambda_1'$ and $1/Y_o^2 = \lambda_2'$, the equation of each ellipse making an angle θ to the principal extensions (x direction) is given by

$$(\lambda_1' \cos^2 \theta + \lambda_2' \sin^2 \theta)x^2 - 2(\lambda_2' - \lambda_1') \sin \theta \cos \theta \, xy$$
$$+ (\lambda_1' \sin^2 \theta + \lambda_2' \cos^2 \theta)y^2 = 1 \quad (3\text{-}72)$$

After a tectonic deformation where the principal strains are X_t and Y_t and have reciprocal quadratic extensions $\lambda_i' = 1/X_t^2$ and $\lambda_{ii}' = 1/Y_t^2$, the modified ellipse (Figs. 5-21, 5-22) is given by

$$(\lambda_1' \cos^2 \theta + \lambda_2' \sin^2 \theta)\lambda_i'x^2 - 2(\lambda_i'\lambda_{ii}')^{\frac{1}{2}}(\lambda_2' - \lambda_1') \sin \theta \cos \theta \, xy$$
$$+ (\lambda_1' \sin^2 \theta + \lambda_2' \cos^2 \theta)\lambda_{ii}'y^2 = 1 \quad (5\text{-}18)$$

On the basis of the principle established in Sec. 3-11 it follows that we can

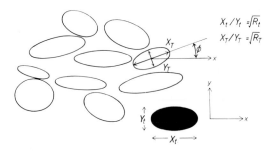

Figure 5-21

The ellipses of Fig. 5-20 deformed by a homogeneous strain with axial ratio $X_t/Y_t = R_t^{\frac{1}{2}}$. The angle θ that the major axis of the ellipses made with the x direction is modified to the angle ϕ. The maximum value of ϕ defines the fluctuation.

Figure 5-22

The ellipses of Fig. 5-20 modified by a greater homogeneous strain than that of the deformation of Fig. 5-21. The resulting elliptical forms show a great variation in actual axial ratio $X_T/Y_T = R_T^{\frac{1}{2}}$ and the fluctuation is decreased.

establish the shape and orientation of the resulting ellipse. Figure 5-23 shows the type of Mohr diagram solution to the problem. The original ellipse is represented by a circle of center C_1, radius C_1X_1.

$$OA_1 = \lambda_1' \cos^2 \theta + \lambda_2' \sin^2 \theta$$

$$OB_1 = \lambda_1' \sin^2 \theta + \lambda_2' \cos^2 \theta \tag{5-19}$$

and $$A_1X_1 = (\lambda_2' - \lambda_1') \sin \theta \cos \theta$$

and the deformed ellipse is represented by the circle of center C_2, radius C_2X_2, where

$$OA_2 = (\lambda_1' \cos^2 \theta + \lambda_2' \sin^2 \theta)\lambda_i' \tag{5-20a}$$

$$OB_2 = (\lambda_1' \sin^2 \theta + \lambda_2' \cos^2 \theta)\lambda_{ii}' \tag{5-20b}$$

$$A_2X_2 = (\lambda_i'\lambda_{ii}')^{\frac{1}{2}}(\lambda_2' - \lambda_1') \sin \theta \cos \theta \tag{5-20c}$$

The lengths of the axes of the final ellipse are recorded as the reciprocal

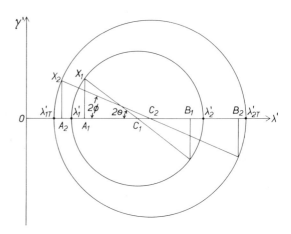

Figure 5-23

Mohr diagram for the solution of the problem of determining the shape and orientation of an ellipse formed by an initial elliptical object homogeneously.

principal quadratic extensions λ'_{1T} and λ'_{2T} and the angle ϕ between its major axis and the x axis is given by

$$\tan 2\phi = \frac{A_2 X_2}{A_2 C_2} \tag{5-21}$$

but $A_2 C_2 = OC_2 - OA_2 = \dfrac{OA_2 + OB_2}{2} - OA_2 = \dfrac{OB_2 - OA_2}{2}$

Therefore

$$\tan 2\phi = \frac{2(\lambda'_i \lambda'_{ii})^{\frac{1}{2}}(\lambda'_2 - \lambda'_1)\sin\theta\cos\theta}{(\lambda'_1 \sin^2\theta + \lambda'_2 \cos^2\theta)\lambda'_{ii} - (\lambda'_1 \cos^2\theta + \lambda'_2 \sin^2\theta)\lambda'_i}$$

Dividing by $\lambda'_1 \lambda'_i$ and replacing the ratios λ'_2/λ'_1 by R_o and λ'_{ii}/λ'_i by R_t,

$$\tan 2\phi = \frac{2R_t^{\frac{1}{2}}(R_o - 1)\sin\theta\cos\theta}{(\sin^2\theta + R_o \cos^2\theta)R_t - (\cos^2\theta + R_o \sin^2\theta)}$$

Expressing this in terms of 2θ,

$$\tan 2\phi = \frac{2R_t^{\frac{1}{2}}(R_o - 1)\sin 2\theta}{(R_o + 1)(R_t - 1) + (R_o - 1)(R_t + 1)\cos 2\theta} \tag{5-22}$$

This is the basic equation for determining the orientation of the principal axis of the resultant ellipse knowing the square of the ratios of minor and major axes of the original ellipse (R_o) and of the tectonic strain ellipse (R_t). Figures 5-21 and 5-22 illustrate the final ellipse shapes of the originally randomly oriented ellipses shown in Fig. 5-20, and Figs. 5-24 and 5-25 graphically illustrate the relationships that exist between the final ellipse shapes (R_T is the square of the ratios of major to minor axes) and their orientation ϕ. The stronger the deformation (greater value of R_t) the smaller the variation in

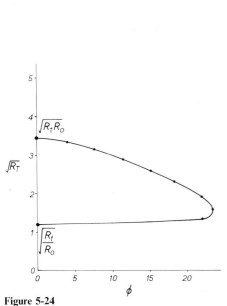

Figure 5-24

Variations in ratio $R_t^{\frac{1}{2}}$ plotted against ϕ for the ellipses shown in Fig. 5-21.

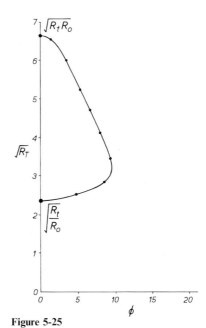

Figure 5-25

Variations in ratio $R_t^{\frac{1}{2}}$ plotted against ϕ for the ellipses shown in Fig. 5-22.

orientation of the principal axes of the ellipses. The maximum departure of the long axis of the ellipse from the principal tectonic extension (maximum value of ϕ, that is, the fluctuation in Cloos' terminology) is smallest in the most strongly deformed material. This maximum angle depends entirely on the ratios of the axes of the original ellipse of Fig. 5-20, and the tectonic strain ellipse. The fluctuation can be found by discovering the maximum value of ϕ from (5-22). Differentiating and equating $d\phi/d\theta$ to zero,

$$0 = [(R_o + 1)(R_t - 1) + (R_o - 1)(R_t + 1) \cos 2\theta] \cos 2\theta$$

$$+ [(R_o - 1)(R_t + 1) \sin 2\theta] \sin 2\theta$$

$$0 = (R_o + 1)(R_t - 1) \cos 2\theta + (R_o - 1)(R_t + 1)$$

Therefore
$$\cos 2\theta = -\frac{(R_o - 1)(R_t + 1)}{(R_o + 1)(R_t - 1)} \tag{5-23}$$

Substituting this condition back in (5-22) with $\sin 2\theta = (1 - \cos^2 2\theta)^{\frac{1}{2}}$,

$$\frac{2R_t^{\frac{1}{2}}(R_o - 1)[(R_o + 1)^2(R_t - 1)^2 - (R_o - 1)^2(R_t + 1)^2]^{\frac{1}{2}}}{(R_o + 1)(R_t - 1)[(R_o + 1)(R_t - 1) - (R_o - 1)^2(R_t + 1)^2/(R_o + 1)(R_t - 1)]}$$

$$= \max \tan 2\phi$$

which simplifies to

$$\frac{R_t^{\frac{1}{2}}(R_o - 1)}{[(R_o R_t - 1)(R_t - R_o)]^{\frac{1}{2}}} = \text{max tan } 2\phi \qquad (5\text{-}24)$$

This function has been plotted for values of R_o between 1.1 and 2.0 in Fig. 5-26. Where $R_t \leqslant R_o$ the function is indeterminate; geologically this means that the fluctuation is the same as the initial random variation, that is, $180°$. Wherever $R_t > R_o$ the fluctuation has a finite value less than $90°$, that is, the major axes of the ellipses will always make an angle of less than $45°$ with the direction of principal tectonic elongation. The theoretical values of the fluctuation in

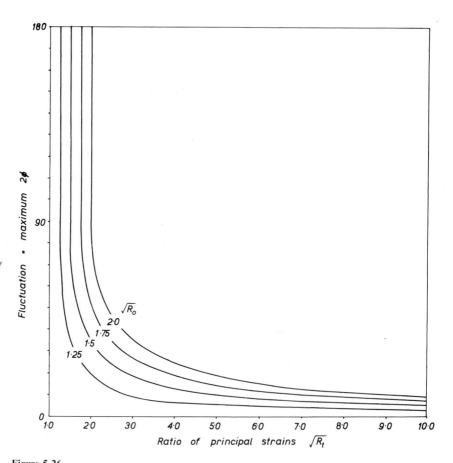

Figure 5-26

Graphical presentation of the fluctuation of the ellipse axes with various initial elliptical shapes ($R_o^{\frac{1}{2}}$) and tectonic strain ($R_t^{\frac{1}{2}}$). This should be compared with the data obtained by Cloos (Fig. 5-19).

Fig. 5-26 should be compared with the observations of Cloos on naturally deformed oolites (Fig. 5-19); the correspondence is a very close one.

Lengths of the principal axes of the final ellipse   We can calculate the ratios of the principal axes of the ellipse formed by the tectonic deformation of an initially elliptical object as follows: If the two principal axes of the final ellipse have actual lengths $X_T > Y_T$ and if $1/X_T^2 = \lambda'_{1T}$ and $1/Y_T^2 = \lambda'_{2T}$, then from the Mohr diagram shown in Fig. 5-23

$$OA_2 = \lambda'_{1T} \cos^2 \phi + \lambda'_{2T} \sin^2 \phi$$

$$OB_2 = \lambda'_{1T} \sin^2 \phi + \lambda'_{2T} \cos^2 \phi$$

(5-25)

Combining (5-20*a,b*) and (5-25),

$$\frac{OA_2}{OB_2} = \frac{\lambda'_{1T} \cos^2 \phi + \lambda'_{2T} \sin^2 \phi}{\lambda'_{1T} \sin^2 \phi + \lambda'_{2T} \cos^2 \phi} = \frac{(\lambda'_1 \cos^2 \theta + \lambda'_2 \sin^2 \theta)\lambda'_i}{(\lambda'_1 \sin^2 \theta + \lambda'_2 \cos^2 \theta)\lambda'_{ii}}$$

(5-26)

Replacing $\lambda'_{2T}/\lambda'_{1T}$ by R_T, λ'_2/λ'_1 by R_o, and λ'_{ii}/λ'_i by R_t,

$$\frac{1 + R_T \tan^2 \phi}{\tan^2 \phi + R_T} = \frac{1 + R_o \tan^2 \theta}{(\tan^2 \theta + R_o)R_t}$$

or

$$R_T = \frac{\tan^2 \phi(1 + R_o \tan^2 \theta) - R_t(\tan^2 \theta + R_o)}{R_t \tan^2 \phi(\tan^2 \theta + R_o) - (1 + R_o \tan^2 \theta)}$$

(5-27)

From (5-22) the angle ϕ can be expressed as a function of θ, R_t, and R_o and therefore using (5-22) and (5-27), R_T can be calculated knowing the values of ϕ, R_t, and R_o.

5-6   DEFORMATION OF ELLIPTICAL OBJECTS WITH RANDOM FABRIC

If the directions of the principal tectonic strains are known, it is possible to compute the principal strain ratio ($R_t = \lambda_i/\lambda_{ii}$) from measurements made on a single ellipse of the ratios of its principal axes ($R_T^{\frac{1}{2}}$) and the angle ϕ between the principal tectonic strains and the longest axis of this ellipse. To do this we have to know the initial shape factor (R_o), and therefore with naturally deformed rock materials this method is not very useful for determining strains. We shall now consider what is the most practical solution to this problem where we do not know either the directions of the principal strains or the original shape of the undeformed ellipses.

One good way of finding the tectonic strains depends on establishing the graphs illustrated in Figs. 5-24 and 5-25 to determine how the ratios of the axes of the deformed ellipses vary with the orientation of their major and minor axes. First a line is drawn on the section as a reference azimuth and then the angle (α) between the longest axis of each ellipse is recorded together with the ratios of its major and minor axes $X/Y = R_T^{\frac{1}{2}}$. These are plotted on a graph (Fig. 5-27). The points should all lie on or about a curve which is symmetrical about a line of some fixed value of α (Fig. 5-27, α'). This value of α' gives the angle between the principal tectonic extension and the reference azimuth. The maximum and minimum values of $R_T^{\frac{1}{2}}$ are now read off along the line $\alpha = \alpha'$. As this line is a principal strain direction, we know from the previous discussion of fluctuation that $\theta = \phi = 0$ or $90°$. From (5-27) we see that where $\theta = \phi = 0°$,

$$\text{maximum } R_T = R_t R_o \qquad (5\text{-}28a)$$

and where $\theta = \phi = 90°$,

$$\text{minimum } R_T = \frac{R_o}{R_t} \text{ or } \frac{R_t}{R_o} \qquad (5\text{-}28b)$$

depending on whether $R_o > R_t$ or $R_t > R_o$, respectively.

The product or quotient of the maximum and minimum values of $R_T^{\frac{1}{2}}$ give values for either the tectonic component R_t alone or the original shape component R_o.

This technique has to be modified slightly if the original ellipses did not all have the same ratios R_o. If R_o varies between two extreme values R_o' and R_o'' such that $R_o' > R_o''$, then the maximum and minimum values of R_T will be given by $R_t R_o'$, $R_t R_o''$ and by R_t/R_o', R_t/R_o''. If $R_t > R_o'$ then $R_t R_o' > R_t R_o''$ and $R_t/R_o' < R_t/R_o''$, and therefore the absolute maximum and minimum values of R_T will be given by $R_t R_o'$ and R_t/R_o'. If the absolute maximum and minimum values of the bounding curve which forms the envelope to all the points in Fig. 5-28 are used for the computation, the tectonic components can still be obtained from the product of these values of $R_T^{\frac{1}{2}}$.

The only difficulties that arise with the practical application of this technique of determining the maximum and minimum values of $R_T^{\frac{1}{2}}$ occur where $R_o > R_t$ and where the fluctuation exceeds $90°$. It will be found that where the longest axis of the original ellipse is at a high angle to the principal tectonic extension, we have to allow the ratio $R_T^{\frac{1}{2}}$ to take a value less than 1 in order to obtain a smoothly continuous curve crossing the line $R_T^{\frac{1}{2}} = 1$.

By using this method on three mutually perpendicular sections through a rock which contains deformed particles that were initially ellipsoidal and variably oriented, it is possible to isolate the tectonic strains in each section. If the initial ellipsoids (with principal lengths $X_o > Y_o > Z_o$) were randomly oriented, then any section through the undeformed rock will show elliptical cross sections of these ellipsoids with values of R_o ranging from 1 (circular section through ellipsoid) to $(X_o/Z_o)^2$. Providing that enough measurements can be made on the deformed material, it should always be possible to separate the tectonic and original parts of the fabric by the method described above, to isolate the strain components in each plane by the method described

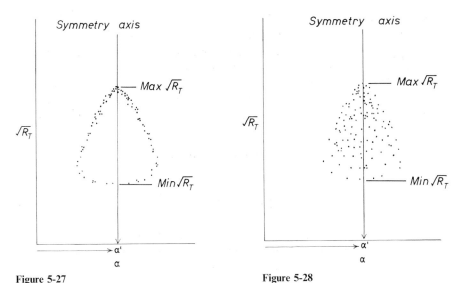

Figure 5-27

Graphical method for solving the values of the tectonic strain ratio $(R_t)^{\frac{1}{2}}$, and orientation of the major axis of the strain ellipse by recording variations in angle α and ratio $R_T^{\frac{1}{2}}$ of deformed initially elliptical objects.

Figure 5-28

Appearance of the graph (Fig. 5-27) where the initial elliptical shapes have a variable ratio $R_o^{\frac{1}{2}}$.

in Sec. 5-3, and to find the principal strains. It is impossible to separate the original and tectonic components if the initial ellipsoids had a perfect plane or linear fabric.

5-7 DEFORMATION OF ELLIPSOIDAL OBJECTS WITH RANDOM FABRIC

It is sometimes possible to isolate deformed particles from their enclosing rock matrix and make direct measurements of their principal axes. Deformed conglomerate pebbles can sometimes be pulled neatly out from the surrounding rock, especially if the matrix has been weathered, and individual ooids can sometimes be isolated from their enclosing limestones by treating the rock with very dilute acid or hydrogen peroxide. Wherever this is possible, the following special method may afford a convenient way of determining the tectonic strains. Let us assume that the particles had an original shape which was approximately ellipsoidal; their deformed shape will then also be ellipsoidal.

The ellipsoidal pebbles in many deformed conglomerates often show an extremely great variation in shape, sometimes much greater than that of the undeformed pebbles (see Figs. 5-29, 5-30), although the orientation of the

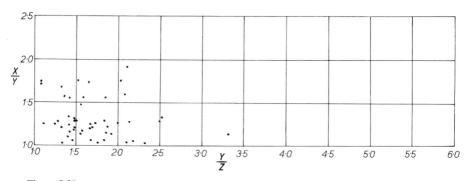

Figure 5-29

Plots of axial ratios of 50 undeformed pebbles from a conglomerate west of Barberton, South Africa.

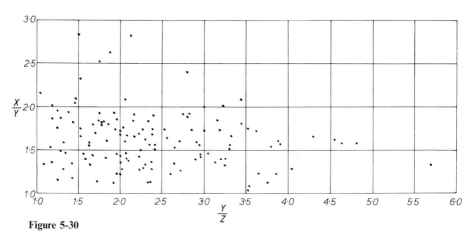

Figure 5-30

Plots of axial ratios of 138 pebbles from the same horizon as those of Fig. 5-29, but in a highly deformed state. Note the greater variation in shapes in the deformed material than in the undeformed conglomerate. (After Ramsay, 1965.)

longest axes of the pebbles in deformed conglomerates generally shows much less variation than that of the undeformed pebbles (Figs. 5-31, 5-32). Some may be pancake-shaped with $k = 0$, while others may have elongated cigar-shape forms with a k value close to ∞. This great range is the result of the variable orientation of the initial pebbles with reference to the fixed directions of the principal tectonic strains. To see how this comes about consider a group of randomly oriented ellipsoidal particles in a rock all with the same axial ratios $X_o/Y_o = a_o$ and $Y_o/Z_o = b_o$ (Fig. 5-33). If these randomly oriented ellipsoids are distorted homogeneously with their matrix by a tectonic deformation with principal quadratic extensions $\lambda_1 = kX_t^2$, $\lambda_2 = kY_t^2$, $\lambda_3 = kZ_t^2$, and with axial ratios $X_t/Y_t = a_t$ and $Y_t/Z_t = b_t$ (Fig. 5-33), we obtain a large number of

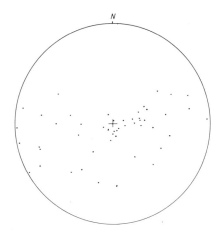

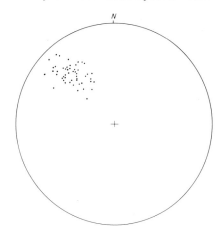

Figure 5-31

*Orientations of the longest axis of 50
undeformed pebbles shown in Fig. 5-29. The
fabric is a weak planar one.*

Figure 5-32

*Orientations of the longest axes of 50
deformed pebbles from the same locality as
Fig. 5-30. The fabric has been modified from
a planar to a linear one.*

different-shaped ellipsoids with axial ratios $X_T/Y_T = a_T$, $Y_T/Z_T = b_T$, depending on the way the original ellipsoids are combined with the tectonic strain ellipsoid. If the original ellipsoids are randomly oriented there are an infinite number of different-shaped final ellipsoids, but six special coaxial combina-

Figure 5-33

*Plots of the elliposid $(b_o,\ a_o)$ representing
the original shape, and $(b_t,\ a_t)$ representing
strain ellipsoid.*

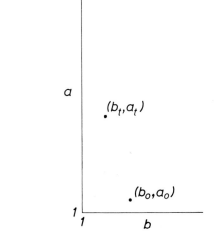

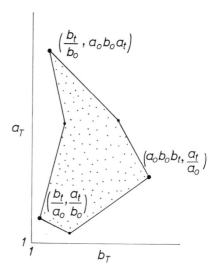

Figure 5-34

Field of possible ellipsoid shapes resulting from the homogeneous straining of randomly oriented initial ellipsoids (b_o, a_o). The coordinates of the three extreme-shaped members of the field are given in terms of the initial and tectonic components.

tions of the original and tectonic ellipsoids give rise to six extreme-shaped ellipsoids given by

$$X_t X_o : Y_t Y_o : Z_t Z_o \qquad X_t X_o : Y_t Z_o : Z_t Y_o$$

$$X_t Y_o : Y_t X_o : Z_t Z_o \qquad X_t Y_o : Y_t Z_o : Z_t X_o$$

$$X_t Z_o : Y_t X_o : Z_t Y_o \qquad X_t Z_o : Y_t Y_o : Z_t X_o$$

The a_T and b_T values for these various extreme ellipsoids can be plotted (Fig. 5-34). On this graph the six points representing these ellipsoids enclose an area which represents the field in which the plots of all the other possible ellipsoids fall. The shape and the size of this field vary with both the original and tectonic shape factor and it will be seen that the shapes of deformed pebbles can never be taken as simple indicators of the strain the rock has suffered. The determination of the strain from a few measurements of deformed pebbles or from arithmetic means of the axial ratios of deformed pebbles is unlikely to produce results of any validity.

In the discussion of the measurement of strain in two dimensions from distorted particles of initially elliptical shape (Sec. 5-6), it was shown how the extreme shapes of the final ellipses brought about by special combinations of original and tectonic components could be used to isolate these two components. We shall now, therefore, examine the points in Fig. 5-34 which have extreme values of a_T and b_T which bound the field of all possible ellipsoids to see if these have any special properties which may be useful for separating the strain components from that of the original shape of the particles. There are always three points which bound the ellipsoid field; for example, if $a_t > a_o$ and $b_t > b_o$, these ellipsoids have axial dimensions

(i) $X_t X_o > Y_t Z_o > Z_t Y_o$

(ii) $X_t Y_o > Y_t X_o > Z_t Z_o$

(iii) $Z_t Z_o > Y_t Y_o > Z_t X_o$

We can calculate the axial ratios a_T and b_T in terms of the axial ratios of the original and tectonic components

(i) $$a_T = \frac{X_t X_o}{Y_t Z_o} = \frac{X_t X_o Y_o}{Y_t Y_o Z_o} = a_t a_o b_o \qquad b_T = \frac{Y_t Z_o}{Z_t Y_o} = \frac{b_t}{b_o}$$

(ii) $$a_T = \frac{X_t Y_o}{Y_t X_o} = \frac{a_t}{a_o} \qquad b_T = \frac{Y_t X_o}{Z_t Z_o} = \frac{Y_t X_o Y_o}{Z_t Y_o Z_o} = b_t a_o b_o$$

(iii) $$a_T = \frac{X_t Z_o}{Y_t Y_o} = \frac{a_t}{b_o} \qquad b_T = \frac{Y_t Y_o}{Z_t X_o} = \frac{b_t}{a_o}$$

If we know the original axial ratios of the particles (a_o, b_o) and we can establish the extreme positions of the deformation field and their six coordinates in terms of a_T and b_T, then there are three independent ways of computing the tectonic axial ratios a_t, and three for b_t. We can also solve this problem without

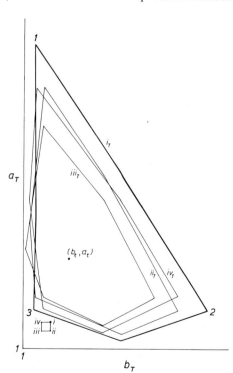

Figure 5-35

The effects of the superposition of a homogeneous tectonic strain (b_t, a_t) on a series of initially variably shaped and randomly oriented ellipsoids delimited by the fields i, ii, iii, and iv. After deformation the resulting shapes fall in the fields i_T, ii_T, iii_T, and iv_T. The ellipsoids which have an original shape iv control the three extreme positions of the deformation field.

knowing the initial shape ratios because by finding values for a_T and b_T we can obtain six separate equations in terms of the four unknown variables a_t, b_t, a_o, and b_o.

It has been assumed that all the initial ellipsoids had the same axial ratios a_o and b_o. Consider ellipsoids of variable initial shape so that their axial-ratio plots fall in the area between the plots *i*, *ii*, *iii*, and *iv* in Fig. 5-35 and deform them by a homogeneous strain (axial ratios a_t and b_t). Each of these four bounding ellipsoids *i*, *ii*, *iii*, and *iv* will give rise to a range of final ellipsoids that lie within the areas enclosed by the lines i_T, ii_T, iii_T, and iv_T, respectively, in Fig. 5-35. It will be seen that all possible ellipsoids fall within the three extreme points 1, 2, and 3 controlled by the initial ellipsoid *i*. The field is therefore bounded by three ellipsoids which have coordinates given by $(b_t/b_{oi}, a_t a_{oi} b_{oi})$, $(b_t a_{oi} b_{oi}, a_t/a_{oi})$, and $(b_t/a_{oi}, a_t/b_{oi})$ and therefore the method of determining the principal tectonic strain ratios from six simultaneous equations as described above is still valid even though the original particles had variable axial ratios.

5-8 DEFORMATION OF ELLIPSOIDAL OBJECTS WITH AN ORIGINAL FABRIC

Until now it has been assumed that the major and minor axes of the ellipsoidal particles before deformation had random orientation. If this is not the case then the extreme combinations of tectonic and original ellipsoids will not be developed and the line which bounds the field of deformed ellipsoids will no longer pass through the points with known a_T and b_T coordinates.

In many conglomerate deposits and in some oolitic limestones the ellipsoidal particles have an initial preferred orientation and the rock has an original fabric. Generally the shortest axes of the ellipsoids are initially perpendicular to the bedding (a planar sedimentary fabric). Sometimes there may be in addition a preferred orientation of the long axes of the particles about a certain linear direction within the bedding surface, or imbricated across the bedding surface. Where an unknown original planar fabric exists, it is generally very difficult or even impossible to calculate the tectonic strains from deformed material.

First, let us find out how we can recognize that the particles in a deformed rock initially had a preferred orientation. Consider a block of rock containing ellipsoidal particles with axial ratios $X:Y:Z = 4:2:1$ all arranged so that their shortest axes (Z) are perpendicular to the bedding surface, and with their X and Y axes arranged in a random fashion within the surface. Let us now consider the shapes and orientations of the various elliptical sections of these ellipsoids that appear on a section (Figs. 5-36 and 5-37, plane A) through the block making an angle of 45° to the bedding surface. As we know the shape of the initial ellipsoid, we can calculate the shapes of the sections. The ratios of the principal elongations of the ellipses ($R_o^{\frac{1}{2}}$) will vary from

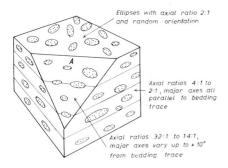

Ellipses with axial ratio 2:1
and random orientation

Axial ratios 4:1 to
2:1, major axes all
parallel to bedding
trace

Axial ratios 32:1 to 14:1,
major axes vary up to ± 10°
from bedding trace

Figure 5-36

*A block containing original ellipsoids all with
the same axial ratio $X:Y:Z = 4:2:1$, and
with a plane fabric so that all have their
Z axis perpendicular to the bedding surface.*

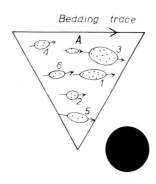

Bedding   trace

Figure 5-37

*Appearance of the elliptical cross sections
on surface A from Fig. 5-36.*

1.4 to 3.2. The longest axes of the ellipsoids also vary in orientation and make
angles of up to 10° on either side of the trace of the bedding surface on plane
A, Fig. 5-37. If the axial ratio $R_o^{\frac{1}{2}}$ of each ellipse is plotted against the angle
that it makes with the bedding trace, all the ellipses fall on the curve shown in
Fig. 5-38. This is the distribution of the undeformed ellipse sections; let us
now subject the particles and their matrix to a homogeneous finite strain.
All the ellipses on this section will change their axial ratios and their orientation depending on the axial ratio of the strain ellipse and its orientation

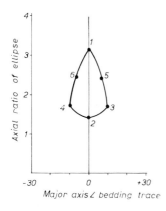

Figure 5-38

*Graphic plot of orientation of long axes of
ellipses on surface A, and their axial
ratio $x_o/y_o = R_o^{\frac{1}{2}}$.*

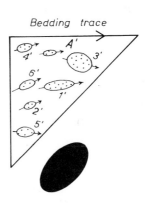

Bedding   trace

Figure 5-39

*Section A from Fig. 5-37 deformed by a
homogeneous tectonic strain.*

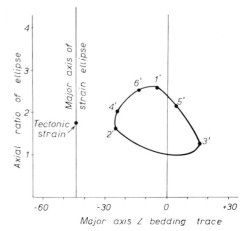

Figure 5-40

*Graphic plot of orientation of long axis
of ellipses on deformed surface A
(Fig. 5-39) and their axial ratios
$x_T/y_T = R_T^{\frac{1}{2}}$.*

(Fig. 5-39). These changed shapes have been calculated using the methods described in Sec. 5-5; and the various axial ratios $(R_T^{\frac{1}{2}})$ and orientations (α) of the new ellipses are plotted in Fig. 5-40. It will be seen that the distribution of the ellipse plots is asymmetric about both axes of this graph. This arrangement should be compared with the symmetrical distribution that results from the deformation of randomly oriented ellipse sections (Figs. 5-27 and 5-28); an asymmetric distribution of ellipse plots always implies an initial fabric or preferred orientation of the ellipse sections. In order to determine the tectonic strains from an arrangement like this, we have to know both the original fabric and the orientation of the principal elongations of the strain ellipse; it follows that it is generally impossible to obtain a solution to this problem. The only section on which the two-dimensional strains may be determined simply is one that is parallel to the bedding surface, always provided there is no linear orientation of the X axes within this plane.

With some special arrangements of tectonic and original fabrics, it may, however, be possible to use direct three-dimensional measurement techniques like those described for random fabrics. These techniques can be used where there is some special coincidence of the bedding planes parallel to a principal plane of the strain ellipsoid. Probably the most frequent coincidence in strongly deformed rocks is where the bedding planes are parallel to, or lie very close to, the XY principal plane, i.e., bedding is parallel to the slaty cleavage or schistosity. Where this occurs Z_o and Z_t coincide and all the various extreme forms which have $X_T > Y_T > Z_T$ given by

$$X_tX_o > Y_tY_o > Z_tZ_o \qquad X_tY_o > Y_tX_o > Z_tZ_o$$

The final ellipsoids that are derived from initial ellipsoids plotted as (b_o, a_o) in Fig. 5-41 by a tectonic strain ellipsoid (b_t, a_t) lie on a line bounded by the two ellipsoids given above which have coordinates (b_ob_t, a_oa_t) and $(a_ob_ob_t,$

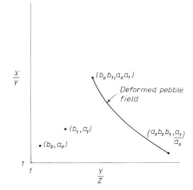

Figure 5-41

Special coincidence of tectonic fabric plane $X_t Y_t$ with the $X_o Y_o$ fabric plane of the initial ellipsoids. As a result of deformation all ellipsoids lie on a curved linear field.

a_t/a_o). If we can obtain values for the coordinates of these end points, we obtain four equations in the variables a_o, b_o, a_t, and b_t, but from the nature of these equations we can only calculate independently a_o and a_t. If, however, we have some knowledge of the initial undeformed ellipsoid shape, we have several independent ways of calculating a_t and b_t. If the particles were of initially variable shape and plots of undeformed ellipsoids fell between the points *i*, *ii*, *iii*, and *iv* in Fig. 5-42, then after deformation each of these bounding ellipsoids would be deformed according to the principles discussed above and would form a range of ellipsoids lying on the lines i_T, ii_T, iii_T, and iv_T, and all the possible ellipsoids initially plotting between *i*, *ii*, *iii*, and *iv* would lie in the stippled area between i_T, ii_T, iii_T, and iv_T. Because of the presence of an initial fabric, most of the deformed ellipsoids have a k_T value ($k_T = a_T/b_T$) quite different from that of the strain ellipsoid ($k_t = a_t/b_t$). This feature is obviously of extreme importance when interpreting the tectonic strains from deformed conglomerates yet it appears to have been overlooked by practically every study so far made in this subject. It completely invalidates the method of

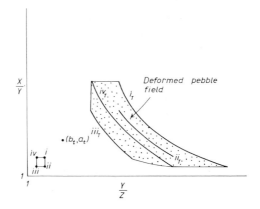

Figure 5-42

Special coincidence of tectonic fabric plane $X_t Y_t$ with the $X_o Y_o$ fabric plane of initially variably shaped ellipsoids i, ii, iii, and iv. The final deformation field occupies the stippled area between i_T, ii_T, iii_T, and iv_T.

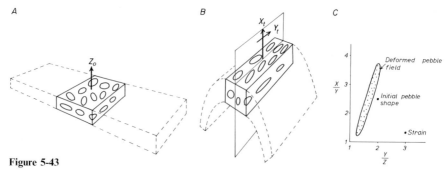

Figure 5-43

A. Ellipsoidal particles with ratios 5:2:1 with a preferred orientation such that all the shortest axes (Z_o) are arranged perpendicular to the bedding planes. B. The effect of deforming these particles so that the greatest tectonic extension (X_t) coincides with the Z_o direction. This is the type of arrangement that might occur near the hinge zone of a fold. C. The deformation field occupied by deformed pebbles; all their longest axes are arranged parallel or subparallel to the Y_t tectonic direction.

taking arithmetic means of the axial ratios of the pebbles to determine the strain.

In folded conglomerate layers the pebbles near the hinge zones of the folds often have an elongation parallel or subparallel to the hinge lines (Runner, 1934; Sander, 1948; Fairbairn, 1949). This feature has been explained in two ways: first, the principal extensive finite strain is parallel to the fold hinges and the pebbles are behaving as strain ellipsoids, or second, the pebbles have undergone a rolling motion as a result of lamellar movement in the rocks and in consequence do not give the shape (or correct orientation) of the strain ellipse. There is yet a third way to account for these pebble elongations using the ideas evolved above of special coincidence of tectonic and sedimentary fabrics producing rather abnormally shaped objects.

Let us consider a conglomeratic bed in its undeformed state with pebbles of uniform axial ratio $X_o:Y_o:Z_o$ and with Z_o axes perpendicular to the bedding planes. Subject this layer to a homogeneous strain with ratio $X_t:Y_t:Z_t$ such that X_t becomes perpendicular to the bedding surface. The pebbles would change shape, and the special coaxial combinations which limit the ranges of shape would have axes of lengths

$$Z_o X_t \quad Y_o Y_t \quad X_o Z_t$$
$$Z_o X_t \quad Y_o Z_t \quad X_o Y_t$$

Consider now an actual example with undeformed pebble ratios 5:2:1 (Fig. 5-43A) deformed with tectonic strains 4:3:1 (Fig. 5-43B). The resulting pebble shapes range from nearly spherical to apparently strongly constricted types (6:5:4 to 15:4:2) and occupy the field shown in Fig. 5-43C. Because the two fabrics have been combined in a particular manner, the longest axes of all the pebbles are aligned *parallel or subparallel to the Y_t direction*, even though the maximum finite tectonic elongation is perpendicular to this line. The individual deformed pebbles nearly all have k values greater than 1, even though the tectonic strain is a flattening type of deformation ($k = 0.17$).

The interpretation of deformed conglomerates is therefore no simple routine matter; the shapes of the deformed pebbles in different parts of the same folded layer are likely to be quite different, depending on the nature of the original and tectonic fabrics and the manner in which they coincide.

5-9   THE BEHAVIOR OF RIGID OR COMPETENT OBJECTS IN A DUCTILE MATRIX

The discussion on the shapes of deformed particles has until now assumed that both the particle and the rock material which surrounds it are deformed in a homogeneous manner. Although this may be correct in certain specialized geological environments, it is unlikely to be generally true. Studies of deformed layers containing pebbles of differing composition often indicate that there is a range of pebble shapes (Fig. 5-44). In a study of a metamorphosed, deformed conglomerate in Saxony, Mehnert (1939) showed that the shapes of the deformed pebbles had the following variation in Y/Z ratios; quartz 1:1 to 2:1; quartzite 8:1; graywacke 10:1 to 12:1; schistose graywacke and shale 18:1. Some of this variation may have been due to initial shape variation of the different types of pebble, and some to competence difference during

Figure 5-44

Deformed pebbles with a range in pebble shapes on account of their different compositions. Cristallina, Pennine Alps, Switzerland.

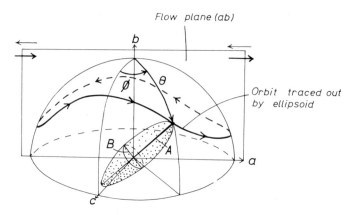

Figure 5-45

Orbit traced out by the end of the unique axis of a rigid uniaxial ellipsoid in a fluid undergoing lamellar flow, flowplane ab and flow direction a.

deformation. The use of deformed pebble shapes as indicators of the strain in the bulk of the rock is therefore not always possible, even if the restrictions resulting from variable initial shape and initial fabric could be overcome.

If there is a competence difference between the particle and its matrix, the particle undergoes a rotary motion which is not the same as that of the matrix. Jeffery (1923) has investigated the motion of an ellipsoidal particle carried along in a fluid undergoing slow lamellar flow (simple shear). He found that the forces which acted on the particle could be reduced into two couples, one tending to make the ellipsoid adopt the same rotation as the surrounding fluid, and another tending to set the ellipsoid so that its axes moved toward those of the principal distortion axes in the fluid. Jeffery found that the equations defining the motion made by an ellipsoid could be readily solved if the ellipsoid were of either a uniaxial prolate or uniaxial oblate type. If ϕ and θ are the two Euler angles shown in Fig. 5-45, then the unique axis of the ellipsoid moves according to the equation

$$\tan \phi = \frac{A}{B} \tan \frac{\dot{\gamma}ABt}{A^2 + B^2} \tag{5-29}$$

$$\tan^2 \theta = \frac{k^2 A^2 B^2}{A^2 \cos^2 \phi + B^2 \sin^2 \phi} \tag{5-30}$$

where A is the length of the unique axis, B the length of the other principal axes, $\dot{\gamma}$ the rate of shear, t time, and k an integration constant. The motion is periodic with a period given from (5-29) by

$$t = \frac{2\pi(A^2 + B^2)}{AB\dot{\gamma}} \tag{5-31}$$

The ends of the A axis of the ellipsoid trace out orbits which are a pair of spherical elliptic cones whose eccentricity is defined by k.

The angular rotation ϕ made by a particle for various finite shear strains γ has been graphically recorded (Fig. 5-46) for variously shaped ellipsoids with different A/B ratios ($= R$), starting at a position $\phi = 0$ using (5-29) in the form

$$\tan \phi = R \tan \frac{\gamma}{R + 1/R} \tag{5-32}$$

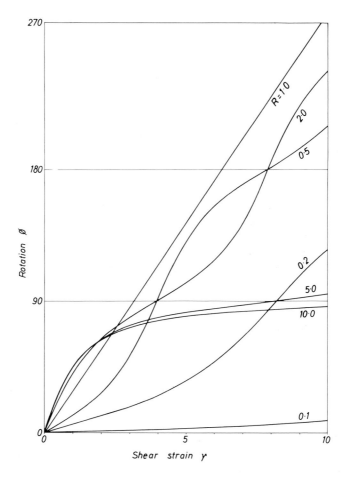

Figure 5-46

Graphical illustration of the amount of finite rotation ϕ of the axes of rigid ellipsoidal particles in a fluid undergoing lamellar flow.

From Eq. (5-30), if $\phi = 0$ the integration constant may be evaluated:

$$k = \tan \theta \qquad (5\text{-}33)$$

The elliptic orbits are projected in Fig. 5-47 from the equation of Trevelyan and Mason (1951, p. 356)

$$\tan \theta = k \left(\cos^2 \phi + \frac{\sin^2 \phi}{R} \right)^{-\frac{1}{2}} \qquad (5\text{-}34)$$

Flow of a fluid material containing rigid ellipsoidal particles leads to the development of a preferred orientation of the ellipsoids. If the ellipsoids are prolate types (Fig. 5-46, ellipsoids with $R > 1$), then those oriented at a high angle to the shear plane rotate with a rapid but decelerating motion along

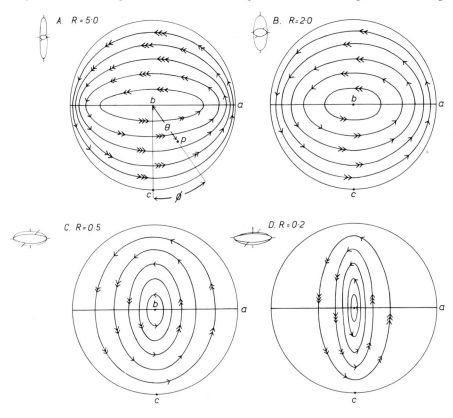

Figure 5-47

Elliptical orbits traced out by the unique axes of ellipsoids with various A/B axial ratios ($= R$) as defined by Eq. (5-33). The arrows indicate the relative speeds of motion at different stages on the elliptic orbit. The shear plane is ab; a is the direction of fluid flow and c is perpendicular to ab.

their orbit, and those lying close to the flow plane move more slowly (Fig. 5-46; curves show a slope decrease as ϕ approaches 90°). If the ellipsoidal particles are initially fairly uniformly distributed in the fluid, a type of linear fabric will be developed with the longest axes of the particles concentrated around the flow direction (*a*).

If the ellipsoids have an oblate form (Fig. 5-46, $R < 1$), those ellipsoids with their unique (shortest) axis at a high angle to the shear plane rotate very slowly (curves have a low slope where ϕ is close to 0 and 180°), but accelerate and rotate most rapidly where this axis becomes more nearly parallel to the flow plane (ϕ near to 90°). Motions of this type tend to produce planar fabrics, with the equatorial diameter of the ellipsoid tending to adopt a position parallel to the plane of flow (*ab*).

If the particles were originally randomly dispersed, then the degrees of development of these linear and planar fabrics depend on the value of R and the amount of shear. The symmetry of the fabric formed depends on the orientation of the *a*, *b*, and *c* directions defining the flow (cf. Sec. 4-10).

Although each ellipsoid periodically rotates in a single orbit, some of the orbits dissipate less energy than others and Jeffery suggested that the rotating ellipsoid might move over from an orbit of high-energy dissipation into one that dissipates less energy. With prolate ellipsoids the minimum energy orbit is found when the longest axis lies parallel to the *b* direction, and the ellipsoid rotates about this axis. With oblate ellipsoids the minimum energy orbit occurs where one of the equatorial long axes coincides with the *b* axis, and the disk rotates with a variable angular velocity about the *b* direction. Experiments carried out by Taylor (1923) verify Jeffery's conclusions; but he found that it required about 85 revolutions for the oblate spheroid to take up the minimum energy orbit, and 370 revolutions for the prolate ellipsoid to do the same.

This basic theory seems to have very considerable applications in a number of deformation processes found in geological phenomena. The most obvious application is in the study of fabrics shown by the crystal components of igneous rocks where platy and acicular crystals are carried along by the flow of the surrounding magma. Under these conditions linear and planar fabrics are formed, and it seems likely that there might occur shear sufficient to cause the rotating particles to move into the orbits of least dissipation of energy. More recently the problems of aggregate formation have been the subject of a number of detailed studies (e.g. Mason, 1950), for it appears that if the fluid contains many rotating particles then there is a fairly high probability that some will collide and form aggregates. Aggregation of large porphyritic crystals is a fairly well-known phenomenon seen in igneous rocks crystallized from melts. Jeffery's theory has also been applied to explain the fabric of certain glacial tills, and it has been suggested that the behavior of rock boulders contained in moving ice sheets might approximate to that of a rigid particle enclosed in a fluid (e.g., Holmes, 1941; Glenn, Donner, and West, 1957).

It is difficult to know whether the theory is directly applicable to the behavior of conglomerate pebbles deforming in a ductile matrix under conditions of tectonic flow. The pebbles themselves generally undergo a change in shape, although this change is generally of a different type and extent from that of their matrix. It seems that it might be possible to explain the behavior of competent but ductile pebbles by considering the types of changing rotations produced by a particle of constantly changing axial ratio. The mechanism of flow investigated by Jeffery is, however, a rather specialized type. Jeffery pointed out that even assuming completely rigid particles he was unable to obtain a general solution to the series of equations which would give the particle motions for the most general type of fluid flow. The problem of determining the behavior of progressively deforming particles in a more ductile matrix undergoing general flow is a problem of considerable complexity which still awaits solution.

It has often been suggested that the long axes of deforming pebbles adopt an attitude perpendicular to the shear direction (*a*) and are rotated about this axis in the manner described by Jeffery's minimum energy dissipation orbit. From Taylor's experiments, however, it seems very unlikely that a sufficiently strong shear strain is developed in real rock materials to accomplish the number of revolutions of the pebbles that is required to produce such an arrangement. It seems more likely that this phenomenon is explained by strong tectonic elongation of the pebbles, or by the superposition of two types of fabric in the manner shown in Fig. 5-43.

To summarize the discussion on pebble shapes and orientations in deformed rocks it seems that these features are likely to be controlled by a number of variable factors interacting in a very complex manner:

1. The state of finite strain in the rock
2. The original shape and fabric of the pebble association in the undeformed state
3. The difference in ductility between the pebbles and their matrix
4. The amount of differential rotation between pebbles and matrix. This depends on their ductility contrast, but it also depends on the actual mechanism of deformation and the deformation paths followed by the components of the strain tensor.

5-10   PRESSURE SOLUTION AND ITS EFFECTS IN DEFORMED CONGLOMERATES

In Sec. 5-2 it was shown how pressure sometimes leads to the removal of material along certain zones in the deforming rock, a feature which has been described in certain clastic and conglomeratic sediments. The Upper Tertiary conglomerates of the Alpine molasse zone of Europe are made up of pebbles

Figure 5-48

Pitted pebbles from the Upper Tertiary Alpine molasse. Grenoble, France.

Figure 5-49

Thin section of the contacts between pebbles from the Upper Tertiary Alpine molasse. Rigi, Switzerland.

showing a considerable variety of composition of which limestone, marl, chert, granite, and other crystalline materials are the most abundant. Near to the Alpine tectonic front these conglomerates have been involved in thrusting and folding, and the limestone pebbles frequently show pit-like depressions in their margins (Fig. 5-48). When these pitted pebbles are examined further, it is found that they are caused by the penetration of immediately adjacent pebbles (Fig. 5-49). This penetration appears to be the result of pressure solution. The pebbles have been pressed together and the material making up the most soluble pebbles has gone into solution, thus enabling the least soluble pebble to penetrate into it. Considerable changes of the bulk rock shape are accomplished by this solution, but few attempts have yet been made to assess the shape change. Such a study might be possible, using the technique described in Sec. 5-2 (method 3), by determining the relative displacements of the centers of individual pebbles.

5-11   DETERMINATION OF STRAIN FROM DEFORMED FOSSILS

One of the most valuable methods for the determination of the state of finite strain in sedimentary strata involves an analysis of the shapes of fossil impressions contained in rocks distorted by the various deformations which have occurred since the fossils were originally interred in the sediment. By measuring the changes in angles and lengths of lines within the fossil impression, it is generally possible to compute the values of the principal elongations (or their ratio λ_1/λ_2) and their orientations. By combining the two-dimensional strain data obtained in this way from three differently oriented sections, the values and orientations of the principal strains in three dimensions can be found (Sec. 4-7). The actual technique used for determining the two-dimensional strain state depends on how much is known about the original shape of the fossil and how many deformed specimens are available at any one locality. If the initial shape of the fossils is not known, it is generally impossible to determine the strained state from a single specimen unless the orientations of the principal axes of strain are known. If two or more specimens of deformed fossils are available and their relative position is known, the ratios and orientations of the principal axes of strain can be calculated. Most deformed fossils afford excellent material for the computation of strain, being essentially patternlike impressions in the rock. If they deform homogeneously with the surrounding rock material, they can be used as almost perfect strain gauges. Care must be taken in using larger fossils in this way since there may be a competence difference between the fossil and its rock matrix. Special techniques have to be devised to use these types for strain measurement.

The deformations recorded by the distorted fossils reflect the total change in shape of the enclosing rock since the dead organism was interred in the sediment. Some of these changes in shape may be the result of processes that went on before the development of tectonically produced strains (Ferguson, 1963). In some sediments (especially clays and marls) considerable volume loss often takes place as a result of compaction and loss of water during diagenetic changes, and these processes lead to a uniaxial strain ($k = 0$ type ellipsoid) with the shortest axis of the strain ellipsoid arranged perpendicular to the bedding surfaces. If the tests of dead organisms lie on the bedding surfaces there will be little or no change of shape as a result of this compaction; but if they are obliquely inclined to the bedding surfaces, then compaction may give rise to appreciable changes in length of lines and in angles within the fossil impression. In rocks which have suffered no tectonic strain these distortions can be used to calculate the amount of sediment compaction (Allen, 1946; Ferguson, 1962, 1963). Tectonic deformation will lead to further distortion of the fossil impressions; the finite-strain ellipsoid represents both compaction and tectonic strain in the manner described in Sec. 4-11.

5-12   FOSSILS WHICH DEFORM HOMOGENEOUSLY WITH THEIR MATRIX

Many fossils are strained homogeneously with the rock that encloses them, either because they are of an identical or nearly identical composition as the matrix or because they are sufficiently thin not to interfere appreciably with the rock competence. Deformed fossils of this type provide the best material for the computation of rock strain. The techniques that are employed depend on the shape of the fossil.

Originally circular disks   Fossils with this form (e.g., the ossicles of certain species of crinoids) can be used directly to determine the strain ratios and the orientations of the principal strain ellipse. If the disks originally had variable orientations in the sediment, then the states of strain on a number of cross sections through the rock can be determined and the ratios of the three principal strains computed (Sec. 4-7). Some crinoid ossicles have an originally elliptical cross section and, although their deformed shapes will also be ellipses, these forms will not be similar in shape and orientation to the strain ellipse. Where there are a number of deformed ossicles, it is possible to separate the original shape factor from the strain components in the manner described with nonspherical ooids (Sec. 5-6).

Originally circular tubes   Homogeneous deformation changes originally circular, cylindrical tubes into elliptical cylinders which are, unfortunately, less useful for determination of the strains than is sometimes realized. A cross section of the elliptical tube made perpendicular to the tube axis will not in general represent the shape of a strain ellipse (see Fig. 5-50) because it has been derived from an originally elliptical cross section of the circular cylinder. The only section that should be used is that which was the originally circular section of the circular cylinder, and this is only a true cross section of the elliptical cylinder when the tube axis is parallel to one of the three axes of the strain ellipsoid. If, therefore, the elliptical cylinder has no markers to indicate

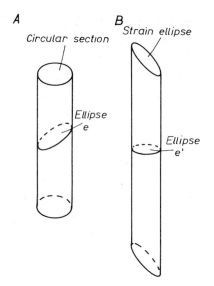

Figure 5-50

*An undeformed cylinder (part A) and the elliptical
cylinder (part B) derived from it by deformation.
The ellipse section e′ of the elliptical cylinder
normal to the tube axis is not a true strain
ellipse because it has been derived from an
originally elliptical cross section e of the
circular cylinder.*

the position of the initial circular section, no information on the strained state
can be inferred.

Fossils with bilateral symmetry   Many organisms found fossilized in rock
material had originally a bilateral symmetry. The line joining any two points
having mirror-image relationships was originally perpendicular to the axis of
bilateral symmetry. After deformation this may no longer be a right angle
and the divergence from the perpendicular state will give a measure of the
angular shear strain ψ for two directions in the deformed material. Where
many bilateral symmetric fossils are deformed by a common homogeneous
strain (Fig. 5-60, page 239), there are always two positions in the material
where deformed fossils still show a bilateral symmetry (Fig. 5-51*A* and *B*,
Fig. 5-60, forms *A* and *B*). At these orientations the line of symmetry in the
fossil must coincide with a line of symmetry in the strain ellipse, and it coincides with the direction of greatest or least extension in the material. In these
positions the form of the deformed fossil is known as *narrow* (Fig. 5-51*A*) or
broad (Fig. 5-51*B*). The recognition of these types is especially valuable for
the rapid determination of these maximum and minimum strain directions
in deformed assemblages of fossils. Where the symmetry line of the fossil does
not coincide with the major and minor axes of the strain ellipse, the deformed
fossil loses its bilateral symmetry and its form is known as *oblique* (Figs. 5-51*C*,
5-60*C*). Where the original line of symmetry (e.g., median line in the trilobite
shown in Fig. 5-51*C*) is deflected clockwise (or to the right) with respect to
the thoracic rings, the form is *right oblique*; if deflected anticlockwise, the
form is *left oblique*. Oblique forms are more common than symmetric forms.

Figure 5-51

Deformed pygidia of trilobites. A (narrow form) and B (broad form) are symmetric deformed forms where the principal elongations have coincided with the axis of bilateral symmetry of the original fossil. C is a right oblique form.

They can be used to determine the amounts and orientations of the greatest and least strains, but the calculations are more involved than with deformed symmetric forms.

The length of lines in fossil impressions is changed as a result of strain, and the actual method of manipulating data depends on what is known about the constancies or variations of lengths, or ratios of lengths, within the undeformed species. Some communities of fossil species may all have had about the same original size (Fig. 5-60), but it is more common for a given population of organisms to show a range of size. For example, a population of brachiopods may show a variation in the original lengths of their hinge lines (h_o) and median lines (m_o) about some mean values as indicated in Fig. 5-52, and this variation will be incorporated into the data from deformed fossils. Where there is such a variation, however, it is often found that the ratios of two perpendicular lengths in the fossil ($h_o/m_o = r_o$) may have a fairly constant value even though the fossils had variable sizes initially (Fig. 5-52). Sometimes this ratio is variable where the measurements are made on small or immature individuals (Fig. 5-53). The actual technique chosen to manipulate the available data on measured lines in deformed fossils depends on how much is known about the original

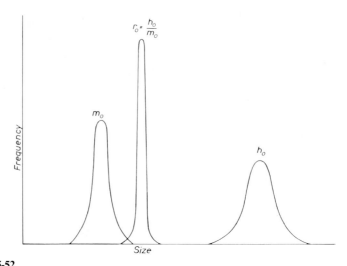

Figure 5-52

Frequency distribution for lengths of lines (h_o), median lines (m_o), and their ratios ($h_o/m_o = r_o$) for an original population of brachiopods.

community, or how much can be deduced from a close scrutiny of these measurements.

The various basic techniques for the investigation of deformed shapes of originally bilaterally symmetric forms will now be examined in detail. The first problem is how few data are essential for a complete determination of the state of strain in two dimensions, and this will be followed by a consideration of the modifications of these techniques where many deformed fossils are available at the same locality. In many of the computations for determining the maximum and minimum strains and their orientations, the Mohr circle enables particularly convenient and rapid solution of the strain equations.

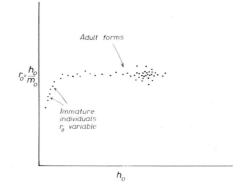

Figure 5-53

Variation in ratio r_o of fossils with size. During a period of growth the organism has a changing r_o ratio, then a more constant value is reached in mature individuals.

1. SINGLE OBLIQUE FORMS   If the directions of the principal axes of strain are known (Fig. 5-54), it is possible to calculate the ratios of the principal extensions by using the measurements of the angular shear strains for the distorted fossil. The angle θ' between the axis of original bilateral symmetry and the principal elongation direction is measured (Fig. 5-55B) and the angular shear strain ψ determined. The Mohr construction can now be applied directly (Sec. 3-8, construction 1, p. 74) to obtain a value for the ratio of the maximum and minimum strains. Arithmetic and graphic solutions to this problem are also simple to apply. Consider a fossil whose axis of symmetry initially makes an angle θ with the line which is to become the major axis of the strain ellipse (Fig. 5-55A). The angle between the perpendicular to the symmetry axis and the major axis is $(90° - \theta)$. From Eq. (3-34) we know how both of these angles become modified by straining: θ changes to θ' and the angle $(90° - \theta)$ changes to $(90° - \theta' - \psi)$. If λ_1 and λ_2 are the principal quadratic extensions in the surface, then

$$\sqrt{\frac{\lambda_2}{\lambda_1}} = \frac{\tan \theta'}{\tan \theta} \qquad \text{and} \qquad \sqrt{\frac{\lambda_2}{\lambda_1}} = \frac{\tan (90° - \theta' - \psi)}{\tan 90° - \theta'}$$

Multiplying these two equations together and using $\tan \theta \tan (90° - \theta') = 1$,

Figure 5-54

Specimens of the deformed trilobite Angelina showing the linear structure developed parallel to the axis of maximum finite elongation. By measuring the shearing strain in each fossil it is possible to make three independent computations of the strain ratio, λ_2/λ_1.

we obtain

$$\frac{\lambda_2}{\lambda_1} = \tan \theta' \tan (90° - \theta' - \psi) = \frac{\tan \theta'}{\tan (\theta' + \psi)} \qquad (5\text{-}35)$$

Arithmetic solutions to this equation to determine the strain ratio are simple and a convenient graphic solution after the method of Breddin (1956) is presented in Fig. 5-56. The numerical value of the angular shear strain for a particular direction is represented by a single point which falls somewhere in the field of curves of different λ_2/λ_1 values and defines the strain ratio (e.g., the data from Fig. 5-55b produce the point x on the graph showing that the strain ratio is 0.5). For each fossil there are two directions at angles θ' and ϕ' from the greatest extension direction which have the same numerical value of ψ, and when plotted (x and x') both should give the same ratio λ_2/λ_1. If they do not, then the direction of greatest extension has been incorrectly positioned and the problem is insoluble.

It will be apparent from the graphic solution that the clearest separation between the curves with different $(\lambda_2/\lambda_1)^{\frac{1}{2}}$ values occurs where the angular

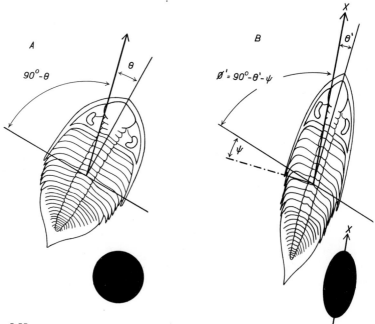

Figure 5-55

Changes in angle within an originally bilaterally symmetric fossil (A) as a result of strain (B). X is the direction of principal extensive strain, and ψ the angular shearing strain.

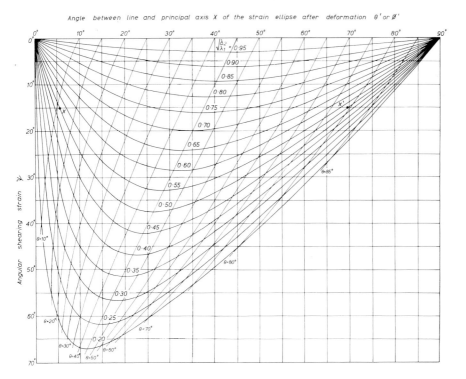

Figure 5-56

Graphical solution to Eq. (5-35) for the determination of the ratio of the minor and major extensions in deformed fossils. Points x and x' represent the plots of the data from the deformed trilobite of Fig. 5-55.

strain ψ is large. The above method is not very precise if the shearing strain is small and it cannot be used at all if the deformed fossil is symmetric ($\psi = 0°$).

2. SINGLE FORMS OF KNOWN RELATIVE SHAPE   The original size of a deformed fossil can never be determined precisely, but with many species the original ratio of the lengths of certain lines within the fossil may have had a fixed value. If this is so, it is possible to determine both the ratio and the orientations of the maximum and minimum extensions from a single deformed fossil. For example, let the ratio of the lengths of the hinge line and median line of an undeformed brachiopod species be $h_o/m_o = r_o$, while in a deformed fossil the measured ratio is $h/m = r$. The reciprocal quadratic extensions for the directions of the hinge and median lines are, respectively,

$$\left(\frac{h_o}{h}\right)^2 = \lambda'_h \qquad \left(\frac{m_o}{m}\right)^2 = \lambda'_m$$

Dividing one by the other and replacing h_o/m_o by r_o and h/m by r, we obtain the relationship between these two strain parameters:

$$\lambda'_h = \left(\frac{r_o}{r}\right)^2 \lambda'_m \tag{5-36}$$

Although we do not know the absolute reciprocal quadratic extensions of these lines, we can express one in terms of the other. In a similar way we can express the strain parameters λ'_h and λ'_m in terms of the same unknown, for

$$\gamma'_m = \lambda'_m \tan \psi \tag{5-37}$$

$$\gamma'_h = -\lambda'_m \left(\frac{r_o}{r}\right)^2 \tan \psi \tag{5-38}$$

The states of strain in the two directions may therefore be recorded as two points on a Mohr diagram with coordinates (λ'_m, γ'_m) and (λ'_h, γ'_h), all expressed in terms of the unknown multiplier λ'_m. The Mohr circle can now be constructed, the strain ratio λ_1/λ_2 determined from $\lambda'_2\lambda'_m/\lambda'_1\lambda'_m$, and the angle θ' of the hinge line to the principal extension determined from the angle $2\theta'$ in the Mohr diagram.

3. TWO DEFORMED FOSSILS Where two differently oriented, originally bilaterally symmetric forms lie in the same surface and have been strained homogeneously together, it is possible to compute the ratios and orientations of the maximum and minimum strains, even if the fossils are of different species. The two angular shearing strains ψ_A and ψ_B are measured for the two fossils (Fig. 5-57). If α' is the angle between the directions of like parts

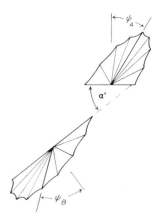

Figure 5-57

Two brachiopods deformed in a bedding surface.

(e.g., hinge lines), it follows from (5-35) that if θ' is the angle between one of these hinge lines and the principal elongation, then

$$\sqrt{\frac{\lambda_2}{\lambda_1}} = \frac{\tan \theta'}{\tan (\theta' + \psi_A)} \qquad (5\text{-}39a)$$

$$\sqrt{\frac{\lambda_2}{\lambda_1}} = \frac{\tan (\theta' + \alpha')}{\tan (\theta' + \alpha' + \psi_B)} \qquad (5\text{-}39b)$$

These two equations can be expanded and solved numerically for the two unknown quantities λ_2/λ_1 and θ'. The most convenient graphical solution employs a Mohr circle in the following way.

Consider two deformed brachiopods A and B illustrated in Fig. 5-57; determine the shearing strains ψ_A and ψ_B and the angle between the hinge lines α'.

1. Lines representing the λ' and γ' axes are drawn on a sheet of tracing paper (Fig. 5-58). Two lines are now set off, making angles of ψ_A and ψ_B with this λ' axis. The states of strain represented along the two hinge directions must lie somewhere along these two lines.

2. On a separate sheet of paper construct a circle of center C of any radius r (Fig. 5-58). Any two points A and B are selected on this circle such that ACB is $2\alpha'$. These two points will represent the states of strain along the hinge direction of the two fossils.

3. The next part consists of making the two earlier parts consistent. By placing the tracing paper over the Mohr circle and moving it around, it will be found that there is only one position [the one solution to Eq. (5-39)] where (1) C lies on the λ' axis, (2) the line ψ_A degrees away from the λ' axis passes through the point A on the Mohr circle, and (3) the line ψ_B degrees from the λ' axis passes through point B. It is now possible to determine the intersections of the circle on the λ' axis to define λ_1/λ_2, and the angle $2\theta'$

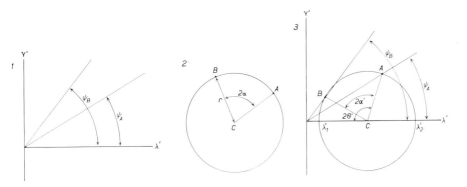

Figure 5-58

Mohr construction for determination of the strain state in Fig. 5-57.

will define the direction of the principal extension θ' degrees from the hinge line of fossil A (Fig. 5-58).

Another convenient graphical solution to the strain equations can be found using Fig. 5-56. On a piece of tracing paper two points are placed in their correct angular separation (α') so as to represent the shearing strains ψ_A and ψ_B of the two fossils (Fig. 5-59A). This tracing sheet is now placed over the graph (Fig. 5-56) so that the orientation axes coincide and is then slid along until the two points have the same $(\lambda_2/\lambda_1)^{\frac{1}{2}}$ ratio. There is only one position where this can be done (Fig. 5-59B), and the angle θ' defines the orientation of the principal extension from fossil B.

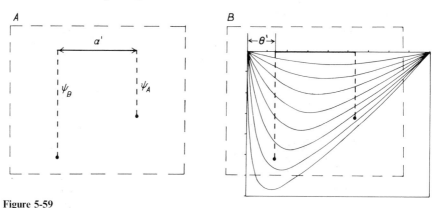

Figure 5-59

Graphical method for determining the state of strain in Fig. 5-57.

Once the values of the orientations of the maximum and minimum strains and their ratios have been calculated, it is possible to reconstruct the shape of the fossil in its undeformed state and the original ratio $h_o/m_o = r_o$. On the Mohr circle (Fig. 5-58) a line through C at an angle of $2(\theta' + \psi_A + 90°)$ from the λ' axis is plotted and the point where this cuts the circle records the state of strain of the median line of fossil A. The ratio λ'_h/λ'_m is now determined from the diagram. It follows that if $h/m = r$ is the ratio of hinge length to median line length in fossil A, the undeformed ratio r_o is given from (5-36)

$$r_o = r\left(\frac{\lambda'_h}{\lambda'_m}\right)^{\frac{1}{2}} \tag{5-40}$$

4. SEVERAL DEFORMED FOSSILS OF THE SAME OR DIFFERENT SPECIES   If a number of homogeneously deformed fossils of originally bilateral symmetry are contained in a single surface, it is possible to use the individuals two at a time to make a number of separate determinations of the maximum and minimum strain orientations and their ratios. It is often more convenient to analyze

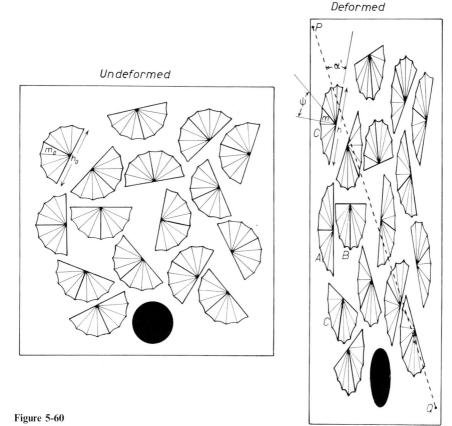

Figure 5-60

The shape changes in an assemblage of equal-sized, bilaterally symmetric fossils as a result of homogeneous strain. A is in the broad form, B the narrow form, and C the oblique form.

the distortion in another way by combining all the measurements into a synoptic diagram. These can be prepared in several ways, depending on the nature of the shape variation of the undeformed fossils.

If the fossils were all about the same size, then the variation in length of one particular feature with direction is graphically recorded. In the deformed material of Fig. 5-60 an azimuth line PQ is first drawn, and for each fossil the length of the hinge h is plotted against the angle α' between the hinge and the direction of PQ (Fig. 5-61). The same method has been applied to the median lines of the same fossil. The plots should fall on or about a particular line given from Eq. (3-31)

$$\frac{1}{\lambda} = \left(\frac{h_o}{h}\right)^2 = \frac{\cos^2 (\alpha' - \theta')}{\lambda_1} + \frac{\sin^2 (\alpha' - \theta')}{\lambda_2}$$

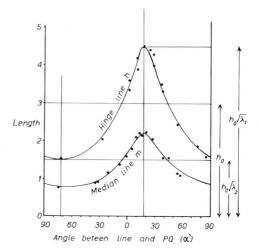

Figure 5-61

Graph of the changes in dimensions h and m in the fossils of Fig. 5-60 with change in orientation angle α'.

where θ' is the angle of the principal extension direction from PQ. If R is the ratio λ_1/λ_2, this equation becomes

$$h = h_o \left[\frac{\lambda_1}{1 + (R - 1) \sin^2 (\alpha' - \theta')} \right]^{\frac{1}{2}} \tag{5-41}$$

This result has maximum and minimum values where $\alpha' = \theta'$ and $\alpha' = \theta' + 90°$, respectively, and at these points the values of h are $h_o(\lambda_1)^{\frac{1}{2}}$ and $h_o(\lambda_2)^{\frac{1}{2}}$.

If h_o is variable this method does not provide a satisfactory means of determining the principal strain ratios and more accurate computations are made by determining the variations in the shearing strains in different directions within the surface, or by determining the changes in the ratios of lengths of originally perpendicular lines.

The use of variations in shear is valuable for determining the principal strains and their orientations, and the method is independent of the size or species of the deformed fossil. In Fig. 5-62 variation in angular shear ψ is plotted against the angle α' between the hinge lines and PQ. The points all fall on a curve whose equation is given by combining (3-31) and (3-42)

$$\tan \psi = \frac{\gamma'}{\lambda'} = \frac{(\lambda'_2 - \lambda'_1) \sin (\theta' - \alpha') \cos (\theta' - \alpha')}{\lambda'_1 \cos^2 (\theta' - \alpha') + \lambda'_2 \sin^2 (\theta' - \alpha')}$$

where θ' is the angle between the principal extension direction and PQ. Dividing top and bottom of the right-hand side by $\lambda'_1 \sin (\theta' - \alpha') \cos (\theta' - \alpha')$ and replacing λ'_2/λ'_1 by R, this simplifies to

$$\tan \psi = \frac{(R - 1) \tan (\theta' - \alpha')}{1 + R \tan^2 (\theta' - \alpha')} \tag{5-42}$$

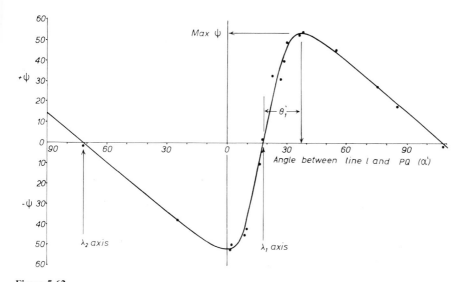

Figure 5-62

Graph of the variation in angular shear strain ψ with change in orientation angle α' from Fig. 5-60.

The two positions where this curve cuts the α' axis give the orientation of the maximum and minimum extensions. The strain ratio R can be found by discovering the curve of best fit to the observed data by standard statistical procedures or by comparison with the graphs of Fig. 5-56. Rapid approximations of the value of R can be found by determining the maximum angular shearing strain in the system and then applying the methods given on p. 198, employing the functions

$$R = \cot^2 \theta'_1 \tag{5-43}$$

$$R = 1 + 2 \tan^2 \psi_{max} + 2 \tan \psi_{max} \sec \psi_{max} \tag{5-3}$$

Another method uses the shapes of deformed fossils which still have bilateral symmetry. This technique can be used for the case in which the fossils were of an initially variable size providing their shape was constant. The ratio of the lengths of two originally perpendicular lines in the fossil is plotted against the angular shear of that fossil (Fig. 5-63) and the points plotted on a curve. Even if the truly bilateral symmetric forms are absent in the deformed community of fossils, extrapolation of this curve enables their shape to be estimated. The h/m ratios of the narrow and broad bilaterally symmetric forms have values, respectively,

$$\frac{h_o}{m_o} \left(\frac{\lambda_1}{\lambda_2} \right)^{\frac{1}{4}} \quad \text{and} \quad \frac{h_o}{m_o} \left(\frac{\lambda_2}{\lambda_1} \right)^{\frac{1}{4}}$$

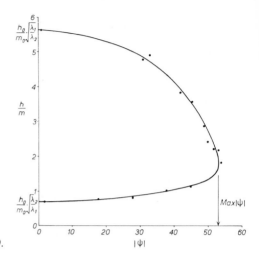

Figure 5-63

Plot of variation in h/m ratio against the numerical value of the angular shear in the deformed fossil assemblage of Fig. 5-60.

If the h_o/m_o value is constant, it follows that the tectonic strain ratio λ_1/λ_2 can be obtained by dividing one by the other, while the square of the original ratio h_o/m_o can be found by multiplying the two together.

An intriguing method of reconstructing the shape and orientation of the strain ellipse which involves no measurements or calculations has been described by Wellman (1962) and its application to the deformed fossil assemblage of Fig. 5-60 is illustrated in Fig. 5-64. Any line *pq* of finite length

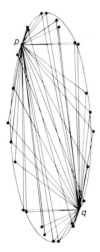

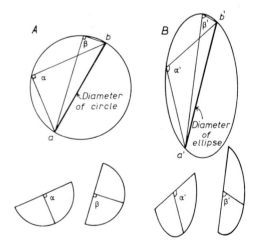

Figure 5-64

Wellman's graphical method of constructing the strain ellipse from the deformed fossils of Fig. 5-60.

Figure 5-65

The theoretical basis of Wellman's strain-ellipse construction.

is drawn so that it has the same orientation as line *PQ* drawn on the surface of the specimens. The directions of the median and hinge lines of each deformed fossil are projected through *p* and *q* so that they have their correct angular orientation as seen on the deformed surface. For each fossil two points of intersection of these lines are found, one in the mirror image of the other through *pq* (see dots in Fig. 5-64). This process is repeated for every deformed fossil. The resulting intersections define an ellipse which has the same shape and orientation as the strain ellipse. The theory that accounts for this surprising geometric fact is illustrated in Fig. 5-65. Consider the geometric relations of the two right angles α and β in two undeformed brachiopods as projected through the end points *a* and *b* of a diameter of a circle. Now let these fossils be deformed in the same way as the circle. The circle becomes an ellipse with *a'b'* as a diameter and the angles α' and β' depart from the perpendicular so that the apices of the two triangles lie on the ellipse.

Fossils without original bilateral symmetry Fossils without an originally bilateral symmetry are more difficult to use for determining the state of strain. The technique employed depends on how much is known about the shape of the undeformed fossil impression, and when undeformed material is not available a number of deformed specimens are needed before a complete solution of the state of strain can be made. Most of the techniques employed here involve determining the change in angles between linear features of the fossil, but methods involving the determination of changes in lengths and ratios of lengths similar to those described above may be applicable.

1. SINGLE SPECIMENS OF DEFORMED FOSSILS   If the changes in angle between two linear features in a single deformed fossil are known, together with the

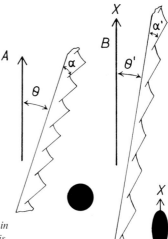

Figure 5-66

Change in the angle α as a result of strain in deformed Graptolites. A is undeformed, B is in the deformed state.

orientation of the principal extension, then the ratio of the maximum and minimum strains can be computed directly. Consider the changes between the directions of the stipe and thecae of the graptolite shown in Fig. 5-66; these can be expressed using the basic form of Eq. (3-34), where $\lambda_1/\lambda_2 = R$, by

$$\tan \theta = R^{\frac{1}{2}} \tan \theta' \tag{5-44}$$

$$\tan (\theta + \alpha) = R^{\frac{1}{2}} \tan (\theta' + \alpha') \tag{5-45}$$

Expanding (5-45) and expressing θ in terms of θ' from (5-44)

$$\frac{R^{\frac{1}{2}} \tan \theta' + \tan \alpha}{1 - R^{\frac{1}{2}} \tan \theta' \tan \alpha} = \frac{R^{\frac{1}{2}} (\tan \theta' + \tan \alpha')}{1 - \tan \theta' \tan \alpha'} \tag{5-46}$$

If α, α', and θ' are known, this equation is soluble for R.

If the values of two angles α and β between three directions in an undeformed fossil are known, together with their deformed derivatives α' and β', it is possible to calculate the orientations and ratio of the principal extensions. From Fig. 5-67 it follows that in addition to the expression (5-46)

$$\frac{R^{\frac{1}{2}} \tan \theta' + \tan (\alpha + \beta)}{1 - R^{\frac{1}{2}} \tan \theta' \tan (\alpha + \beta)} = \frac{R^{\frac{1}{2}} [\tan \theta' + \tan (\alpha' + \beta')]}{1 - \tan \theta' \tan (\alpha' + \beta')} \tag{5-47}$$

If α and β are known, Eqs. (5-46) and (5-47) involve only the two unknowns R and θ' and can therefore be solved. This is generally done most easily with a Mohr diagram. We construct a triangle ABC of any size but with angles

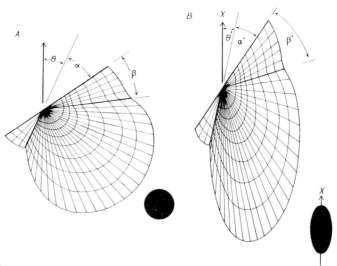

Figure 5-67

Changes in the angles α and β in a deformed lamellibranch as a result of strain. A is undeformed, B is deformed.

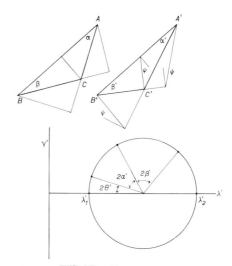

Figure 5-68

Mohr construction for the determination of the state of strain from the deformed lamellibranch of Fig. 5-67.

$BAC = \alpha$ and $ABC = \beta$, and another triangle $A'B'C'$ representing the deformed state of ABC with angles $B'A'C' = \alpha'$ and $A'B'C' = \beta'$. By dropping perpendiculars from A onto BC, B onto CA, and C onto AB, and discovering how they are distorted in triangle $A'B'C'$ using the methods described in Sec. 3-8, it is possible to calculate the shearing strains for the three directions $A'B'$, $B'C'$, and $C'A'$ (Fig. 5-68). Because the relative alteration in length of the sides of the triangle AB to $A'B'$, etc., are known, their relative reciprocal quadratic extensions can be determined. Thus it is possible to calculate a λ' and γ' value for each direction and then construct a Mohr circle from the three points representing the relative strain states along the three directions (Fig. 5-68).

2. TWO OR MORE DEFORMED FOSSILS   Where two differently oriented specimens of a single fossil species are found deformed on the same surface, a complete strain solution is possible if we know the undeformed angle (α) between two line elements, the two deformed derivatives (α' and α''), and the angle ϕ' between the fossils. This solution is possible because a further equation can be added to that of the single fossil example (5-46),

$$\frac{R^{\frac{1}{2}} \tan (\theta' + \phi') + \tan \alpha}{1 - R^{\frac{1}{2}} \tan (\theta' + \phi') \tan \alpha} = \frac{R^{\frac{1}{2}} \tan (\theta' + \phi') + \tan \alpha''}{1 - \tan (\theta' + \phi') \tan \alpha''} \tag{5-48}$$

These two equations can be solved completely for θ' and R. If only deformed material is available (that is, α is unknown), it is necessary to have three differently oriented fossils with known relative orientations before the strain state can be completely determined.

It has been shown that if the undeformed angles α and β between three line elements in a fossil are known, it is possible to determine the state of strain

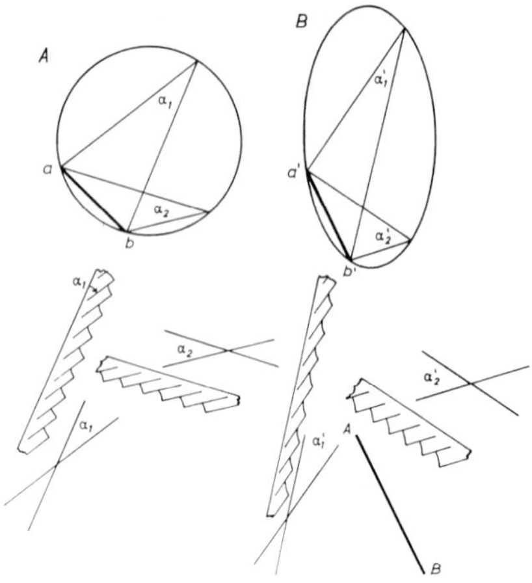

Figure 5-69

Wellman's graphical construction for determining the strain ellipse from deformed angles α_1 and α_2 not originally right angles. A represents the relationships in the undeformed state and B those in the deformed state.

from a single deformed fossil. If only deformed material is available and α and β are not known, it is necessary to have at least two fossils before the four equations essential for the complete solution of the four unknown quantities R, θ', α, and β can be obtained.

A graphical solution for the determination of the major and minor axes of the strain ellipse from a number of deformed specimens of fossils has been described by Wellman (1962); this solution requires no computation. An azimuth line AB is drawn on the specimen surface and a line $a'b'$ is drawn parallel to it on a sheet of paper (Fig. 5-69). The angles α'_1, α'_2, etc., are then transferred from the deformed fossils and projected through the points a' and b'. The two lines from each deformed fossil intersect in a point, and the series of points from a number of the fossils constructed in this way outline an ellipse with the line $a'b'$ as a chord of the ellipse. This ellipse is similar in shape to the strain ellipse and it has the same orientation of major and minor axes. The geometric principle used here is that a chord of a circle subtends equal angles α_1, α_2 at the circumference, and after deformation the

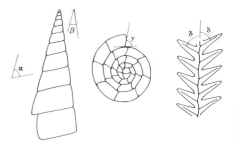

Figure 5-70

Types of angular features which may be used for strain computation in fossils which do not have original bilateral symmetry.

angles α_1' and α_2' subtended from this deformed chord vary in a way which describes the deformed circle, i.e., the strain ellipse.

Because there is a very great range of possible morphologic forms of fossil impressions, it is not possible to discuss all the various adaptations of the basic techniques discussed above to cover their application to all types of fossils. However, most of these adaptations depend on the analysis of the changes in angles between line elements within the impressions. Some of the types of features that might be used for distortion parameters are shown in Fig. 5-70, but great care must be taken that these angular features are constant in the undeformed species being used.

5-13 FOSSILS WHICH DEFORM INHOMOGENEOUSLY WITH THEIR MATRIX

Certain large fossils made up of material of a different composition or structure from that of the surrounding rock may not undergo a deformation homogeneous with their matrix. Deformed fossils of this type may be extremely useful for determining the *actual linear changes* that have gone on in the rock, and it may be possible to compute the values of the three principal strains and so determine volume change. The disrupted belemnites in certain deformed Mesozoic strata in the Swiss and French Alps exemplify these (Fig. 5-71) (Daubrée, 1876; Heim, 1878, 1919; Brace, 1960; Goguel, 1962; Badoux, 1963). The rostrum of a belemnite is made up of a radiating fibrous aggregate of calcite crystals with the long axes of the crystals arranged perpendicular to the length of the test. When a belemnite is subjected to an elongation, the

Figure 5-71

Extended and broken belemnite in a deformed Jurassic limestone. Fernigen, Switzerland. The strong linear fabric runs parallel to the direction of greatest elongation in the rock.

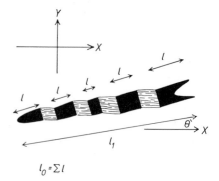

Figure 5-72

Measurements that are made on deformed belemnites in order to determine the quadratic extension ($\lambda = l_1^2/l_0^2$) in the direction θ' from the direction of greatest extension (X).

weakness in the structure enables the test to break up. The spaces between the separated parts often become infilled by interstitial liquids, and quartz and calcite crystallize from these liquids in long fibers arranged parallel to the direction of extension (Figs. 5-71, 5-72). The quadratic elongation (λ) in certain directions can be calculated by reconstructing the separate pieces of the belemnite (Fig. 5-72, $l_o = \Sigma l$) and comparing this length with their separated positions in the deformed rock (l_1), $\lambda = (l_1/l_0)^2$. If the quadratic extensions in two directions in the rock can be determined in this way, and if the orientation of the maximum elongation is known (e.g., from a "stretching" lineation, see Fig. 5-71), then a Mohr diagram (see Sec. 3-8, construction

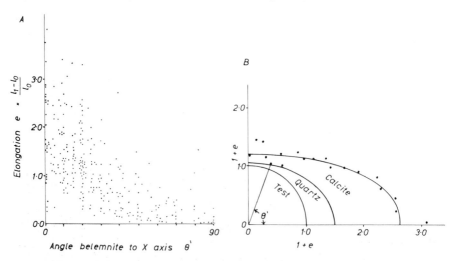

Figure 5-73

A. Variation in elongation of deformed belemnites with variation in angle θ' from the direction of greatest extension X. B shows how the data from A have been averaged to give the mean shape of the strain ellipse. (After Badoux, 1963.) The sectors labelled "calcite" and "quartz" show the proportions of these minerals in the extended zones between the broken belemnites.

3, for details) will enable the two principal extensions to be absolutely deter-mined. If the quadratic extensions for three different directions can be found, then the orientations of the strain-ellipse axes can be determined as well as the values of the maximum and minimum strains (Sec. 3-8, construction 4). Badoux (1963) has described the results of measurements of 300 deformed belemnites in Liassic slates from the Morcles nappe near Leytron, Switzerland (Fig. 5-73). He found that the extensions shown by these belemnites varied from 0 to 375 percent depending on their orientation; by determining the mean extension for certain directions, he was able to reconstruct the shape and orientation of the strain ellipse in these rocks.

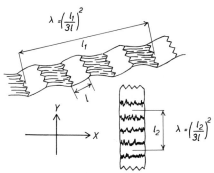

Figure 5-74

Deformation of ossicles of crinoid stems as a result of tectonic strain. Where the stems are aligned in a direction of extension, the ossicles are pulled apart; whereas if the stems are aligned along a direction of contraction, pressure solution takes place along the articulating surfaces between the ossicles.

A similar technique can be applied to deformed crinoid stems built up of independent ossicles (Figs. 5-74, 5-75) since the ossicles become separated one from the other if the crinoid stem is subjected to an extensive strain. When a compressive strain acts along the length of the crinoid stem, the stem may contract by the removal of calcite by pressure solution of the articulating surfaces between adjacent ossicles (Fig. 5-76). This process may lead to the development of stylolitic structures along the contacts of any two ossicles (Figs. 5-74).

Figure 5-75

Thin section of extended crinoid ossicles from Devonian limestones. Ilfracombe, N. Devon.

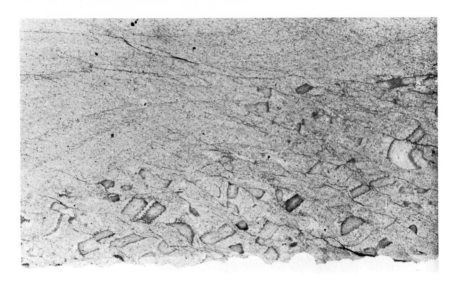

Figure 5-76

Crinoid ossicles which have undergone pressure solution along zones parallel to the slaty cleavage in a limestone. Ilfracombe. N. Devon.

5-14 DETERMINATION OF STRAIN FROM FOLDED AND BOUDINAGED VEINS AND OTHER FEATURES

Although a deformed rock may not contain ooids, fossils, or other objects useful for computing the state of strain, it may possess other features which can be used. Many metamorphic rocks contain ptygmatic veins, boudinage structures, etc., which can sometimes be used to determine the extensions in certain directions within the rock material (Flinn, 1961, p. 424). With ptygmatic veins, care must be taken to see that the folded layer does have a constant orthogonal thickness and that the shortening is measured in a true profile section perpendicular to the fold hinge (Morris and Fearnsides, 1926, p. 266). Measurements of the shortening shown by ptygmatic veins arranged in several directions in the rock (Fig. 5-77) might be used to estimate the amount of strain since their emplacement provided that all were developed simultaneously. When measuring extensions using boudinaged competent layers, allowance must be made for ductile extensions along the layer. Veins with pinch and swell should be avoided since the measurements of extensive strain on them will generally be too small.

The orientations of the primary structural features of igneous and sedimentary rocks such as polygonal columnar jointing in lava flows and dikes, mud cracks, worm tubes, and cross-bedding in deformed rocks may also be used to calculate the amount of angular distortion (Brace, 1961).

Figure 5-77

Synchronous ptygmatic veins in two directions in a gneiss.

REFERENCES AND FURTHER READING

Oolitic limestones, pisolites, vesicles, spots, concretions

Bucher, W. H.: On Oolites and Spherulites, *J. Geol.*, **26**: 593-609 (1918).

Chapman, F.: On Oolitic and Other Limestones with Sheared Structure Form, Ilfracombe, *Geol. Mag.*, **10**: 100-104 (1893).

Cloos, E.: Oolite Deformation in South Mountain Fold, Maryland, *Geol. Soc. Am. Bull.*, **58**: 843-918 (1947).

Cloos, E., and A. Hietenan: Geology of the Martic Overthrust, *Geol. Soc. Am. Spec. Papers* 35, 1941.

Green, J. F. N.: The Age of the Chief Instrusions of the Lake District, *Geol. Assoc. Proc.*, **28**: 1-30 (1917).

Heim, A.: "Untersuchungen über den Mechanismus der Gebirgsbildung," Schwabe, Basel, 1878.

Heim, A.: "Geologie der Schweiz," Tauchnitz, Leipzig, 1921.

Jones, O. T., and W. J. Pugh: A Multilayered Dolerite Complex, *Quart. J. Geol. Soc.*, **104**: 43-70 (1948).

Neuvonen, K. J., and A. S. I. Matisto: Some Observations on the Tectonics in the Tampere Schist Area, *Comm. Geol. Finland Bull.*, **142**: 79-86 (1948).

Sorby, H. C.: On the Origin of Slaty Cleavage, *New Phil. J. Edinburgh*, **55**: 137-148 (1853).

Sorby, H. C.: On the Application of Quantitive Methods to the Study of the Structure and History of Rocks, *Quart. J. Geol. Soc.*, **64**: 171-232 (1908).

Williams, H.: The Igneous Rocks of the Capel Curig District, *Liverpool Geol. Soc.*, **13**: 166-202 (1922).

Wilson, G.: The Tectonics of the Tintagel Area, North Cornwall, *Quart. J. Geol. Soc.*, **106**: 393-432 (1951).

Conglomerates. Original fabric and deformed fabric

Binder, R. C.: The Motion of Cylindrical Particles in Viscous Flow, *J. Appl. Phys.*, **10**: 711-713 (1939).

Brace, W. F.: Quartzite Pebble Deformation in Central Vermont, *Am. J. Sci.*, **253**: 129-145 (1955).

Bretherton, F. P.: The Motion of Rigid Particles in Shear Flow at Low Reynold's Number, *J. Fluid Mech.*, **14**: 284-304 (1962).

Burckhardt, C. E.: Geologie und Petrographie des Basodino-Gebietes (NW Tessin), *Schweiz. Mineral. Petrog. Mitt.*, **22**: 99-186 (1942).

Drescher, F. K.: Über Quartz Gefügereglung im Dattelquartzit von Krummendorf (Schlesien), *Mineral. Petrog. Mitt.*, **42**: 217-263 (1932).

Dziedzie, K.: The Geological Significance of the Orientation of Pebbles, *Geol. Sudetica*, **1**: 301-307 (1964).

Eirich F., H. Margaretha, and M. Bunzl: Untersuchungen über die Viskosität von Suspensionen und Lösungen, *Kolloid-Z.*, **75**: 20-37 (1936).

Elwell, R. W. D.: The Lithology and Structure of a Boulder Bed in the Dalradians of Mayo, Ireland, *Quart. Geol. Soc.*, **111**: 71-84 (1955).

Fairbairn, H. W.: "Structural Petrology of Deformed Rocks," Addison-Wesley Publishing Company, Inc., Reading, Mass., 1949.

Flinn, D.: On the Deformation of the Funzie Conglomerate, Fetlar, Shetland, *J. Geol.*, **64**: 480-505 (1956).

Flinn, D.: On Deformation at Thrust Planes in Shetland and the Jotunheim Area of Norway, *Geol. Mag.*, **98**: 245-256 (1961).

Glenn, J. W., J. J. Donner, and R. G. West: On the Mechanism by which Stones in Till Become Oriented, *Am. J. Sci.*, **255**: 194-205 (1957).

Goldschmidt, V. M.: Konglomeraterne in den Høifjeldskvartsen, *Norg. Geol. Undersokelse*, 77 (1916).

Günthert, A.: Beiträge zur Petrographie und Geologie des Maggia-Lappens (NW Tessin), *Schweiz. Mineral Petrog. Mitt.*, **34**: 1-159 (1954).

Hitchcock, E., C. H. Hager, and A. D. Hager: "Report on the Geology of Vermont," Claremont, New York, 1861.

Holmes, C. D.: Till Fabric, *Geol. Soc. Am. Bull.*, **52**: 1299-1354 (1941).

Jeffery, G. B.: "On the Motion of Ellipsoidal Particles Immersed in a Viscous Fluid," *Proc. Roy. Soc. (London)*, *Ser. A*, **102**: 161-179 (1922).

Johnson, W. A.: Imbricated Structure in River Gravels, *Am. J. Sci.*, **4**: 387-390 (1922).

Krumbein, W. C.: Preferred Orientation of Pebbles in Sedimentary Deposits, *J. Geol.*, **47**: 673-706 (1939).

Kvale, A.: Petrofabric Analysis of a Quartzite from the Bergsdalen Quadrangle, Western Norway, *Norsk Geol. Tidsskr.*, **25**: 193-215 (1945).

Mason, S. G.: The Flocculation of Pulp Suspension, *Tappi*, **33**: 440-444 (1950).

McCallie, S. W.: Stretched Pebbles from Ocee Conglomerate, *J. Geol.*, **14**: 55-59 (1906).

Mehnert, K. R.: Die Meta-Konglomerate des Wiesenthaler Gneiszuges im sächsischen Erzgebirge, *Mineral. Petrog. Mitt.*, **50**: 194-272 (1939).

Oftedahl, C.: Deformation of Quartz Conglomerates in Central Norway, *J. Geol.*, **56**: 476-487 (1948).

Peach, B.: The Geology of Beinn Wyvis, Carn Chuinneag, Inchbae and the Surrounding Country, *Geol. Surv. Scotland Mem.*, 1912.

Pettijohn, F. J.: Conglomerate of Abram Lake, Ontario, and Its Extensions, *Geol. Soc. Am. Bull.*, **45**: 479-506 (1934).

Ramsay, A. C.: The Geology of North Wales, *Geol. Surv. Gr. Brit. Mem.*, 1881.

Ramsay, D. M.: Deformation of Pebbles in Lower Old Red Sandstone Conglomerates Adjacent to the Highland Boundary Fault, *Geol. Mag.*, **101**: 228-248 (1964).

Ramsay, J. G.: Structural Investigations in the Barberton Mountain Land, Eastern Transvaal, *Geol. Soc. S. Africa Trans.*, **66**: 353-401 (1965).

Runner, J. J.: Pre-Cambrain Geology of the Nemo District, Black Hills, South Dakota, *Am. J. Sci.*, **28**: 353-372 (1934).

Sander, B.: "Einführung in die Gefügekunde der geologischen Korpu," Springer-Verlag OHG, Vienna, 1948.

Scheumann, K. K.: Konglomerattektonite und ihre Begleitgesteine in der epizonalen Schiefer-schotte südlich von Strehlen in Schlesien, *Mineral. Petrog. Mitt.*, **48**: 325-372 (1936).

Strand, T.: Structural Petrology of the Bygdin Conglomerate, *Norsk Geol. Tidsskr.*, **24**: 14-31 (1944).

Tavener-Smith, R.: A Deformed Katanga Conglomerate in Northern Rhodesia, *Geol. Soc. S. Africa Trans.*, **65**: 177-192 (1962).

Taylor, G. I.: The Motion of Ellipsoidal Particles in a Viscous Fluid, *Roy. Soc. (London)*, *Ser. A*, **103**: 58-61 (1923).

Trevelyan, B. J., and S. G. Mason: Particle Motion in Sheared Suspensions, 1-Rotation, *J. Colloid Sci.*, **6**: 354-367 (1951).

Turner, F. J.: Lineation, Symmetry and Internal Movement in Monoclinic Tectonite Fabrics, *Geol. Soc. Am. Bull.*, **68**: 1-18 (1957).

Twenhofel, W. H.: "Principles of Sedimentation," McGraw-Hill Book Company, New York, 1939.

Zingg, T.: Beitrag zur Schotteranalyse, *Schweiz. Mineral. Petrog. Mitt.*, **15**: 39-140 (1935).

Fossil deformation

Allen, P.: Notes on Wealden Fossil Soil Beds, *Proc. Geologists' Assoc. (Engl.)*, **57**: 303-314 (1946).

Badoux, H.: Les bélemnites tronçonnées de Leytron (Valais), *Bull. Lab. Geol., Mineral. Geophys. Musee Geol. Univ. Lausanne*, **138**: 1-7 (1963).

Baumberger, E.: Bivalven aus dem Stampien des Vorarlbergs mit besonderer Berücksichtigung des Deformationsproblems, *Eclogae Geol. Helv.*, **30**: 361 (1937).

Brace, W. F.: Analysis of Large Two-dimensional Strain in Deformed Rocks, 21*st Intern. Geol. Congr., Copenhagen*, **18**: 261-269 (1960).

Brace, W. F.: Mohr Construction in the Analysis of Large Geological Strain, *Bull. Geol. Soc. Am.*, **72**: 1059-1080 (1961).

Breddin, H.: Die tektonische Deformation der Fossilien im Rheinischen Schiefergebirge, *Deut. Geol. Ges. Z.*, **106**: 227-305 (1956*a*).

Breddin, H.: Tektonische Gesteinsdeformation im Karbongürtel Westdeutschlands und Süd-Limburgs, *Deut. Geol. Ges. Z.*, **107**: 232-260 (1956*b*).

Breddin, H.: Tektonische Fossil und Gesteinsdeformation im Gebiet von St. Goarhausen, *Decheniana*, **110**: 289-350 (1957).

Breddin, H.: Funde tektonisch deformierter Fossilien von der Zeche Mathias Stinnes in der Emscher Mulde und ihre Bedeutung für die Tektonik und Paläontologie des Ruhrkarbons, *Glückauf*, **94**: 1095-1101 (1958).

Breddin, H.: Die tektonische Deformation der Fossilien und Gesteine in der Molasse von St. Gallen (Schweiz), *Geol. Mitt. Aachen*, **4**: 1-68 (1964).

Bryan, W. H., and O. A. Jones: Radiolaria as Critical Inductors of Deformation, *Univ. Queensland Papers, Dept. Geol.*, **4**: 1-6 (1955).

Cloos, E.: Oolite Deformation in the Southern Mountain Fold, Maryland, *Geol. Soc. Am. Bull.*, **58**: 843-918 (1947).

Dale, T. N.: Structural Details in the Green Mountain Region in Eastern New York, *U.S. Geol. Surv. Bull.*, **195**: 22-44 (1902).

Daubrée, G. A.: Expériences sur la schistosité des roches et sur les déformations des fossiles, corrélatives de ce phénomène, *Compt. Rend. Acad. Sci., Paris*, **82** (13): 710 and **82** (15): 798 (1876).

Dufet, H.: Note sur les déformations des fossils contenues dans les roches schisteuses et sur la détermination de quelques espèces du genre Ogygia, *Ann. Sci. Ecole Normale Superieure, Paris*, 2nd Ser., **4**: 183-190 (1875).

Engels, B.: Zum Problem der tektonishen Verformung der Fossilien im Rheinischen Schiefergebirge, *Deut. Geol. Ges. Z.*, **106**: 306-307 (1956).

Ferguson, L.: Distortion of *Crurithyris urei* (Fleming) from the Visean Rocks of Fyfe, Scotland, by Compaction of the Containing Sediment, *J. Paleontol.*, **36**: 115-119 (1962).

Ferguson, L.: Estimation of the Compaction Factor of a Shale from Distorted Brachiopod Shells, *J. Sediment, Petrol.*, **12**: 796-798 (1963).

Furtak, H., and E. Hellermann: Die tektonische Verformung von pflanzlichen Fossilien des Karbons, *Geol. Mitt., Aachen*, **2**: 49-69 (1961).

Goguel, J.: "Tectonics," W. H. Freeman and Company, New York, 1962.

Gräf, I.: Tektonisch deformierte Fossilien aus dem Westfal der Bohrung Rosenthal im Erkelenzer Steinkohlenrevier, *N. Jb. Geol. Pal. Mh.*, **2**: 68-95 (1958).

Harker, A.: On Slaty-cleavage and Allied Rock Structures, *Rept. Brit. Assoc. Sci.*, 813-852 (1884).

Haughton, S.: On Slaty Cleavage and the Distortion of Fossils, *Phil. Mag., 4th Ser.*, **12**: 409-421 (1856).

Heim, A.: "Geologie der Schweiz," Tauchnitz, Leipzig, 1919.

Hellmers, H.: Crinoidenstillglieder als Indikatoren der Gesteinsdeformation, *Geol. Rundsch.*, **44**: 87-92 (1955).

Hills, E. S., and D. E. Thomas: Deformation of Graptolites and Sandstones in Slates, Victoria, Australia, *Geol. Mag.*, **81**: 216-222 (1944).

Jannettaz, E.: Mémoir sùr les clivages des roches, *Geol. Soc. France Bull.*, **12**: 211-236 (1884).

Ladurner, J.: Zur Kenntnis des Gefüges "gestreckter" Belemniten, *Mineral. Petrog. Mitt.*, **44**: 479-494 (1933).

Lake, P.: Restoration of the Original Form of Distorted Species, *Geol. Mag.*, **80**: 139-147 (1943).

Phillips, J.: On certain Movements in the Parts of Stratified Rocks, *Brit. Assoc. Adv. Sci.*, 60-61 (1843).

Phillips, J.: Report on Cleavage and Foliation in Rocks, *Brit. Assoc. Sci.*, **269**: 60-61 (1857).

Renevier, E.: Résumé des travaux de Mr. D. Sharpe sur le clivage et la foliation des roches, *Bull. Soc. Vaudoise Sci. Nat.*, **4**: 379-384 (1856).

Rutsch, R. F.: Die Bedentung der Fossil-Deformation, *Verhandl. Schweiz. Petrog. Geol. Ingr. Bull.*, **15**: 5-18 (1949).

Sdzuy, K.: Uber das Entzerren von Fossilien, *Palaont*, **36**: 275-284 (1962).

Sharpe, D.: On Slaty Cleavage, *Quart. J. Geol. Soc.*, **3**: 74-104 (1846).

Sorby, H. C.: On Slaty Cleavage as Exhibited in the Devonian Limestones of Devonshire, *Phil. Mag.*, **11**: 20-37 (1855).

Voll, G.: New Work on Petrofabrics, *Liverpool–Manchester Geol. J.*, **2**: 503-567 (1960).

Wellman, H. W.: A Graphical Method for Analysing Fossil Distortion Caused by Tectonic Deformation, *Geol. Mag.*, **99**: 348-352 (1962).

Wettstein, A.: Über die Fischfauna des Tertiären Glarner Schiefers, *Schweiz. Paläont. Ges. Abhandl.*, **13**: 1-101 (1886).

6 Relationship between stress and strain

To study the development of the structures in naturally deformed rocks, the geologist must consider how rocks behave when they are subjected to stress. The behavior of real rock materials is often complex for a number of reasons. Most rocks are generally initially heterogeneous (the individual parts dissimilar) and anisotropic (properties not the same in all directions). Even if they are initially homogeneous and isotropic, a fabric anisotropy is likely to develop when they suffer large strains. Most of the classical theoretical principles of the stress-strain relationships have been built up on the assumption that the deforming materials are both homogeneous and isotropic. The results of these analyses may not, therefore, be exactly applicable to the behavior of real materials although they probably offer a reasonable first approximation.

During the prolonged tectonic activity of orogenic processes, the values of the principal stresses and the relative orientation of their axes change during the progress of time. These changes mean that the incremental strains resulting from the various stress states are generally superimposed to build up the finite strains in a complex manner.

Many factors can influence the stress-strain relationships and it seems likely that most of these parameters play a significant role in geological processes. The most important are (1) the confining pressure, (2) the rate of strain, (3) the temperature, and (4) the nature of the chemical environment. It is well known that under the temperatures and confining pressures

Figure 6-1

Folds produced by ductile flow. Pennine Alps, Ticino, Switzerland.

that exist even at relatively moderate depths in the crust, the strongest of rocks will flow and undergo large permanent strain (Fig. 6-1), while rocks near to the surface or those subjected to rapid strain rates may buckle and be fractured (Fig. 6-2).

One of the most profitable methods of investigating the rheological (from the Greek *rheo* meaning *flow*) properties of rocks under a variety of environmental conditions has involved laboratory experiments as an extension of a rigorous theoretical analysis. Experiments on these lines started at the end of the nineteenth century by determining such elementary properties as the crushing strength of rocks. The importance of the information accruing from these techniques soon became apparent, and with improved apparatus and instruments a considerable amount of data on the behavior of rocks resulted. The methods and results of these experiments will now be outlined, while a more analytical approach to the theoretical aspects of the deformation will be undertaken in later sections of the chapter.

6-1 BEHAVIOR OF ROCKS UNDER EXPERIMENTAL CONDITIONS

Experiments of short duration Most of the data presently available on the behavior of deformed rocks under stress have been determined using a

Figure 6-2

Brittle fracture in folded chert layers. Barberton area, Transvaal, South Africa.

triaxial testing rig. With this rig it is possible to control the confining pressure during the application of a uniaxial compressive or tensile stress. Thus the three principal stresses are all under the control of the operator, but two of these stresses are always of equal value. Cylinders of rock are jacketed in copper to prevent the external pressurizing liquid from entering the specimen; then they are placed in a thick-walled bomb, subjected to a known confining pressure, and loaded by means of a piston. The temperature is controlled by means of an external furnace, and various fluids may be incorporated inside the jacket around the specimen. The relationship between the applied stress and the strains can be easily determined and graphed (Fig. 6-3), although the sensitivity of the stress-strain record is not uniformly accurate throughout the course of the experiment. Using the triaxial rig it has been possible to record the behavior of all the main rock types of interest to the geologist over a great range of temperature and pressure. The behavior is found to vary considerably with rock type, temperature, confining pressure, and strain rate.

The types of stress-strain curve obtained for a number of rock materials are rather similar to those obtained during the testing of metals, and the metallurgical terminology for the description of deformation has been taken over almost in its entirety.

The stress-strain curves from the first stages of deformation are generally almost straight and with steep slopes, implying that only small strains are

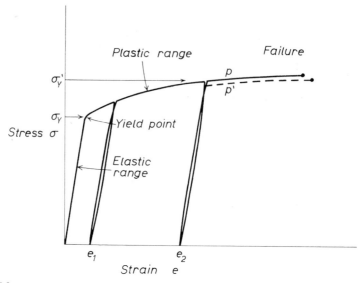

Figure 6-3

Generalized stress-strain relationships in an elastic-plastic material.

developed for a large increment in stress (Fig. 6-3). These stages of deformation are the *elastic region* in which there exists a linear or nearly linear stress-strain relationship. If the stresses are removed, the original dimension of the material is completely recovered. In a number of porous sedimentary rocks, however, it is often found that this initial strain is not fully recoverable because of the irreversible collapse of the pore spaces. The rock may suddenly fracture while in the elastic range; this effect is known as *brittle failure* and the value of the stress at this point is known as the *brittle strength*. Brittle strength varies with the type of stress applied to the specimen (whether tensile, compressive, or shear). The criteria that determine the brittle strength and the orientations of the fracture surfaces will be discussed in Sec. 6-10.

   If the material is not brittle, then the slope of the stress-strain curve diminishes. If the stresses are now removed, the stress-strain curve does not return to the origin but intersects the e axis at e_1 (Fig. 6-3). The specimen has therefore been *permanently strained* because the elastic limit has been exceeded. The point where this limit is first exceeded is known as the *yield point*, and the stress σ_Y at which this occurs is known as the *yield stress* or *yield-point stress*. The deformation which occurs at stresses at or above the yield point and which leads to permanent strain is known as *plastic deformation*. The yield point may be sharp, situated at a sudden change of slope of the stress-strain curve, but in practice it is often difficult to locate with precision. The magnitude of the yield stress in any one rock type depends on a number of environmental

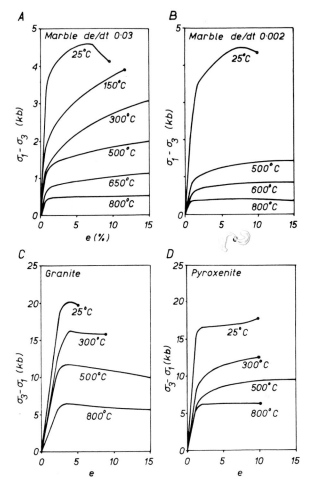

Figure 6-4

Stress-strain curves from triaxial tests made on various rock materials. A and B, Yule marble in extension; C, granite in compression; D, pyroxenite in compression: all with 5 kilobars confining pressure. (After Griggs, Turner, and Heard, 1960.)

conditions, it decreases with increasing temperature (Fig. 6-4), increases with increasing confining pressure (Fig. 6-5), and also increases with increasing strain rate (Fig. 6-4A, B). At the yield stress the material may deform continuously in a ductile manner with a zero slope of the stress-strain curve (Fig. 6-5B). The rock is then known as a *perfectly plastic material*. At low to moderate temperatures, however, the stress-strain curves from most deformed rocks have a small positive slope (Figs. 6-3, 6-4A, B). This means that if plastic deformation is to proceed, the stresses have to be increased above the initial yield stress,

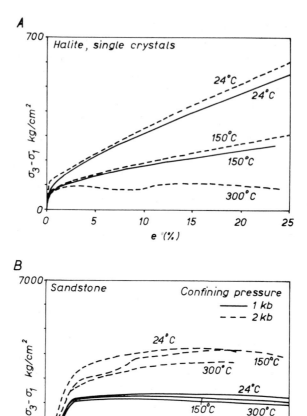

Figure 6-5

Stress-strain curves from triaxial tests on halite and sandstone, deformed under dry conditions at different temperatures and confining pressures. (After Handin and Hager, 1958.)

a process known as *strain hardening* or *work hardening*. If, after a certain amount of strain-hardened plastic deformation, the stresses are removed, the stress-strain curve falls in an almost linear fashion to the *e* axis at a point which indicates the total permanent strain e_2 (Fig. 6-3). If the same stresses are immediately reapplied, the stress-strain curve almost retraces its previous path back to the position of the plastic deformation curve. If they are now increased, the stress-strain relations continue to be those of the original plastic curve (Fig. 6-3, curve *p*); i.e., the rock behaves as if it has an increased elastic range and increased yield stress σ'_Y. Strain hardening may therefore be envisaged as a continuous increase in the yield strength with the progress of deformation. If after the removal of stresses some time elapses before they are reapplied, the new yield point generally lies beneath the previous plastic curve (Fig. 6-3, curve p'). Under experimental conditions the material generally ruptures some way along the plastic curve; the stress at which this occurs has been termed the *ultimate strength* of the material.

Other terms are used to describe the state of the crystal fabric in deformed polycrystalline aggregates. If the permanent deformation takes place at low temperatures and involves relative displacement of individual grains accompanied by crushing, fracturing, and mechanical granulation, the process is known as *cataclastic flow*. If the individual crystals do not become crushed, but undergo distortion, bending, twinning, and the development of slip or deformation bands, the process is called *cold working*. As a result of cold working, the fabric of the original rock becomes changed because of the realignment of the crystal particles, but no new grains are formed. *Hot working* occurs if the temperature is sufficiently high for the material to behave in an almost perfectly plastic manner. The internal distortions (dislocations) that exist inside the crystal grains of a cold-worked material may be cleared if the rock is heated to a moderate temperature. This process of *recovery* and *polygonization* leads to the formation of a number of unstrained *subgrains* from the original distorted crystals which have a slightly different optical orientation from the parent crystal. At higher temperatures unstrained crystals develop around new nuclei and replace the deformed grains of the cold-worked fabric. Prolonged *annealing* of this type may lead to the growth of large new grains and the complete transformation of the original fabric.

The process of polygonization and annealing recrystallization exert important controls on the reconstruction of the crystalline fabric of deformed metamorphic rocks (Griggs, Turner, and Heard, 1960).

In laboratory experiments of short duration it is found that practically all the rock materials are brittle at conditions of room temperature and atmospheric confining pressure. As the confining pressure is increased the rocks become stronger and fracturing is inhibited, but they also become much more brittle. Some sedimentary rocks can flow at relatively low confining pressures (Goguel, 1948). Heard (1960) has made a very thorough investigation of the conditions which control the transition from brittle to ductile behavior over a considerable range of pressure and temperature (0 to 5 kilobars, 24 to 600°C). The behavior of sedimentary rocks generally varies greatly with their composition, texture, and purity (Handin and Hager, 1957, 1958). Coarse-grained igneous rocks (peridotite, pyroxenite, granite) are generally much stronger than sedimentary rocks, and increases in temperature have a very marked effect in reducing the yield stress (Fig. 6-4*C*, *D*) (Griggs, Turner, and Heard, 1960).

Creep tests of long duration   One of the main difficulties in interpreting the results of the triaxial tests in geological terms is that the experiments only take a few minutes or hours to perform. In geological processes time is an extremely important variable, and orogenic deformations may take many millions of years to complete. With the triaxial tests it has already been pointed out that a decrease in the strain rate lowers the stress at which yielding takes place. It is therefore important to discover how this lowering of the yield stress goes on at very low rates of strain. It is well known that many apparently solid materials deform very slowly at room temperature if subjected to a stress which may be only a fraction of their brittle strengths. This time-dependent deformation known as *creep* often sets up important problems for the engineer, and the investigation of the behavior of rocks under these conditions is of vital significance in the interpretation of natural deformation. There is a great deal of scope for further experimental work on creep in rocks. For these results to be applicable to geological problems the duration of the tests must be as long as possible, preferably in the order of months or years.

Creep experiments are generally performed by measuring the deflection of beams of rock subjected to a constant stress or by loading cylindrical samples of rock by using a system of levers (Price, 1964). The strains recorded in these experiments are generally very small and are measured in units of microstrain

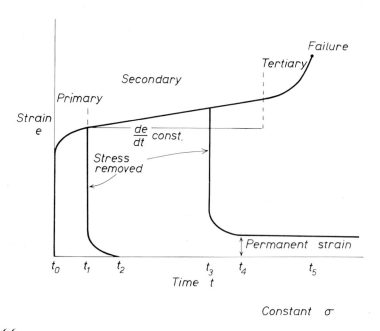

Figure 6-6

Generalized strain-time curve from a creep test in an elastic-plastic material.

$(e/10^6)$. In consequence the strain-measuring devices must be extremely sensitive, and generally optical methods are used. The experiments also have to be performed under carefully controlled conditions of temperature ($\pm 1.5°C$) and humidity (specimens coated with bitumen).

The results of these experiments are generally presented in graphical form, plotting the variation in strain against time under constant-stress conditions (Fig. 6-6). At the first application of the stress (time t_0) the rock undergoes an immediate elastic strain. Then follows a stage known as *primary* or *transient* creep in which the strain rate is fairly rapid at first but gradually decreases. This stage represents a *delayed elastic creep* deformation because if the stress is removed (Fig. 6-6, time t_1) there is an instantaneous (but incomplete) recovery followed by a decelerating recovery which is complete at time t_2. After the initial stage of primary creep there follows a stage where the rate of strain de/dt is approximately constant. This is known as the stage of *secondary steady state* or *pseudoviscous creep*. The rock deforms plastically, and if during this stage the stress is removed (time t_3), then, after an instantaneous and delayed elastic recovery (time t_4), there remains a permanent strain. The final stage of creep is known as *tertiary* or *accelerating* creep; the strain rate increases and eventually the material fails (time t_5). If experiments are performed at constant temperature but at various different levels of applied stress, the series of creep curves so obtained indicates that the rate of strain in the stage of steady-state creep varies but that there is some limiting stress at which the strain rate is zero (Fig. 6-7). This indicates that rocks have a

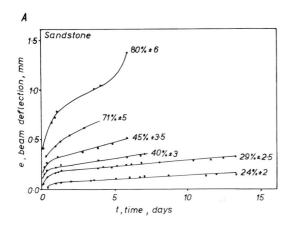

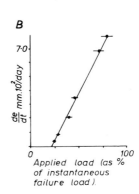

Figure 6-7

Strain-time curves from creep tests on sandstone made at different stress levels expressed in terms of the instantaneous failure stress. Part B illustrates the extrapolation from Part A for determination of the limiting stress below which the rock behaves as an elastic solid. (After Price, 1964.)

stress threshold or *creep strength* below which they behave as solids and above which they flow plastically, behaving as elastoviscous materials.

One of the most remarkable features of creep curves is that they are broadly similar for a great range of materials of quite different makeup (e.g., rock, asphalt, metals, polymers, rubber). The precise mechanism of deformation must vary with each of these different materials but the bulk behavior takes a much more restricted form.

A number of attempts have been made to fit equations to the creep curves (Kennedy, 1962). Most of these are empirical; the simplest equation that can represent the curve with reasonable precision is that given by Weaver (1936) and Griggs (1939),

$$e = a + b \log t + ct \qquad (6\text{-}1)$$

where e is the total strain, t is time, and a, b, and c are constants depending on the material, the temperature, and the value of the applied stress. Other equations based on exponential functions will be derived from model theory (Sec. 6-7).

Controlled strain-rate experiments   Experiments on deformation of marble with a constant rate of strain have been described by Heard (1963). These tests were carried out at temperatures of 25 to 500°C and at strain rates varying from 10^{-1} to 10^{-8} per second. The stress on the rock specimen was recorded throughout the deformation, and the results presented by plotting

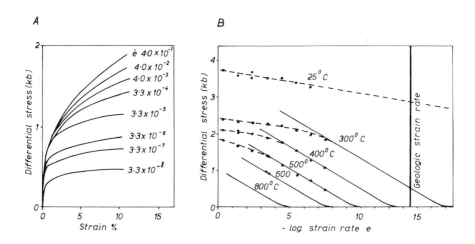

Figure 6-8

Stress-strain curves for deformation of Yule marble performed under a constant strain rate ($\dot{e}$). Part B illustrates the method of extrapolation to determine the behavior at geologically representative strain rates. (After Heard, 1963.)

stress against total strain (Fig. 6-8*A*). These curves show first an initial elastic deformation followed by a period of strain-hardening plastic flow.

The slowest strain rate used in these experiments (10^{-8} per second) was considerably faster than the representative geological rate of 3×10^{-14} per second determined by Whitten (1956) from measurements of the San Andreas fault zone. Heard used the steady-state flow equations proposed by Ree, Ree, and Eyring (1960) in order to extrapolate the experimental data into the field of geologically likely strain rates:

$$\dot{e} = \dot{e}_o \exp\left(\frac{-E}{RT}\right) \sinh \frac{\sigma}{\sigma_o}$$

where $\dot{e}$ is the strain rate, E the activation energy of diffusion, R the gas constant, T the absolute temperature, σ the stress difference applied to the rock specimen, and $\dot{e}_o$ and σ_o constants with the dimensions of strain rate and stress, respectively. At low temperatures where the rocks showed strain hardening, the experimental data diverged from these curves; but at higher temperatures a steady-state flow was attained and the experimental data closely fitted the theoretical curves (Fig. 6-8*B*). Extrapolation indicates that at geologically representative strain rates Yule marble at 300°C has a fundamental strength of 70 to 500 bars depending on the orientation of the principal axes of stress to the original fabric, and that at stresses above yield, slow steady flow would go on at equivalent viscosities ranging from 10^{23} poises at 25°C to 10^{16} poises at 500°C.

6-2 DEFORMATION OF POLYCRYSTALLINE AGGREGATES

The theoretical principles of elastic and plastic behavior which will be developed in later sections of this chapter are built up on an essentially macroscopic view of the material, a view which is broad enough for all the various structural discontinuities in the rock to be smoothed out. These theories do not attempt to explain the various deformation phenomena in terms of the properties of the various components which make up the rock aggregate. It is therefore important to have some idea of these actual processes, but as yet no one has been able to explain in quantitative terms the relationship between stress and strain in the rock bulk and the deformation mechanisms of the individual crystal components. Many problems of the growth of crystals under stress still remain uninvestigated.

Great advances have been made over the past decades in the study of the behavior of stressed crystals and crystal aggregates from a metallurgical viewpoint (Cottrell, 1948, 1953; Barrett, 1952; Read, 1951). A completely convincing application of these ideas to the structure of silicate minerals awaits development, but interesting and pertinent discussions on the application of these principles to the interpretation of the textures of deformed metamorphic rocks has been made by Voll (1960). The difficulty of direct applications of these principles is that they have been developed for the rather simple structural forms of metals (body-centered cubic and hexagonal systems), whereas most of the rock-forming minerals of interest to the geologist have a much lower symmetry.

All crystalline materials possess the property of cohesion resulting from the bonding forces that exist between the atoms of the crystal lattice. When the crystal lattice is subjected to an externally applied stress, then the various atomic spacings are slightly changed by an amount which is directly proportional to the value of the stress (Fig. 6-9A), while on removal of the stress the original distances are recovered. The initial elastic behavior of rocks where the stress and strain show a linear relationship probably reflects the stability of the interatomic bonding of the component parts of the crystal aggregate.

The plastic behavior of the material when subjected to stresses equal to or greater than the initial yield stress cannot be explained by simple dilation of the crystal lattices. If the individual crystals in a cold-worked material are viewed under an electron microscope, it is possible to see parallel lines which appear to be traces of regularly oriented slip surfaces. The mechanism of

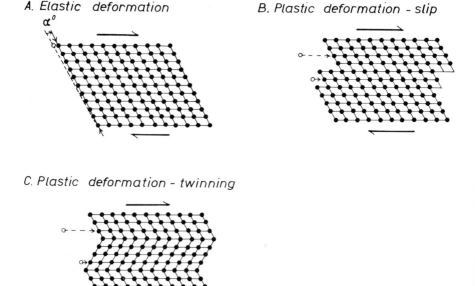

A. Elastic deformation

B. Plastic deformation - slip

C. Plastic deformation - twinning

Figure 6-9

Behavior of crystal lattices under conditions of elastic and plastic deformation. Elastic deformation leads to the distortion of the lattice through an angle α^o, whereas plastic deformation is developed by lattice translation.

deformation is that of slip or gliding of one part of the crystal lattice over another. The atoms become displaced by an amount which is equal to some integral value of the atomic spacing and take up new stable positions with the same interatomic spacing as in the undeformed original lattice (Fig. 6-9B). Because there is no limit to the extent of these displacements, there is no limit to the magnitude of the strain that can be produced by this mechanism. There is also little or no volume change as a result of the slip process.

It is also possible in certain types of lattice for the atomic displacements to move over by some other distance so that the atoms in the deformed lattice take up stable positions which are the mirror image of their arrangement in the undeformed lattice (Fig. 6-9C). The optical orientations of the crystal are changed as a result of the displacements, and the sectors of the crystal show a twinned relationship to one another. Mechanically formed twins of this type are common in a number of deformed rock materials (e.g., calcite marbles).

Shearing stresses are necessary before the crystal lattice can be deformed by these slip mechanisms. This explains the observed fact that hydrostatic pressures do not cause plastic deformation, and that the stress system must have a sufficiently large deviatoric component before the bonds which hold the atomic layers together can be broken. Theoretical calculations have been made to determine the extent of the shearing stresses which are necessary to rupture the interatomic forces in perfect crystals. These values are always about 10^2 to 10^3 times greater than those which are found to produce rupture in crystals deformed under laboratory conditions. This apparent anomaly is explained by the fact that crystal lattices are rarely, if ever, perfect. Atoms may be missing or in the wrong place, while other important defects in the crystal structure arise from disorder or out-of-stepness of the elementary

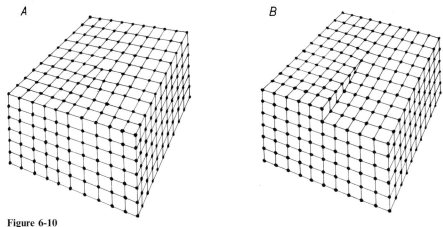

Figure 6-10

Types of dislocations or imperfections that may exist in a crystal lattice. A, edge dislocation; B, screw dislocation.

lattice units. Structural discontinuities of this latter type are known as *dis-locations* and they can be of various types. An *edge dislocation* is an incomplete, wedgelike layer of lattice units within the regular structure (Fig. 6-10*A*), while in a *screw dislocation* the individual planes of atoms are curved to form a spiral ramplike structure (Fig. 6-10*B*). These dislocations act as stress concentrators; when small stresses are applied to the crystal, they are caused to move and migrate through the more ordered part of the lattice. During plastic deformation this mobilization of the dislocations tends to initiate or liberate other dislocations. When several sets of dislocations sweep through the crystal lattice, they interfere with one another and may become locked. It is thought that these obstructions inhibit free gliding of the lattice and so account for the property of strain hardening. With the application of thermal energy the atoms may be reactivated to a sufficient level for the lattice defects and locked dislocations to be removed and the crystal annealed.

The slip mechanisms of polycrystalline aggregates are very complex. The movement on a slip zone in one grain may be transferred to another across their common boundary. If the boundaries are to remain closely fitting, it has been shown theoretically that at least five slip directions must be developed in each grain. In some rock materials slip probably occurs along grain boundaries. Generally, however, displacements of this type tend to produce openings and fissures between the grains. These fissures may be infilled by grain growth of the adjacent crystals or by the development of crystals from new nuclei.

6-3   ROCK BEHAVIOR IN TERMS OF MECHANICAL ANALOGS

One way of viewing the behavior of rock materials under stress is by using analogous mechanical models (Eirich, 1956; Bland, 1960; Reiner, 1960*a*, 1960*b*; Jaeger, 1962; Price, 1964; Sperry, 1964). The fundamental model elements are those which represent an ideal solid, a perfect liquid, and a yield stress. The elastic model is a perfect spring obeying the linear elastic stress-strain laws (Fig. 6-11*A*). The model representing the perfect liquid is sometimes known as a *dashpot* (Fig. 6-11*B*), and consists of a loosely fitting piston which moves through a cylindrical tube filled with a viscous liquid at a rate which is proportional to the stress on the piston. The yield stress (σ_Y) is represented by a block resting on a surface and exerting a frictional resistance which must be overcome before it will move (Fig. 6-11*C*).

These elementary models are combined in various ways to form compounded models whose properties can be investigated. It is found that the behavior of many of these compounded models under stress may be very like the behavior of real rock materials in laboratory experiments. The compound models can be grouped into two main types. The first are the *viscoelastic models*, which, for any given stress have a certain specific limiting value for the strain. The

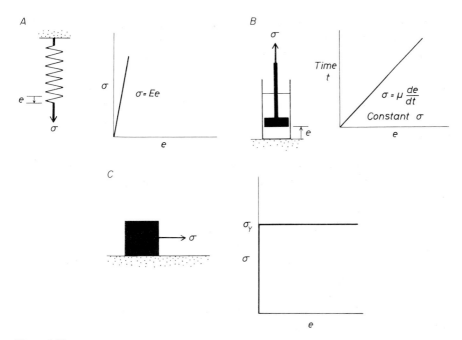

Figure 6-11

Basic model analogs. A, elastic spring (Hookean model); B, liquid model (Newtonian model); C, plastic model (St. Venant model).

material takes some time to take up this value; it does so exponentially and the time required to take up $1/e$ (e is the exponential constant 2.7183) of its final value is defined as the *retardation time*. These materials are basically solids. The second group of models are the *elastoviscous* types which are basically liquids in that for any given stress there is no limiting value of strain and any differential stress produces a continuous deformation. If the strain is held at a constant value, then the internal stresses are gradually dissipated, and the time taken for these stresses to diminish by $1/e$ of their original value is termed the *relaxation time*. The friction block representing the yield stress can be inserted between elastic and viscous components and divide the behavior of the model into two parts; for example, up to a certain stress the material may have viscoelastic properties, and above the yield strength they may show elastoviscous behavior.

The representation of actual phenomena by models is essentially a late nineteenth century approach to physical behavior. The principal disadvantage of thinking in these terms is that the models offer only a qualitative description of the phenomena. Sometimes the elementary components making up the model may have some sort of reality in fact. For example, a dry aggregate of particles may behave elastically; whereas if the pore spaces are filled with liquid, then the liquid has to escape during deformation and in consequence there is a viscous resistance to the migration in a way which directly compares with the movement of the liquid past the piston in the viscous model. However, internal deformation in rocks proceeds in a number of ways: by the development of crystal dislocations and internal translations, by grain boundary movements, and by recrystallization, and in general, the identification of each of the model elements with these individual processes going on in the rock is impossible.

The models generally offer only phenomenological descriptions of rock behavior under stress. These rules for deformation may sometimes be usefully extrapolated over time so that the long-term geological effects of the processes can be computed.

6-4   THE KELVIN OR VOIGT VISCOELASTIC MODEL

Real solids very rarely behave exactly according to the laws of elasticity. Elastic strain should develop instantaneously with the application of stress, and these strains should immediately disappear on removal of the stress. In real materials, however, it is often found that the elastic strain is time dependent and that the application (or removal) of an applied stress takes some time to make its effect. It is as if there is some damping mechanism which slows down

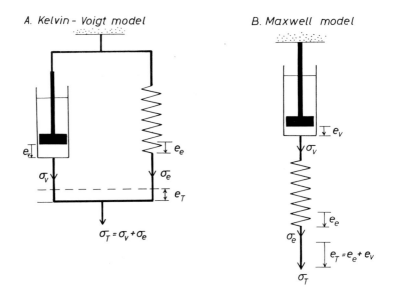

Figure 6-12

Compound models. A, viscoelastic (Kelvin-Voigt) model; B, elastoviscous (Maxwell) model.

the immediate reaction of material. William Thomson (later Lord Kelvin) (1890) and Voigt (1892) both proposed a model employing viscous and elastic elements which could describe this damping mechanism and the property known as *viscosity of solids.*

The model (Fig. 6-12*A*) is built up on the assumption that the displacement on the elastic spring is viscously damped by the interaction of a dashpot built in parallel with the spring.

Let σ_v and σ_e be the stresses acting on the viscous and elastic elements which react by strain e_v and e_e, respectively. Then the stress-strain relationships are given by

$$\sigma_v = \mu \frac{de_v}{dt} \tag{6-2}$$

and

$$\sigma_e = Ee_e \tag{6-3}$$

where μ and E are the viscous and elastic moduli. Because the framework of the model is assumed to be rigid, it follows that the strains e_v and e_e are identical and equal to the total strain e_T, and also that the total stress σ_T is the sum of the stresses acting on the two elements

$$e_T = e_v = e_e \tag{6-4}$$

and

$$\sigma_T = \sigma_v + \sigma_e \tag{6-5}$$

Combining (6-2), (6-3), and (6-4) with (6-5),

$$\sigma_T = \mu \frac{de_T}{dt} + Ee_T \tag{6-6}$$

This simple first-order differential equation is solved as follows: Replacing μ/E by r (the significance of r will be seen later), rearranging and multiplying throughout by a term f (known as an integrating factor),

$$f \frac{de_T}{dt} + \frac{fe_T}{r} = \frac{f\sigma_T}{\mu} \tag{6-7}$$

Now

$$\frac{d(fe_T)}{dt} = f \frac{de_T}{dt} + e_T \frac{df}{dt}$$

or

$$f \frac{de_T}{dt} = \frac{d(fe_T)}{dt} - e_T \frac{df}{dt} \tag{6-8}$$

Substituting this in (6-7),

$$\frac{d(fe_T)}{dt} + e_T \left(\frac{f}{r} - \frac{df}{dt} \right) = \frac{f\sigma_T}{\mu} \tag{6-9}$$

We now choose a value for f so that the coefficient of e_T becomes zero, that is,

$$\frac{f}{r} - \frac{df}{dt} = 0 \quad \text{or} \quad \frac{1}{f}\frac{df}{dt} = \frac{1}{r}$$

Integrating with respect to t,

$$\log_e f = \frac{t}{r}$$

or
$$f = \exp\left(\frac{t}{r}\right) \tag{6-10}$$

Replacing (6-10) in (6-9) so that the coefficient of e_T becomes zero,

$$\frac{d}{dt}\left[e_T \exp\left(\frac{t}{r}\right)\right] = \frac{\sigma}{\mu}\exp\left(\frac{t}{r}\right) \tag{6-11}$$

Integrating with respect to t,

$$e_T \exp\left(\frac{t}{r}\right) = \frac{1}{\mu}\int \sigma(t)\exp\left(\frac{t}{r}\right)dt + C \tag{6-12}$$

The integration constant C is determined by applying the conditions $t = 0$, $\sigma = 0$, then

$$e_T \exp(0) = C$$

or $C = e_T = e_o$, the total strain where $t = 0$.

The equation can now be rearranged to express the total strains

$$e_T = \exp\left(\frac{-t}{r}\right)\left[e_o + \frac{1}{\mu}\int \sigma(t)\exp\left(\frac{t}{r}\right)dt\right] \tag{6-13}$$

If the stress is constant $\sigma(t) =$ a constant σ, and the initial strain (e_o) is zero, this result simplifies to

$$e_T = \frac{\sigma}{E}\left[1 - \exp\left(\frac{-t}{r}\right)\right] \tag{6-14}$$

The behavior of the Kelvin-Voigt material under a sudden stress is as follows: There is no instantaneous, elastic strain, but an initially rapid and exponentially decreasing strain rate (Fig. 6-13). For any given stress the strains asymptotically approach a finite value σ/E. The time taken for the material to strain by an amount equal to $1/2.718$ of the final value σ/E is known as the *retardation time* which has a value $\mu/E = r$.

If after a time t_1 the stress is removed, then the strains exponentially decrease to a zero value. The strain e_1 at a time t_1 is given from (6-14) by

$$e_1 = \frac{\sigma}{E}\left[1 - \exp\left(\frac{-t_1}{r}\right)\right] \tag{6-15}$$

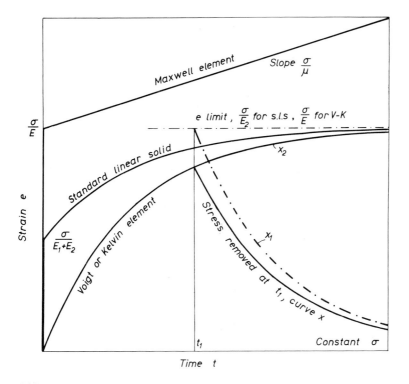

Figure 6-13

Strain-time relationships for viscoelastic and elastoviscous models.

If the stress is now removed ($\sigma = 0$ at $t = t_1$), then the material recovers the strain according to (6-13),

$$e_r = \frac{\sigma}{E}\left[\exp\left(\frac{t - t_1}{r}\right)\right]\left[1 - \exp\left(\frac{-t_1}{r}\right)\right] \qquad (6\text{-}16)$$

where e_r is the strain recovery, or

$$e_r = \frac{\sigma}{E}\left\{\left[1 - \exp\left(\frac{-t}{r}\right)\right] - \left[1 - \exp\left(-\frac{t_1 + t}{r}\right)\right]\right\} \qquad (6\text{-}17)$$

Thus the recovery strain is the difference between the original strain starting from time t_o and an identical curve starting from time t_1 (Fig. 6-13, curve x_1 − curve x_2). This is known as *Boltzmann's principle of superposition* and can be stated in words: If the equations relating stress, strain, and their time derivatives are linear, then the strain that results from a series of stress states is the sum of the individual reactions to the several stress increments. The material behaves almost as if it could remember its past history. Boltzmann developed this interesting theory from a study of glass fibers which had been subjected to a series of repeated torques; after the last twist, instead of returning gradually to their original length, the fibers underwent a series of strain oscillations about their original value.

Although the Kelvin-Voigt model does not exactly describe the results of creep experiments made on rock materials, the features of dampened elastic strain do accord with the type of behavior seen in these experiments. If this model is combined with other elements (see the standard linear solid below), certain parts of the creep deformation field are described fairly exactly.

6-5   THE MAXWELL ELASTOVISCOUS MODEL

Some materials (e.g., pitch) which cannot withstand any shearing stress of long duration (and are therefore in effect liquids) also have the property that if they are suddenly stressed they strain in an elastic manner. If the applied stress is *immediately* removed, the material regains its original dimensions; but if some time elapses before the stress is removed, the elastic recovery is not quite as great as the initial elastic strain—the material has suffered a permanent strain. In a truly elastic material the stress is held within the material and can be recovered instantly and completely, but elastoviscous materials behave as if the stresses are dissipated with time—the energy is taken up by the substance as a permanent strain. This phenomenon is known as *stress relaxation.* Maxwell (1868) was the first to describe these stress relaxations and propose a descriptive mathematical basis for the property.

The Maxwell model is made up of a viscous and elastic component (dashpot and spring) arranged in series (Fig. 6-12B). Let σ_v and σ_e be the stresses acting on the two elements, and the strains produced, e_v and e_e, respectively. Then the stress-strain relations are those expressed in (6-2) and (6-3). Because of the arrangement in series, it follows that the total stress σ_T is equal to each of the individual stresses, and that the total strain e_T is given by the sum of the component strains.

$$\sigma_T = \sigma_e = \sigma_v \tag{6-18}$$

$$e_T = e_e + e_v \tag{6-19}$$

From (6-2) and (6-18)

$$e_v = \frac{1}{\mu} \int_0^t \sigma_T \, dt \tag{6-20}$$

Substituting (6-3) and (6-20) in (6-19), we obtain the basic equation expressing the stress-strain relationships in this elastoviscous model.

$$e_T = \frac{\sigma_T}{E} + \frac{1}{\mu} \int_0^t \sigma_T(t) \, dt \tag{6-21}$$

If the applied stress is constant (σ), this reduces to

$$e_T = \frac{\sigma}{E} + \frac{\sigma t}{\mu} \qquad (6\text{-}22)$$

Thus when the stress is applied at time t_0 the material shows an instantaneous elastic strain of σ/E (Fig. 6-13), and then the material flows at a constant rate given by $de/dt = \sigma/\mu$. On removal of the stress at any time t, the elastic strain σ/E is completely recovered, but a permanent strain of $\sigma t/\mu$ remains.

To determine the stress relaxation properties and the behavior of the stresses when the material is held in a constant state of strain (e_T constant), we differentiate (6-22) with respect to time and obtain

$$0 = \frac{1}{E}\frac{d\sigma}{dt} + \frac{\sigma}{\mu} \qquad \text{or} \qquad \frac{d\sigma}{\sigma} = -\frac{E}{\mu}\,dt$$

Integrating both sides

$$\int_{\sigma_0}^{\sigma}\frac{d\sigma}{\sigma} = -\frac{E}{\mu}\int_{0}^{t}dt$$

where σ_0 is the stress level at time t_0.

$$\log_e \sigma - \log_e \sigma_0 = -\frac{Et}{\mu} \qquad \text{or} \qquad \sigma = \sigma_0 \exp -\frac{Et}{\mu} \qquad (6\text{-}23)$$

This means that the stresses relax exponentially with time (Fig. 6-14) and the *relaxation time* when the stresses are reduced by $1/2.718$ of their original value is given by μ/E. In rocks the viscosity may be very high and the relaxation times in consequence very large. Price (1959) discussed the results of Grigg's creep experiments on Solenhofen limestone in these terms. The viscosity of this rock has been calculated at 2.2×10^{22} poises, and the relaxation

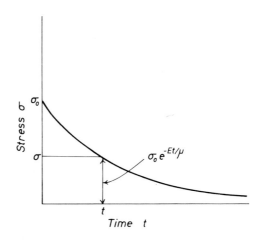

Figure 6-14

Stress relaxation in the Maxwell elastoviscous model.

time could not be less than 100,000 years and could easily exceed 10,000,000 years. With this latter value the time for the residual stresses to relax to 1 percent of their original value would be of the order of 50,000,000 years. Thus it seems likely that small remnants of the last orogenic stresses may be preserved in the rocks long after the geological events which initiated the stress state have ceased. Price (1964) has described the results of a series of compressive creep tests on nodular limestones which have a bearing on these long relaxation times in naturally deformed rocks. Cylinders of limestone were subjected to compressive loads of 4100 and 5250 lb/in.^{2} and held at that stress level for a number of weeks. The specimens behaved in a variety of ways. Only one strained in the manner that is normally expected in creep experiments and a number of specimens actually *expanded* under the compressive load (Fig. 6-15). After the specimen had been placed in the testing rig and measured for a few weeks in an unloaded condition to check its stability, it was subjected to a compressive load of 5250 lb/in.^{2} After the first instantaneous deformation, the material expanded rapidly at first, and then at a fairly constant rate of about 0.7 microstrains per day. After eight weeks it was unloaded; the limestone showed an instantaneous elastic recovery but the length was *greater* than that of the original even though it had been subjected to compression over a long period of time. With the load completely removed, the material continued to expand but with a decreasing rate. It was loaded a second time; again there was an elastic change which was then followed by a period of more normal creep, i.e., contraction under a compressive load.

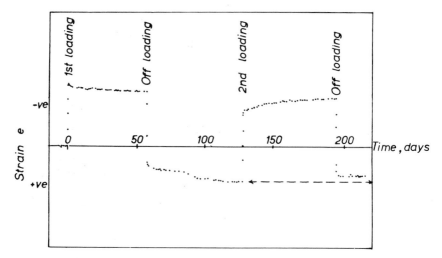

Figure 6-15

Strain-time relationships in a nodular limestone deformed during a creep test. The dashed line gives the amount of expansion after the first off loading. (After Price, 1964.)

The behavior of a number of other specimens investigated by Price showed similar anomalous strain-time relationships and he suggested that all these rocks contained stresses stored from some previous strain history of orogenic antiquity which had not been completely relaxed.

The relaxation time of these rocks must be of the order of tens or hundreds of millions of years, implying a viscosity of something like 10^{26} poises. When these nodular limestones were subjected to a series of alternating cycles of compression and tension before creep testing, all behaved normally. This cycling presumably led to the working out of the stored strain energy.

Price (1959) has suggested that these unrelaxed elastic strains may be released at a later stage in the orogenic history of a region during periods of crustal uplifting, and this might account for the development of shear and tension joints in the rocks. The stress trajectory orientations suggested by the joint patterns are therefore related to the previous tectonic history of the rock and not necessarily to the later uplift. The "movement picture" of the joints is therefore often related to the major structures (folds and faults) formed during preceding orogenic phases.

6-6 THE STANDARD LINEAR SOLID (VISCOELASTIC)

The model with stress-strain-time properties most closely analogous to those of real solids is known as the *standard linear solid* (s.l.s.) and consists of a parallel arrangement of a Maxwell elastoviscous unit with a single elastic

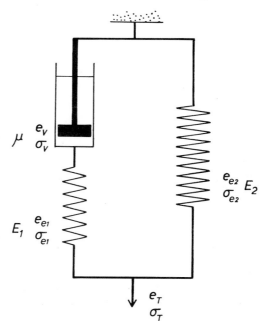

Figure 6-16

Model for the standard linear viscoelastic material.

spring (Fig. 6-16). If suddenly stressed, this model responds in an elastic way; but slowly the elastic stresses stored in the Maxwell unit undergo relaxation. In mathematical terms the stress-strain properties are determined as follows: Let the stresses and strains in the viscous and elastic elements of the Maxwell unit be σ_v, σ_{e1} and e_v, e_{e1}, respectively, with viscosity μ and elastic modulus E_1 so that

$$\sigma_{e1} = E_1 \, e_{e1} \tag{6-24a}$$

$$\frac{d\sigma_{e1}}{dt} = E_1 \frac{de_{e1}}{dt} \tag{6-24b}$$

$$\sigma_v = \mu \frac{de_v}{dt} \tag{6-25}$$

and
$$\sigma_{e1} = \sigma_v \tag{6-26}$$

In the parallel elastic element with stress, strain, and elastic modulus σ_{e2}, e_{e2}, and E_2, respectively, we have

$$\sigma_{e2} = E_2 e_{e2} \tag{6-27a}$$

Therefore
$$\frac{d\sigma_{e2}}{dt} = E_2 \frac{de_{e2}}{dt} \tag{6-27b}$$

If the total stress acting on the unit is σ_T and the total strain is e_T, then

$$\sigma_T = \sigma_v + \sigma_{e2} \tag{6-28}$$

$$e_T = e_{e1} + e_v \tag{6-29}$$

$$e_T = e_{e2} \tag{6-30}$$

Differentiating (6-29) with respect to time t, and using (6-24b) and (6-25),

$$\frac{de_T}{dt} = \frac{1}{E_1} \frac{d\sigma_{e1}}{dt} + \frac{\sigma_v}{\mu}$$

Therefore
$$\sigma_v = \mu \left(\frac{de_T}{dt} - \frac{1}{E_1} \frac{d\sigma_{e1}}{dt} \right) \tag{6-31}$$

Using (6-31) and (6-27a) in (6-28)

$$\sigma_T = \mu \left(\frac{de_T}{dt} - \frac{1}{E_1} \frac{d\sigma_{e1}}{dt} \right) + E_2 e_{e2} \tag{6-32}$$

Replacing (6-26) in (6-28) and differentiating with respect to t,

$$\frac{d\sigma_{e1}}{dt} = \frac{d\sigma_T}{dt} - \frac{d\sigma_{e2}}{dt} \tag{6-33}$$

Replacing (6-33) in (6-32) and using the relationship of (6-27b) and (6-30),

$$\sigma_T = \mu \left(\frac{de_T}{dt} + \frac{E_2}{E_1} \frac{de_T}{dt} - \frac{1}{E_1} \frac{d\sigma_T}{dt} \right) + E_2 e_T \tag{6-34}$$

This is the basic first-order differential equation which expresses the behavior of the s.l.s. We can investigate the change in strain that occurs under conditions of constant stress by letting $d\sigma_T/dt = 0$ and $\sigma_T =$ constant σ. Under these conditions (6-34) simplifies to

$$\frac{de_T}{dt} + \frac{e_T}{r_1 + r_2} = \frac{\sigma}{E_2(r_1 + r_2)} \tag{6-35}$$

where $r_1 = \mu/E_1$ and $r_2 = \mu/E_2$. This is the same type of differential equation as (6-6) and is solved using identical methods employing the integrating factor $f = \exp [t/(r_1 + r_2)]$. The solution is

$$e_T = \frac{\sigma}{E_1 + E_2} + \sigma \left[\frac{1}{E_2} - \frac{1}{E_1 + E_2} \right] \left[1 - \exp \left(-\frac{t}{r_1 + r_2} \right) \right] \tag{6-36}$$

Thus if the material is stressed rapidly so that no relaxation occurs, the material behaves as an elastic substance with a modulus $E_1 + E_2$ and it strains by an amount $\sigma/(E_1 + E_2)$ (Fig. 6-13). If the stresses are held constant the material undergoes elastic creep to approach the strain value of σ/E_2 with a relaxation time given by $\mu(E_1 + E_2)/E_1 E_2$.

The strain-time behavior of the s.l.s. simulates quite well the initial instantaneous elastic and primary creep parts of the curves of the experimental data (Figs. 6-6, 6-7). The other rheological models discussed below are an attempt to adjust the viscous and elastic elements further so that the secondary creep stage may be accommodated in the equations.

6-7   ELASTOVISCOUS AND PLASTIC MODELS FOR CREEP STRAIN

The model with properties which give the best fit for the normal creep curves derived experimentally from deformed rocks is made by compounding in series (1) the elastic spring, (2) a Kelvin-Voigt unit, and (3) a viscous dashpot (Fig. 6-17A). If such a system is subjected to a constant stress, then the total strain e_T is given by the sum of the individual strains of the competent elements,

$$e_T = e_{e1} + e_{K-V} + e_{V2}$$

or

$$e_T = \frac{\sigma}{E_1} + \frac{\sigma}{E_2} \left[1 - \exp \left(-t \frac{E_2}{\mu_1} \right) \right] + \frac{\sigma t}{\mu_2} \tag{6-37}$$

The creep curve for this material is compounded of three curves representing the three component terms of Eq. (6-37) (Fig. 6-18, curve $1 + 2 + 3$). With this function there should always be a viscous (secondary) creep because the

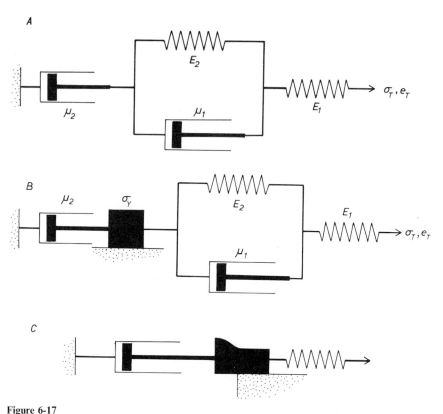

Figure 6-17

Various model analogs for behavior of materials during creep tests.

final term of the equation always has a positive value providing that $\sigma > 0$. In many experiments performed at normal temperatures, however, it is found that if the applied stress falls below a certain value, then only instantaneous-elastic and delayed-elastic (primary creep) deformation occurs (Fig. 6-7). Under these conditions the material behaves essentially as a viscoelastic solid and the material has an inherent strength. The stresses have to reach a certain value (σ_Y = yield stress) before viscous flow will occur and in order to amend our model to include this behavior we add a friction block into the series of elements (Fig. 6-17B). The time-strain behavior can now be expressed

$$e_T = \frac{\sigma}{E_1} + \frac{\sigma}{E_2}\left[1 - \exp\left(-t\frac{E_2}{\mu_1}\right)\right] \qquad \text{if } \sigma < \sigma_Y \quad (6\text{-}38a)$$

or $\quad e_T = \frac{\sigma}{E_1} + \frac{\sigma}{E_2}\left[1 - \exp\left(-t\frac{E_2}{\mu_1}\right)\right] + \frac{(\sigma - \sigma_Y)t}{\mu_2} \qquad \text{if } \sigma > \sigma_Y \quad (6\text{-}38b)$

Finally if the rock material is capable of storing strain energy in the manner

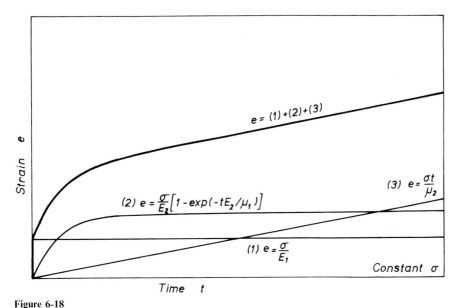

Figure 6-18

Strain-time behavior of model A, Fig. 6-17, representing the sum of curves (1), (2), and (3).

discussed under the heading of the Maxwell model, we ought to include in the model another dashpot with a very high viscosity μ_3 between the Kelvin-Voigt element and the friction block so that it will have a very long relaxation time.

General plastic or Bingham material The model which represents the behavior of many simple plastic materials consists of a spring, friction block, and dashpot arranged in series. At low stresses the material is a solid and behaves elastically; but once the yield stress σ_Y is reached, the frictional resistance of the block is overcome and the material flows like a liquid. On removal of the stress the elastic spring recovers, but there is a permanent deformation as a result of the flow. Materials which show strain hardening (i.e., a progressive increase in the yield stress with deformation) may be represented by changing the thickness of the friction block and allowing only a part of it to rest initially on the surface (Fig. 6-17C).

6-8 RELATIONSHIPS BETWEEN THE STRESS AND STRAIN TENSORS

Because both stress and strain are complex tensor quantities, the problem of connecting the nature of the deformation with the state of stress that exists in a material is that of relating these tensors. The stress state that exists at any one time must control the *increment* of strain. If the deforming material is both isotropic and homogeneous, then the directions of the principal axes of incremental (infinitesimal) strain coincide with those of the principal stresses; it is generally assumed that there is some linear relationship between the components of the two tensors, or of their rates of change with time. If the materials are anisotropic and heterogeneous, these relationships do not hold and the mathematical analysis becomes extremely complex.

The nature of the stress and infinitesimal-strain tensors will be briefly summarized. It has been shown already how the asymmetric tensor of infinitesimal strain (T_e) can be decomposed into symmetric and skew-symmetric parts (Sec. 4-12) which relate, respectively, to the irrotational and rotational components of the deformation. This decomposition may be continued further, for the irrotational component itself can be considered as being made up of a volume change (Δ) and a distortion.

$$
T_e =
\begin{vmatrix}
\dfrac{\Delta}{3} & 0 & 0 \\[2mm]
0 & \dfrac{\Delta}{3} & 0 \\[2mm]
0 & 0 & \dfrac{\Delta}{3}
\end{vmatrix}
+
\begin{vmatrix}
e_x - \dfrac{\Delta}{3} & \tfrac{1}{2}\gamma_{xy} & \tfrac{1}{2}\gamma_{xz} \\[2mm]
\tfrac{1}{2}\gamma_{xy} & e_y - \dfrac{\Delta}{3} & \tfrac{1}{2}\gamma_{yz} \\[2mm]
\tfrac{1}{2}\gamma_{xz} & \tfrac{1}{2}\gamma_{yz} & e_z - \dfrac{\Delta}{3}
\end{vmatrix}
+
\begin{vmatrix}
0 & -\omega_3 & \omega_2 \\[2mm]
\omega_3 & 0 & -\omega_1 \\[2mm]
-\omega_2 & \omega_1 & 0
\end{vmatrix}
$$

$$(6\text{-}39)$$

This breakdown shows that strains can be considered as being composed of three effects. The first part has no shearing strain; its trace (sum of the component across the diagonal) is Δ, the invariant quantity $e_1 + e_2 + e_3 = e_x + e_y + e_z$, and therefore represents a uniform dilation in all directions. The second part has a trace equal to zero showing that it represents a deformation without volume change. The third part has only skew-symmetric terms ω_1, ω_2, and ω_3 and therefore relates only to rigid-body rotations of the material. Thus

$$T_e = \text{dilation} + \text{distortion} + \text{rigid rotation}$$

The stress tensor (T_p) can also be decomposed in a similar manner. The state of stress in a body in equilibrium can be described in terms of a symmetrical tensor which itself can be subdivided into a hydrostatic stress $\bar{\sigma}$ and a stress deviator from the hydrostatic state. When in equilibrium,

$$
T_p =
\begin{vmatrix}
\bar{\sigma} & 0 & 0 \\
0 & \bar{\sigma} & 0 \\
0 & 0 & \bar{\sigma}
\end{vmatrix}
+
\begin{vmatrix}
\sigma_x - \bar{\sigma} & \tau_{xy} & \tau_{xz} \\
\tau_{yx} & \sigma_y - \bar{\sigma} & \tau_{yz} \\
\tau_{zx} & \tau_{zy} & \sigma_z - \bar{\sigma}
\end{vmatrix}
\qquad (6\text{-}40)
$$

If the body is not in equilibrium, however, the second of these components will not be a symmetric tensor for $\tau_{xy} \neq \tau_{yx}$ [see Eq. (2-1)]. It is therefore possible to subdivide this asymmetric tensor into symmetric and skew-symmetric parts. The state of stress can now be stated in terms of three components, the third of which is zero if the body is in equilibrium.

$$T_p = T_{p1} + T_{p2} + T_{p3} \qquad (6\text{-}41a)$$

where

$$T_{p1} = \begin{vmatrix} \bar{\sigma} & 0 & 0 \\ 0 & \bar{\sigma} & 0 \\ 0 & 0 & \bar{\sigma} \end{vmatrix} \qquad (6\text{-}41b)$$

$$T_{p2} = \begin{vmatrix} \sigma_x - \bar{\sigma} & \frac{1}{2}(\tau_{xy} + \tau_{yx}) & \frac{1}{2}(\tau_{xz} + \tau_{zx}) \\ \frac{1}{2}(\tau_{xy} + \tau_{yx}) & \sigma_y - \bar{\sigma} & \frac{1}{2}(\tau_{yz} + \tau_{zy}) \\ \frac{1}{2}(\tau_{xz} + \tau_{zx}) & \frac{1}{2}(\tau_{yz} + \tau_{zy}) & \sigma_y - \bar{\sigma} \end{vmatrix} \qquad (6\text{-}41c)$$

$$T_{p3} = \begin{vmatrix} 0 & \frac{1}{2}(\tau_{xy} - \tau_{yx}) & \frac{1}{2}(\tau_{xz} - \tau_{zx}) \\ \frac{1}{2}(\tau_{yx} - \tau_{xy}) & 0 & \frac{1}{2}(\tau_{yz} - \tau_{zy}) \\ \frac{1}{2}(\tau_{zx} - \tau_{xz}) & \frac{1}{2}(\tau_{zy} - \tau_{yz}) & 0 \end{vmatrix} \qquad (6\text{-}41d)$$

or

T_p = hydrostatic stress + deviatoric stress + disequilibrium component

Each of these three components making up the state of stress is directly related to one of the three components of the strain tensor. The hydrostatic part of the stress system causes changes in volume of the deformed material, the deviatoric stress components cause distortion (which may be either recoverable or permanent on removal of the stress), and the disequilibrium components cause the material to undergo a rotation in space. The mathematical functions which relate these tensor components depend on the nature of the material, whether it behaves as an elastic solid, a viscous fluid, or a plastic substance.

6-9 ELASTIC SOLIDS

If we take a cylinder of solid material (Fig. 6-19) and subject it to an axial load σ_1, and if then the material shows elastic behavior, its length will be strained by an amount e_1, which is proportional to the amount of the applied stress; on removal of the load, it instantly regains its original length. This linear relationship between stress and strain is known as *Hooke's law*, and the constant E connecting them is known as *Young's modulus of elasticity*,

$$\sigma_1 = Ee_1 \qquad\qquad (6\text{-}3)$$

Because strain is a nondimensional (scalar) term, Young's modulus is expressed in the same dimensions as that of stress $(ml^{-1}t^{-2})$. The strains generally produced in elastic materials are generally small (less than 2 percent) and come under the category of infinitesimal strain.

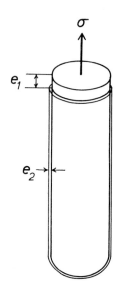

Figure 6-19

A cylinder deformed elastically by the application of an axial tensile stress σ. The cylinder expands by an extension e_1 and contracts laterally by a strain e_2.

The cylinder of material also suffers a lateral contraction (strain e_2) as a result of the longitudinal extension. The ratio of this lateral contraction to the longitudinal extension is generally known as *Poisson's ratio* (v) after the French mathematician who was first to compute its value, while the reciprocal of Poisson's ratio is generally termed *Poisson's number* (m). Both are dimensionless (scalar) quantities like strain

$$v = \frac{e_2}{e_1} = \frac{1}{m} \qquad\qquad (6\text{-}42)$$

As a result of the application of the stress, the cylinder changes shape and volume. The volume change is given by Eq. (4-88) for small strains

$$\Delta = e_1 + e_2 + e_3$$

Combining (6-42) and (6-3) we find that

$$e_2 = e_3 = -\frac{\sigma_1 v}{E}$$

and therefore
$$\Delta = \frac{\sigma_1}{E} - \frac{\sigma_1 v}{E} - \frac{\sigma_1 v}{E} = \frac{\sigma_1}{E}(1 - 2v) \qquad (6\text{-}43)$$

It follows from this that if there is no change in volume, $v = 0.5$. For materials where there is a decrease in volume with compressive strain (increase with tensile), $v < 0.5$.

If a shearing stress τ is applied to the side of a cube of elastic material, it undergoes a shearing strain γ which is directly proportional to the stress, and the elastic constant which connects them is the *rigidity modulus* or *shear modulus* G. Like Young's modulus this also has the dimensions of stress. Another modulus is used to describe the relationship between elastic volume change (dilation $\Delta = e_1 + e_2 + e_3$) and hydrostatic stress $[\bar{\sigma} = (\sigma_1 + \sigma_2 + \sigma_3)/3]$. This is the *bulk modulus* K, and its reciprocal is known as the *compressibility*,

$$\bar{\sigma} = K\Delta \qquad (6\text{-}44)$$

These elastic moduli remain constant throughout isotropic materials, but they are often variable in the anisotropic materials investigated by geologists. For example, the value of E parallel to the bedding in sandstones may be greater than that in a direction perpendicular to the bedding (Price, 1958). The moduli may also show nonlinear variations with changes in the stress conditions (confining pressure) and temperature.

Equations relating stress and strain in elastic materials   If the material is isotropic the directions of the principal strains coincide with the principal axes of stress. It will now be seen why it is more convenient nomenclature to define the principal *tensile* stresses $\sigma_1 \geq \sigma_2 \geq \sigma_3$, for the greatest elongation e_1 is now coincident with the axis of greatest tensile stress σ_1.

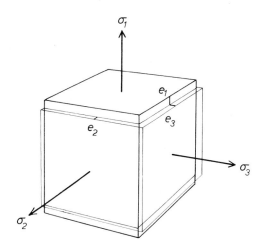

Figure 6-20

Cube of unit dimensions deformed elastically by the three principal stresses σ_1, σ_2, and σ_3, with corresponding strains e_1, e_2, and e_3.

Consider a cube of material acted on by stresses σ_1, σ_2, and σ_3 perpendicular to its sides (Fig. 6-20). It is strained into a rectangular block by amounts e_1, e_2, and e_3. Consider first the strain e_1; this is developed in part by the pull σ_1 which produces an extension σ_1/E. However, the stresses σ_2 and σ_3 also take part in controlling its length, for the stress σ_2 producing a strain σ_2/E in the e_2 direction also contracts the e_1 and e_3 directions by an amount $v\sigma_2/E$. Similarly, stress σ_3 also produces a contraction along the e_1 direction by an amount $v\sigma_3/E$. It therefore follows that the total strain e_1 is given by

$$e_1 = \frac{\sigma_1}{E} - \frac{v\sigma_2}{E} - \frac{v\sigma_3}{E}$$

Similar arguments for e_2 and e_3 lead to the three equations which express the strains in terms of the stresses and the elastic moduli E and v.

$$e_1 = \frac{1}{E}\left[\sigma_1 - v(\sigma_2 + \sigma_3)\right] \tag{6-45a}$$

$$e_2 = \frac{1}{E}\left[\sigma_2 - v(\sigma_1 + \sigma_3)\right] \tag{6-45b}$$

$$e_3 = \frac{1}{E}\left[\sigma_3 - v(\sigma_1 + \sigma_2)\right] \tag{6-45c}$$

and in general

$$e_x = \frac{1}{E}\left[\sigma_x - v(\sigma_y + \sigma_z)\right] \tag{6-45d}$$

$$\gamma_{xy} = \frac{\tau_{xy}}{G} \tag{6-45e}$$

$$\gamma_{yz} = \frac{\tau_{yz}}{G} \tag{6-45f}$$

$$\gamma_{zx} = \frac{\tau_{zx}}{G} \tag{6-45g}$$

The remaining three equations that complete the stress-strain relationships are simply derived from the previous discussion of shear modulus G. Plane stress ($\sigma_2 = 0$) does not generally lead to a plane strain deformation ($e_2 = 0$).

Because the stress-strain relationships can be completely expressed using only two of the elastic moduli, it follows that of the four moduli E, v, G, and K, only two are independent. By adding (6-45a, b, c) we find

$$e_1 + e_2 + e_3 = \frac{\sigma_1 + \sigma_2 + \sigma_3}{E}(1 - 2v)$$

or

$$\Delta = \frac{3\bar{\sigma}}{E}(1 - 2v)$$

and from (6-44)

$$K = \frac{\bar{\sigma}}{\Delta} = \frac{E}{3(1 - 2v)} \tag{6-46}$$

It can also be shown that

$$G = \frac{E}{2(1 + v)} \tag{6-47}$$

Equations (6-45a, b, c) may be reorganized so that the stresses are expressed in terms of the strains into three equations of this type:

$$\sigma_x = \frac{E}{(1 + v)(1 - 2v)}\left[e_x(1 - v) + v(e_y + e_z)\right]$$

These are usually simplified using new elastic moduli λ and $\mu = G$ (Lamé's parameters),

$$\sigma_x = (\lambda + 2\mu)e_x + \lambda e_y + \lambda e_z = 2\mu e_x + \lambda\Delta \tag{6-48a}$$

$$\sigma_y = \lambda e_x + (\lambda + 2\mu)e_y + \lambda e_z = 2\mu e_y + \lambda\Delta \tag{6-48b}$$

$$\sigma_z = \lambda e_x + \lambda e_y + (\lambda + 2\mu)e_z = 2\mu e_z + \lambda\Delta \tag{6-48c}$$

where $\quad \lambda = \dfrac{vE}{(1 + v)(1 - 2v)} \quad$ and $\quad \mu = G = \dfrac{E}{2(1 + v)}$

Elastic strain energy   As a result of elastic deformation, energy is stored up inside the deformed solid. If the restraining stresses are removed, this energy can be recovered. In the simple graphical representation of the elastic relationships of stress and strain (Fig. 6-21), the work done (W) is given by the area under the curve,

$$W = \tfrac{1}{2}\sigma e \tag{6-49}$$

In three dimensions the total stored energy equals the total work done during deformation, which is found by summing the work in each of the three principal directions of the stress (and strain) axes,

$$W = \tfrac{1}{2}(\sigma_1 e_1 + \sigma_2 e_2 + \sigma_3 e_3) \tag{6-50}$$

This total strain energy may be expressed in terms of the stresses alone by using the stress-strain equation (6-45) in (6-50)

$$W = \frac{1}{2E}\left[\sigma_1^2 + \sigma_2^2 + \sigma_3^2 - 2v(\sigma_1\sigma_2 + \sigma_2\sigma_3 + \sigma_3\sigma_1)\right] \tag{6-51}$$

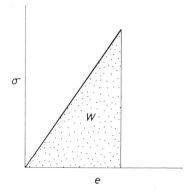

Figure 6-21

*Work done during elastic deformation W
is the area under the stress-strain curve.*

This total stored energy consists of two parts. One part (W_Δ) is that used in changing the volume and is caused by the hydrostatic part of the stress tensor; the second (W') is used in distortion of the solid and is produced by the deviatoric stresses,

$$W_\Delta = \tfrac{1}{2}\bar{\sigma}\Delta$$

Replacing the volumetric strain Δ in terms of stresses, using (6-44),

$$W_\Delta = \frac{\bar{\sigma}^2}{2K} \quad \text{or} \quad \frac{1 - 2v}{6E}(\sigma_1 + \sigma_2 + \sigma_3)^2 \qquad (6\text{-}52)$$

The energy stored by distortion apart from volumetric changes is

$$W' = W - W_\Delta$$

This can be expressed using (6-51), (6-52), and (6-47),

$$W' = \frac{1}{6G}\left[\sigma_1^2 + \sigma_2^2 + \sigma_3^2 - (\sigma_1\sigma_2 + \sigma_2\sigma_3 + \sigma_3\sigma_1)\right] \qquad (6\text{-}53a)$$

or

$$W' = \frac{1}{12G}\left[(\sigma_1 - \sigma_2)^2 + (\sigma_2 - \sigma_3)^2 + (\sigma_3 - \sigma_1)^2\right] \qquad (6\text{-}53b)$$

These extremely important relationships giving the various components of stored energy can be expressed as functions of the invariants of the total stress tensor [using (2-29), (2-30)] or deviatoric stress tensor [using (2-48)],

$$W = I_1^2 - 2(1 + v)I_2 \qquad (6\text{-}54)$$

$$W_\Delta = \frac{I_1^2}{2K} \qquad (6\text{-}55)$$

$$W' = \frac{I_1^2 - 3I_2}{6G} = \frac{I_2'}{2G} \qquad (6\text{-}56)$$

Huber (1904) and Hencky (1924) have suggested that solid material can only take up a certain amount of energy during elastic deformation, and that as soon as this limit is exceeded the solid will deform plastically. It therefore follows that the yield criteria for the elastic limit may be a function of the stored elastic energy.

6-10 BRITTLE FAILURE; THE DEVELOPMENT OF FAULTS AND FRACTURES

When the internal cohesion of rock materials which are deforming in their elastic range is broken, the rock is said to be brittle, and the stress conditions at the moment of failure define the *stress criteria of brittle strength*. The engineer is particularly interested in these conditions of failure, for the stability of rock constructions depends on keeping the applied stresses well within these limits. These conditions also govern the development of faulting during orogenic processes.

Rocks may be subjected to an unlimited compressive hydrostatic stress; they suffer only a volume change. If subjected to a tensile hydrostatic stress, they break apart with explosive violence when the stress becomes equal in value to the cohesive stresses which hold the particles together. It is shearing stresses which lead to the formation of fracturing and fault development of rocks, and therefore the stress system has to have deviatoric components of a sufficiently large value before these shearing stresses become large enough for rupture to occur. Although the state of hydrostatic stress cannot initiate fractures, it does control the values of the stress difference that must be reached before failure under shear will occur. It is well known that rock will fail more easily under a tensile stress than under compressive stress.

When rocks fail it is found that they break on two sets of planar shear surfaces which intersect in lines parallel to the directions of intermediate stress σ_2, and that the acute angle between these planes is always bisected by the maximum compressive stress (σ_3). Although the failure is controlled by shear, it does not generally occur on those surfaces on which the maximum shearing stress acts and which are situated at 45° to the principal compressive stress (see Sec. 2-11).

The stress states which lead to failure are best represented by means of the *Mohr diagram for stress*. Normal stress σ is plotted as abscissa, and shearing stress τ as ordinate. Any state of stress can be represented as a circle with center $(\sigma_1 + \sigma_3)/2$ and radius $(\sigma_1 - \sigma_3)/2$ intersecting the σ axis at two points with values σ_1 and σ_3 of the principal stresses.

Subject a rock to an increasing compressive stress, keeping the other principal stresses zero. The various changing stress states are represented by a series of circles with radii r^i, r^{ii}, r^{iii} (Fig. 6-22). The rock deforms elastically when subjected to principal stress σ_3^i and σ_3^{ii}, but as soon as the compression

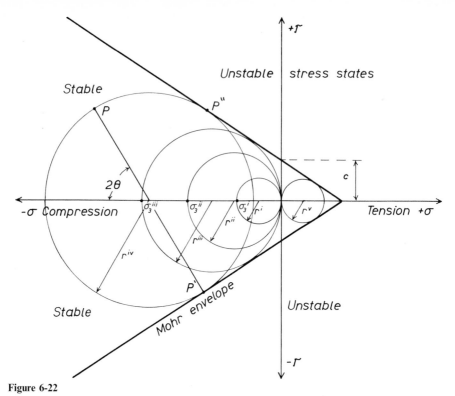

Figure 6-22

Mohr's stress circles for stress states leading to brittle failure, and the Mohr envelope separating unstable and stable stress conditions.

reaches a value of σ_3^{iii} it breaks. If failure tests are repeated but with a different value of the minimum compressive stress σ_1, other limiting circles are obtained for each value of σ_1 with radii r^{iv} and r^v. In this diagram we may therefore distinguish those stable stress states which lead to failure from those which do not. The line forming the tangent to a whole series of limiting failure circles is known as the *Mohr envelope* and gives a clear separation of the stable and unstable stress fields. The Mohr envelope may be a pair of straight lines, or it sometimes has a curved form (Fig. 6-23C). It always opens toward the compressive stress side of the diagram, showing that rocks are stronger under compression than under tension. If it terminates at the origin, it signifies that the material can withstand no tensile stress; i.e., it has no cohesion (e.g., dry sand, Fig. 6-23B). Most envelopes intersect the τ axis at a value c, sometimes called the *cohesion* of the material. In some clays so deformed that pore fluids do not escape, the Mohr envelope is two parallel lines of constant shear stress $(= c)$ (Fig. 6-23A).

The angle θ of the shear planes to the principal axis of compressive stress

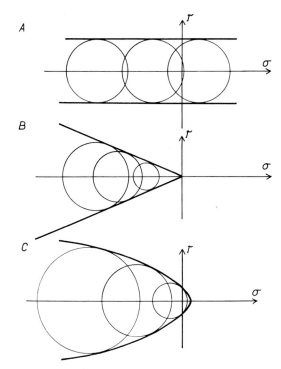

Figure 6-23

Types of Mohr envelopes. A, wet clay; B, dry sand; C, general type of Mohr envelope for rock materials.

may be recorded on the diagram as the angle 2θ (Fig. 6-22). The points P' and P'' represent the failure points and lie on the Mohr envelope. An excellent discussion of this type of simple Mohr envelope is given by Hubbert (1951).

Mohr showed that the conditions of failure given by this envelope must connect the shearing and normal stress in some functional way

$$\tau = f(\sigma) \qquad (6\text{-}57)$$

The simplest envelope is given by the pair of straight lines of slopes $\pm\mu$ (*Coulomb criterion* of failure).

$$|\tau| = c - \mu\sigma \qquad (6\text{-}58)$$

The fracture will take place where the shear stress attains a value given by the sum of the cohesive resistance of the rock and a resistance across the surface equal to μ times the compressive stress. It has been suggested that μ is analogous to the ordinary coefficient of friction resisting the sliding of one block over another. The term μ is known as the *coefficient of internal friction*.

The angle θ that the shear planes make with the direction of principal stresses can be determined by substituting the values of τ and σ in terms of the principal stresses (2-11) and (2-13) into (6-58).

$$c = \mu\left(\frac{\sigma_1 + \sigma_2}{2} + \frac{\sigma_1 - \sigma_2}{2}\cos 2\theta\right) \pm \frac{\sigma_1 - \sigma_2}{2}\sin 2\theta$$

The maximum value of the right-hand side of this equation is found by differentiating it with respect to θ and equating it to zero,

$$0 = -\mu(\sigma_1 - \sigma_2)\sin 2\theta \pm (\sigma_1 - \sigma_2)\cos 2\theta$$

$$\tan 2\theta = \pm\frac{1}{\mu} \tag{6-59}$$

If the internal friction μ is zero, $\theta = \pm 45°$ and the shear planes coincide with the planes of maximum shearing stress. As μ takes up small positive values so θ becomes reduced; in many rock materials μ is 0.5-0.6 and θ has a value around 30°. This two-dimensional theory accords reasonably well with experimental results. It appears that the value of the intermediate stress does affect the conditions of failure, but it is of secondary importance.

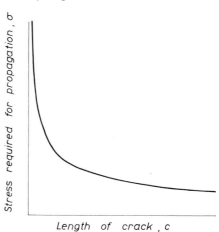

Figure 6-24

Stress required for propagation of a crack of given length. (After Griffith, 1924.)

Another interpretation of rock failure has been put forward by Griffith (1920, 1924). It is found that the cohesive strength of materials calculated from a theoretical basis of molecular structure exceeds the actual fracture strength by as much as three orders of magnitude. Griffith suggested that this was due to inherent defects in the structure caused by the presence of microscopic and submicroscopic cracks. The stresses in materials containing small flaws are very strongly concentrated around the ends of the cracks where the curvature is greatest, and if the cracks are larger than a certain critical value they spread spontaneously. Griffith experimented on stressed glass containing cracks of known length and found that for cracks of length c there was an initial stress σ above which they developed rapidly and that

$$\sigma(c)^{\frac{1}{2}} = \left(\frac{4E\omega}{\pi}\right)^{\frac{1}{2}} = \text{constant} \tag{6-60}$$

where E is Young's modulus and ω is the surface tension energy of the material. Materials with very small cracks can therefore sustain a much higher tensile stress than those with cracks of moderate or large length (Fig. 6-24).

The Mohr envelope developed from this theory has a parabolic form (Odé, 1960; Jaeger, 1962) which closely approximates to the curved envelope obtained under experimental conditions given by

$$|\tau| = 2[\sigma_t(\sigma - \sigma_t)]^{\frac{1}{2}} \tag{6-61}$$

where σ_t is the tensile strength of the material.

Types of faults   Because the shearing stresses along the surface of the earth are zero, it follows that two of the principal stresses are always contained in the surface and the third is perpendicular to it. Anderson (1951) developed the consequences of this, showing that at or near the surface the faults can be classified into three types depending on the orientations of the principal stresses:

NORMAL FAULTS (Fig. 6-25*A*)   Maximum compressive stress σ_3 vertical, and the intermediate σ_2 and least σ_1 compressive stresses contained in a horizontal plane

LOW-ANGLE REVERSE FAULTS OR THRUSTS (Fig. 6-25*B*)   Maximum compressive stress horizontal, least compressive stress vertical

STRIKE-SLIP FAULTS (WRENCH OR TEAR FAULTS) (Fig. 6-25*C*)   Maximum and minimum compressive stresses, both horizontal

At deeper levels in the crust, more complex orientations of the stress tra-

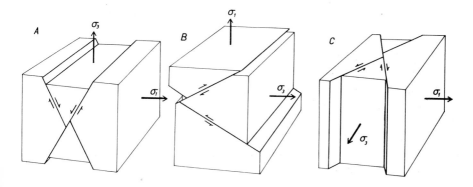

Figure 6-25

Three principal types of faults. A, normal; B, thrust; C, strike-slip.

jectories may be developed; in consequence, the faulting and fracturing systems may be obliquely and asymmetrically inclined to the surface.

Second-order faults   As a result of the formation of a major fault, the stress distributions in the rocks immediately adjacent to the fault plane become altered (Anderson, 1951). The stress trajectories become adjusted in order to come more nearly parallel and perpendicular to the fault plane (Fig. 6-26A). Stress conditions build up around the ends of the fault surface in the same way as they are built up at the ends of the Griffith cracks and the fault surface is propagated at these points. If the redistributed stresses along the sides of the fault surface attain sufficiently large values, the rock may fail again on surfaces which are at 30° to the modified principal compressive strength (Fig. 6-26B). This situation leads to the development of fault systems along the margins of the main faults known as *second-order faults* (Moody and Hill, 1956). It has been suggested that these new second-order faults may themselves lead to further modifications of the stress trajectories and to the development of third-order faults.

This concept has been strongly challenged by Chinnery (1966) who has recalculated the stress distribution in the rocks around a major fault plane. He found that there are fallacies in the arguments put forward by McKinstry, and Moody and Hill, and that after formation of the main fault the shear stresses are reduced on either side of the fault surface. He suggested that secondary faulting is an effect confined to the ends of a master fracture and that second-order fractures will not develop along the walls of the main fault.

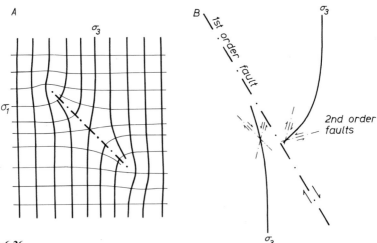

Figure 6-26

The stress trajectories around a fault (after Anderson, 1951), and the possible second-order faults that may result from this redistribution of the stresses.

The stress concentrations at the ends of the master fracture are likely to give rise to various types of secondary fault patterns, of which the most commonly developed is the *splay fault* pattern.

The effects of internal fluid pressure on fault development Rubey and Hubbert (1959) investigated the conditions of formation of thrust faults, taking into account the effects of fluids which are commonly contained within the pores of rocks. They first examined the apparent paradox offered by the sliding of a thrust sheet for a distance of many kilometers over a static basement. With usual values of the coefficient of friction of rock on rock, it seems almost impossible for such sheets to move, for they cannot sustain the stresses which are required to move them without crushing. Yet without any doubt some of the nappe sheets of the Alps have been transported for a horizontal distance of at least 40 km although they are only a few kilometers thick (e.g., Glarus nappe, pre-Alpine nappes). Even if they slid down an inclined surface, this would have to have a tilt of the order of 30°, and the geological evidence precludes such a high inclination. Thus it seems almost impossible to account for the observed thrust sheets on either the hypothesis of a push from behind or that of gliding down a slope.

Rubey and Hubbert suggested a solution to this problem based on the recognition of high fluid pressures in rocks. The vertical stresses that exist in the crust are made up of, first, the pressure of the fluids contained in the rock, known as the *normal* or *true hydrostatic pressure*

$$\sigma_f = -\rho_f gz$$

where ρ_f is the fluid density, g the gravity acceleration, and z the depth, and second, the *geostatic* or *lithostatic pressure* which is the weight of the rock material above the point of observation,

$$\sigma_r = -\rho_r gz$$

where ρ_r is the rock density. The fluid pressure has the effect of supporting some of the weight of the overburden of material and at any point the *effective normal stress* is given by the difference between the geostatic and fluid pressures. The geostatic pressure is a fixed quantity, but the pressure of the fluids in the rock may be changed. As the fluid pressure approaches a value equal to that of the geostatic pressure, so the value of the effective stress decreases to zero. Let the fluid pressure be expressed as a proportion of the normal stress acting on the rock if no fluid were present,

$$\sigma_f = \lambda\sigma$$

Now reexamine the Mohr-Coulomb criterion of rock failure. This states that failure should occur when the shear stress reaches a value equal to the cohesion and the friction times the *effective* normal compressive stress,

$$|\tau| = c - \mu(\sigma - \sigma\lambda)$$

$$|\tau| = c - \sigma\mu(1 - \lambda) \tag{6-62}$$

It therefore follows that if λ increases in value with increasing fluid pressure, the rock fails more easily. Once the rock has begun to move, then the frictional conditions necessary for continued movement are

$$|\tau| = \mu\sigma \tag{6-63}$$

or, where the rocks have a fluid pressure λ, this is modified to

$$|\tau| = \mu\sigma(1 - \lambda) \tag{6-64}$$

Again it can be seen that, as the fluid pressure increases, the block can move at a lower critical value of the shearing stress without in any way altering the value of the friction constant.

From these equations it follows that blocks may be moved by the application of relatively small forces (Table 6-1), and they will slide down slopes of quite

Table 6-1 *Maximum length (km) of horizontal overthrust for various thicknesses and values of λ (after Rubey and Hubbert, 1959)*

Thickness, km	λ						
	0.0	0.465	0.5	0.6	0.7	0.8	0.9
1	8.0	13.4	14.2	17.3	22.5	32.9	64.0
2	10.6	16.7	17.6	21.2	27.1	39.0	74.4
3	13.2	20.1	21.1	25.1	31.8	45.1	84.8
4	15.8	23.5	24.6	29.0	36.4	51.2	95.2
5	18.4	26.7	28.0	32.9	41.0	57.3	106.0
6	21.0	30.2	31.5	36.8	45.6	63.4	116.0
7	23.6	33.6	34.9	40.7	50.3	69.5	126.0
8	26.2	36.9	38.4	44.6	54.9	75.6	137.0

Table 6-2 *Critical inclination necessary for gravity gliding (after Rubey and Hubbert, 1959)*

λ	Critical slope for gravitational gliding, deg
0.46	17.2
0.60	13.0
0.70	9.8
0.80	6.6
0.85	5.0
0.90	3.3
0.94	2.0
0.97	1.0

low inclination (Table 6-2) providing that the pressure of the internal fluid is sufficiently high.

Rubey and Hubbert list an impressive number of occurrences of known abnormally high fluid pressures. In some oil wells it is common for λ to approach the value 0.9 (Fig. 6-27), although the normal value should be 0.46.

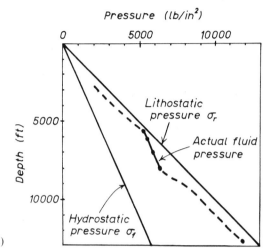

Figure 6-27

Occurrence of abnormally high fluid pressure in an oil well at Chia-Surkh, Iraq. (After Hubbert and Rubey, 1959.)

They suggest that these high pressures result, for the most part, from the compaction of water-filled porous sediments. They might also develop from chemical changes in the clay minerals. In sands and limestones these high pore pressures diminish as the water slowly migrates away from the compacted region, but in clays high pore pressures may be retained for some time. It is the zones of abnormally high fluid pressure, such as are found in clay layers, that will control the formation of faults and the instability of the rock layers. In the overthrust region of West Wyoming (Fig. 6-28) Rubey and Hubbert suggested that the eastward migration of the principal axis of sedimentation may have led to the formation of states of high fluid pressure in the compacted strata at depth and in consequence to the successive eastward development of the thrust sheets.

6-11 STRESS FUNCTIONS AND THEIR APPLICATION TO FAULTING PROBLEMS

The development of structures in the rock materials making up the crust of the earth depend to a large extent on stress distribution, on the values taken up by the principal stresses, and on how these both change with time. As yet very little information is available on measurements of the stress

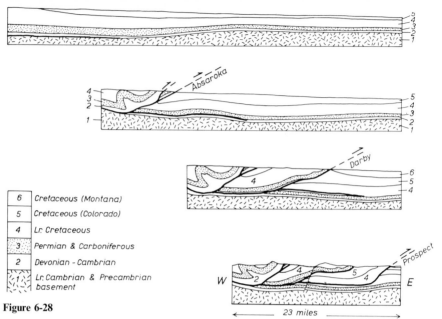

6	Cretaceous (Montana)
5	Cretaceous (Colorado)
4	Lr. Cretaceous
3	Permian & Carboniferous
2	Devonian - Cambrian
1	Lr. Cambrian & Precambrian basement

Figure 6-28

Migration of the zone of high fluid pressure leading to the successive eastward development of thrust faults in West Wyoming. The heavy lines represent thrust faults. (After Rubey and Hubbert, 1959.)

distribution that exists at present in the crust. With a rock mass of known shape and with certain fixed stress values along its boundaries, there is only one possible pattern of stress distribution in the material. The reason for this is that the laws relating stress distribution and displacements of the particles can only be satisfied in one way. The stress distributions that exist in circumstances of likely geological significance have been investigated from a theoretical viewpoint by a number of researchers (Anderson, 1951; Hafner, 1951; Odé, 1957; Sanford, 1959). It seems likely that the comparison of theoretically predicted structural patterns with the pattern of disposition of faults and fractures determined on the field will be of prime importance in determining the stress distributions of the past.

Stress functions for the solution of two-dimensional problems   One of the most convenient ways of determining the internal stress distributions in a body of given shape in static equilibrium is by using particular expressions known as *stress functions*. These are functions of x and y chosen so that the three stress components σ_x, σ_y, and τ_{xy} at any point can be expressed as their second partial derivatives and so that they automatically satisfy the conditions of static equilibrium. The equilibrium conditions (2-53) for three dimensions can be simplified for the two-dimensional problems without body forces as

$$\frac{\partial \sigma_x}{\partial x} + \frac{\partial \tau_{xy}}{\partial y} = 0 \tag{6-65a}$$

$$\frac{\partial \sigma_y}{\partial y} + \frac{\partial \tau_{xy}}{\partial x} = 0 \tag{6-65b}$$

The stress function ϕ is chosen so that

$$\sigma_x = \frac{\partial^2 \phi}{\partial y^2} \tag{6-66a}$$

$$\sigma_y = \frac{\partial^2 \phi}{\partial x^2} \tag{6-66b}$$

$$\tau_{xy} = -\frac{\partial^2 \phi}{\partial x\, \partial y} \tag{6-66c}$$

Substitution will show that these stress components satisfy the conditions of equilibrium (6-65). In order to give a valid stress distribution, however, the stress function is required to fit other laws. First the material is assumed to obey Hooke's law, and therefore the stress-strain relationships must accord with equations (6-45). In two dimensions the stress-strain equations are

$$e_x = \frac{1}{E}\left(\sigma_x - v\sigma_y\right) \tag{6-67a}$$

$$e_y = \frac{1}{E}\left(\sigma_y - v\sigma_x\right) \tag{6-67b}$$

$$\gamma_{xy} = \frac{1}{G}\left(\tau_{xy}\right) \tag{6-67c}$$

Expressing these equations in terms of the stress function using (6-66),

$$e_x = \frac{1}{E}\left(\frac{\partial^2 \phi}{\partial y^2} - \frac{v\partial^2 \phi}{\partial x^2}\right) \tag{6-68a}$$

$$e_y = \frac{1}{E}\left(\frac{\partial^2 \phi}{\partial x^2} - \frac{v\partial^2 \phi}{\partial y^2}\right) \tag{6-68b}$$

$$\gamma_{xy} = -\frac{1}{G}\frac{\partial^2 \phi}{\partial x\, \partial y} \tag{6-68c}$$

The strains e_x, e_y, and γ_{xy} must accord with the compatibility equation (3-79) given by

$$\frac{\partial^2 e_x}{\partial y^2} + \frac{\partial^2 e_y}{\partial x^2} = \frac{\partial^2 \gamma_{xy}}{\partial x\, \partial y}$$

This equation signifies that the material is made up of a connected series of particles and that the longitudinal and shearing displacements have to be internally consistent. The stress function must therefore also accord with this strain compatibility, and the necessary conditions for this to hold are found by substituting (6-68) in (3-79),

$$\frac{1}{E}\left(\frac{\partial^4 \phi}{\partial y^4} - v\frac{\partial^4 \phi}{\partial x^2\, \partial y^2} + \frac{\partial^4 \phi}{\partial x^4} - v\frac{\partial^4 \phi}{\partial x^2\, \partial y^2}\right) + \frac{1}{G}\frac{\partial^4 \phi}{\partial x^2\, \partial y^2} = 0$$

using the relationship between the elastic constants $G = E/2(1 + v)$, and clearing we obtain

$$\frac{\partial^4 \phi}{\partial x^4} + \frac{2\partial^4 \phi}{\partial x^2\, \partial y^2} + \frac{\partial^4 \phi}{\partial y^4} = 0 \tag{6-69}$$

Thus the stress function has to satisfy this fourth-order partial differential equation if it is to represent a valid stress distribution in the material. This is sometimes written in a simpler notation using the symbol ∇ (del) such that "del squared," that is, ∇^2, defines the operation

$$\nabla^2 = \frac{\partial^2}{\partial x^2} + \frac{\partial^2}{\partial y^2}$$

and $$\nabla^4 = \nabla^2\nabla^2 = \left(\frac{\partial^2}{\partial x^2} + \frac{\partial^2}{\partial y^2}\right)^2 = \frac{\partial^4}{\partial x^4} + \frac{2\partial^4}{\partial x^2\, \partial y^2} + \frac{\partial^4}{\partial y^4}$$

Equation (6-69) can now be written

$$\nabla^4 \phi = 0 \tag{6-70}$$

Any function ϕ which satisfies (6-70) (known as a *biharmonic equation*) automatically satisfies the equilibrium, stress-strain and compatibility equations and is therefore a solution of *some* problem of stress distribution. There are a number of types of expressions which are possible stress functions. For example, any polynomial of the type

$$\phi = ax^2 + bxy + cy^2 + dx^3 + ex^2y + fxy^2 \cdots$$

with terms of powers of xy up to three will automatically satisfy $\nabla^4 \phi = 0$, and expressions with terms higher than the third power may do so if they have special relationships between their coefficients. Other types of stress functions may have the form

$$\phi = \exp(x)\, f(y)$$

where from (6-69) $f(y)$ satisfies the differential equation

$$f(y) + \frac{2\partial^2 f(y)}{\partial y^2} + \frac{\partial^4 f(y)}{\partial y^4} = 0 \tag{6-71}$$

Similarly there are many other complex functions of x and y which give valid solutions of $\Delta^4\phi = 0$ (Timoshenko, 1936).

Once the correct stress function has been found for the particular problem in hand, the stress components σ_x, σ_y, and τ_{xy} may be determined directly by differentiation according to Eq. (6-66). The values of the principal stresses σ_1 and σ_2 are given from these by (2-7)

$$\sigma = \tfrac{1}{2}\{\sigma_x + \sigma_y \pm [(\sigma_x - \sigma_y)^2 + 4\tau_{xy}^2]^{\frac{1}{2}}\}$$

and the direction of the stress trajectories from (2-5) by

$$\tan \alpha = \frac{\sigma_y - \sigma_x}{2\tau_{xy}} \pm \left[\frac{(\sigma_x - \sigma_y)^2}{\tau_{xy}^2} + 4 \right]^{\frac{1}{2}}$$

or more simply,

$$\tan 2\alpha = \frac{2\tau_{xy}}{\sigma_x - \sigma_y}$$

where α is the angle from the positive x axis.

The methods by which stress functions may be used to solve stress-distribution problems is most readily shown with an actual example. Hafner (1951) investigated some of the possible stress distributions that arise in a layer subjected to various types of horizontal compressive stress, and also with various combinations of vertical and shearing stress. As a result he predicted the type of fault and fracture pattern that might be found in the crustal layers of the earth.

A reference coordinate system is chosen: x and y contained in the surface, and z vertical. The most important boundary condition that controls the working of the problem is that existing at the surface. Because air is a fluid and cannot withstand any shearing stresses, it follows that at the surface $\tau_{zx} = \tau_{zy} = 0$. The vertical pressure σ_z at the surface is equal to that of the atmosphere which is very small indeed in comparison with the stresses existing in the crust, $\sigma_z = 0$.

In geological problems the stresses that arise through gravity are often very important. At any depth $-z$ there is a vertical stress because of the weight of the superincumbent rock load. Assuming that the surface is flat and that the strata have a uniform density (ρ), this vertical stress is given by

$$\sigma_z = -\rho g z$$

In any problem of stress distribution we can consider that the stresses at any point in the crust represent the combination of two systems, (1) gravitation stress variations, and (2) the applied tectonic stresses. In order to compute the stress distributions, it is most convenient to rearrange these into two other systems (Anderson, 1951): (1) a state of stress which is hydrostatic at every point and which increases with depth in the same manner as the vertical gravitational effect $\sigma_x = \sigma_y = \sigma_z = -\rho g z$ (Anderson termed this the *standard state*), and (2) the stress deviator from the standard state, or *supplementary stress system*.

Because the standard state is a hydrostatic stress, it exerts no influence on the position of the stress trajectories or on the values of the principal stress differences; these depend entirely on the nature of the supplementary stress system.

The standard stress state can be expressed by means of a stress function ϕ_1; and similarly, another stress function ϕ_2 can be fitted to the supplementary stress system. Because both of these functions are independent solutions to the biharmonic equation (6-70), it follows that their sum will also satisfy this equation and represent the stress function ϕ for the total stresses

$$\nabla^4 \phi = \nabla^4 \phi_1 + \nabla^4 \phi_2 = 0 \tag{6-72}$$

We shall now examine the type of stress distribution that arises when the supplementary stress consists of an additional horizontal stress, but no deviation from the normal hydrostatic stress component in the vertical direction. If ϕ_2 is the stress function of the supplementary system, σ_z and σ_x the supplementary normal stresses in the vertical and horizontal directions, and τ_{xz} the supplementary shearing stress, then, because $\sigma_z = 0$,

$$\sigma_z = \frac{\partial^2 \phi_2}{\partial x^2} = 0 \quad \text{for all values of } z$$

Integrating twice with respect to x,

$$\frac{\partial \phi_2}{\partial x} = a\, f_1(z) + b$$

$$\phi_2 = ax\, f_1(z) + bx + c\, f_2(z) + d \tag{6-73}$$

where $f_1(z)$ and $f_2(z)$ are two functions of z.

This stress function has to satisfy the biharmonic equation to be a valid solution to the problem of stress distribution, so putting (6-73) in (6-69) we obtain the condition

$$ax\frac{\partial^4 f_1(z)}{\partial x^4} + \frac{c\partial^4 f_2(z)}{\partial z^4} = 0 \tag{6-74}$$

It follows that the fourth-order derivatives of $f_1(z)$ and $f_2(z)$ are zero, and therefore that their second-order derivatives are linear functions of z, constants or zero. The stress components become

$$\sigma_x = \frac{\partial^2 \phi_2}{\partial z^2} = ax\frac{\partial^2 f_1(z)}{\partial z^2} + c\frac{\partial^2 f_2(z)}{\partial z^2}$$

$$\sigma_z = \frac{\partial^2 \phi_2}{\partial x^2} = 0 \tag{6-75}$$

$$\tau_{xy} = -\frac{\partial^2 \phi_2}{\partial x \, \partial y} = -a\frac{\partial f_1(z)}{\partial z}$$

Three groups of equations can be found which satisfy the boundary conditions at the surface $[z = 0, \ \tau_{xz} = 0, \ \partial f_1(z)/\partial z = 0]$.

Group I $\qquad \dfrac{\partial f_1(z)}{\partial z} = 0 \qquad \dfrac{\partial^2 f_2(z)}{\partial z^2} = z + e$

$$\sigma_x = cz + f \qquad (\text{where } f = ce) \qquad \sigma_z = 0 \qquad \tau_{xz} = 0 \tag{6-76}$$

Group II $\qquad \dfrac{\partial f_1(z)}{\partial z} = z \qquad \dfrac{\partial f_2(z)}{\partial z} = 0$

$$\sigma_x = ax \qquad \sigma_z = 0 \qquad \tau_{xz} = -az \tag{6-77}$$

Group III $\qquad \dfrac{\partial f_1(z)}{\partial z} = \dfrac{z^2}{2} \qquad \dfrac{\partial f_2(z)}{\partial z} = 0$

$$\sigma_x = axz \qquad \sigma_z = 0 \qquad \tau_{xz} = -\frac{az^2}{2} \tag{6-78}$$

The constants a, c, and f are of any value (including zero). Because each of these groups individually satisfies the stress distribution, it follows that any linear combination will also do so.

Consider first Group I where the constant c is zero. Here the supplementary stresses are everywhere constant; likewise, the maximum shearing stress $(\sigma_x - \sigma_z)/2$, and because $\tau_{xz} = 0$, the stress trajectories are everywhere vertical and horizontal. The faults that arise in this instance are two sets of constantly oriented thrusts. If c is not zero, the supplementary horizontal stresses increase in a linear fashion with depth; the maximum shearing stresses also increase with depth, but the stress trajectories remain horizontal and vertical.

Consider now the possible solutions that result from adding Groups I and II but with constant f set at zero as its effects have already been considered above. Adding the stresses of the standard state to those of the supplementary system, we obtain the general components

$$\sigma_x = ax + cz - \rho gz \qquad \sigma_z = -\rho gz \qquad \tau_{xz} = -az$$

Because σ_x is controlled by linear functions of x and z, there are constant stress gradients in the horizontal and vertical directions.

The stress trajectories for these components where the horizontal supple-

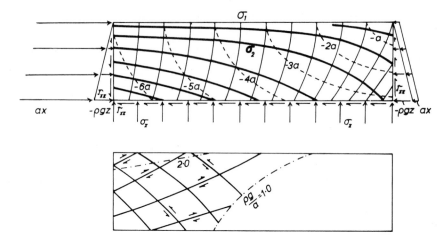

Figure 6-29

Stress trajectories and fault pattern developed in a block with constant horizontal supplementary stress system. The dotted lines are contours of equal maximum shear stress, and the dash-dot lines refer to the stability front for different values of pg/a. (After Hafner, 1951.)

mentary stresses are constant with depth ($c = 0$) are shown in Fig. 6-29, and those where the horizontal supplementary stress gradient is equal to the vertical one ($a = c$) in Fig. 6-30. The trajectories form a series of curves, and the potential faults (drawn at angles of 30° to the line of maximum compressive stress) form sets of curved thrusts. The set which dips toward the region of

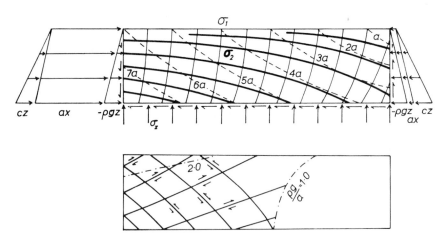

Figure 6-30

Stress trajectories and fault pattern developed in a block with horizontal supplementary stresses increasing in a linear fashion with depth. (After Hafner, 1951.)

maximum pressure is slightly concave upward, while the conjugate set is concave downward. Thrusts of the first type are well known along mountain fronts, but it appears that the complementary set is rather uncommon in nature. The lines of equal maximum shearing stress have been recorded in these diagrams expressed in multiples of the constant a. The maximum shearing stress increases toward the region of greatest horizontal pressure and also with depth. Hafner also calculated the boundaries of the stable and unstable zones in the stresses block. Because this boundary depends on the shearing stresses and on the total confining pressure, its position is influenced by the standard stress state. The position of the boundary limit depends on the constants a and ρg. It dips toward the area of greatest horizontal pressure, depending on the ratio $\rho g/a$, and is steep if $\rho g/a$ is near to unity.

If the supplementary horizontal stresses have a small lateral gradient (about half the vertical pressure gradient), then thrusting is confined to a shallow, gently dipping wedge, although the zone of thrusting may be exposed over a wide area of the surface. If the supplementary horizontal gradient approaches

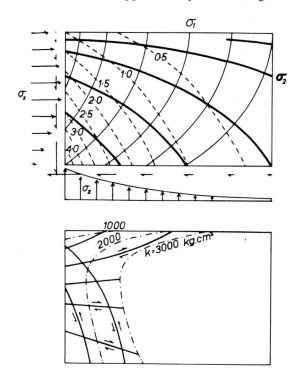

Figure 6-31

Stress trajectories and fault pattern developed in a block where the supplementary horizontal stresses decrease exponentially in a horizontal direction. (After Hafner, 1951.)

that of the vertical gradient (ρgz), then thrusting takes place in a zone which is narrow but which may extend to considerable depth.

These results of Hafner's work have been presented in detail because they offer a particularly clear illustration of the simple use of stress functions for the determination of likely fracture patterns in the crust. Hafner also investigated the stress distribution in a region where the supplementary horizontal stress decreases exponentially using the exponential stress function

$$\phi_2 = k \exp (x) f(z) \qquad (6\text{-}79)$$

The stress trajectories form a series of curves (Fig. 6-31) and the zone of potential thrust faults consists of a rather narrow wedge with the stable boundary curving away from the zone of greatest horizontal pressure as it approaches the surface in a manner which depends on the value of the constant k in the stress function. Stress distributions of this type produce thrust faults dipping at low angles toward the region of greatest horizontal stress and becoming nearly horizontal at a rather shallow depth.

Sanford (1959) utilized the stress function analysis to determine the effects of vertical movements in a basement complex on an overlying homogeneous group of sedimentary rocks which were deforming elastically. The fault patterns predicted by the analysis were of the *horst* and *graben* type, with curving fault surfaces.

6-12   VISCOUS FLUIDS

When a fluid substance is subjected to a hydrostatic stress, it may undergo a small volume change but it will not flow. If, however, the stress system has a deviatoric component, the shearing stresses lead to a continuous deformation by lamellar flow. The rate of this flow is proportional to the amount of shearing stress, the constant connecting the terms is known as the *viscosity coefficient*,

$$\tau = \mu \frac{\partial \gamma}{\partial t} = \mu \dot{\gamma} \qquad (6\text{-}80)$$

The notation of a dot above a symbol will be used throughout the following section to indicate the rate of change of that function with time t, (d/dt), as it greatly simplifies the appearance of the equations which describe the properties of viscous flow. The dimensional of the coefficient of viscosity is stress $\times$ time $(ml^{-1}t^{-1})$, and in the cgs system the coefficient is measured in units of the poise (dyne/cm^{-2}).

If the viscosity is constant for all rates of strain, the fluid is known as a *true Newtonian fluid*. In these fluids viscosity generally decreases with increase in temperature because the molecular activity increases and this leads to less internal cohesion. The fluids known as *generalized Newtonian fluids* have a viscosity coefficient which varies with the strain rate (Fig. 6-32). This variation is generally accredited to structural changes of the particles within the fluid

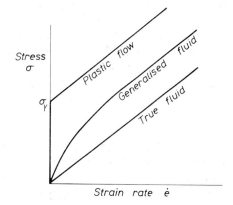

Figure 6-32

Stress-strain rate relationships for various types of fluid flow.

during flow. Plastic materials (Fig. 6-32) are not to be confused with true fluids because at stresses below the yield stress σ_Y they behave as solids and will not flow. *Thixotropy* is the name given to special fluid properties shown by some muds (e.g., bentonite). These materials flow under a low stress only after they have been subjected to a sudden high stress, and return to a stable solid state when the flow ceases.

It seems possible that at moderate or deep levels in the crust many rock materials behave as very slow-moving fluids of high viscosity, and so the principles which govern slow or creeping flow will be outlined below. In many respects this theory has a close mathematical correspondence with that of an elastic deformation but with the linear relation of components of stress and strain tensors replaced by stress and *strain-rate* tensors. These strain-rate components can be defined in terms of rate of change (i.e., velocities) of the displacements u, v, and w of infinitesimal strain (Sec. 4-12). By differentiating Eqs. (4-65) and (4-66) with respect to time, we obtain the standard components of the strain-rate tensor.

$$\dot{e}_x = \frac{\partial e_x}{\partial t} = \frac{\partial^2 u}{\partial x\, \partial t} \quad \text{or} \quad \frac{\partial \dot{u}}{\partial x} \tag{6-81a}$$

$$\dot{e}_y = \frac{\partial \dot{v}}{\partial y} \tag{6-81b}$$

$$\dot{e}_z = \frac{\partial \dot{w}}{\partial z} \tag{6-81c}$$

$$\dot{\gamma}_{xy} = \frac{\partial \dot{u}}{\partial y} + \frac{\partial \dot{v}}{\partial x} \tag{6-82a}$$

$$\dot{\gamma}_{yz} = \frac{\partial \dot{v}}{\partial z} + \frac{\partial \dot{w}}{\partial y} \tag{6-82b}$$

$$\dot{\gamma}_{zx} = \frac{\partial \dot{w}}{\partial x} + \frac{\partial \dot{u}}{\partial z} \tag{6-82c}$$

The rates of change of the rotations derived from (4-67) define the *vorticity* of the flow,

$$\dot{\omega}_x = \frac{1}{2}\left(\frac{\partial \dot{w}}{\partial y} - \frac{\partial \dot{v}}{\partial z}\right) \tag{6-83a}$$

$$\dot{\omega}_y = \frac{1}{2}\left(\frac{\partial \dot{u}}{\partial z} - \frac{\partial \dot{w}}{\partial x}\right) \tag{6-83b}$$

$$\dot{\omega}_z = \frac{1}{2}\left(\frac{\partial \dot{v}}{\partial x} - \frac{\partial \dot{u}}{\partial y}\right) \tag{6-83c}$$

These equations define the flow rates at any point with reference to static coordinate axes. In a general account of fluid flow it is often more valuable to refer these changes to a particular particle in the fluid which is being carried along by the fluid motion; the process known as *differentiation following the fluid motion* (see Rutherford, 1959, p. 6). If the flow is rapid, the motion of the particle is governed by significant acceleration components which act in addition to those defined by $\partial^2 u/\partial t^2$, etc., and under these circumstances the stress equations of small motion (2-52) require additional terms. In this section, only the slow creeping flows such as are likely to occur in rocks will be considered. Because the accelerations in the fluid are extremely small under these circumstances, the stress equations of small motion are still valid.

The volume change produced by infinitesimal strain was derived in Sec. 4-16,

$$\Delta = e_1 + e_2 + e_3 \tag{4-88}$$

and because this is the quantity K_1, the first invariant of the infinitesimal-strain tensor, it follows that

$$\Delta = e_x + e_y + e_z$$

This result can be adapted to express the rate of volume change in a fluid

$$\dot{\Delta} = \dot{e}_x + \dot{e}_y + \dot{e}_z = \frac{\partial \dot{u}}{\partial x} + \frac{\partial \dot{v}}{\partial y} + \frac{\partial \dot{w}}{\partial z} \tag{6-84}$$

Fluids are practically incompressible, and under these conditions this equation simplifies into the important *continuity equation*

$$\frac{\partial \dot{u}}{\partial x} + \frac{\partial \dot{v}}{\partial y} + \frac{\partial \dot{w}}{\partial z} = 0 \tag{6-85}$$

Equations relating stress and strain in viscous fluids   The equations of slow fluid flow will be derived in a similar way to that used to discover the stress-strain relationships for solids. There are important analogs and differences between the two sets of equations.

Consider a cylindrical element of fluid, subject it to an axial tension σ_1 and allow it to flow in the direction of the applied stress at a constant rate $\dot{e}_1$ which is proportional to the stress,

$$\sigma_1 = \eta \dot{e}_1 \tag{6-86}$$

where η is the viscous tensile modulus. As a result of this axial flow, material will move in from the sides at a rate $\dot{e}_2 (= \dot{e}_3)$. Let the ratio $\dot{e}_2/\dot{e}_1 = V$ (cf. Poisson's ratio v for solids. If there is no volume change during flow, it follows from (6-84) that

$$0 = \frac{\sigma_1}{\eta} - \frac{\sigma_1 V}{\eta} - \frac{\sigma_1 V}{\eta} = \frac{\sigma_1}{\eta}(1 - 2V)$$

or

$$V = 0.5 \tag{6-87}$$

Because of this fixed value of V it follows that the stress-strain rate relationships of an incompressible fluid can be expressed in terms of one modulus (cf. elastic deformation where two elastic moduli are required).

Consider a small cubic element within a fluid which is subjected to the stresses σ_x, σ_y, and σ_z perpendicular to its sides. Using the same reasoning as was employed for the elastic deformation, the principal strain rates can be expressed in terms of the stresses:

$$\dot{e}_x = \frac{\sigma_x}{\eta} - \frac{\sigma_y}{2\eta} - \frac{\sigma_z}{2\eta} = \frac{(2\sigma_x - \sigma_y - \sigma_z)}{2\eta} \tag{6-88}$$

But the expression in parentheses represents three times the deviatoric principal stress σ'_x (from 2-44). It therefore follows that the stress-strain relationships can be expressed very simply as

$$\sigma'_x = \frac{2\eta \dot{e}_x}{3} \qquad \sigma'_y = \frac{2\eta \dot{e}_y}{3} \qquad \sigma'_z = \frac{2\eta \dot{e}_z}{3} \tag{6-89}$$

The equations expressing stresses in terms of strain rates can also be presented in another way [cf. the development of the elastic deformation equation (6-48)] using viscous moduli λ and μ (cf. Lamé's parameters),

$$\sigma_x = (\lambda + 2\mu)\dot{e}_x + \lambda\dot{e}_y + \lambda\dot{e}_z = 2\mu\dot{e}_x + \lambda\dot{\Delta} \tag{6-90a}$$

$$\sigma_y = \lambda\dot{e}_x + (\lambda + 2\mu)\dot{e}_y + \lambda\dot{e}_z = 2\mu\dot{e}_y + \lambda\dot{\Delta} \tag{6-90b}$$

$$\sigma_z = \lambda\dot{e}_x + \lambda\dot{e}_y + (\lambda + 2\mu)\dot{e}_z = 2\mu\dot{e}_z + \lambda\dot{\Delta} \tag{6-90c}$$

Adding these three equations and using (6-84),

$$\sigma_x + \sigma_y + \sigma_z = 3\bar{\sigma} = (2\mu + 3\lambda)\dot{\Delta}$$

This equation expressing volume change in terms of hydrostatic stress $\bar{\sigma}$ is similar to the equation for the elastic state (6-44) but with the elastic bulk modulus K equivalent to a viscosity modulus $(2\mu + 3\lambda)$. Because this modulus is unimportant in fluids where volume changes are very small, it follows that

$$2\mu + 3\lambda \approx 0$$

or
$$\lambda \approx -\frac{2}{3}\mu \qquad (6\text{-}91)$$

For a very slightly compressible fluid it is therefore possible to express Eqs. (6-90) in terms of one viscous modulus μ.

$$\sigma_x = \sigma'_x - \bar{\sigma} = 2\mu\dot{e}_x - \frac{2}{3}\mu\dot{\Delta}$$

$$\sigma_y = \sigma'_y - \bar{\sigma} = 2\mu\dot{e}_y - \frac{2}{3}\mu\dot{\Delta} \qquad (6\text{-}92)$$

$$\sigma_z = \sigma'_z - \bar{\sigma} = 2\mu\dot{e}_z - \frac{2}{3}\mu\dot{\Delta}$$

The deviatoric and hydrostatic parts of the stress can be related directly to the deformation and dilation in the fluid. If the fluid is incompressible, $\dot{\Delta} = 0$; and comparing Eqs. (6-92) with (6-89), it follows that

$$\eta = 3\mu \qquad (6\text{-}93)$$

Table 6-3 *Comparison of moduli for elastic solids and viscous fluids.*

Type of modulus	Elastic solid	Fluid		
		Incom-pressible	Slightly compressible	Compressible
Tensile modulus	Young's modulus $E = 2G(1 + v)$	$\eta \approx 3\mu$	$\eta \approx 3\mu$	$\eta = 2\mu(1 + V)$
Shear modulus	G	μ	μ	μ
Compressibility	$1/K = \dfrac{3(1 - 2v)}{2G(1 + v)}$	0	≈ 0	$\dfrac{3(1 - 2V)}{2\mu(1 + V)}$
Ratio of latitudinal to longitudinal strains (or strain rates)	Poisson's ratio v	0.5	≈ 0.5	$V(< 0.5)$

The correspondence of the various elastic and viscous moduli should be clearly apparent from this parallel development of the stress-strain equations (Table 6-3). If the fluid is compressible a second viscosity modulus must be introduced in order to define the stress-strain relationships completely.

Equation of fluid motion   The equation of motion of a very viscous, practically incompressible fluid can be found from the stress equations of small motion (2-52). Because the fluid movement is very small, the accelerations $\partial^2 u/\partial t^2$, $\partial^2 v/\partial t^2$, and $\partial^2 w/\partial t^2$ can be assumed to have a zero value. Substituting the deviatoric and hydrostatic components for the total stress σ_x, (2-52a) becomes

$$0 = X + \frac{\partial \bar{\sigma}}{\partial x} + \frac{\partial \sigma'_x}{\partial x} + \frac{\partial \tau_{yx}}{\partial y} + \frac{\partial \tau_{zx}}{\partial z} \tag{6-94}$$

Substituting for the stresses in terms of strain rate using (6-80) and (6-90a),

$$0 = X + \frac{\partial \bar{\sigma}}{\partial x} + 2\mu \frac{\partial \dot{e}_x}{\partial x} - \frac{2}{3}\mu \frac{\partial \dot{\Delta}}{\partial x} + \mu \left(\frac{\partial \dot{\gamma}_{yx}}{\partial y} + \frac{\partial \dot{\gamma}_{zx}}{\partial z} \right)$$

Expressing $\dot{e}$ and $\dot{\gamma}$ in terms of their velocity components $\dot{u}$, $\dot{v}$, and $\dot{w}$ and using (6-84),

$$0 = X + \frac{\partial \bar{\sigma}}{\partial x} + \frac{\mu}{3}\frac{\partial \dot{\Delta}}{\partial x}$$

$$+ \mu\left(\frac{2\partial^2 \dot{u}}{\partial x^2} - \frac{\partial^2 \dot{u}}{\partial x^2} - \frac{\partial^2 \dot{v}}{\partial x\,\partial y} - \frac{\partial^2 \dot{w}}{\partial x\,\partial z} + \frac{\partial^2 \dot{u}}{\partial y^2} + \frac{\partial^2 \dot{v}}{\partial x\,\partial y} + \frac{\partial^2 \dot{u}}{\partial z^2} + \frac{\partial^2 \dot{w}}{\partial x\,\partial z} \right)$$

which simplies to

$$0 = X + \frac{\partial \bar{\sigma}}{\partial x} + \frac{\mu}{3}\frac{\partial \dot{\Delta}}{\partial x} + \mu\left(\frac{\partial^2 \dot{u}}{\partial x^2} + \frac{\partial^2 \dot{u}}{\partial y^2} + \frac{\partial^2 \dot{u}}{\partial z^2} \right) \tag{6-95}$$

Or, employing the notation $\nabla^2 = \partial^2/\partial x^2 + \partial^2/\partial y^2 + \partial^2/\partial z^2$,

$$0 = X + \frac{\partial \bar{\sigma}}{\partial x} + \frac{\mu}{3}\frac{\partial \dot{\Delta}}{\partial x} + \mu\nabla^2 \dot{u} \tag{6-96a}$$

Similarly,

$$0 = Y + \frac{\partial \bar{\sigma}}{\partial y} + \frac{\mu}{3}\frac{\partial \dot{\Delta}}{\partial y} + \mu\nabla^2 \dot{v} \tag{6-96b}$$

$$0 = Z + \frac{\partial \bar{\sigma}}{\partial z} + \frac{\mu}{3}\frac{\partial \dot{\Delta}}{\partial z} + \mu\nabla^2 \dot{w} \tag{6-96c}$$

These are the fundamental equations of flow for slow-moving, slightly compressible fluids known as the *Navier-Stokes* equations. For two-dimensional incompressible flow with no body force, they simplify to give

$$\frac{1}{\mu}\frac{\partial\bar{\sigma}}{\partial x} + \frac{\partial^2\dot{u}}{\partial x^2} + \frac{\partial^2\dot{u}}{\partial z^2} = 0 \qquad (6\text{-}97a)$$

$$\frac{1}{\mu}\frac{\partial\bar{\sigma}}{\partial z} + \frac{\partial^2\dot{w}}{\partial x^2} + \frac{\partial^2\dot{w}}{\partial z^2} = 0 \qquad (6\text{-}97b)$$

6-13   STREAM FUNCTIONS AND THE SOLUTION OF PROBLEMS OF VISCOUS FLOW

In Sec. 6-11 it was shown how certain problems of stress distribution could be solved by choosing a stress function which satisfied the basic laws of elasticity and from which the stress components could be derived. Some problems of slow viscous flow may be investigated by use of a similar method. A function ψ known as a *stream function* is chosen such that the velocities of the movement parallel to the x and z axes may be expressed as partial derivatives:

$$\dot{u} = -\frac{\partial\psi}{\partial z} \qquad (6\text{-}98a)$$

$$\dot{w} = \frac{\partial\psi}{\partial x} \qquad (6\text{-}98b)$$

This choice means that the stress function automatically satisfies the continuity equation (6-85) for incompressible flow,

$$\frac{\partial\dot{u}}{\partial x} + \frac{\partial\dot{w}}{\partial z} = 0 \qquad (6\text{-}99)$$

We now investigate what further limitations must be put on the stress function for it to satisfy the Navier-Stokes equations and therefore be a correct solution to some problem of fluid flow. Substituting (6-98) in (6-97),

$$\frac{1}{\mu}\frac{\partial\bar{\sigma}}{\partial x} - \frac{\partial^3\psi}{\partial x^2\,\partial z} - \frac{\partial^3\psi}{\partial z^3} = 0 \qquad (6\text{-}100a)$$

$$\frac{1}{\mu}\frac{\partial\bar{\sigma}}{\partial x} + \frac{\partial^3\psi}{\partial x^3} + \frac{\partial^3\psi}{\partial z^2\,\partial x} = 0 \qquad (6\text{-}100b)$$

Differentiating (6-100a) with respect to z, (6-100b) with respect to x, and subtracting,

$$\frac{\partial^4\psi}{\partial x^4} + \frac{2\partial^4\psi}{\partial x^2\,\partial z^2} + \frac{\partial^4\psi}{\partial z^4} = 0 \qquad (6\text{-}101)$$

or

$$(\nabla^2\psi)^2 = 0 \qquad (6\text{-}102)$$

This is a biharmonic equation similar to that derived for the stress function ϕ.

If a solution to this equation can be found which satisfies the specific boundary conditions of the deformation, then this stream function gives the complete solution of velocity components, strain rates, and stress variations for that particular problem. Ramberg (1961, 1963) has employed this method to discover the nature of the buckles which develop in compressed viscous layers by using stream functions of the type

$$\psi = f(y) \cos \omega x$$

From these stream functions Ramberg has calculated the stresses which resist the development of the buckles, and determined the characteristic wavelength of the most easily formed folds.

6-14   PLASTIC FLOW

A material is said to behave plastically if, at stresses below a critical level, it reacts as a solid; while at stresses at or above the critical level (or *yield stress*), it flows continuously without rupture to become permanently strained. The principal difference between plastic flow and true-fluid flow is that real fluids do not possess a fundamental strength—their yield stress is zero.

The mathematical theory of plastic flow is mostly built up around the concept of idealized isotropic and homogeneous materials, and their special models known as *rigid-plastic* and *elastic-plastic* types (Fig. 6-33). Various rules for the behavior of strain-hardening materials have been proposed, but these are rather complex and difficult to use. The mathematical theories of plastic deformation in heterogeneous and anisotropic materials, such as are commonplace in geological problems, are practically insoluble with present methods of analysis (Hill, 1950; Olszak and Urbanowski, 1959). These problems are of great importance because the process of plastic deformation itself leads to inhomogeneity of fabric even in initially homogeneous materials.

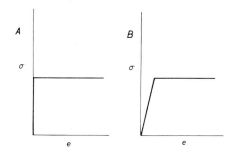

Figure 6-33

Generalized stress-strain curves for A, rigid-plastic and B, elastic-plastic materials.

The critical stress level required for yield   Consider the stresses which act on a material as represented in a three-dimensional stress diagram (Fig. 6-34).

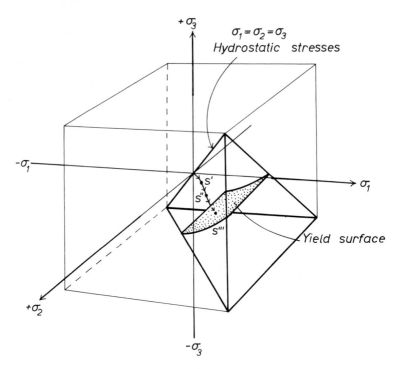

Figure 6-34

Changing stress states s′ and s″ leading to yield of the material at state s‴.

Because of the restrictions placed on the values of the principal stresses $(\sigma_1 \geqslant \sigma_2 \geqslant \sigma_3)$, only one-sixth of the space around the coordinate axes is taken up by points representing possible states of stress. From the initial state of zero stress, the stress history can be shown as a *stress path*. This is a line passing through points representing successive states of stress $s′$, $s″$, $s‴$, etc. The material may behave elastically over the initial part of the stress path $(s′, s″)$ but at a certain critical stress level $s‴$ the material begins to yield plastically. Other states of stress on other stress paths may also cause yielding, and the loci of all possible yield stresses defines a surface in the diagram known as the *yield surface*.

The stress conditions which initiate plastic yield must be independent of the coordinate system used to define the stresses, and it therefore follows that the shape of the yield surface (the yield criterion) must be some function of the three invariants of the stress tensor (see Sec. 2-9):

$$I_1 = \sigma_1 + \sigma_2 + \sigma_3 \tag{6-103a}$$

$$I_2 = \sigma_1\sigma_2 + \sigma_2\sigma_3 + \sigma_3\sigma_1 \tag{6-103b}$$

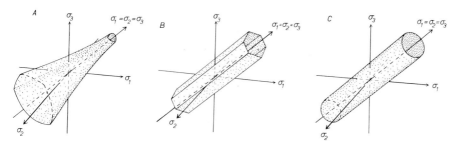

Figure 6-35

Yield surfaces. A, generalized surface; B, Tresca hexagon; C, Von Mises cylinder.

$$I_3 = \sigma_1\sigma_2\sigma_3 \tag{6-103c}$$

that is,
$$f(I_1, I_2, I_3) = \text{constant} \tag{6-104}$$

It is found that the hydrostatic state $(\bar{\sigma})$ does not influence the yield stress, and because $I_1 = 3\bar{\sigma}$ (2-42) it follows that the yield criteria must be a function only of I_2 and I_3, or a function of the invariants of the deviatoric stress system I_2' and I_3', where

$$I_1' = \sigma_1' + \sigma_2' + \sigma_3' = 0 \tag{6-105a}$$

$$I_2' = -(\sigma_1'\sigma_2' + \sigma_2'\sigma_3' + \sigma_3'\sigma_1') \tag{6-105b}$$

$$I_3' = \sigma_1'\sigma_2'\sigma_3' \tag{6-105c}$$

The yield criteria can therefore be expressed as

$$f(I_2, I_3) = \text{constant} \tag{6-106a}$$

$$f(I_2', I_3') = \text{constant} \tag{6-106b}$$

From the symmetry of these invariants, where σ_1, σ_2, and σ_3 may be interchanged with each other without altering the expression, it follows that the yield surface can be represented as a surface with threefold symmetry about the line $\sigma_1 = \sigma_2 = \sigma_3$, that is, the line of hydrostatic stress states (Fig. 6-35A). If restrictions are placed on the stress values, then only one-sixth of this surface will cut the zone of possible stress states shown in Fig. 6-34. The principal yield criteria that have been proposed will now be examined further.

Tresca yield criterion This, the simplest workable criterion for plastic yield, was put forward by Tresca in 1864 following his studies on the extrusion of metals through variously shaped dies. It was suggested that, because the flow in a body is governed by the values of the shear stresses, yielding will occur when these shearing stresses reach some prescribed maximum value. From Sec. 2-11 this criterion may be expressed as

$$\frac{\sigma_1 - \sigma_3}{2} = \frac{\sigma_1' - \sigma_3'}{2} = \frac{k}{2} \tag{6-107}$$

Put in this form it is an essentially two-dimensional concept. The significance of the constant k may be found by subjecting the material to simple tension $(\sigma_1 = \sigma_t, \sigma_3 = 0)$; then k represents the yield stress in tension (or simple compression).

The yield surface, being independent of values of σ_2, is a plane parallel to σ_2 inclined at 45° to the σ_1 and σ_3 axes, and cutting them at points σ_t and $-\sigma_t$, respectively. If no restrictions are placed on the relative magnitudes of the principal stresses, then the yield surface is a regular hexagonal prism about the central line $\sigma_1 = \sigma_2 = \sigma_3$, known as the *Tresca hexagonal prism* (Fig. 6-35B).

If one of the principal stresses is kept constant, the values of the others which limit the yield condition fall on a hexagon which represents a section through the hexagonal prism (Fig. 6-37A).

Von Mises yield criterion In 1913 Von Mises put forward what is now the most generally accepted criterion of yielding. He suggested that only the invariant I_2' was involved in the yield function:

$$I_2' = -(\sigma_1'\sigma_2' + \sigma_2'\sigma_3' + \sigma_3'\sigma_1') = c^2 \tag{6-108a}$$

This can be expressed in a number of forms. From (6-105a) it follows that

$$0 = I_1'^2 = \sigma_1'^2 + \sigma_2'^2 + \sigma_3'^2 + 2(\sigma_1'\sigma_2' + \sigma_2'\sigma_3' + \sigma_3'\sigma_1')$$

or

$$-(\sigma_1'\sigma_2' + \sigma_2'\sigma_3' + \sigma_3'\sigma_1') = \tfrac{1}{2}(\sigma_1'^2 + \sigma_2'^2 + \sigma_3'^2) = c^2 \tag{6-108b}$$

Or using (2-48a),

$$c^2 = I_2' = 3\left(\frac{\sigma_1 + \sigma_2 + \sigma_3}{3}\right)^2 - \left(\sigma_1\sigma_2 + \sigma_2\sigma_3 + \sigma_3\sigma_1\right)$$

$$3c^2 = \sigma_1^2 + \sigma_2^2 + \sigma_3^2 - \sigma_1\sigma_2 - \sigma_2\sigma_3 - \sigma_3\sigma_1$$

$$6c^2 = (\sigma_1 - \sigma_2)^2 + (\sigma_2 - \sigma_3)^2 + (\sigma_3 - \sigma_1)^2 \tag{6-108c}$$

The significance of the constant c can be found by considering yielding under simple tensions $\sigma_1 = \sigma_t, \sigma_2 = \sigma_3 = 0$. Then from (6-108c)

$$6c^2 = 2\sigma_t^2$$

$$c = \frac{\sigma_t}{(3)^{\frac{1}{2}}} \tag{6-109}$$

Every state of stress can be considered as a combination of a hydrostatic state and a stress deviator (Fig. 6-36). If D is the length of the diagonal of the "deviatoric box," its direction cosines (l_1, m_1, and n_1) are given by

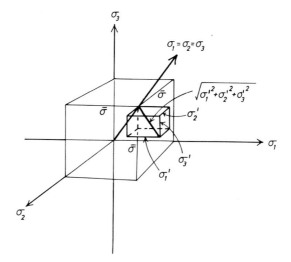

Figure 6-36

Graphic representation of the hydrostatic and deviatoric components of a stress system.

$$l_1 = \frac{\sigma_1'}{D} \qquad m_1 = \frac{\sigma_2'}{D} \qquad n_1 = \frac{\sigma_3'}{D}$$

The direction cosines l_2, m_2, and n_2 of the hydrostatic line are all equal to $1/(3)^{\frac{1}{2}}$. The angle between the diagonal of the deviatoric box and the hydrostatic line is given by (2-15),

$$\cos \alpha = l_1 l_2 + m_1 m_2 + n_1 n_2$$

$$\cos \alpha = \frac{1}{(3)^{\frac{1}{2}} D} (\sigma_1' + \sigma_2' + \sigma_3')$$

$$\cos \alpha = 0 \qquad \text{(from 6-105a)}$$

The two lines are always perpendicular. The yield criterion expressed in the form of (6-108b) implies that this diagonal D has a constant length, and it therefore follows that the yield surface is a circular cylinder at a distance $(2/3)^{\frac{1}{2}} \sigma_t$ from the hydrostatic line $\sigma_1 = \sigma_2 = \sigma_3$.

The physical significance of the Von Mises yield criterion is open to a number of interpretations. As expressed in (6-108c) it is identical to the expression of stored elastic deformation energy (6-53b), and Huber and Hencky have suggested that yielding occurs when the stored energy of deformation reaches some critical value. Nadai (1950) has noted that the criterion expressed in the form (6-108c) is similar to that of the shearing stress on a plane which is inclined at the same angle to all three principal stress axes [Eq. 2-38, where $l = m = n = 1/(3)^{\frac{1}{2}}$, the *octahedral plane*], and he suggested that yielding occurs when the shear stresses on these surfaces reach some critical limiting value.

A number of experiments made on deformed metals confirm that the Von Mises criterion provides a more accurate account of the yielding conditions than does the Tresca criterion (Taylor and Quinney, 1931; Siebel, 1953; Hundy and Green, 1954; Lianis and Ford, 1957). Little work has been done to see if these conditions of yielding hold in rock materials.

Any section through the yield cylinder will give a two-dimensional yield ellipse. Consider the ellipses which represent the yield locus where σ_2 has some constant value which can be defined in terms of the simple tensile yield strength σ_t,

$$\sigma_2 = n\sigma_t$$

The Von Mises criterion for yield is given by

$$(\sigma_1 - n\sigma_t)^2 + (n\sigma_t - \sigma_3)^2 + (\sigma_3 - \sigma_1)^2 = 2\sigma_t^2$$

or $$\sigma_1^2 + \sigma_3^2 - n\sigma_t\sigma_1 - n\sigma_t\sigma_3 - \sigma_1\sigma_3 + \sigma_t^2(n^2 - 1) = 0 \qquad (6\text{-}110)$$

This is the equation for a series of ellipses and because σ_1 and σ_3 are interchangeable it follows that the line $\sigma_1 = \sigma_3$ is a line of symmetry through all the ellipses. Three ellipses of this family are plotted in Fig. 6-37B with different values of n. All have ratios of length of major and minor axes of $(3)^{\frac{1}{2}}$ because all represent parallel sections through the same yield cylinder.

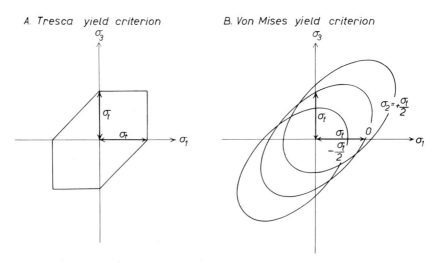

Figure 6-37

Two-dimensional sections through A, the Tresca hexagonal yield surface; B, the Von Mises cylindrical yield surface; the three ellipses represent yield conditions for values of $n = +\frac{1}{2}, 0,$ and $-\frac{1}{2}$.

An excellent confirmation of the validity of two-dimensional yield ellipses of this type has been given by Kehle (1964) from a study of the deformation of ice on the Ross ice shelf of Antarctica. The strain rates at different points of the glacier surface were determined by observing the displacement of a series of markers, and from these measurements the principal stresses were calculated assuming that the principal strain increments were proportional to the principal stresses. The principal stresses σ_1 and σ_3 act parallel to the glacier surface, and close to the surface where the measurements were made the intermediate stress σ_2 was assumed to take a fairly constant value. The values of the surface stresses were plotted graphically (Fig. 6-38). Stresses in the zones where the ice was flowing and crevasses actively forming were termed *unstable*, and those in regions without active crevasse formation were termed *stable*. The limit of stable and unstable stresses defines the two-dimensional yield ellipse of (6-110).

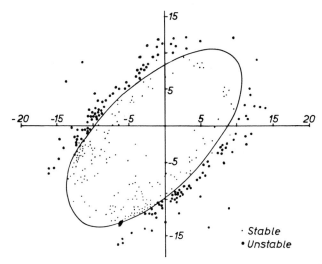

Figure 6-38

Plots of unstable and stable stress states (values in bars) in ice from the Ross ice sheet. (After Kehle, 1964.)

Laws of plastic flow The basic law governing the flow of plastic material was originally suggested by St. Venant in 1870. He proposed that the principal axes of the *strain increments* were parallel to the principal axes of stress. Lévy (1871) and Von Mises (1913) independently suggested that the ratios of the incremental strains were directly proportional to the ratios of the magnitude of the stresses. With reference to coordinate axes x, y, and z, the *Lévy-Mises equations* are

$$\frac{\delta e_x - \delta e_y}{\sigma_x - \sigma_y} = \frac{\delta e_y - \delta e_z}{\sigma_y - \sigma_z} = \frac{\delta e_z - \delta e_x}{\sigma_z - \sigma_x} = \psi \tag{6-111}$$

where δe_x, δe_y, and δe_z are the strain increments and ψ is a factor of proportionality.

It is found experimentally that plastic flow proceeds under conditions of nearly constant volume even when the permanent deformations are very large. From (4-88)

$$\delta e_x + \delta e_y + \delta e_z = \delta e_1 + \delta e_2 + \delta e_3 = 0 \tag{6-112}$$

From (6-111) it follows that

$$\delta e_x - \delta e_y = \psi(\sigma_x - \sigma_y)$$

$$\delta e_x - \delta e_z = \psi(\sigma_x - \sigma_z)$$

Adding these equations and using (6-112) in the form $-(\delta e_y + \delta e_z) = \delta e_x$,

$$\delta e_x = \frac{2\psi}{3} \left[\sigma_x - \tfrac{1}{2}(\sigma_y + \sigma_z) \right] \tag{6-113}$$

This is the basic equation relating incremental extensions to the principal stresses; its form is very similar to the equation defining elastic strain (6-45), where $2\psi/3$ replaces $1/E$ and $1/2$ replaces Poisson's ratio v. The simplest form of the stress-strain relationships in plastic materials comes from the recognition that the terms in square brackets have the value $3/2$ of the deviatoric stress σ_x' [see Eq. (2-44)],

$$\delta e_x = \psi\sigma_x' \tag{6-114a}$$

also

$$\delta e_y = \psi\sigma_y' \tag{6-114b}$$

$$\delta e_z = \psi\sigma_z' \tag{6-114c}$$

The incremental shearing strains can be expressed in terms of the principal strain increments and the direction cosines l, m, and n from (4-87),

$$\left(\frac{\delta\gamma}{2}\right)^2 = (\delta e_1 - \delta e_2)^2 l^2 m^2 + (\delta e_2 - \delta e_3)^2 m^2 n^2 + (\delta e_3 - \delta e_1)^2 n^2 l^2$$

From (6-111), $\delta e_1 - \delta e_2 = \psi(\sigma_1 - \sigma_2)$, etc., and therefore

$$\left(\frac{\delta\gamma}{2}\right)^2 = \psi^2 [(\sigma_1 - \sigma_2)^2 l^2 m^2 + (\sigma_2 - \sigma_3)^2 m^2 n^2 + (\sigma_3 - \sigma_1)^2 n^2 l^2]$$

The function in the square brackets is equal to τ^2 from (2-38) and therefore

$$\delta\gamma = 2\psi\tau$$

The six basic equations for plastic flow may therefore be expressed as

$$\delta e_x = \psi\sigma'_x \tag{6-114a}$$

$$\delta e_y = \psi\sigma'_y \tag{6-114b}$$

$$\delta e_z = \psi\sigma'_z \tag{6-114c}$$

$$\delta\gamma_{xy} = 2\psi\tau_{xy} \tag{6-115a}$$

$$\delta\gamma_{yz} = 2\psi\tau_{yz} \tag{6-115b}$$

$$\delta\gamma_{zx} = 2\psi\tau_{zx} \tag{6-115c}$$

There are close analogies in the form of these equations and those for incompressible viscous flow [(6-80), (6-89)]. However, there is no direct correspondence between the viscosity constants and the proportionality factor ψ. The factor ψ is *not* a physical constant; its value is dependent on the yield criterion and it varies from point to point in the material during deformation depending on the equilibrium and boundary conditions.

The equations established above refer to a rigid-plastic material, one where the elastic components of deformation are absent, or are so small in relation to the plastic strains that they can be neglected. If the elastic strains produce a significant contribution to the strain, then these equations are amended to the form known as the *Prandtl-Reuss equations*. The elastic-strain increment depends on the changes that go on in the stress state, that is, $\delta\sigma_x$, $\delta\sigma_y$, and $\delta\sigma_z$:

$$\delta e_x = \frac{2}{3}\psi\left[\sigma_x - \tfrac{1}{2}(\sigma_y + \sigma_z)\right] + \frac{1}{E}\left[\delta\sigma_x - v(\delta\sigma_y + \delta\sigma_z)\right] \tag{6-116}$$

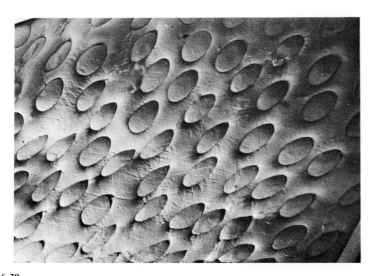

Figure 6-39

Slip lines in deformed clay.

Slip lines   The internal adjustments in plastically deforming material are sometimes accomplished by slip on shear surfaces which may appear as visible discontinuities in the material (Fig. 6-39). These surfaces are planes of maximum rate of shearing stress, or maximum incremental shearing stress. These surfaces form orthogonal sets oriented at 45° to the principal stresses σ_1 and σ_3, and are known as *slip lines* (Prager and Hodge, 1951; Hill, 1950; Ford, 1963). The study of the development of these shear planes, known as the *theory of slip-line fields* has been used to solve a number of problems in plasticity. These studies indicate that there may exist different velocity components across the slip lines, even though the stress states remain continuous across the lines. The mathematical theory of the development of these *velocity discontinuities* will not be considered here, but can be found in Prager and Hodge, 1951 and Ford, 1963. Odé (1960) has suggested that faulting in plastically deforming material might be the result of the development of velocity discontinuities, and also it seems possible that the discontinuities that develop in rocks during the process of folding (microlithon structure and strain-slip cleavage) might be explained in this way.

6-15   PROGRESSIVE DEFORMATION

One of the characteristic features of the deformation of liquid or plastic materials is that flow leads to the development of large permanent strains. The process whereby a large finite strain can be built up by the superposition of a number of strain increments will now be examined.

Consider the deformation history of a cross section of a small unit of rock material (Fig. 6-40). At time t_o flow begins, the incremental strain (increment i_1) between time t_o and t_i depends on the stress states that exist through this period, and is recorded by the strain ellipse $i_1 = f_1$. From time t_1 to t_2 stress states lead to the superposition of increment i_2 on the strained state i_1. This leads to the formation of a finite-strain ellipse f_2 according to the principles discussed in Sec. 3-11 and 4-11. The progress of deformation continues by the superposition of increment i_3 on the finite strain f_2 to give the ellipse f_3. The successive three-dimensional changes in strain can be recorded graphically (Sec. 4-6) and define a *strain path* or *deformation path*. The problem of relating the finite strains to the strain increments (and consequently to the stress states) is a complex one; we have to examine the general process of the continuous oblique superposition of two strain tensors through time, the finite-strain tensor and the strain-increment tensor.

Progressive strain in two dimensions   In Chap. 3 it was shown how a homogeneous strain can be defined in terms of an algebraic rule by which points with coordinates (x,y) are transferred to new positions (x_1,y_1):

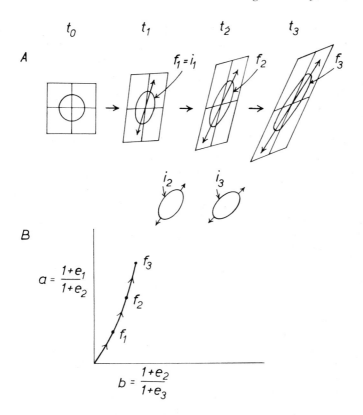

Figure 6-40

Progressive deformation, illustrating how the successive finite-strain states f_1, f_2, and f_3 are developed at times t_1, t_2, and t_3 by incremental strains i_1, i_2, and i_3. B is a deformation path illustrating the changes in principal strain ratios of the states of finite strain in three dimensions.

$$x_1 = a_1 x + b_1 y$$

$$y_1 = c_1 x + d_1 y$$

This transformation defining a finite strain f_1 is more simply written in matrix form

$$Mf_1 = \begin{vmatrix} a_1 & b_1 \\ c_1 & d_1 \end{vmatrix} \qquad (6\text{-}117)$$

The process of progressive deformation consists of transposing (x_1, y_1) to a new position (x_2, y_2) by the superposition of some similar law which defines the incremental strain:

$$x_2 = a_2 x_1 + b_2 y_1$$

$$\text{or} \quad Mi = \begin{vmatrix} a_2 & b_2 \\ c_2 & d_2 \end{vmatrix} \tag{6-118}$$

$$y_2 = c_2 x_1 + d_2 y_1$$

The finite strain produced by superposition of the increment defined by Mi on the existing strain defined by Mf_1 is found by combining the transformations (6-118) and (6-117),

$$x_2 = a_2(a_1 x + b_1 y) + b_2(c_1 x + d_1 y)$$

$$y_2 = c_2(a_1 x + b_1 y) + d_2(c_1 x + d_1 y)$$

or

$$x_2 = x(a_1 a_2 + c_1 b_2) + y(b_1 a_2 + d_1 b_2)$$

$$y_2 = x(a_1 c_2 + c_1 d_2) + y(b_1 c_2 + d_1 d_2)$$

This total transformation can be written in matrix form Mf_2 as the product of the two matrices Mf_1 and Mi:

$$Mf_2 = Mf_1 \times Mi = \begin{vmatrix} a_1 a_2 + c_1 b_2 & b_1 a_2 + d_1 b_2 \\ a_1 c_2 + c_1 d_2 & b_1 c_2 + d_1 d_2 \end{vmatrix} \tag{6-119}$$

This important rule for matrix multiplication is noncommutative; the multiplication of the first matrix by the second does not in general produce the same answer as multiplication of the second by the first.

The effect of progressive deformation can now be investigated by successive multiplication of matrices representing the units of incremental strain. We shall now examine three simple examples to illustrate the process.

PROGRESSIVE SIMPLE SHEAR   The shear transformation matrix where γ is the shear increment is

$$Mi = \begin{vmatrix} 1 & \gamma \\ 0 & 1 \end{vmatrix}$$

Successive multiplications of the original undeformed state represented by the unit matrix M_o give

$$M2i = \begin{vmatrix} 1 & 2\gamma \\ 0 & 1 \end{vmatrix} \qquad M3i = M2i \times Mi = \begin{vmatrix} 1 & 3\gamma \\ 0 & 1 \end{vmatrix}$$

$$Mni = M(n-1)i \times Mi = \begin{vmatrix} 1 & n\gamma \\ 0 & 1 \end{vmatrix}$$

ROTATION   The body-rotation matrix is

$$Mi = \begin{vmatrix} \cos\theta & \sin\theta \\ -\sin\theta & \cos\theta \end{vmatrix}$$

The product of two matrices of this value is

$$Mi \times Mi = M2i = \begin{vmatrix} \cos^2 \theta - \sin^2 \theta & 2 \sin \theta \cos \theta \\ -2 \sin \theta \cos \theta & -\sin^2 \theta + \cos^2 \theta \end{vmatrix}$$

$$= \begin{vmatrix} \cos 2\theta & \sin 2\theta \\ -\sin 2\theta & \cos 2\theta \end{vmatrix}$$

and further products give

$$Mni = \begin{vmatrix} \cos n\theta & \sin n\theta \\ -\sin n\theta & \cos n\theta \end{vmatrix}$$

that is, rotation through an angle $n\theta$.

IRROTATIONAL STRAIN WITH PRINCIPAL AXES PARALLEL TO x AND y.   If the principal incremental strains are e_1 and e_2, the incremental matrix is

$$M_1 = \begin{vmatrix} 1 + e_1 & 0 \\ 0 & 1 + e_2 \end{vmatrix}$$

and successive applications of the rule of matrix multiplication of (6-119) give

$$Mni = \begin{vmatrix} (1 + e_1)^n & 0 \\ 0 & (1 + e_2)^n \end{vmatrix}$$

If e_1 and e_2 are small and the number of increments is not large, powers of e_1 and e_2 can be neglected and the matrix becomes

$$Mni = \begin{vmatrix} 1 + ne_1 & 0 \\ 0 & 1 + ne_2 \end{vmatrix}$$

For large numbers of increments, however, this approximation leads to inaccuracies, for terms like $n(n - 1)e^2/2$ have significant value.

These simple results are easily derived in other ways; they have been included here to illustrate the general method of matrix multiplication. The calculation of the finite-strain transformation from the general two-dimensional strain-increment transformation given by (3-80) looks more elaborate but is identical to the three examples above.

$$Mi = \begin{vmatrix} 1 + \dfrac{\partial u}{\partial x} & \dfrac{\partial u}{\partial y} \\ \dfrac{\partial v}{\partial x} & 1 + \dfrac{\partial v}{\partial y} \end{vmatrix}$$

$$
Mi = \begin{vmatrix} 1 + \dfrac{\partial u}{\partial x} & \dfrac{1}{2}\left(\dfrac{\partial u}{\partial y} + \dfrac{\partial v}{\partial x}\right) + \dfrac{1}{2}\left(\dfrac{\partial u}{\partial y} - \dfrac{\partial v}{\partial x}\right) \\[2ex] \dfrac{1}{2}\left(\dfrac{\partial v}{\partial x} + \dfrac{\partial u}{\partial y}\right) + \dfrac{1}{2}\left(\dfrac{\partial v}{\partial x} - \dfrac{\partial u}{\partial y}\right) & 1 + \dfrac{\partial v}{\partial y} \end{vmatrix}
$$

Or, put in terms of the strain components,

$$
Mi = \begin{vmatrix} 1 + e_x & \dfrac{\tau_{xy}}{2} - \omega \\[2ex] \dfrac{\gamma_{xy}}{2} + \omega & 1 + e_y \end{vmatrix} \tag{6-120}
$$

$$
M2i = \begin{vmatrix} (1 + e_x)^2 + \dfrac{\gamma_{xy}^2}{4} - \omega^2 & \left(\dfrac{\gamma_{xy}}{2} - \omega\right)(2 + e_x + e_y) \\[2ex] \left(\dfrac{\gamma_{xy}}{2} + \omega\right)(2 + e_x + e_y) & (1 + e_y)^2 + \dfrac{\gamma_{xy}^2}{4} - \omega^2 \end{vmatrix} \tag{6-121}
$$

The matrices representing successive products of $M2i$ with Mi, etc., and which represent the successive states of finite strain are the results of a very intricate interaction of all four components of the strain increment. There is no general simple rule for Mni, and it follows that there is *no simple relationship between the strain increments and the finite strain, or between the finite strain and stresses which produced the flow.*

Progressive strain in three dimensions   Exactly similar methods can be used for computing the successive states of three-dimensional finite strain to those described above, using 3×3 matrix representations. Any finite strain can be represented by the matrix Mf, and any strain increment by the matrix Mi, where

$$
Mf_1 = \begin{vmatrix} a_1 & b_1 & c_1 \\ d_1 & e_1 & f_1 \\ g_1 & h_1 & i_1 \end{vmatrix} \qquad Mi = \begin{vmatrix} a_2 & b_2 & c_2 \\ d_2 & e_2 & f_2 \\ g_2 & h_2 & i_2 \end{vmatrix}
$$

and the product of $Mf1 \times Mi$, which gives the new strain state, can be represented by

$$
Mf_2 = \begin{vmatrix} a_2a_1 + b_2d_1 + c_2g_1 & a_2b_1 + b_2e_1 + c_2h_1 & a_2c_1 + b_2f_1 + c_2i_1 \\ d_2a_1 + e_2d_1 + f_2g_1 & d_2b_1 + e_2e_1 + f_2h_1 & d_2c_1 + e_2f_1 + f_2i_1 \\ g_2a_1 + h_2d_1 + i_2g_1 & g_2b_1 + h_2e_1 + i_2h_1 & g_2c_1 + h_2f_1 + i_2i_1 \end{vmatrix}
$$

$$\tag{6-122}$$

Thus, according to the rules of (6-112), successive matrix multiplication of the general strain-increment matrix Mi

$$Mi = \begin{vmatrix} 1 + e_x & \dfrac{\gamma_{xy}}{2} - \omega_3 & \dfrac{\gamma_{xz}}{2} + \omega_2 \\ \dfrac{\gamma_{xy}}{2} + \omega_3 & 1 + e_y & \dfrac{\gamma_{yz}}{2} - \omega_1 \\ \dfrac{\gamma_{xz}}{2} - \omega_2 & \dfrac{\gamma_{yz}}{2} + \omega_1 & 1 + e_z \end{vmatrix}$$

will determine the progressive changes in finite-strain state. A digital computer is necessary if the number of strain increments is large. It will again be apparent that there are no simple rules for determining the finite strain from the strain increments even where the value of the components of the strain-increment matrix keep constant throughout the deformation. This fact has a number of very important implications. For example, it is possible to produce a finite strain of a constriction type $(1 + e_2) < 1$ by successive superposition of strain increments of the flattening types $(1 + e_2) > 1$ with certain values of the strain components γ and ω.

The various types of shape change that go on in materials subjected to different stress conditions will now be examined.

CONSTANT STRESS STATE WITH NO ROTATION Where the orientations and values of the principal deviatoric stresses remain constant through the flow history, the incremental strain matrix simplifies to

$$Mi = \begin{vmatrix} 1 + e_1 & 0 & 0 \\ 0 & 1 + e_2 & 0 \\ 0 & 0 & 1 + e_3 \end{vmatrix}$$

where $e_1 = \psi\sigma_1'$, $e_2 = \psi\sigma_2'$, and $e_3 = \psi\sigma_3'$, and where ψ is either a viscosity constant or the plastic proportionality factor. Progressive deformation gives states of finite strain represented by the matrix

$$Mni = \begin{vmatrix} (1 + e_1)^n & 0 & 0 \\ 0 & (1 + e_2)^n & 0 \\ 0 & 0 & (1 + e_3)^n \end{vmatrix}$$

$$= \begin{vmatrix} (1 + \psi\sigma_1')^n & 0 & 0 \\ 0 & (1 + \psi\sigma_2')^n & 0 \\ 0 & 0 & (1 + \psi\sigma_3')^n \end{vmatrix}$$

that is, a finite-strain ellipsoid with principal axes $(1 + e_1)^n > (1 + e_2)^n$

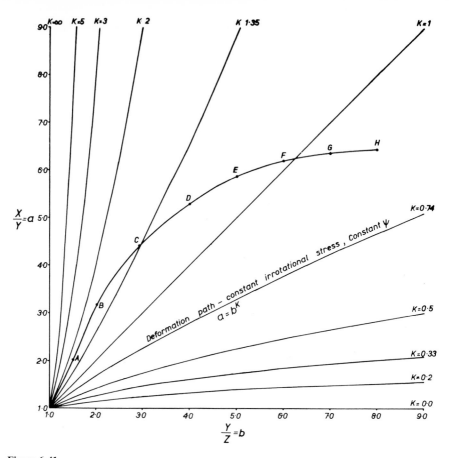

Figure 6-41

Deformation paths illustrating changes in ratios of the three principal strains. A,B,C, . . ., H represent a series of ellipsoids on a generalized deformation path.

$> (1 + e_3)^n$. The deformation path can be plotted using the ratio method with b as abscissa and a as ordinate, where

$$a = \left(\frac{1 + e_1}{1 + e_2}\right)^n \qquad b = \left(\frac{1 + e_2}{1 + e_3}\right)^n$$

If ψ remains constant during the deformation the quantities represented in the parentheses keep constant values k_1 and k_2, respectively. The deformation paths given by progressive irrotational strains of this type are given by eliminating n from the equations,

$$a = k_1^n \qquad b = k_2^n$$

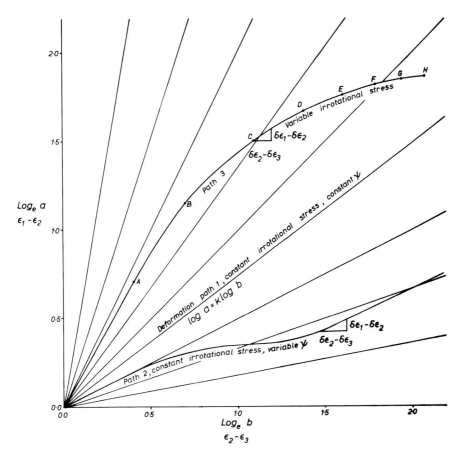

Figure 6-42

Deformation paths illustrating changes in the logarithmic ratios of the three principal strains.

Taking logarithms of both sides, dividing one by the other,

$$\log a = \log b \, \frac{\log k_1}{\log k_2}$$

or $$a = b^K \qquad \text{where } K = \frac{\log k_1}{\log k_2} \qquad (6\text{-}123)$$

These are a series of power curves (Fig. 6-41) true for deforming liquids or for plastically deforming materials at points where ψ is constant throughout the deformation. Flinn (1962) suggested that deformation paths under these conditions were straight lines. This follows from his assumption that incremental strains are added to the previously established finite ellipsoid axes, an assumption which is only true where the number of increments n is small (see Ramsay, 1964). The simplest way of representing these deformation paths is by using logarithmic ratios, plotting $\log_e [(1 + e_1)/(1 + e_2)] = \log_e a$ against $\log_e b$. These represent the expressions of natural strains $\varepsilon_1 - \varepsilon_2$ and $\varepsilon_2 - \varepsilon_3$, respectively. With this method the deformation paths become straight lines through the origin (Fig. 6-42; path 1).

$$\varepsilon_1 - \varepsilon_2 = K(\varepsilon_2 - \varepsilon_3) \tag{6-124}$$

If the function ψ does not remain constant, k_1, k_2, and K vary during the deformation even though the stresses remain constant. Thus the incremental natural strain differences represented by $\delta\varepsilon_1 - \delta\varepsilon_2$ and $\delta\varepsilon_2 - \delta\varepsilon_3$ are no longer proportional to the finite strains $\varepsilon_1 - \varepsilon_2$ and $\varepsilon_2 - \varepsilon_3$ respectively. The deformation path diverges from a straight line (Fig. 6-42, path 2).

VARIABLE STRESS STATES WITH NO ROTATION   If the values of the principal stresses vary, then the incremental strains will also vary in proportion to the values of the stress deviators. Because of the limitations placed on the deviatoric stresses given by

$$\sigma_1' + \sigma_2' + \sigma_3' = 0 \tag{6-125}$$

it follows that only certain combinations of deviatoric stress ratios are possible and in consequence only certain incremental strain ratios are possible. Dividing (6-125) by σ_1',

$$1 + \frac{\sigma_2'}{\sigma_1'} + \frac{\sigma_3'}{\sigma_2'}\frac{\sigma_2'}{\sigma_1'} = 0$$

or
$$\frac{\sigma_2'}{\sigma_1'} = -\frac{1}{1 + \sigma_3'/\sigma_2'} \tag{6-126}$$

The various types of deviatoric stress (and strain increment) systems defined by (6-126) are plotted in Fig. 6-43.

The principal strain increments e_1, e_2, and e_3 control the values of $\delta\varepsilon_1$, $\delta\varepsilon_2$, and $\delta\varepsilon_3$, and therefore the increments $\delta\varepsilon_1 - \delta\varepsilon_2$, $\delta\varepsilon_2 - \delta\varepsilon_3$ shown on Fig. 6-42 (path 3) will vary along the deformation path. The slope of the deformation path therefore reflects the nature of the deviatoric stress system at the time; slopes greater than 1 indicate a constricting type of strain increment, slopes less that 1 are produced by a flattening type of increment.

If the proportional constant ψ changes its value during deformation, further complications will develop in the deformation path. Provided that the deformation is irrotational, however, variations of this type will not alter the general rule about the interpretation of the slope of the deformation path.

VARIABLE STRESS STATE WITH ROTATION   This must be the most common type of stress history during practically all geological processes of flow, and it may lead to a strain history of considerable complexity.

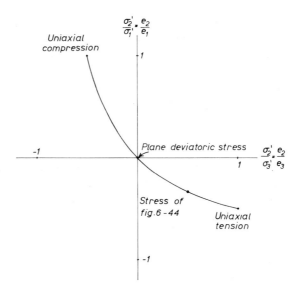

Figure 6-43

Possible types of deviatoric stress ratios.

Consider a certain point on a deformation path representing the state of finite strain at a certain time in the history. This ellipsoid has principal axes measured in logarithmic strain $\varepsilon_1 > \varepsilon_2 > \varepsilon_3$ (Fig. 6-44). Consider now the possible changes in shape of this ellipsoid as the result of deformation under a stress system $\sigma_1' > \sigma_2' > \sigma_3'$ as represented in Fig. 6-43. This gives strain increments

$$e_1 = \ \ \ 0.003\psi$$

$$e_2 = -0.001\psi$$

$$e_3 = -0.002\psi$$

The incremental strain ellipsoid has axes $\varepsilon_{i1} = \log_e (1 + 0.003\psi)$, $\varepsilon_{i2} = \log_e (1 - 0.001\psi)$, $\varepsilon_{i3} = \log_e (1 - 0.002\psi)$. These increments can be superimposed on the previously established finite-strain ellipsoid in an infinite number of ways. Six extreme combinations formed by coaxial superposition of the incremental strain ellipsoid on the finite-strain ellipsoid are given in Table 6-4. These increments have been plotted in Fig. 6-44 for values of the plastic proportionality factor $\psi = 0.5$, 1.0, and 2.0. There are an infinite number of changes of shape of the finite ellipsoid, and even though these incremental strains are of a constrictional type, under certain circumstances it is possible to make the deformation path move toward the finite field of flattened ellipsoids.

The interpretation of the shapes of naturally produced finite-strain ellipsoids is a matter of great difficulty. It is perhaps not surprising that the shapes of these strain ellipsoids are often variable even at one locality, for it is likely

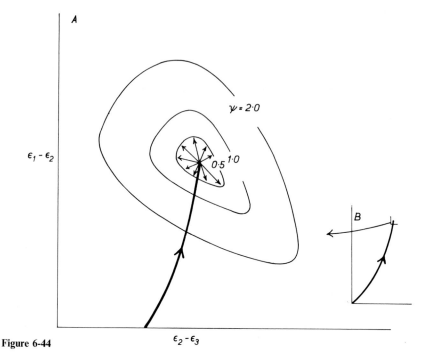

Figure 6-44

Progress of a deformation path as a result of an increment in strain. B, shows the complete deformation path in a material and A shows the possible continuation of this path as the result of the stress state that exists at that time. Depending on the orientation of the principal stresses it is possible for the path to continue along any one of the directed lines by an amount depending on the plastic proportionality factor ψ.

that the stresses, rotations, and plastic proportionality factors change during the deformation history from point to point in the rocks. The interpretation of the range of finite-strain ellipsoid shapes and their deformation fields offers an interesting but complicated problem for future research.

Table 6-4   *Values of the natural principal strains of six extreme combinations of strain state ε_1, ε_2 and ε_3, with increments ε_{i1}, ε_{i2}, and ε_{i3}*

| | Natural principal strains of new finite-strain ellipsoid | | | Increment | |
				$\delta\varepsilon_1 - \delta\varepsilon_2$	$\delta\varepsilon_2 - \delta\varepsilon_3$
i	$\varepsilon_1 + \varepsilon_{i1}$	$\varepsilon_2 + \varepsilon_{i2}$	$\varepsilon_3 + \varepsilon_{i3}$	$\varepsilon_{i1} - \varepsilon_{i2}$	$\varepsilon_{i2} - \varepsilon_{i3}$
ii	$\varepsilon_1 + \varepsilon_{i1}$	$\varepsilon_2 + \varepsilon_{i3}$	$\varepsilon_3 + \varepsilon_{i2}$	$\varepsilon_{i1} - \varepsilon_{i3}$	$\varepsilon_{i3} - \varepsilon_{i2}$
iii	$\varepsilon_1 + \varepsilon_{i2}$	$\varepsilon_2 + \varepsilon_{i3}$	$\varepsilon_3 + \varepsilon_{i1}$	$\varepsilon_{i2} - \varepsilon_{i3}$	$\varepsilon_{i3} - \varepsilon_{i1}$
iv	$\varepsilon_1 + \varepsilon_{i2}$	$\varepsilon_2 + \varepsilon_{i1}$	$\varepsilon_3 + \varepsilon_{i3}$	$\varepsilon_{i2} - \varepsilon_{i1}$	$\varepsilon_{i1} - \varepsilon_{i3}$
v	$\varepsilon_1 + \varepsilon_{i3}$	$\varepsilon_2 + \varepsilon_{i1}$	$\varepsilon_3 + \varepsilon_{i2}$	$\varepsilon_{i3} - \varepsilon_{i1}$	$\varepsilon_{i1} - \varepsilon_{i2}$
vi	$\varepsilon_1 + \varepsilon_{i3}$	$\varepsilon_2 + \varepsilon_{i2}$	$\varepsilon_3 + \varepsilon_{i1}$	$\varepsilon_{i3} - \varepsilon_{i2}$	$\varepsilon_{i2} - \varepsilon_{i1}$

6-16   THE CONCEPT OF SYMMETRY AND TECTONIC AXES

The component parts of a deformed material may be arranged in some irregular or regular order relative to each other which defines what is known as the *rock fabric*. Fabrics are nearly always heterogeneous and anisotropic on a small scale, but are often in some sort of orderly anisotropic arrangement when a larger unit is considered. If the fabric is anisotropic, the ordered arrangements may be defined in terms of *fabric symmetry*. These symmetrical representations are rarely of the same order of perfection as is found in the nearly ideal symmetry of the atomic components of crystal lattices, but are more nearly a type of *statistical symmetry*, a term which describes the average order of the particular fabric element (Paterson and Weiss, 1961; Turner and Weiss, 1963). The symmetry shown by different fabric elements (e.g., crystallographic axes, bedding surfaces, cleavage surfaces, linear structures) may be of different types and of different orders of perfection.

The nature of the symmetry of deformed rocks was recognized by geologists around the end of the nineteenth century and the geometrical forms of rock structures were described in terms of orthogonal reference axes (e.g., Heim, 1878, 1921). It was Sander (1911, 1930) and Schmidt (1926) who took up this concept and developed it into the form it is generally recognized in today. Sander suggested that many rock fabrics produced as the result of tectonic processes, termed *tectonite fabrics*, appear to have a monoclinic symmetry. Simply folded rock strata often have a plane of symmetry perpendicular to the hinge lines of the folds, and he suggested that the symmetry of such structures could be referred to the *fabric symmetry axes a, b*, and *c*, defined as follows:

ac   the symmetry plane
b   normal to the symmetry plane
a   perpendicular to *b* in the principal fabric plane (e.g., cleavage)
c   normal to *ab*

The importance of the *b* axis of the fabric parallel to the fold axes led some investigators to elevate its status, naming it the *principal fabric axis* and denoting it by the capital letter *B*.

The physical quantities of stress and strain also show a symmetry. For example, the equilibrium stress tensor has an orthorhombic symmetry and the infinitesimal- and finite-strain tensors generally have triclinic symmetry; but if the strain is irrotational the tensor is orthorhombic, and if the deformation is simple shear the tensor has monoclinic symmetry.

Sander used orthogonal axes to define the process of flow by simple shear, known as the *kinematic symmetry axes a, b*, and *c*,

a   direction of flow
ab   plane of flow with *b* perpendicular to *a*

c   normal to *ab*

Anderson (1948) used *p*, *q*, and *r* as coordinate axes equivalent to the kinematic axes *a*, *b*, and *c* to describe the same process, thus making clearer the distinction between kinematic axes and fabric axes. These sets of axes take a restricted view of the process of flow; they imply that flow is the result of progressive simple shear and so are essentially two-dimensional in their outlook. In its simplest form the finite-strain matrix for progressive simple shear given by *n* increments of the incremental matrix *Mi* is

$$
Mni = \begin{vmatrix} 1 & 0 & n\gamma \\ 0 & 1 & 0 \\ 0 & 0 & 1 \end{vmatrix}
$$

This is a highly specialized matrix and is one of the few whose finite transformations can be computed easily. Although this matrix may offer an approximate indication of the behavior of some deforming materials, it is most unlikely that it is generally applicable to the deformation process.

There is considerable confusion of usage of the kinematic *a* direction or movement direction in published works on tectonic structures. The simple shear direction *a* is sometimes confused with the finite translation (the shortest distance between the original position of point *P* and the position after deformation *P'*), and sometimes with the actual movement path of *P* to *P'*. These are both impossible to determine in naturally deformed rocks because the original position *P* can never be accurately known. The term *a* direction is often used in a rather indefinite way to indicate the approximate orogenic translation. These usages are to be avoided. The *a* direction is also frequently confused with the direction of maximum relative separation of particles of rock material, that is, the principal finite elongation or *X* direction of the finite-strain ellipsoid with axes $X > Y > Z$. These various incorrect usages are indicated in Fig. 6-45. The study of rock structure sometimes appears to be in danger of being confounded by the use of poorly defined terms that may

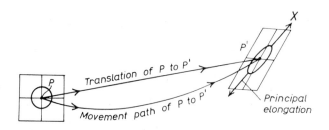

Figure 6-45

The relationship between translation, movement path, and principal elongation in a deformed material.

be both restricting to thought and confusing to the reader. In the past the study of rock deformation has often avoided the precision of real mathematical formulation to escape into a specialized type of obscure symbolic jargon. If precise description of rock structures necessitates the use of reference axes, it seems advisable to define their use clearly. Clarification of these descriptions might be attained by restricting the reference frames to three:

X Y Z principal finite strains
A B C monoclinic fabric symmetry axes equivalent to Sander's a, b, and c
a b c kinematic axes for simple shear equivalent to Sander's kinematic
        a, b, and c

Sander used the symmetry concept to correlate the fabric of a deformed rock with the symmetry of the movements which produced the fabric. This is a generally accepted principle, although Sander's limitation on types of kinematic axes often needlessly restricts its useful application. Because the deformation process generally has a triclinic symmetry, the rock fabrics that are produced also have a triclinic symmetry.

Another important principle limiting the simple application of the symmetry principle is that the rock material before deformation usually has some original fabric. The symmetry of the final product of deformation is therefore a combination of the kinematic fabric and the original fabric, and is therefore generally of a lower order than either of the components (Weiss, 1955; Ramsay, 1960, 1962; Turner and Weiss, 1963). This important modification to the symmetry principle of Sander will be developed further in the later chapters.

REFERENCES AND FURTHER READING

Elasticity, viscosity, and plasticity

Andrade, E. N. da C.: "Viscosity and Plasticity," Chemical Publishing Company, Inc., New York, 1951.

Bland, D. R.: "The Theory of Linear Viscoelasticity," Pergamon Press, London and New York, 1960.

Boltzman, L.: Zur Theorie der elastichen Nachwirkung, *Pogg. Ann. Erganzungsband*, **7**: 624-654 (1871).

Crandall, S. H., and N. C. Dahl: "An Introduction to the Mechanics of Solids," McGraw-Hill Book Company, New York, 1959.

Drysdale, C. V.: "The Mechanical Properties of Fluids," Blackie and Son, Ltd., London and Glasgow, 1946.

Eirich, F.: "Rheology, Theory and Application," Academic Press Inc., New York, 1956.

Flinn, D.: On Folding during Three-dimensional Progressive Deformation, *Quart. J. Geol. Soc.*, **118**: 385-433 (1962).

Ford, H.: "Advanced Mechanics of Materials," Longmans, Green & Co., Ltd., London, 1963.

Frankel, J. P.: "Principles of the Properties of Materials," McGraw-Hill Book Company, New York, 1957.

Goldstein, S.: "Modern Development in Fluid Mechanics," Oxford University Press, London, 1938.

Green, A. E., and W. Zerna: "Theoretical Elasticity," Oxford University Press, London, 1954.

Hencky, H.: Zur Theorie plasticher Deformationen, Z. *Angew. Math. Phys.*, **4**: 323-324 (1924).

Hill, R.: "The Mathematical Theory of Plasticity," Oxford University Press, London, 1950.

Hundy, B. B., and A. P. Green: A Determination of Plastic Stress-Strain Relations, *J. Mech. Phys. Solids*, **3**: 16-21 (1954).

Jaeger, J. C.: "Elasticity, Fracture and Flow," 2d ed., Methuen & Co., Ltd., London and John Wiley & Sons, Inc., New York, 1962.

Jameson, A. H.: "An Introduction to Fluid Mechanics," Longmans, Green & Co., Ltd., London, New York, 1937.

Johnson, W., and P. B. Mellor: "Plasticity for Mechanical Engineers," I., Van Nostrand Company, Inc., Princeton, N.J., 1962.

Kehle, R. O.: Deformation of the Ross Ice Shelf, Antarctica, *Geol. Soc. Am. Bull.*, **75**: 259-286 (1964).

Koerber, G. G.: "Properties of Solids," Prentice-Hall, Inc., Englewood Cliffs, N.J., 1962.

Lamb, H.: "Hydrodynamics," Dover Publications, Inc., New York, 1932.

Levy, M.: Extrait du Memoire sur les equations des Mouvements interieurs des corps Solides-ductiles, *J. Math. Pure Appl.*, **16**: 369-372 (1871).

Lianis, G., and H. Ford: An Experimental Investigation of the Yield Criterion and the Stress-Strain Laws, *J. Mech. Phys. Solids*, **5**: 215-222 (1957).

Long, R. R.: "Mechanics of Solids and Fluids," Prentice-Hall, Inc., Englewood Cliffs, N.J., 1961.

Love, A. E. H.: "The Mathematical Theory of Elasticity," Dover Publications, Inc., New York, 1944. (Reissue of book first published in 1892.)

Maxwell, J. C.: On the Dynamical Theory of Gases, *Phil. Mag.*, **35**: (4) 129-185 (1868).

Michelson, A. A.: Elastic Viscous Flow, *J. Geol.*, **25**: 405-410 (1917), **28**: 18-24 (1920).

Mises, R. von: Mechanik der festen Körper in plastisch-deformablen Zustand, *Nachr. Akad. Wiss. Goettingen Math. Physik Kl. IIa*, 582-592 (1913).

Nadai, A.: "The Theory of Flow and Fracture of Solids," McGraw-Hill Book Company, New York, 1950.

Olszak, W., and Urbanowski: Non-homogeneity in Elasticity and Plasticity, pp. 259-298 in Plastic Non-homogeneity Symp., Proc. of I.U.T.A.M., Pergamon Press, London and New York, 1959.

Paterson, M. S.: Lüders Bands and Plastic Deformation in the Earth's Crust, *Geol. Soc. Am. Bull.*, **68**: 129-130 (1957).

Pearson, C. E.: "Theoretical Elasticity," Harvard University Press, Cambridge, Mass., 1959.

Phillips, A.: "Introduction to Plasticity," The Ronald Press Company, New York, 1956.

Popov, E. P.: "Mechanics of Materials," Prentice-Hall, Inc., Englewood Cliffs, N.J., 1952.

Prager, W., and P. G. Hodge: "The Theory of Perfectly Plastic Solids," Chapman & Hall, Ltd., London, 1951.

Prescott, J.: "Applied Elasticity," Dover Publications, Inc., New York, 1946.

Ramberg, H.: Contact Strain and Folding Instability of a Multilayered Body under Compression, *Geol. Rundschau*, **51**: 405-439 (1961).

Ramberg, H.: Fluid Dynamics of Viscous Buckling Applicable to Folding of Layered Rocks, *Bull. Am. Assoc. Petrol. Geologists*, **47**: 485-505 (1963).

Ramsay, J. G.: Progressive Deformation in Tectonic Processes, *Am. Geophys. Union Trans.*, **45**: 106 (1964).

Reiner, M.: "Lectures of Theoretical Rheology," North Holland Publishing Company, 3d ed., Amsterdam, 1960*a*.

Reiner, M.: "Deformation, Strain and Flow," 2d ed., H. K. Lewis and Co., London, 1960*b*.

Rogers, G. L.: "Mechanics of Solids," John Wiley & Sons, Inc., New York, 1964.

Rosenthal, D.: "Introduction to Properties of Materials," D. Van Nostrand Company, Inc., Princeton, N.J., 1964.

Rouse, H.: "Elementary Mechanics of Fluids," John Wiley & Sons, Inc., New York, 1946.

Rutherford, D. E.: "Fluid Dynamics," Oliver & Boyd Ltd., Edinburgh and London, 1959.

Siebel, M. P. L.: Combined Bending and Twisting of Thin Cylinders in the Plastic Range, *J. Mech. Phys. Solids*, **1**: 189-206 (1953).

Sokolnikoff, I. S.: "Mathematical Theory of Elasticity," McGraw-Hill Book Company, New York, 1956.

Sommerfeld, A.: "Mechanics of Deformable Bodies," Academic Press Inc., New York, 1950.

Streeter, V. C.: "Fluid Mechanics," McGraw-Hill Book Company, New York, 1958.

Sperry, W. C.: Rheological Model Concept, *J. Acoust. Soc. Am.*, **36**: 376-385 (1964).

Taylor, G. I., and H. Quinney: Plastic Distortion of Metals, *Phil. Trans.*, *A***230**: 323-362 (1931).

Thomson, W.: On the Elasticity and Viscosity of Metals, Math. and Phys. Papers III, Cambridge University Press, London, 1890.

Timoshenko, S. P.: "Theory of Elastic Stability," McGraw-Hill Book Company, New York, 1936.

Timoshenko, S. P.: "Strength of Materials," D. Van Nostrand Company, Inc., Princeton, N.J., 1955.

Tresca, M. H.: Memoires sur l'ecoulement des Corps solides soumis a de Fortes Pressions, *Compt. Rend. Acad. Sci. (Paris)*, **59**: 754-758 (1864), **64**: 809-812 (1867).

Voight, W.: Ueber innere Reinbung fester Körper insbesondere der Metalle, *Ann. Physik*, **47**: 671-693 (1892).

Westergaard, H. M.: "Theory of Elasticity and Plasticity," Harvard University Press and John Wiley & Sons, Inc., New York, 1952.

Young, J. F.: "Materials and Processes," John Wiley & Sons, Inc., New York, 1954.

Zener, C. M.: "Elasticity and Anelasticity of Metals," The University of Chicago Press, Chicago, 1948.

Creep

Benioff, H.: Earthquakes and Rock Creep, *Seismol. Soc. Am. Bull.*, **41**: 31-62 (1951).

Finnie, J., and W. R. Heller: "Creep of Engineering Materials," McGraw-Hill Book Company, New York, 1959.

Griggs, D. T.: Creep in Rocks, *J. Geol.*, **47**: 225-251 (1939).

Hardy, H. R.: Time Dependent Deformation and Failure of Geological Materials, *3rd Symp. on Rock Mechanics*, Colorado School of Mines, 1959.

Kennedy, A. J.: "Processes of Creep and Fatigue in Metals," Oliver & Boyd, Ltd., Edinburgh and London, 1962.

Lomnitz, C.: Creep Measurements in Igneous Rocks, *J. Geol.*, **64**: 473-479 (1956).

Murrell, S. A. F., and A. K. Misra: Time Dependent Strain or "Creep" in Rocks and Similar Non-metallic Materials, *Trans. Inst. Mining Met.*, **71**: 353 (1962).

Orowan, E.: Creep in Metallic and Non-metallic Materials, *Proc. 1st Nat. Congr. Appl. Mech.*, *ASME*, 453-472 (1952).

Price, N. J.: A Study of the Time-Strain Behaviour of Coal-measure Rocks, *Intern. J. Rock Mech. Mining Sci.*, **1**: 277-303 (1964).

Ree, F. H., T. Ree, and H. Eyring: Relaxation Theory of Creep of Metals, *J. Eng. Mech. Div.*, *ASCE*, **86**: 41-59 (1960).

Robertson, E. C.: Creep of Solenhofen Limestone under Moderate Hydrostatic Pressure, in Rock Deformation, *Geol. Soc. Am. Mem.*, **79**: 227-244 (1960).

Tocher, D.: Creep of the San Andreas Fault; Creep Rate and Related Measurements, *Seis. Soc. Am. Bull.*, **50**: 396-404 (1960).

Weaver, S. H.: The Creep Curve and Stability of Steel at Constant Stress and Temperature, *Trans. ASME*, **58**: 745-751 (1936).

Fracture and brittle strength

Anderson, E. M.: The Dynamics of the Formation of Cone-sheets, Ring-dykes, Caldron-substances, *Proc. Roy. Soc. Edinburgh*, **56**: 128-157 (1936).

Anderson, E. M.: "The Dynamics of Faulting," Oliver & Boyd Ltd., Edinburgh, 1951.

Adams, F. D., and J. A. Bancroft: Internal Friction during Deformation and Relative Plasticity of Rocks, *J. Geol.*, **25**: 597-637 (1917).

Balmer, G.: A General Analytical Solution for Mohr's Envelope, *Amer. Soc. Test Mater. Proc.* **52**: 1260 (1952).

Brace, W. F.: An Extension of Griffith Theory of Fracture to Rocks, *J. Geophys. Res.*, **65**: 3477-3480 (1960).

Brace, W. F.: Dependence of Fracture Strength of Rocks on Grain Size, *Mineral Exp. Sta. Penn. State Univ. Bull.*, **76**: 99-103 (1961).

Brace, W. F.: Brittle Fracture of Rocks, *RAND Corp. Mem.*, RM-35-83, 1963.

Bucher, W. H.: The Mechanical Interpretation of Joints, *J. Geol.*, **28**: 707-730, **29**: 1-28 (1920).

Cloos E.: Experimental Analysis of Fracture Patterns, *Geol. Soc. Am. Bull.*, **66**: 241-256 (1955).

Chinnery, M. A.: Stress Changes That Accompany Strike-Slip Faulting, *Seis. Soc. Am. Bull.*, **53**: 921-932 (1963).

Chinnery, M. A.: Secondary Faulting, *Can. J. Earth Sci.*, **3**: 163-190 (1966).

Donath, F. A.: Experimental Study of Shear Failure in Anisotropic Rocks, *Geol. Soc. Am. Bull.*, **72**: 985-990 (1961).

Donath, F. A.: Analysis of Basin Range Structure, S-Central Oregon, *Geol. Soc. Am. Bull.*, **73**: 1-66 (1962a).

Donath, F. A.: Role of Layering in Geological Deformation, *Trans. N.Y. Acad. of Sci.*, **24**: 236-249 (1962b).

Donath, F. A.: Strength Variation, *RAND Corp. Mem.*, RM-35-83, 1963.

Garson, M. S.: Stress Patterns of Carbonatitic and Alkaline Dykes at Tundulu Ring Structure, Southern Nyasaland, *20th Intern. Geol. Congr., Mexico, Assoc. de Serv. Geol. Africanos*, 309-323 (1956).

Griffith, A. A.: The Phenomena of Rupture and Flow in Solids, *Phil. Trans. Roy. Soc. London*, *Ser. A* **221**: 163-197 (1920).

Griffith, A. A.: The Theory of Rupture, *Proc. (1st) Intern. Congr. Appl. Mech.*, 55-63 (1924).

Hafner, W.: Stress Distributions and Faulting, *Geol. Soc. Am. Bull.*, **62**: 373-398 (1951).

Hodgson, J. H.: Nature of Faulting in Large Earthquakes, *Geol. Soc. Am. Bull.*, **68**: 611-644 (1957).

Hubbert, M. K.: Mechanical Basis for Certain Familiar Geologic Structures, *Geol. Soc. Am. Bull.*, **62**: 355-372 (1951).

Hubbert, M. K., and W. W. Rubey: Role of Fluid Pressure in Mechanics of Overthrust Faulting, *Geol. Soc. Am. Bull.*, **70**: 115-166 (1959).

Jaeger, J. C.: Shear Failure of Anisotropic Rocks, *Geol. Mag.*, **97**: 65-72 (1960).

McKinstry, H. E.: Shears of the Second Order, *Am. Jour. Sci.*, **251**: 401-414 (1953).

Moody, J. D., and M. J. Hill: Wrench Fault Tectonics, *Geol. Soc. Am. Bull.*, **67**: 1207-1246 (1956).

Odé, H.: Mechanical Analysis of the Dike Pattern of the Spanish Peaks Area, Colorado, *Bull. Geol. Soc. Am.*, **68**: 567-568 (1957).

Odé, H.: "Faulting as a Velocity Discontinuity in Plastic Deformation," in Rock Deformation, *Geol. Soc. Am. Mem.*, **79**: 293-321 (1960).

Orowan, E.: Fracture and Strength of Solids, *Rept. Progr. Phys.*, **12**: 185-232 (1949).

Parker, J. M.: Regional Systematic Jointing in Slightly Deformed Sedimentary Rocks, *Geol. Soc. Am. Bull.*, **69**: 465-476 (1958).

Price, N. J.: A Study of Rock Properties in Conditions of Triaxial Stress, in "Mechanical Properties of Non-metallic Brittle Materials," Butterworth Scientific Publications, London, pp. 106-122, 1958.

Price, N. J.: Mechanics of Jointing in Rocks, *Geol. Mag.*, **96**: 149-167 (1959).

Price, N. J.: The Influence of Geological Factors on the Strength of Coal Measure Rocks, *Geol. Mag.*, **100**: 428-443 (1963).

Price, N. J.: "Fault and Joint Development," Pergamon Press, New York, 1966.

Raleigh, G. B., and D. T. Griggs: Effect of the Toe in the Mechanics of Overthrust Faulting, *Geol. Soc. Am. Bull.*, **74**: 819-830 (1963).

Robertson, E. C.: Experimental Study of the Strength of Rocks, *Geol. Soc. Am. Bull.*, **66**: 1275-1314 (1955).

Rubey, W. W., and M. K. Hubbert: Role of Fluid Pressure in Mechanics of Overthrust Faulting. II, *Geol. Soc. Am. Bull.*, **70**: 167-206 (1959).

Sanford, A. R.: Analytical and Experimental Study of Simple Geologic Structure, *Geol. Soc. Am. Bull.*, **70**: 19-52 (1959).

Walsh, J. B.: The Effect of Cracks on the Uniaxial Elastic Compression of Rocks, *J. Geophys. Res.*, **70**: 381-384 (1965).

Crystal dislocation and growth

Barrett, C. S.: "Structure of Metals," McGraw-Hill Book Company, New York, 1952.

Bragg, W. H., and W. L. Bragg: "The Crystalline State," G. Bell & Sons, Ltd., London, 1949.

Cohen, M.: Dislocations in Metals, *AIMME*, New York, 1954.

Cottrell, A. H.: "Theoretical Structural Metallurgy," Edward Arnold (Publishers) Ltd., London, 1948.

Cottrell, A. H.: "Dislocations and Plastic Flow in Crystals," Clarendon Press, Oxford and New York, 1953.

MacDonald, G. J. F.: Thermodynamics of Solids under Non-hydrostatic Stress with Geological Applications, *Am. J. Sci.*, **255**: 266-281 (1957).

MacDonald, G. J. F.: Orientation of Anisotropic Minerals in a Stress Field, *Geol. Soc. Am. Mem.*, **79**: 1-8 (1960).

Read, W. T.: "Dislocations in Crystals," McGraw-Hill Book Company, New York, 1953.

Rosenhain, W.: "An Introduction to the Study of Physical Metallurgy," Constable & Co., Ltd., London, 1935.

Verma, A. R.: "Crystal Growth and Dislocation," John Wiley & Sons, Inc., New York, 1952.

Voll, G.: New Work on Petrofabrics, *Liverpool Manchester Geol. J.*, **2**: 503-567 (1960).

Experimental rock deformation

Griggs, D. T.: Deformation of Rocks under High Confining Pressure, *J. Geol.*, **44**: 541-577 (1936).

Griggs, D. T.: Experimental Flow of Rocks under Conditions Favoring Recrystallization, *Geol. Soc. Am. Bull.*, **51**: 1001-1022 (1940).

Griggs, D. T., F. J. Turner, and N. C. Heard: Deformation of Rocks at 500° to 800°, *Geol. Soc. Am. Mem.*, **79**: 39-104 (1960).

Goguel, J.: "Introduction a l'étude méchanique des déformations de l'écorce terrestre," Paris, France, Imprimerie Nationale, 1948.

Handin, J., and R. V. Hager: Experimental Deformation of Sedimentary Rocks under Confining Pressure. Tests at Room Temperature on Dry Samples, *Bull. Am. Assoc. Petrol. Geologists*, **41**: 1-50 (1957); Tests at High Temperature, **42**: 2892-2934 (1958).

Heard, H. C.: Transition from Brittle to Ductile Flow in Solenhofen Limestone, *Geol. Soc. Am. Mem.*, **79**: 193-226 (1960).

Heard, H. C.: Effect of Large Changes in Strain Rate in the Experimental Deformation of Yule Marble, *J. Geol.*, **71**: 162-195 (1963).

Robertson, E. C.: Experimental Study of the Strength of Rocks, *Geol. Soc. Am. Bull.*, **66**: 1275-1314 (1955).

Serdengecti, S., and G. D. Boozer: The Effects of Strain Rate and Temperature on the Behaviour of Rocks Subjected to Triaxial Compression, *4th Symp. on Rock Mechanics, Penn. State Univ.*, 83-98 (1961).

Symmetry concept

Anderson, E. M.: On Lineation and Petrofabric Structure and the Shearing Movement by Which They Have Been Produced, *Quart. J. Geol. Soc.*, **104**: 99-132 (1948).

Flinn, D.: On the Symmetry Principle in the Deformation Ellipsoid, *Geol. Mag.*, **102**: 36-45 (1965).

Heim, A.: "Untersuchungen über den Mechanismus der Gebirgsbildung," Schwabe, Basel, 1878.

Heim, A.: "Geologie der Schweiz," 2, Tauchnitz, Leipzig, 1921.

Knopf, E. B., and E. Ingerson: Structural Petrology, *Geol. Soc. Am. Mem.*, **6** (1938).

Knopf, E. B.: Study of Experimentally Deformed Rocks, *Science*, **103**: 99-103 (1946).

Kvale, A.: Linear Structures and Their Relation to Movement in the Calidonides of Scandinavia, *Quart. J. Geol. Soc.*, **109**: 51-74 (1953).

Oertel, G.: Extrapolation in Geological Fabrics, *Geol. Soc. Am. Bull.*, **73**: 325-342 (1962).

Paterson, M. S., and L. E. Weiss: Symmetry Concepts in the Structural Analysis of Deformed Rocks, *Geol. Soc. Am. Bull.*, **72**: 841-882 (1961).

Ramsay, J. G.: The Deformation of Early Linear Structures in Areas of Repeated Folding, *J. Geol.*, **68**: 75-93 (1960).

Ramsay, J. G.: The Geometry of Conjugate Fold Systems, *Geol. Mag.*, **99**: 516-526 (1962).

Sander, B.: Uber Zusammenhänge zwischen Teilbewegung und Gefüge in Gesteinen, *Tschermaks Mineral. Petrog. Mitt.*, 281-301 (1911).

Schmidt, W.: Gefügesymmetrie und Tektonik, *Jahrb. Geol. Bundesanstalt*, **76**: 407-430 (1926).

Schmidt, W.: "Tektonik und Verformungslehre," Gebrüder Borntraeger, Berlin-Nikolassee, 1932.

Sander, B.: "Gefügekunde der Gesteine," Springer-Verlag OHG, Vienna, 1930.

Sander, B.: "Einführung in die Gefügekunder Geologischen Korper," Springer-Verlag OHG, Vienna, 1948 and 1950.

Turner, F. J.: Lineation, Symmetry and Internal Movement in Monoclinic Tectonite Fabrics, *Geol. Soc. Am. Bull.*, **68**: 1-18 (1957).

Turner, F. J., and L. E. Weiss: "Structural Analysis of Metamorphic Tectonites," McGraw-Hill Book Company, New York, 1963.

Weiss, L. E.: Fabric Analysis of a Triclinic Tectonite and Its Bearing on the Geology of Flow in Rocks, *Am. J. Sci.*, **253**: 225-236 (1955).

Geotectonics

Birch, F.: Physics of the Crust, in Poldervaart and Arie, "Crust of the Earth," *Geol. Soc. Am. Spec. Paper* **62**: 101-118 (1955).

Carey, S. W.: The Rheid Concept in Geotectonics, *Geol. Soc. Austr. J.*, **1**: 67-117 (1953).

Goguel, J.: "Introduction à l'étude mécanique des déformations de l'écorce terrestre," Imprimerie Nationale, Paris, 1948.

Jeffreys, H.: "The Earth," Cambridge University Press, London, 3d ed., 1952.

Scheidegger, A. E.: Examination of the Physics of Theories of Orogenesis, *Geol. Soc. Am. Bull.*, **64**: 127-150 (1953).

Scheidegger, A. E.: Rheology of the Earth, the Basic Problems of Geodynamics, *Can. J. Phys.* (1957).

Scheidegger, A. E.: "Principles of Geodynamics," Academic Press Inc., New York, 1963.

Whitten, C. A.: Crustal Movement in California and Nevada, *Am. Geophys. Union Trans.*, **37**: 393-398 (1956).

7 ‖ Folds and folding

ONE of the most intriguing features of layered rocks deformed during natural orogenic processes is that their surfaces are often curved to form folds. The methods of classification of these folds and the determination of their mechanisms of formation have long been controversial subjects in the study of rock structure. It seems worthwhile at the start of this chapter to examine carefully the process of fold formation to see what may be deduced from the geological observations, and also to determine what features are likely to offer the most logical basis for fold classification.

Let us now examine the geological effects of a progressive deformation in two dimensions, remembering that the features we observe in two dimensions must, in reality, be extended to cover the third dimension. Consider first a cross section of a layered material (Fig. 7-1, diagram 1) which contains circular markers that can be used to measure the total strain at different times during the progress of the deformation. We subject this material to various states of stress and as a result it becomes strained. The relationship of the increments of strain to the stress state depend on:

1. The nature and orientation of the stress tensor and how the tensor is modified during the history of formation of the fold
2. The initial rheological properties of the material and how these properties are changed during progressive deformation

The distortion of the material generally leads to the development of a new fabric and often to fabric inhomogeneity in the folded layer. The material properties vary greatly in different directions and the axes of symmetry (if any) of these material properties rarely coincide with the principal stresses. Under these conditions the relationship between the stress tensor and the incremental strain tensor is complex (see Chap. 6). As deformation continues, the geometrical form of the material changes; Fig. 7-1, diagrams 2 and 3, show two stages during the fold development which preceded the formation of the final stage shown in diagram 4. Let us examine this process of progressive deformation to see what the observer may legitimately determine from the final product.

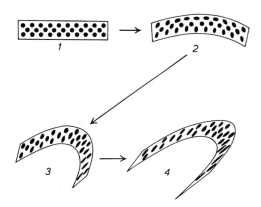

Figure 7-1

Progressive development of a fold.

1. The state of finite strain in the material varies from point to point. Folded rock strata (formed from initially plane-parallel stratification) are always the result of inhomogeneous finite strain (see definitions of homogeneous and inhomogeneous strain, Chap. 3, page 53). If there are markers of known original shape in the deformed rocks, we can completely determine the states of strain throughout the folded material. A classification of folds from this knowledge would be unambiguous and complete, but unfortunately such information is seldom available in naturally deformed rocks, so the method is not practicable.

2. Even if the finite-strain states throughout the structure are known at every point within the fold, there are an infinite number of possible intermediate deformation paths linking the undeformed initial state to that in the final fold (although not all of these deformation paths are equally likely to occur). It is therefore impossible to determine with precision the mechanism of formation of the fold, as we have no knowledge of the nature of the strain increments. It may, however, be possible to get some information about the nature of the intermediate stages of fold formation from a study of the minor structures that were produced at different stages of the fold development. A particularly interesting example of progressive fold development is seen in Fig. 7-2. Here the pegmatite sheet was intruded during the fold formation,

Figure 7-2

Progressive deformation in fold development. The pegmatite was intruded into the rocks as they were undergoing folding and has therefore taken up only the later strains of the finite fold in the surrounding metamorphic sediments. Glen Cannich, Northern Highlands of Scotland.

and so it was affected by only a part of the deformation suffered by the surrounding rock.

3. Because the exact nature of the strain increments and the rheological state of the rock material during deformation are both unknown, we can never determine accurately the states of stress that existed during the formation of the fold.

4. We can determine the relative orientation of adjacent folded surfaces in the structure, but from this it is impossible to deduce the variable states of finite strain in the rock masses between these two bounding surfaces. In fact, there are an infinite number of ways of accommodating the various states of strain between any two boundaries of known shape. It therefore follows that if the methods of classification are based on the geometrical form of the folded layers (and these methods are the most practical ones to apply to folds), folds which come under the same classification can have quite distinct variations of finite strain within the layers, and can be formed in many ways.

Most of these conclusions appear at first sight disheartening to the geologist bent on understanding folding, but it is most important at an early stage in our investigations to realize exactly our limitations so as not to waste effort attempting to solve problems which are insoluble with the data at our disposal.

The geometrical features to be used for a classification should be *invariant features of the folds* and should not depend on any external reference co-ordinates (e.g., orientation with reference to a horizontal datum plane); i.e., the fold should not change its position in a classification just because its position in space is changed.

7-1   DESCRIPTION AND CLASSIFICATION OF FOLDS

The following sections will establish an exact and unambiguous nomenclature that will make it possible to describe folds accurately. There is still much work to be done in discovering the geometrical forms of naturally formed folds. It is important to classify these folds so that the structure produced by experiment on deformed layered models, and the shape of folds produced by theoretical models of rock deformation, can be compared with the results of geological processes.

The description and classification of the geometry of folded surfaces should not involve terms dependent on the genetic development of the fold structure. Some terms that have been used to classify folds are unsatisfactory in this respect, as the mechanism which led to the formation of the fold may be still under debate. The classification established below relies only on the fold morphology and it provides a complete account of all folds. The strength of this new classification is primarily that it offers a very sensitive method for determining slight changes in the fold geometry from layer to layer in a fold structure.

Folds are developed in rocks containing visible surfaces. Before folding, these surfaces may be plane or curved and adjacent surfaces may or may not be parallel. The terms that describe the parts and shapes of folds can be divided into two groups: those which refer to a single folded surface within the structure, and those which describe the relationships of adjacent surfaces.

7-2   DESCRIPTION OF SINGLE FOLDED SURFACES

If the cross section of a single folded surface is examined (Fig. 7-3), it will be seen that there are certain points at which the curved surface generally reaches its highest position above a horizontal datum plane, and other points where it reaches its lowest position. These points are known as *crest points* and *trough points*, respectively. The join of successive crest points on a number of cross sections defines a *crest line*, and the join of trough points defines a *trough line*. These lines may be either rectilinear or curved in space and they may or may not be parallel to one another. The point where a crest line or trough line reaches a maximum height above the horizontal datum plane is known as the *culmination point*. The point where the crest or trough has a minimum height is known as the *depression point*.

The *curvature* of the surface at any one point is described by determining

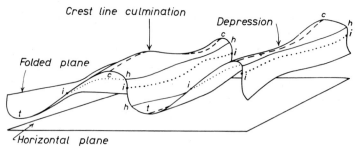

Figure 7-3

The location of crest lines (c), trough lines (t), hinge lines (h), and inflection lines (i) on a single folded surface.

the two-dimensional curvature of sections of the surface in planes drawn so that they contain the normal at this point. These planes are known as *normal planes,* and the curvature as *normal plane curvature.* Generally there is one normal section in which the curvature reaches a maximum value and another in which it has a minimum value, and these are known as the *principal curvatures* of the surface. If the folded surface has a constant curvature in all directions (part of a spherical surface), there is a single crest point or *apex,* or a single trough point or *nadir,* and crest and trough lines do not exist on the surface. These fold forms are known as *domes* and *basins,* respectively, and although not always distinguished, it is probably advisable to differentiate a dome defined in this way from a fold crest showing a marked culmination.

The parts of a fold defined above depend on the orientation of the horizontal datum plane and as such are not fundamental features of the folded surface. The more important *invariant features of the folds* will now be defined.

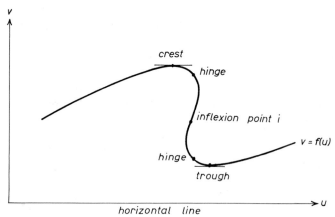

Figure 7-4

The position of crest, trough, hinge, and inflection points on a cross section (profile) of a folded surface.

In general the folded surface will have a variable curvature. Along certain lines on the surface the principal normal curvatures will attain maximum values and these lines are known as *hinge lines*. Hinge lines, like crest and trough lines, may be rectilinear or curved and adjacent fold hinges may be parallel or convergent. The cross section of the folded surface perpendicular to the hinge line at any point in the fold is known as the *fold profile* (Fig. 7-4). Within this section the various fold features may be expressed more exactly in terms of the function of the curve $v = f(u)$ representing the line of intersection of the folded surface with the profile plane:

Crest points: $\qquad \dfrac{du}{dv} = 0 \qquad \dfrac{d^2u}{dv^2} = -ve$

Trough points: $\qquad \dfrac{du}{dv} = 0 \qquad \dfrac{d^2u}{dv^2} = +ve$

Hinge point where curvature c has a maximum value:

$$c = \frac{d^2u/dv^2}{[1 + (du/dv)^2]^{3/2}}$$

If the maximum curvature is constant over a sector of the curve (part of a circular arc), the hinge point cannot be simply defined in the terms given above. In folds of this type the hinge point is best located at the midpoint of the circular arc (see Turner and Weiss, 1963, p. 106; Fleuty, 1964, p. 462).

On this profile there are generally *points of inflection i*, where the rate of change of the slope is zero ($d^2u/dv^2 = 0$). The points where inflections occur are the *limits of individual folds* in the cross section, and the lines joining these points in adjacent profiles are *lines of inflection* (see Fig. 7-3) delimiting the separate folds in the surface. With some types of fold (e.g., Fig. 7-5, fold 1) the slope may remain constant over certain parts of the cross section, and along these sectors all points satisfy the condition $d^2u/dv^2 = 0$. The particular inflection point defining the limits of individual folds is most conveniently chosen as the midpoint of each straight sector.

Let us now consider the shape of an individual fold on the profile contained between two lines of inflection i_1 and i_2 (Fig. 7-5). If the curvature of the surface is constant between the two inflection points (i.e., the profile section is part of a circular arc, Fig. 7-5, folds 4 and 5), then the fold has no true hinge, as defined on location of maximum curvature. Between i_1 and i_2 the curvature may increase and reach a maximum value; the fold then has a *single hinge* (Fig. 7-5, folds 1, 2, 3, 6). It is possible, however, for the curvature to vary in such a way that it reaches two (Fig. 7-5, fold 7) or more (Fig. 7-5, fold 8) maxima and the fold has two or more hinge lines. With these *multiple hinge folds* the curvature between the individual hinge lines does not decrease to zero; if it did, we would be dealing with several individual single hinge folds. The

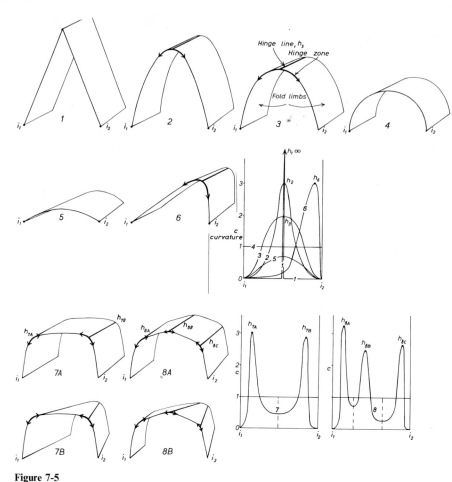

Figure 7-5

Types of single folded surfaces based on curvature. Folds 1 to 6 are single-hinge types, and 7 and 8 multiple-hinge types. The graphs illustrate the changes in curvature c in the various folds.

several hinge lines in multiple hinge folds may be rectilinear or curved, and the adjacent hinge lines may be either parallel (Fig. 7-5, folds 7*A* and 8*A*) or nonparallel (Fig. 7-5, folds 7*B* and 8*B*). If they are not parallel, it is impossible to construct a single-plane profile through the structure.

If every surface in the fold structure contains a line parallel to the hinge line, the fold is known as a *cylindrical fold*; the line is known as the *fold axis* and its orientation is the *axial direction*. A cylindrical fold can be defined as the surface generated by moving a line parallel to itself through space. Although many natural folds are cylindrical or closely approximate to this form, in those known as *noncylindrical* the folded surfaces do not all contain a line parallel to the hinge. Although noncylindrical folds have hinge lines, they do not have fold axes. Some noncylindrical folds have their surfaces arranged so as to form part of a cone and are known as *conical folds* (Dahlstrom, 1954; Evans, 1963; Stauffer, 1964). Structures of this type are very rare, however, and noncylindrical folds generally have a much more complex surface form.

The part of the fold in the vicinity of the hinge is generally known as the *hinge zone*. Although the limits of this zone have not been strictly defined, it is probably useful to have some exact definition of this part of the fold because the appearance of the fold depends to a large measure on its extent. Probably the simplest practical way of defining the limits of the hinge zone is by comparing the curvature of the fold surface with that (unity) of a circular arc drawn with $i_1 i_2$ as diameter. That part of the fold where the curvature of the fold surface exceeds that of the circular arc can be defined as the *hinge zone*, and the parts of the fold on either side of the hinge zone where the curvature is less than that of the circular arc are defined as the *fold limbs*. Fleuty (1964) has used the term "fold limb" for the part of the folded surface between adjacent fold hinges on the same surface. Where the principal normal curvatures of the surfaces on the fold are all equal or less than unity (Fig. 7-5, folds 4 and 5), there is neither a hinge zone nor limbs.

It seems likely that future work on the geometry of folds will require some basic parameters to describe the change of curvature of the surfaces and the general shape of the fold. Various terms have been suggested to describe the tightness of the fold and the angularity of the hinge zones (Knill, 1960; Fleuty, 1964). Fleuty has suggested that the *interlimb angle* of the fold be measured (Fig. 7-6) and that the tightness be defined with the terms listed in Table 7-1.

Table 7-1 *Terms used to describe the tightness of a fold (After Fleuty, 1964)*

Description of fold	Angle between surface inclinations measured at the two inflection points, deg
Gentle	180 to 120
Open	120 to 70
Close	70 to 30
Tight	30 to 0
Isoclinal	0
Elasticas	Negative value

However, the appearance of "tightness" of the fold also depends to some extent on the changes in curvature of the folded surface. The three folds represented in Fig. 7-6 have the same interlimb angle but need some other parameter to describe their shapes in cross section. It is possible to use the ratio of the

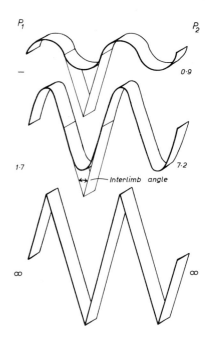

Figure 7-6

Folds with the same interlimb angle but with different values of parameters P_1 and P_2.

extent of the fold limbs to that of the hinge zone to obtain a useful parameter (P_1) describing the fold shape.

$$P_1 = \frac{\text{length of projections of limbs on the join of } i_1 i_2}{\text{length of projection of hinge zone on the join of } i_1 i_2}$$

With the fold styles generally described as *zigzag, chevron,* or *accordion folds* (Fig. 7-5, fold 1), the hinge zone is very small in proportion to the extent of the fold limbs, and P_1 approaches infinity. If the folds have "subangular" or "subrounded" hinge zones, $\infty > P_1 > 0$. Where the fold has no hinge zones, the ratio is indeterminate. Another useful descriptive parameter (P_2) is obtained by expressing the maximum curvature of the fold surface as a ratio of the unit curvature of the circle drawn with $i_1 i_2$ as a diameter. The average curvature of the fold surface at any position between two points x and y on the fold surface is determined by first drawing normals to the surface at x and y (Fig. 7-7). The point of intersection of these two normals is

Figure 7-7

Calculation of mean curvature of a segment of a fold between points x and y.

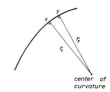

the center of curvature, and the average radius of curvature is $r = (r_x + r_y)/2$. The curvature is the inverse of the radius of curvature, $c = 1/r$. Examples of parameters P_1 and P_2 applied to folds with identical interlimb angles are given in Fig. 7-6, and parameter values for the folds of Fig. 7-5 are given in Table 7-2.

Where the folds have more than one hinge, each hinge has its own characteristic parameters.

Table 7-2 *Values of parameters P_1 and P_2 for folds illustrated in Fig. 7-5*

Fold	Hinge	P_1	P_2
1		∞	∞
2		2.0	3.0
3		1.0	2.0
4		—	1.0
5		—	0.7
6		2.0	3.0
7	A	2.0	3.0
7	B	2.8	2.8
8	A	1.8	3.3
8	B	1.5	2.5
8	C	2.4	2.7

Fold dimensions and symmetry   The folded layer may form either a periodic or an irregular waveform that oscillates between two limiting surfaces called the *enveloping surfaces* (Turner and Weiss, 1963). The surface which joins the successive lines of inflection will be called the *median surface m* (Fig. 7-8). The enveloping and median surfaces may be planar or curved, the median surface may or may not lie midway between the enveloping surfaces, and the distance between the enveloping surfaces may change from place to place. The orientation of the median surface defines the mean orientation of the folded layer or *sheet-dip* (Fig. 7-9) and more closely approximates the attitude of main lithological units in the major structures than do the individual limbs of the smaller folds. The orientation of the sheet-dip should, whenever possible, be determined in the field, for it is not possible to compute this from the mean of the values of isolated dips measured on the small folds.

If the median surface is planar, if the projections of the fold hinges onto this plane lie midway between the inflection points, and if the fold curves between adjacent hinges are identical (or mirror image) forms, then the folds are *symmetric* (Fig. 7-8A). If these conditions do not hold, the fold is *asymmetric*. The symmetry or asymmetry of a fold depends on the relative lengths of the two limbs. If the fold is viewed down its plunge, its symmetry may be recorded on a map by one of three symbols: *M* (symmetric), *S* or *Z* (asymmetric). For example, the fold symmetry of Fig. 7-33 is *S*, that of Fig. 7-87 is

A. Periodic symmetrical waves

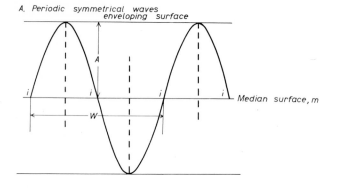

B. Periodic asymmetrical waves (i)

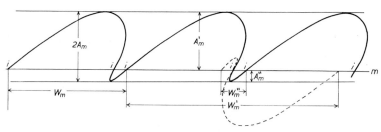

C. Periodic asymmetrical waves (ii)

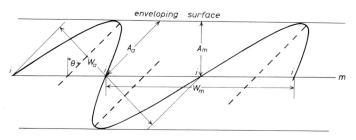

D. Non-periodic asymmetric waves with curved median surface

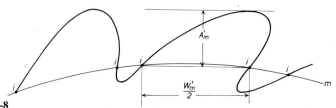

Figure 7-8

Amplitude and wavelength measurements of a folded surface.

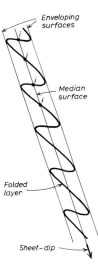

Figure 7-9

Enveloping surfaces and median surfaces of a folded layer.

M. The relationships of the symmetry of minor fold structures to the large-scale fold structures is illustrated in Fig. 7-10. In regions where the major structures are plunging, tight, or isoclinal folds, this method of recording the symmetry patterns of small-scale folds sometimes affords an excellent method of recognizing the individual fold limbs and hinge zones of the large folds.

If the folds show a regular periodic repetition and their median surface lies midway between the two enveloping surfaces (Fig. 7-8*A* and *B*), their dimensions may be expressed in terms of wavelength and amplitude. The *wavelength* of a fold is given by the length of the periodic unit W (or W_m where the folds are asymmetric) and the *amplitude* A (or A_m) is half the perpendicular distance between the two enveloping surfaces. Sometimes a more fundamental measure is referred to other directions in the folds (e.g., parallel and perpendicular to the axial surface *a*). It may be more significant to make measurements of the periodicity W_a and the half distance between the enveloping surfaces A_a with reference to some other coordinate directions making an angle $(90° - \theta)$ with the median surface (Fig. 7-8*B*). There is a simple relationship between these measurements:

$$W_m = W_a \sec \theta \tag{7-1}$$

$$A_m = A_a \cos \theta \tag{7-2}$$

Where the fold curves are periodic but the median surface does not lie exactly midway between the enveloping surfaces, it is possible to define the total wavelength as the distance between alternating inflection points W_m, and to define the average amplitude A_m as half the distance between the

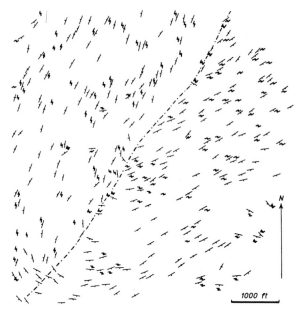

Figure 7-10

Method of mapping the symmetry of minor folds and determination of the trace of a major fold. The dash-dot line represents the major fold trace. South side of Loch Monar, Inverness-shire, Scotland.

enveloping surfaces. Two separate and perhaps more significant measures refer to the parts of the fold on either side of the median surface, imagining each part to be half of a more complete waveform (see Fig. 7-8C). Then

$$\text{Total } W_m = \frac{W_m'}{2} + \frac{W_m''}{2} \tag{7-3}$$

$$\text{Total } A_m = \frac{A_m'}{2} + \frac{A_m''}{2} \tag{7-4}$$

If the folding is not periodic, there will be no truly characteristic wavelength and amplitude. However, using a technique similar to that in Fig. 7-8C, it is possible to express the apparent wavelength and amplitude of each fold between adjacent inflection points by imagining that it forms half of a complete wave (see Fig. 7-8D).

In a single folded layer it is sometimes found that various orders of fold size exist (Fig. 7-11). In folded ore deposits the determination and correct extrapolation of these orders of periodicity may have great economic consequence. The method of determining the various orders of periodicity depends on first establishing the position of the median surface from the observed small-scale folds. If the median surface has a folded form, then its character-

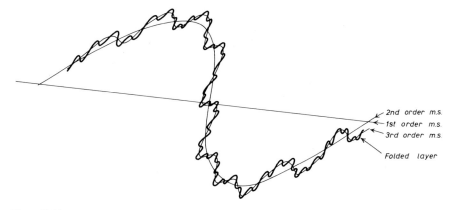

Figure 7-11

First-, second-, and third-order folds derived from the curvature of the median surfaces.

istic wavelength and amplitude are determined. A line is now drawn through the inflection points of the folded median surface (i.e., the median surface of the folded original median surface). This line may itself show a folded form. By proceeding in this way, drawing a succession of median surfaces, the successive orders of folding can be accurately determined.

7-3   RELATIONS OF ADJACENT SURFACES IN THE FOLD

Certain geometrical features of folds exist which are the result of the relations of the different folded surfaces of the features that we have established on each surface. The successive positions of the fold hinges form an important curved or plane surface known as the *axial surface* (or axial plane) (Figs. 7-12 and 7-13), and the successive positions of crest and trough lines in a number of folded surfaces define the *crestal surface* and *trough surface*. The positions of

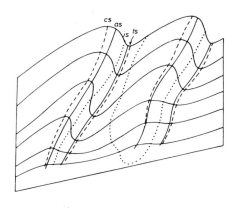

Figure 7-12

Traces of crest surface (cs), trough surface (ts), axial surface (as), and inflection surface (is) through a number of folded layers.

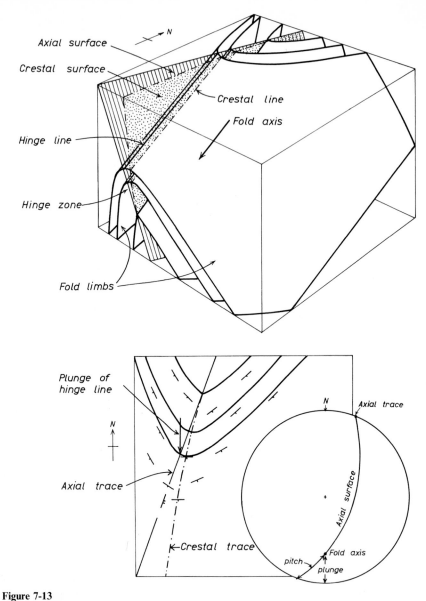

Figure 7-13

Position of axial surface and crestal surface on folded layers, and position of the axial trace and crestal trace.

the crest and trough surfaces depend on the reference frame used to describe the folds and, in particular, the orientation of the horizontal datum; but the positions of the hinge lines and axial surfaces are invariant and do not depend

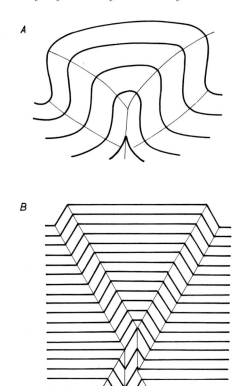

Figure 7-14

*Characteristic form of a box fold (A) and
conjugate kink fold (B).*

on any external reference axes. The lines of inflection on each surface connect
to form *inflection surfaces* limiting the domains of adjacent folds: these are
generally curved; and adjacent inflection surfaces delimiting a specific fold
diverge or converge where the fold amplitude in successive layers decreases
(Fig. 7-12). These various surfaces in the fold may be intersected by other
surfaces (profile plane or ground surface), and the line of intersection is termed
crest-surface trace, trough-surface trace, or *axial-surface trace* (generally
shortened to *axial trace*) (Fig. 7-13).

The axial surface need not bisect the interlimb angle. Although symmetric
folds have been defined with respect to a single fold surface, this description
must be extended. For a complete definition of *symmetric folds* the axial
surface must be the right bisector of a line drawn between two adjacent
inflection points and this bisector must divide the fold into two identical
(and mirror image) parts.

Where individual folds have *multiple hinge lines,* they possess two or more
axial surfaces generally inclined to each other. These folds are known as

polyclinal folds. Where there are two axial surfaces, the folds are generally called *conjugate folds* or *box folds.* There are two principal differences between conjugate and box folds (Fig. 7-14):

1. The hinge zone of a conjugate fold is generally small (P_1 large or ∞), while that of a box fold has a smaller curvature and is wider (P_1 small to zero).

2. Successive surfaces in a conjugate fold have a nearly identical form (except for the complication that occurs where the inclined axial surfaces— most commonly planes—cross one another); while the folded surfaces in a box fold generally show a rapid change of shape along the axial surface (often curved), and the polyclinal shape changes into a single-hinge type.

7-4   TERMS USED TO DESCRIBE THE ATTITUDE OF FOLDS

Many of the terms used to describe the orientation of folds have never been satisfactorily defined and some are of doubtful value (see Fleuty, 1964, p. 481).

The fold limbs are said to converge or *close* as they are traced toward the fold hinge. If the fold closes upward it is known as an *antiform*; if downward, a *synform.* If the folded layers retain their correct depositional sequence in the structure, the same folds are termed *anticline* and *syncline,* respectively, or more exactly, *antiformal anticline* and *synformal syncline.* If the rock strata are in inverted sequence, the antiform and synform would be known as *antiformal syncline* and *synformal anticline,* respectively (Fig. 7-15).

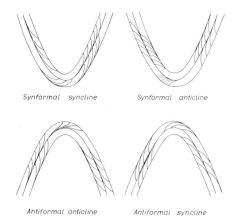

Synformal syncline   Synformal anticline

Antiformal anticline   Antiformal syncline

Figure 7-15

Definitions of antiform and synform, anticline and syncline. The cross-bedded sediments illustrate the order of succession of the strata.

In some folds the hinge line is parallel to the dip direction of the axial surface; the fold closes sideways and is known as a *neutral* fold. If the axial surface is horizontal (or has a dip not exceeding 10°, see Fleuty 1964, p. 483),

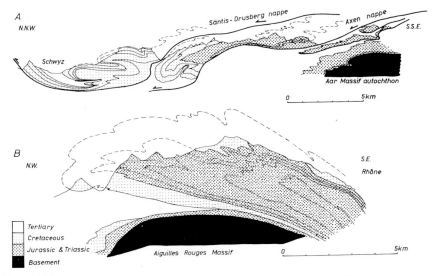

Figure 7-16

Helvetic nappes. A, a thrust nappe (after Heim, 1921), and B, a fold nappe (after Lugeon, 1940).

the fold is described as *recumbent*. The sheet of rock making up a complete, large, recumbent fold with an overturned limb exceeding about 5 km in length is termed a *fold nappe* (Fig. 7-16B). In many nappes the inverted limb of the fold is partially or completely cut by a thrust fault and the structure is termed a *thrust nappe* (Fig. 7-16A). *Upright* folds have vertical or nearly vertical axial surfaces, and *vertical* folds are those upright folds with a hinge that plunges steeply (80 to 90°). All folds whose axial surfaces dip at angles of from 10 to 80° may be termed *inclined* folds, and *overturned* folds or *overfolds* are those inclined folds in which the limbs dip in the same direction. Fleuty has also suggested that inclined folds with hinge lines which have a pitch of 80 to 100° on the axial surface should be termed *reclined* folds.

Redefinitions of the terms used to describe fold attitude based on the dip of the axial surface and on the plunge of the hinge line have been proposed by Fleuty (1964). The strict application of this revised scheme (Fig. 7-17) should do much to clarify fold descriptions and make the terms really useful to the geologist.

7-5 THE GEOMETRICAL CLASSIFICATION OF FOLDS

The morphological classification of folds to be developed below depends on the shape of the profile section of the folded layers. The form of any one layer in the structure depends on the relationships of the bounding surfaces of the layer, and in particular, the relative rates of change of the inclination of these

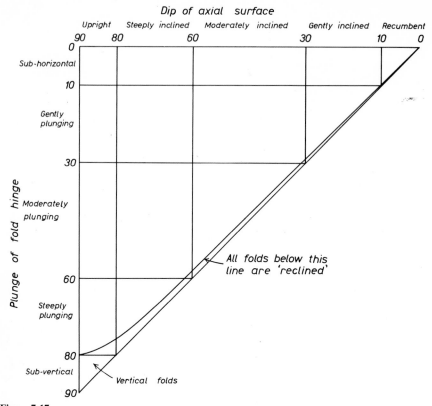

Figure 7-17

Definitions of fold attitude after Fleuty (1964).

bounding surfaces. There are a number of ways by which we can measure these relative rates of change and hence describe the shape of the folded layer.

Orthogonal thickness t If we examine a cross section (profile) of any fold (fold 1, Fig. 7-18), we can determine the thickness t_α of the material between two folded surfaces by constructing tangents to each surface making an angle of $90° - \alpha$ with the axial trace of the fold. In general, the value of t_α varies with the angle of inclination α, and at any given inclination α it is possible to express the proportional thickness t_α with that measured at the hinge zone of the fold (t_o). This ratio gives a valuable parameter $t'_\alpha = t_\alpha/t_o$ defining the proportional change in orthogonal thickness throughout the fold with variation in α. This variation in t'_α relates to the fold shape and can be recorded graphically (Fig. 7-19). Every folded layer can be recorded as a single line on this graph, and every line drawn on the graph records the change of shape of a possible fold, although not all are equally likely in naturally folded

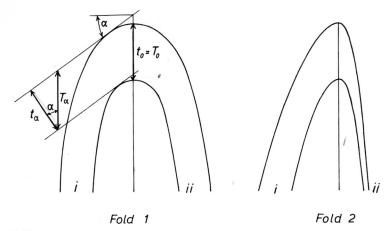

Figure 7-18

Cross-section shapes of folded layers in two folds.

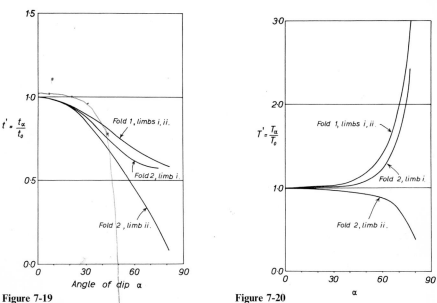

Figure 7-19

Plots of t'_α in folds 1 and 2 of Fig. 7-18.

Figure 7-20

Plots of T'_α in folds 1 and 2 of Fig. 7-18.

material. For a given set of values of t'_α (Fig. 7-21) and a known shape of one surface of the fold, an envelope of tangential lines can be constructed which defines the other surface of the layer. With every fold examined in this way two sets of values of t'_α can be established, one for the sector of the fold between the hinge and one inflection point, another for the other half of the fold.

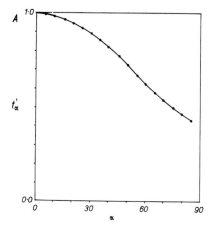
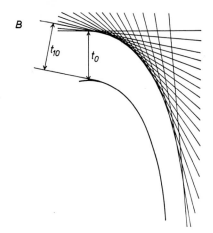

Figure 7-21

The determination of the envelope of a folded layer from values of t'_α through the structure. A shows a plot of t'_α for a folded layer and B illustrates that when one surface in the fold is given this graph defines the envelope of the upper surface of a folded layer.

Sometimes these plots of t'_α coincide (symmetric fold 1, Fig. 7-18, and some asymmetric folds), while others will form two separate curves (some asymmetric folds—fold 2, Fig. 7-18).

   Thickness parallel to the axial surfaces T Another useful parameter for describing the variation in thickness of the folded layers records the distance between the bounding surfaces measured in a direction parallel to the axial surface of the fold (Fig. 7-18, fold 1, distance T_α). For the purpose of describing the variation in this distance, this thickness is expressed as a proportion of the same measurement made at the fold hinge (T_o); this is the parameter $T'_\alpha = T_\alpha/T_o$. The variations in T'_α with α are best expressed graphically with T'_α as ordinate and α as abscissa (Fig. 7-20). Variations of T'_α throughout the fold appear on this graph as a continuous curve; again every fold can be represented by a single (or double) continuous curve, and every line on the graph expresses the change of shape of a layer in a possible fold.
   The two parameters t'_α and T_α are not independent. In Fig. 7-18, fold 1, it will be seen that

$$t_\alpha = T_\alpha \cos \alpha \tag{7-5}$$

and because $t_o = T_o$, it follows that

$$t'_\alpha = T'_\alpha \cos \alpha \tag{7-6}$$

for any position in any fold.
   Although these parameters are dependent, they are both useful, for one may show certain special properties of a fold better than the other.

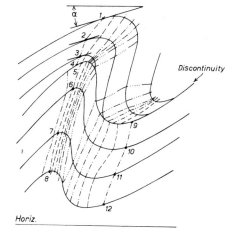

Figure 7-22

Construction of dip isogons in folded surfaces. Points 1, 2, . . ., 12 all have the same inclination α. A discontinuity in the structure produces a discontinuity in the dip isogons.

Lines of equal dip, or dip isogons On the profile section of the fold it is possible to determine points on the folded surface which have the same slope (Fig. 7-22: the points (1,2,3, . . . , 12) are all situated on the folded surface where the tangent to the surface makes an angle of α with the horizontal line in the profile plane). By joining points of the same slope on adjacent surfaces, we obtain lines of equal slope known as *dip isogons* (Elliott, 1965). These isogons generally form a series of continuous curved lines on the section; however, where there is some planar discontinuity in the structure, individual isogonal lines stop against this structural break. The inclination of the isogons reflects the variations in the slope of adjacent surfaces.

The following rules govern the interpretation of dip isogons: *if the isogons are traced from the outer arc of a fold to the inner arc of that fold,* i.e., toward the concavity of the surface, then:

1. (*a*) If the isogons dip toward the axial trace, the mean curvature of the outer arc is less than that of the inner arc.

(*b*) If the isogons are parallel to the axial trace, the mean curvatures of the outer and inner arcs of the fold are equal.

(*c*) If the isogons dip away from the axial trace, the mean curvature of the outer arc is greater than that of the inner arc.

2. (*a*) If adjacent isogons are inclined toward each other, the *curvature* of the outer arc between the isogons is less than that of the inner arc.

(*b*) If adjacent isogons are parallel, the curvature of the two surfaces is equal.

(*c*) If adjacent isogons are inclined away from each other, the curvature of the outer arc exceeds that of the inner arc.

Because dip isogons can be constructed simply and rapidly, they afford an excellent practical method for comparing the variations in slope of the surfaces in the fold. For the purposes of classifying folds it is found that the isogons are best referred to the datum direction given by the inclination of the axial surface of the fold (Fig. 7-24).

In Sec. 7-2 it was shown how the various features of a single folded surface could be recorded. In each fold we can determine the various parameters of amplitude, wavelength, rates of curvature, and the size and curvature of the hinge zone for *each* surface, and hence we could make a complete and system-atic description of the fold. Such a system would be lengthy and it is doubtful whether the geometrical properties of the structure could be easily appreciated from such a list. Classification of the folds in this way is unlikely to have any practical use for the geologist. However, much of what we mean by "fold style" depends on the cross-sectional shape of individual layers of material between two folded surfaces, and in particular, the changes and rates of change of the inclination of the two bounding surfaces. The parameters t' and T' of the fold sum the effects of the various changes of inclination of the upper and lower surfaces. The inclinations of the dip isogons with respect to the axial trace of the fold enable us to compare the mean curvature of the bounding surfaces, and these are related to the rates of change of t'_α and T'_α with respect to the angle α (that is, $dt'_\alpha/d\alpha$ and $dT'_\alpha/d\alpha$). For example, if the dips on one of the folded surfaces shown in Fig. 7-23 at positions A, B, C, D, etc., are equal to those on the adjacent surface at A', B', C', D', respectively, then the dip isogons in this sector of the fold will run parallel to the axial trace. Through this sector T'_α will remain constant, that is, $dT'/d\alpha = 0$. If the angle of dip of the outer arc of the fold is less than that of the inner arc

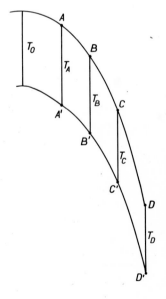

Figure 7-23

Variations in values of T through a fold.

(dip isogons converge onto the axial trace), then T'_α must increase with α, that is, $dT'/d\alpha > 0$; if the dip of the outer arc exceeds that of the inner arc (isogons diverge from axial trace), T'_α decreases with the increase in α, that is, $dT'/d\alpha < 0$.

The inclinations of adjacent dip isogons with respect to each other (when traced toward the inner fold arc) reflect the change of the curvature of adjacent surfaces, and these are related to the rates of change of $dt'_\alpha/d\alpha$ and $dT'_\alpha/d\alpha$ with respect to α, that is, $d^2t'_\alpha/d\alpha^2$ and $d^2T'_\alpha/d\alpha^2$. This means that if the rates of change of curvature of the adjacent surfaces at dip α are the same, then $d^2T'_\alpha/d\alpha^2 = 0$.

It is possible to express the changes in shape within the fold and to classify the folded layer using any of the following parameters:

1. Orthogonal thickness t
2. Thickness parallel to the axial surface T
3. The inclination of the dip isogons

In any folded layer the rates of change of the inclinations of adjacent surfaces may vary in a number of ways, providing that the maximum curvature occurs at the hinge zone and that the curvature reaches a zero value at the points of inflection on the folded surfaces. There are three types of fold which we can consider as fundamental classes; their definition is based on the comparison of curvature changes on adjacent surfaces in the folded layer:

Class 1 Curvature of the inner fold arc *always exceeds* that of the outer arc.

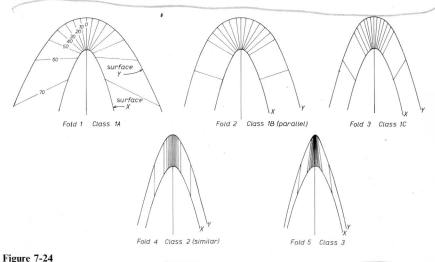

Fold 1   Class 1A   Fold 2   Class 1B (parallel)   Fold 3   Class 1C

Fold 4   Class 2 (similar)   Fold 5   Class 3

Figure 7-24

Fundamental types of fold classes. Dip isogons have been drawn at 10° intervals from the lower to the upper surfaces X and Y.

Class 2 Curvature of the inner and outer arcs is *equal*.
Class 3 Curvature of the inner fold arc is *always less than* that of the outer
   arc.

All other folds, where the rates of change vary in more complex ways, can
be envisaged as combinations of these three classes. Let us first consider the
geometrical properties of these classes. Consider a layer bounded by a lower
surface (X) and an upper surface (Y) (Fig. 7-24) and let X have a specific
folded form. The rate of change of slope of surface Y varies according to the
principles of one of the three fold classes, and variously shaped folded layers
result. The values of the parameters t' and T' of the main classes of folds are
illustrated in Figs. 7-25 and 7-26, respectively.

Class 1 Folds with convergent dip isogons   If the curvature of the outer
arc is always less than that of the inner arc, the dip isogons converge on each
other and on the axial trace of the fold *as they are followed in toward the inner
arc of the fold* (Fig. 7-24, folds 1, 2, and 3). With respect to a vertical axial
trace, the inclinations of the outer arc are always less than those of the inner
arc and T'_α always exceeds unity (Fig. 7-26). Folds with these properties can
be subdivided into one of three subclasses depending on the strength of the

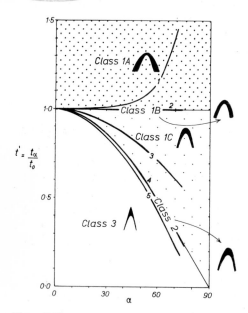

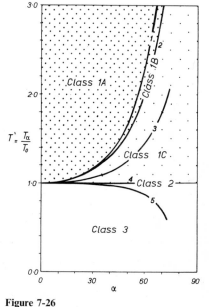

Figure 7-25

*Position of t'_α plots for the various
fundamental fold classes.*

Figure 7-26

*Position of T'_α plots for the various
fundamental fold classes.*

convergence of the dip isogons and values of the parameters t'_α, T'_α and their derivatives.

SUBCLASS 1A FOLDS WITH STRONGLY CONVERGENT DIP ISOGONS (Fig. 7-24, fold 1) In these folds the orthogonal thickness of the beds (t_α) always exceeds that at the hinge of the structure (t_o) and $t'_\alpha > 1$. It also follows from (7-6) that $T'_\alpha > \sec \alpha$. From these two values the rates of change of dips of the surfaces on either side of the folded layer must vary so that $d^2t'/d\alpha^2 > 0$ and $d^2T'/d\alpha^2 > (2 \sec^3 \alpha - \sec \alpha)$ throughout the structure. The dip isogons drawn on surfaces inclined at an angle of $(90 - \alpha)$ degrees from the axial surface of the fold always make an angle with the axial surface which is greater than α.

SUBCLASS 1B PARALLEL FOLDS (Fig. 7-24, fold 2) This is a well-known geometrical model in which the layer keeps a constant orthogonal thickness throughout the fold ($t'_\alpha = 1$ and all derivatives which express changes in t' are zero). The thickness of any bed measured parallel to the axial trace of the fold is always greater than that at the hinge such that $T'_\alpha = \sec \alpha$ [from (7-6)]. Dip isogons are always perpendicular to the surfaces of the folded layer and converge with the axial trace to make an angle of α with the trace.

SUBCLASS 1C FOLDS WITH WEAKLY CONVERGENT ISOGONS (Fig. 7-24, fold 3) Folds that fit into this subclass differ from 1A and 1B in that the orthogonal thickness of the folded layer on the flank of the fold is always less than that at the fold hinge ($t'_\alpha < 1$), but t'_α always exceeds $\cos \alpha$ (the condition for the next class of folds). The dip isogons converge with the axial trace but make an angle of less than α with the fold trace.

Class 2 Folds with parallel isogons (Fig. 7-24, fold 4) Where the changes in curvature of the surfaces of the layer are the same throughout the fold, the variations in dip of the surfaces are the same and the geometrical forms of the sides of the layer are identical. The distances between the folded boundaries of the layer in a direction parallel to the axial surface are equal, $T'_\alpha = 1$. The orthogonal thickness of the layer always has a minimum value on the flank of the fold and a maximum value at the hinge of the structure. Folds which show relative thinning on the fold limbs have sometimes been loosely grouped together as folds of general similar type (Ramsay, 1962), but this is an unsatisfactory grouping since folds of Classes 1C, 2, and 3 all have this property. In a true similar fold the orthogonal thickness of the layer should change according to the relationship $t'_\alpha = \cos \alpha$.

Class 3 Folds with divergent isogons (Fig. 7-24, fold 5) With folds of this class the rate of change of the slope of the outer arc always exceeds that of the inner arc. The inclinations α of the outer surface are always greater than those on the inner surface, and dip isogons diverge from each other and from the axial trace when traced from outer to inner arc of the structure. These

Table 7-3 Properties of the three fold classes

Fold class	1. Curvature on adjacent surfaces				2. Dip values on adjacent surfaces				3. Integration of dip change through fold	
	Comparison of inner and outer arcs	Adjacent isogons	$\dfrac{d^2T'}{dx^2}$	$\dfrac{d^2t'}{dx^2}$	Comparison of inner and outer arcs	Relation of dip of isogons and axial trace	$\dfrac{dT'}{dx}$	$\dfrac{dt'}{dx}$	Orthogonal thickness ratio t'_α	Thickness ratio parallel to axial trace T'_α
1A	$i > 0$	Converge strongly	$> 2\sec^3\alpha - \sec\alpha$	> 0	$i > 0$	Converge strongly. Angle with axial trace $>$ angle of dip α	$> \tan\alpha \sec\alpha$	> 0	> 1	$> \sec\alpha$
1B parallel	$i > 0$	Converge	$2\sec^3\alpha - \sec\alpha$	0	$i > 0$	Converge. Angle with axial trace = angle of dip	$\tan\alpha \sec\alpha$	0	1	$\sec\alpha$
1C	$i > 0$	Converge slightly	$(2\sec^3\alpha - \sec\alpha) > \dfrac{d^2T'}{dx^2} > 0$	$0 > \dfrac{d^2t'}{dx^2} > -\cos\alpha$	$i > 0$	Converge slightly. Angle with axial trace $<$ angle of dip	$\tan\alpha\sec\alpha > \dfrac{dT'}{dx} > 0$	$0 > \dfrac{dt'}{dx} > -\sin\alpha$	$1 > t'_\alpha > \cos\alpha$	$\sec\alpha > T'_\alpha > 1$
2 similar	$i = 0$	Parallel	0	$-\cos\alpha$	$i = 0$	Parallel	0	$-\sin\alpha$	$\cos\alpha$	1
3	$0 > i$	Diverge	< 0	$< -\cos\alpha$	$0 > i$	Diverge	< 0	$< -\sin\alpha$	$0 > t'_\alpha > \cos\alpha$	< 1

changes mean that measurements of the thickness of the layer on the flank of the structure made in a direction parallel to the axial trace are always less than that at the hinge of the fold, that is, $T'_\alpha < 1$; and for this to occur, the rates of change of the upper and lower surfaces of the layer must accord with the condition that $d^2 T'/d\alpha^2 < 0$. As a consequence of these features the relative orthogonal thickness t'_α of the layer on the fold limbs is always severely reduced such that $0 < t'_\alpha < \cos \alpha$.

The geometric characteristics of these fundamental classes are completely listed in Table 7-3. Although most of the folds that the geologist encounters belong to one of the classes described above, more complex fold shapes may be found that do not fit simply into the broad classification so far established. Although these complex folds are probably very rare, the classification can be expanded to include them. These structures are geometrically possible because any variation in curvature of the folded surfaces is possible, providing that the curvature of the surface attains a single maximum value at the fold hinge and a zero value on the limb of the fold (inflection point on the folded surface).

Figure 7-27 illustrates a part of a complex fold where the curvatures do not conform to the simple relations of the three principal classes established above and where the isogons converge and diverge with each other and with the axial trace of the fold. Let us investigate in more detail the geometry of this structure to see how it can be classified. First, it is possible to construct a graph to illustrate the variation in relative orthogonal thickness of the layer through the structure (Fig. 7-28A). The continuous line exactly represents the geometrical form of the fold and lies partly in the fields of fold Classes 3, 2, and 1C of Fig. 7-25. It will be apparent, however, that the slopes of this line

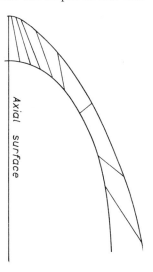

Figure 7-27

Cross section of a complex folded layer with dip isogons constructed.

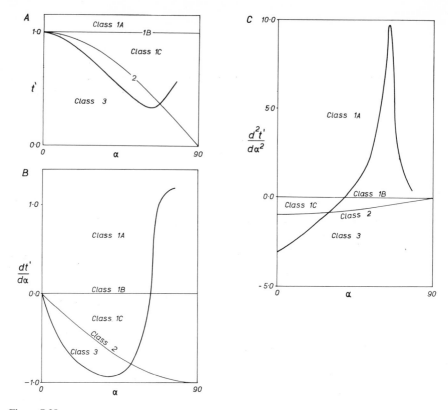

Figure 7-28

Various plots for the fold illustrated in 7-27. A. t'_α, B. $dt'_\alpha/d\alpha$, C. $d^2t'_\alpha/d\alpha^2$.

in these three fields do not always accord with the conditions for these fold classes (see Table 7-3). Another graph has been constructed (Fig. 7-28B) which gives the slope of graph 7-28A; that is, it represents the variation of $dt'/d\alpha$ with α. The part of the fold with t' values indicating a Class 3 fold can be divided into five sections depending on whether the values of $dt'/d\alpha$ are characteristic of fold Classes 3, 2, 1C, 1B, or 1A; and the parts of the fold with t' values of types 2 and 1C both have $dt'/d\alpha$ values characteristic of Class 1A folds. In a similar way to that by which Fig. 7-28B was prepared from Fig. 7-28A, it is possible to represent graphically the function $d^2t'/d\alpha^2$ which shows the comparative rates of change of curvature on the two surfaces (Fig. 7-28C). The part of the fold which has $dt'/d\alpha$ values of a Class 3 fold shows $d^2t'/d\alpha^2$ values of fold Classes 3, 2, 1C, 1B, and 1A. We can best express all the complex geometrical changes that go on in this fold in tabular form (Table 7-4).

Only a small part of the fold, that near the hinge, has the true characteristic

Table 7-4   *Classification of the fold of Fig. 7-27 according to the values of t′ and its derivatives with respect to α*

α	0-29	29	29-40	40	40-52	52	52-63	63	63-67	67	67
t'	3	3	3	3	3	3	3	3	3	2	1C
$\dfrac{dt'}{d\alpha}$	3	3	3	3	3	2	1C	1B	1A	1A	1A
$\dfrac{d^2t'}{d\alpha^2}$	3	2	1A	1B	1A	1A	1A	1A	1A	1A	1A

of a fold with divergent dip isogons (Class 3). In the rest of the structure this geometrical form is modified and is taking up the form of a fold with convergent dip isogons (Class 1). The structure may therefore be classified as a divergent isogon fold compounded with a convergent isogon fold (Class 3 modified by Class 1A).

One very important feature of this classification is that it is very easy to apply to any folded material. It is a precise and sensitive method of determining geometrical changes, and many of the properties of folds that might be overlooked or impossible to detect with a more traditional classification show up very clearly. It is interesting to see that the two models around which much fold classification has centered in the past are but two unique types in a whole field of possible geometrical models, and it seems hardly surprising that natural folds rarely conform exactly to the morphology of the parallel and similar fold models.

Another especially important property of certain types of folds—that they are unable to affect an unlimited number of layers of material without alteration in shape—is brought out clearly by this method of classification. Where dip isogons either converge or diverge toward the inner fold arc, they cannot continue to do this indefinitely or they will come into contact with each other or with the axial surface. Folds in this category *must* either change their type from layer to layer (e.g., Class 1 to Class 3, or Class 1 or 3 to Class 2) or develop some structural discontinuity in the material (thrusts, slides, shear zones). The only folds that can affect an unlimited number of layers are those of Class 2 (similar folds). In any detailed analysis of the geometry of folded materials, therefore, we must expect rapid and abrupt changes in fold style from layer to layer in the structure, and also perhaps within individual layers.

It is possible to investigate and describe the variations in shape of the fold in the third dimension in the same way as has been done in the two-dimensional profile sections. This can be done by determining the variations in t_α, T_α, and dip isogons in a section drawn along the axial surface of the fold. These variations express the relative changes in orientation of the hinge lines at different positions in the fold, and the folds may be grouped into three types depending on whether the hinge-line plunge isogons converge, diverge, or are parallel.

The main types of folds that are seen in naturally deformed rocks will now be described, and a summary of the theoretical work on fold development will be given, in order to account for some of the geometrical characteristics of folds. The formation of the small-scale features, minor folds, cleavage, etc., that may be found in different parts of the major structure will be related to the strain history of the larger structures.

7-6   FOLDS DEVELOPED BY BUCKLING

Many of the folds that are found in deformed rocks owe their development to a compressive stress which acted along the length of the layers. If the layers have different properties, an instability is produced which leads to the buckling of the more competent layers in the material (see model experiments Fig. 7-29). The folded competent layers which are produced in this way generally have a geometrical form of Class 1 which often closely approximates that of Class 1*B*, the parallel fold model. Other classes of folds are generally developed in the incompetent material.

Figure 7-29

Development of a buckle produced by compressive stress acting along a competent layer embedded in an incompetent matrix.

Fold wavelength   A number of attempts have been made, based on a consideration of the properties of the material involved, to deduce theoretically the wavelength of the folds produced by buckling.

There seems little doubt that the stimulating approaches made by Biot and Ramberg to the problems of the folding of layered materials will greatly assist the mechanical interpretation and understanding of the morphology of folded rock strata. Many of their results which indicate the way layers of different competence behave when strained are very pertinent to naturally folded rocks, and it seems likely that in the future it will be possible to obtain exact information on viscosity ratios of the folded layers in naturally deformed rocks.

Some workers (Smoluchowski, 1909; Goldstein, 1926; Gunn, 1937; Kienow, 1942) have investigated the buckling of elastic layers, restricting the problem to one of static instability which is independent of time. More recent discussions of this problem of elastic instability have been presented by Biot (1961) and Currie, Patnode, and Trump (1962), and for the details of the mathematical discussion the reader is referred to their works. Biot considers first the simple problem of instability of a rod subjected to an axial load F. The rod buckles into a sinusoidal half wave ($W/2$) when F attains the value given by the Euler formula

$$F = 4\pi^2 \frac{EI}{W^2} \tag{7-7}$$

where E = Young's modulus of the rod
I = moment of inertia of the cross section
W = wavelength of the buckle

If the rod is infinitely long and is held at a number of points distance $W/2$ apart (Fig. 7-30), it will be stable with a wavelength given by

$$W = 2\pi \sqrt{\frac{EI}{F}} \tag{7-8}$$

If it is not held at these nodal points, the rod becomes unstable. If the rod is surrounded by a viscous material, the lateral restraint required to prevent this

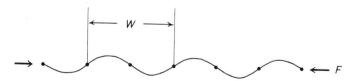

Figure 7-30

Characteristic stable wavelength W of an elastic rod subjected to a force F, and held at nodal points.

instability can be calculated; Biot has shown that the rod will develop a stable wavelength W which is independent of the viscosity of the material surrounding the elastic rod and is given by

$$W = 2\pi \sqrt{\frac{2EI}{F}} \tag{7-9}$$

If the elastic plate has a thickness h, the wavelength is

$$W = \pi h \sqrt{\frac{E}{(1 - v^2) F}} \tag{7-10}$$

where v is Poisson's ratio for the elastic material. Experimental checks on this equation have been made (Biot, Odé, and Roever, 1961) by buckling sheets of aluminum and an elastic cellulose in a viscous matrix (see Fig. 7-31). The observed relationships suggest, however, that factors other than those expressed in (7-10) affect the wavelength of the fold.

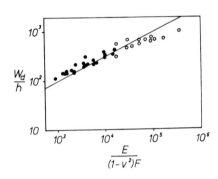

Figure 7-31

Experimental results of dominant wavelength/thickness ratios (W_d/h) of buckling elastic cellulose sheets (filled circles) and buckled aluminum sheets (open circles) in viscous matrix. The straight line represents the relationships of Eq. (7-10). (After Biot, Odé, and Roever, 1961.)

Many of the rock materials in the earth's crust are made up of aggregates of particles, with crystalline silicates probably the most abundant. We have no exact knowledge of the rheological state of these materials during the slow deformations which occur naturally in the crust and which probably lead to the formation of folds, but it seems that as a first approximation we can consider them as fluids (Newtonian) with very high viscosities (perhaps of the order of 10^{17} to 10^{22} poises). Biot (1961) has also pointed out that even if the rock materials do have some elastic properties, these tend to be overshadowed by the viscous properties when folding mechanisms are considered.

Biot (1957, 1961, 1964a, 1965b) and Ramberg (1959, 1960, 1961b, 1963b, 1964b) have investigated the problems of the deformation of layered viscous materials. Only the outlines of the methods and results of these investigations will be presented here; for further details the reader is referred to the original publications. Ramberg's theoretical discussion is built around discovering the particular function which satisfies both the biharmonic equation of the stream function (6-101) and the boundary conditions around the buckled layer (or layers). On the other hand, Biot's method considers the stress-strain relationships in and around the layers and investigates the potential development and amplification of small-scale sinusoidal folds in the competent layer. If a single layer of material with viscosity μ_1, and thickness t embedded in a medium of lower viscosity μ_2, is subjected to a compressive strain at a constant rate parallel to the layer (Fig. 7-32), both Biot and Ramberg have shown that the instability that is set up in the material leads to the production of folds of various wavelengths. The rate of growth of these various wave-

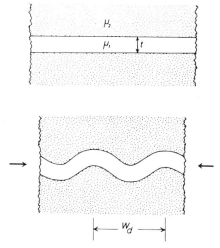

Figure 7-32

Characteristic dominant wavelength W_d produced by buckling a sheet of material of thickness t and viscosity μ_1 in a matrix of viscosity μ_2.

lengths is not uniform; one is generated more rapidly than the others. This is known as the *dominant wavelength* and its value is given by

$$W_d = 2\pi t \sqrt[3]{\frac{\mu_1}{6\mu_2}} \tag{7-11}$$

One rather peculiar feature of this function is that as μ_1 approaches the same value as μ_2, the dominant wavelength approaches a value $3.46t$. The presence of a finite dominant wavelength where $\mu_1 = \mu_2$ shows that this expression breaks down when W_d/t takes small values (less than 5). Another interesting feature of (7-11) is that the fold wavelength is independent of both the amount of compressive load and the strain rate [compare with the elastic layer theory where W_d varies with $F^{-\frac{1}{2}}$ (7-10)]. It must be remembered, however, that if the strain rates are high, the rock is unlikely to behave as a viscous fluid, and the initial conditions to establish (7-11) are invalid. Although experiments performed by McBirney and Best (1961) seem to indicate a large discrepancy between the theoretical and experimental results, those made by Biot, Odé, and Roever (1961) appear to accord well with the theory.

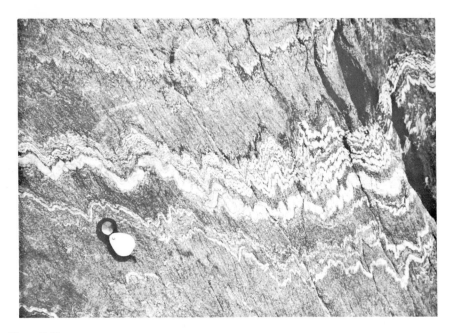

Figure 7-33

Varying wavelengths of folds produced by buckling pegmatite sheets of variable thickness in a mica schist matrix. Pennine Alps, Ticino, Switzerland.

With either elastic or viscous materials, Eqs. (7-10) and (7-11) show that the dominant wavelength is directly proportional to the thickness of the competent layer. This is a particularly important feature which may be seen

Figure 7-34

Relationship of thickness to fold wavelength in buckled layers. (After Currie, Patnode, and Trump, 1962.)

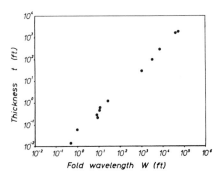

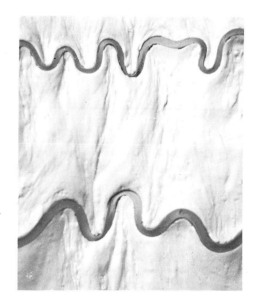

Figure 7-35

Experimental verification of linear dependence of fold wavelength on layer thickness in buckled layers. The buckled layers have the same composition.

in many natural fold systems over a great range of scales (Figs. 7-33, 7-34) and it can easily be verified experimentally (Fig. 7-35).

It will be seen from (7-11) and Fig. 7-36 that there is always a characteristic dominant wavelength of the folds even when the viscosities of the substances are equal. Biot has investigated this problem by considering how folds with a particular dominant wavelength established during the deformation grow in amplitude. He found that, although a dominant wavelength is always established, the fold development depends entirely on the degree of amplification (A_d) of this particular wavelength. The degree of amplification expresses the magnitude of the factor by which the amplitude of folds with a certain

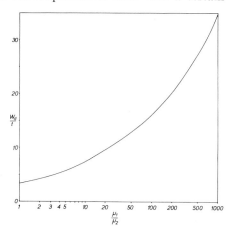

Figure 7-36

Plot of characteristic wavelength/thickness ratio of buckled layers of viscosity μ_1 in a matrix of viscosity μ_2.

wavelength is multiplied after a certain time. The intensity of dominant fold amplification at any one time t_0 during the fold history is referred to a time t_1 when the layer has been shortened by 25 percent of its original length.

$$\log_e A_d = \frac{t_0}{t_1}\left(\frac{\mu_1}{6\mu_2}\right)^{\frac{2}{3}} = \frac{x}{25}\left(\frac{\mu_1}{6\mu_2}\right)^{\frac{2}{3}} \tag{7-12}$$

where x is the percentage of finite strain at time t_0. This function has been graphically recorded in Fig. 7-37 for different values of viscosity ratio. For any value of μ_1/μ_2 (say 100) the fold amplification is at first rather small, then at some particular finite strain (20 to 25 percent) the amplification accelerates very greatly. This great increase in amplification marks a sudden or "explosive" development of folds in the layer. Biot has suggested that this occurs where A_d takes a value of about 1,000, which from (7-12) occurs where

$$\frac{t_0}{t_1} = 6.90\left(\frac{6\mu_2}{\mu_1}\right)^{\frac{2}{3}} \tag{7-13}$$

This is only a function of the viscosity ratio. Where this ratio has a value less than 100, the amplifications are small until the total compressive strain exceeds 25 percent. Large strains exceeding this amount would tend to mask the effects of the folding, and Biot has concluded that clearly marked folding requires that the viscosity ratio must exceed 100. This value seems rather large and may indicate that the assumptions which led to (7-13) might not be completely correct. The absence of folds in some layered complexes of rocks in metamorphic environments may be explained in this way; although there may be great compositional differences in the rock strata under metamorphic conditions, their viscosities might not be sufficiently dissimilar to set up folds.

Ramberg (1964*b*) has also discussed the effects of deformation by buckling and by more homogeneous strain. The greater the competence difference between the layer and matrix, the more rapid the development of buckles within the competent layer. This means that if a rock is made up of several layers of different competence and is subjected to compressive strain, the variable development of folds in each layer might give the appearance that each layer had suffered a different amount of shortening (Fig. 7-38). This would be incorrect, for the phenomenon is entirely the result of the difference in proportions of shortening taken up by buckling and by homogeneous strain in each layer.

Biot, Odé, and Roever (1961) performed a number of model experiments the results of which were in good agreement with Biot's and Ramberg's theoretical predictions (7-11).

In geological problems we commonly have to deal with the folding of not just a single layer but with a multilayered complex. The theory of the behavior

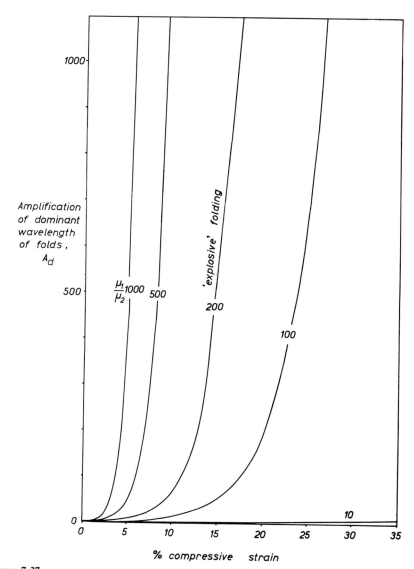

Figure 7-37

Amplification of folds at different values of total compressive strain.

of such materials has been investigated by Biot (1961, 1963*b*, 1963*c*, 1964*a*, 1964*b*) and Ramberg (1961*b*, 1963*b*, 1964*b*). Consider that there are *n* competent layers all of equal thickness *t* and viscosity μ_1 and that the compound slab of high-viscosity material is embedded in a medium of viscosity μ_2. If there is perfect lubrication between the individual layers (probably an unlikely con-

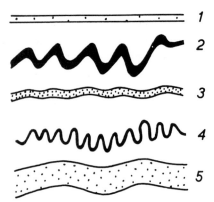

Figure 7-38

Effects of buckling a series of layers of different viscosity (viscosities 4 > 2 > 3 > 5 > 1 > matrix) by an identical compressive strain. (After Ramberg, 1964b.)

dition to hold in geological problems), Biot has shown that the dominant wavelength is larger than that of a single layer by a factor $n^{\frac{1}{3}}$

$$W_d = 2\pi t \left(\frac{n\mu_1}{6\mu_2}\right)^{\frac{1}{3}} \tag{7-14}$$

Thus, comparing the dominant wavelength of the folds in a nonstratified sheet with that of a stratified sheet of the same thickness and made up of n separate layers, that of the stratified sheet is less by a factor $n^{-\frac{2}{3}}$.

Biot (1964b) has also investigated the properties of a model consisting of n alternating layers with viscosities μ_1 and μ_2 and each with a constant thickness t confined between two straight and rigid boundaries. He determined the dominant wavelength W_d and found that this increased as the number of layers in the pile was increased, but that it was rather insensitive to changes in the viscosity contrast between the two types of layers (Fig. 7-39).

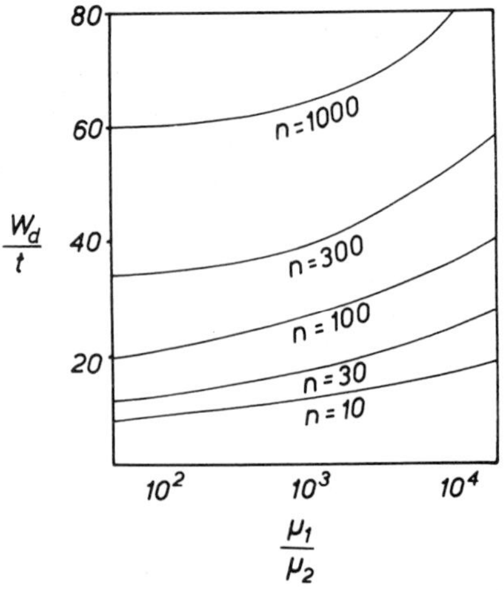

Figure 7-39

Characteristic wavelength/thickness ratio produced by buckling n alternating constant thickness layers of viscosities μ_1 and μ_2. (After Biot, 1964b.)

Ramberg (1961b, 1963b, 1964b) has approached the problems of multi-layered complexes by finding solutions to the plane-strain stream functions which satisfy various imposed boundary conditions on the layers. He considers a multilayered model made up of two materials of thicknesses and viscosities t_1, μ_1 and t_2, μ_2, respectively. Ramberg takes a model made up of a very large number of layers and assumes that each competent layer in the complex buckles such that the boundary deflections for each layer (except

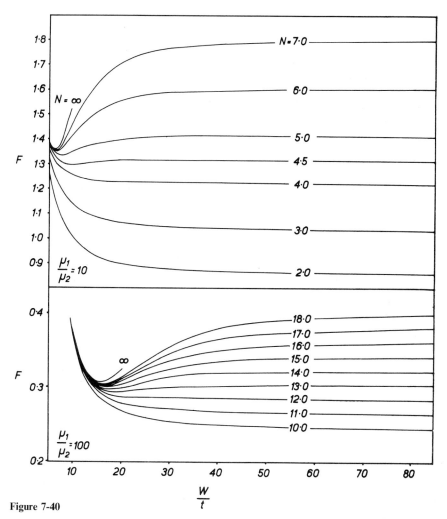

Figure 7-40

The buckling force F neccessary to initiate buckles of a characteristic wavelength/thickness ratio in multilayered complexes of alternating layers of thickness t_1 and t_2 ($t_2/t_1 = N$) and viscosities μ_1 and μ_2. (After Ramberg, 1964b.)

those near the top and bottom of the pile which are ignored) are identical. The buckling force F necessary to initiate the folding of the layers into buckles of a characteristic wavelength-thickness ratio is calculated. This is graphically recorded in Fig. 7-40. For particular values of thickness ratio $t_2/t_1 = N$, F has a specific minimum value which defines the W_d/t_1 ratios of the most easily formed folds and gives the dominant wavelength of the system.

It is found that F decreases with the spacing ratio N. Where $\mu_1/\mu_2 = 10$ and $N \leqslant 4$, no minimum turning value for F exists (Fig. 7-40). This signifies that if the layers are close to one another a minimum resistance to buckling occurs where W_d/t_1 is as large as possible. The layers will therefore form into a half wave whose size is governed only by the length of the layers. Where $N \geqslant 4.5$ a minimum-value turning point occurs in the graphs. This turning point is very pronounced at $N = \infty$, and this gives the wavelength-thickness ratio of folds formed in a single competent layer enclosed in an infinite medium of a lower viscosity given by (7-11). At values of N lying between 4.5 and ∞, the dominant wavelength-thickness ratio is slightly greater than that of the single layer. The changeover from single wave folds in multilayers to a condition where many waves of a characteristic wavelength appear occurs at about $N = 1.5\pi (\mu_1/6\mu_2)^{\frac{1}{3}}$.

Ramberg carried out a series of experimental checks on these theoretical results using elastic materials. He points out that the equations governing elastic instability are very similar to those he has evolved for viscous instability where the elastic rigidity modulus G replaces the viscosity μ. The experiments agree very well with the theoretical deductions, and particularly interesting are the results confirming the predictions about the appearance of folds with either a fixed dominant wavelength or a single half wave (Fig. 7-41). Biot (1964b) has criticized Ramberg's approach to these problems of fluid dynamics, suggesting that they are based on approximations which depend on the incorrect assumption of perfect correspondence of the behavior of elastic and viscous substances.

Another geologically important feature occurs at the contact of two unlike rock types when the rocks suffer a progressive compressive strain. If we take a model in which a single plane surface separates two substances of unlike

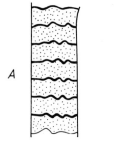
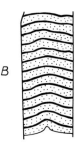

Figure 7-41

Experiments made on compressed multilayered rubber sheets to determine whether a characteristic fold wavelength develops (A), or whether the sheets buckle in a single half wave (B).
(After Ramberg, 1964b.)

A

B

Figure 7-42

Characteristic form assumed by a strongly compressed contact separating material of different viscosities. The dark material has the greatest viscosity.

viscosity $\mu_1 > \mu_2$, and deform this by a principal compressive strain parallel to the boundary surface, an instability is set up and the surface becomes folded. The form of the folded contact during the progress of deformation follows a series of changing shapes very similar to the type of development of the fold shapes on one side of a buckled layer (Fig. 7-42). From an originally near-sinusoidal form whose wavelength depends on the viscosity ratio, the surface develops a series of antiforms of broad, rounded aspect (wavelength large, amplitude small) separated by synforms which have a much tighter cross section (wavelength small, amplitude large) and a characteristic "pinched" appearance. This contrast in shape persists throughout further strain increments. Large-scale geological examples of this type of structure can be found at the unconformable contacts of "basement" and "cover" during a common period of orogenic activity. The pinched-in synclinal zones of Mesozoic sediments found within the external crystalline basement massifs of the Swiss and French Alps afford excellent examples (Fig. 7-43).

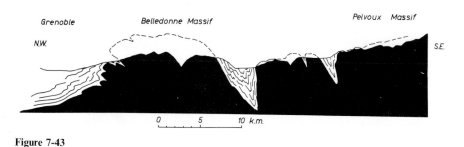

Figure 7-43

Section across the external massifs of the French Alps, illustrating the different form of the tightly pinched synclines and less strongly curved anticlinal zones of basement. (After Ramsay, 1963a.)

This type of structure is also very common in certain folded zones in the African Continent; the cover sediments occupy tight or isoclinal synclines "nipped in" between larger dome-like regions of deformed basement (Fig. 7-44). These

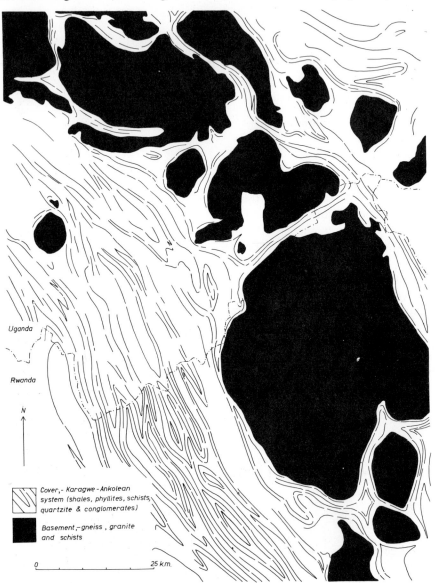

Figure 7-44

Basement-cover relationships in Southern Uganda and Ruanda illustrating the gently curved basement zones and separating synclines of tightly pinched cover sediments. (After Geol. Surv. Uganda Sheets SA, 36-1, 36-5, 1963, and Carte lithologique du Ruanda, 1963.)

synclinal zones sometimes intersect in a complex manner either because the compressive strain was acting in all directions within the surface layers, or

Figure 7-45

Deformed contact of a dike of hornblende schist cutting through granitic gneisses showing characteristic lobate form.

Figure 7-46

Mullion structure formed at a contact of competent sandstone and incompetent slaty materials. North Eifel, Germany. (Photograph by G. Thomas.)

because there were several successive superimposed deformations in which the basement always behaved in a more viscous manner than did the cover. Perhaps the mechanism of formation of the well-known mantled gneiss domes of Scandinavia (Eskola, 1949) can also be accounted for in this way.

On a small scale one may often see structures of this form developed at the contacts of rocks of different competence (Fig. 7-45), and it seems possible that the feature known as *mullion structure* can sometimes be developed in this way (Fig. 7-46) (Wilson, 1953). In many unconsolidated sediments, complex structures often develop at the contacts of rocks of different composition. Graywackes often show peculiar undulations (*bottom structures*) at the base of individual graded units; and although many of these structures are undoubtedly the result of erosion and the sculpting of the sediment surface by marine currents and objects carried by them, others, such as *load casting* and *flame structure* (Fig. 7-47), might sometimes be formed by a shortening of the layer contacts in the manner described above.

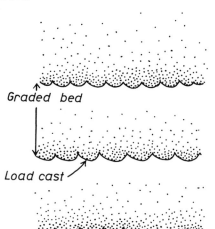

Figure 7-47

Load casts developed at the base of graded graywacke layers where the coarse sediments are in contact with incompetent shales.

7-7   BUCKLING: THE SHAPE OF THE FOLDED LAYERS

It is generally assumed that the shape of the buckles is initially a sine wave of the form $y = A \sin 2\pi x/W$, where A is the amplitude and W the wavelength of the fold. As the fold develops, the shape of the buckled layer departs from this function and the shape it takes up depends on a number of factors.

1. The value of the initial wavelength-thickness ratio W/t
2. Whether the competent bed is a single layer surrounded by less competent material or whether it forms part of a multilayered complex. If it is part of a complex of multilayers, the fold shape depends on the thickness and closeness of the adjacent layers and their viscosity.

3. The rheological properties of the material and the alteration in these properties during the formation of the folds

4. The physical environments of the rocks during folding (nature of the stress tensor, strain rate de/dt).

Consider first the shape of a single layer of material of viscosity μ_1 surrounded by material of a lower viscosity μ_2. As the fold develops, the initial sine function may become modified so that the curvature of the layer becomes more uniform. As a result, the fold curves may closely approximate to circular arcs and the fold is termed a *concentric fold*. If the initial wavelength-thickness ratio is rather small [less than $\pi/2$—such a small ratio is most likely to develop in a layer made up of a number of independent sheets, see (7-14)], there will be a position of maximum shortening of the layer by parallel folding (Class 1*B*) alone. Where the wavelength-thickness ratio is $\pi/2$, this limit occurs where 36 percent shortening has been accomplished, see de Sitter (1958) and Fig. 7-48. With further compressive strain the fold is ruptured,

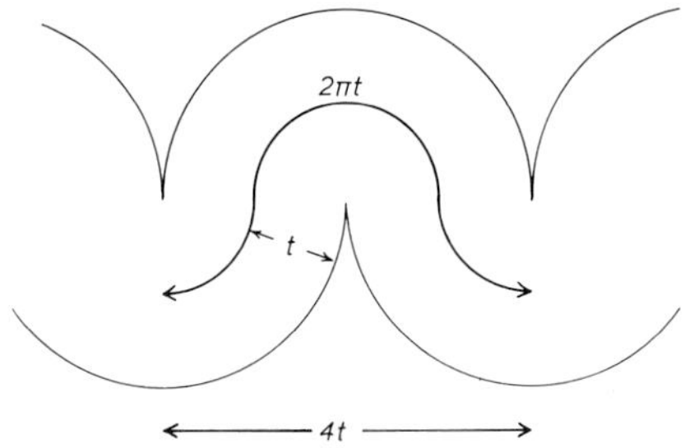

Figure 7-48

Maximum shortening of a concentrically folded competent layer. The length initially $2\pi t$ now occupies a length of $4t$; the shortening is therefore $100(2\pi t - 4t)/2\pi t = 36$ percent.

or its shape becomes modified into a fold of Class 1*C* known as *flattened parallel fold*. The geometry of these folds will be discussed in more detail in a later section. If the initial wavelength-thickness ratio is large in the layer, the compressive strain that can be taken up by folding is much greater than 36 percent and the shape of the structure is frequently modified into a form which resembles the curve known as an *elasticas* (Fig. 7-49) (Love, 1944). Folds of this type are frequently found in strongly deformed migmatites and gneisses, and the folded veins are known as *ptygmatic veins*.

Where the folded layer is one in a complex of multilayers, its development is influenced by the development of the folds in the adjacent competent layers. If the more competent layers are all of about the same thickness and lie fairly close to one another, they buckle together with a common initial wavelength (*harmonic folding* of the layers). As the folds develop they press aside and displace the surrounding less competent material. At later stages in the folding history, however, the bulk strain increments tend to increase the distance between adjacent fold hinges. The curved competent layers be-

Figure 7-49

Elasticas in ptygmatic veins. Isle of Mull, Scotland.

tween adjacent fold hinges may become straightened out and the structures become modified to *chevron folds*. The limbs of the fold may be stretched, thinned, and perhaps ruptured by this process. The hinge zones of adjacent folds may be displaced relative to one another in zones separated by the sheared limbs of the folds (Figs. 7-50, 7-51). This displacement leads to the slicing up of the rock into a series of subparallel slabs sometimes called *microlithons* or *Gleitbretter* (Schmidt, 1932; de Sitter, 1954) separated by surfaces of discontinuity or shear zones. The width of the microlithons is controlled by the dominant wavelength of the initially formed folds and the amount of modification of this wavelength by later compressive strain (cf. Fig. 7-50C and D). The width of the microlithons is therefore controlled by the thickness of the competent layers and is greater in thick layers than in thin layers (Hills, 1963, p. 238). Where microlithons can be traced from thick into thin competent layers, it is found that the large microlithons become subdivided into a number of narrower microlithon compartments. This structure can also be produced in layers which are obliquely inclined to the principal compressive strain, and here the shear structure is asymmetrically inclined to the layering (Figs. 7-52, 7-53). The shear planes that are developed in this way by the deformation of schistosity or slaty cleavage are often termed

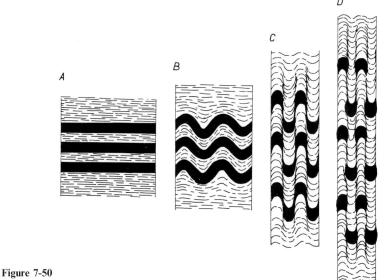

Figure 7-50

Development of Gleitbretter (microlithon) structure by buckling and rupture of competent layers.

shear cleavage or *strain-slip cleavage*. The individual shear surfaces can often be seen as a series of dark lines running subparallel to the axial surfaces of the folds (Fig. 7-33). The micaceous and other flaky minerals in the rock are

Figure 7-51

Development of shear surfaces along the limbs of strongly compressed buckled multilayers. Triassic Quartenschiefer, Brigels, Switzerland.

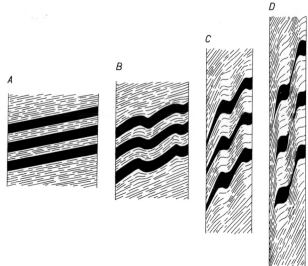

Figure 7-52

Development of asymmetric Gleitbretter (microlithon) structure.

folded and forced into a parallel alignment in these zones, and generally the spacing of the separate shear zones is proportional to the size of the micaceous minerals of the metamorphic rock. In coarsely crystalline rocks the shear zones may be up to 3 cm apart, and the micas are folded into coarse fold pleats between these zones to form what is sometimes called a *crenulation*

Figure 7-53

Asymmetric Gleitbretter in chloritoid schist.
Thin section from Unst, Shetland Islands.

cleavage or *herringbone cleavage*. The planes of slip are commonly the loci of recrystallization of the rock material during the production of the folds. The composition of the rock material within and in the immediate vicinity of the shear zones is changed, and the slip band often appears much darker than the surrounding rock because of the removal of such minerals as quartz and calcite from that zone (Fig. 4-38).

The detached and isolated fold hinges of the more competent layers in the microlithons often form long narrow pieces or *rods* which generally have their longest dimension parallel to the hinges of the larger scale folds. The folding and rupture of quartz veins by this mechanism is a common phenomenon in metamorphic rocks and leads to the production of *quartz rods* (see Wilson, 1953, 1961).

7-8 STATES OF STRAIN WITHIN THE BUCKLED LAYER

The buckled rock layer undergoes various internal distortions as a result of the folding process. These changes of shape can be accommodated in several ways depending on the nature of the material of the layer and the shape of

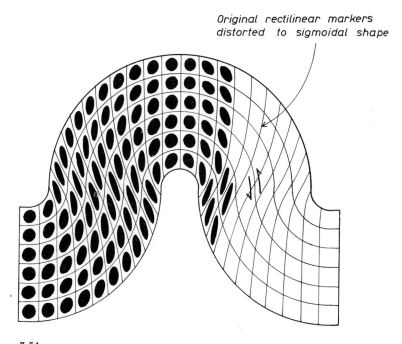

Original rectilinear markers distorted to sigmoidal shape

Figure 7-54

Internal strain in a layer deformed by flexural flow.

the fold. These mechanisms will now be discussed, and the types of small-scale structures that are formed will be described.

Shearing parallel to layer boundaries   Many sedimentary strata have a well-developed plane-parallel stratification, and this inherent weakness in the rocks frequently controls the type of internal deformation during buckling. The individual rock layers are flexed and the outermost layers slip over the inner layers toward the fold hinge zones. The folds developed in this way have a true parallel form (Class 1*B*) and are sometimes known as *flexural folds*. At any point in the fold this simple shear strain may be homogeneously distributed through the strata to produce *flexural-flow folds* (Fig. 7-54); or it may be inhomogeneously distributed with more movement on the layer boundaries than in the center of the layers (Fig. 7-55) and give rise to *flexural-slip folds* (Donath, 1962).

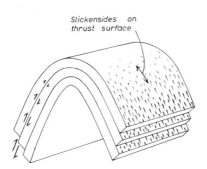

Figure 7-55

Development of thrusts in a flexural-slip fold.

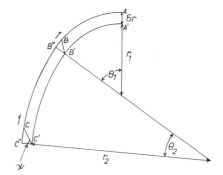

Figure 7-56

Total shear developed in a layer of thickness δr positioned in a flexural fold made up of two circular arcs (internal radii r_1 and r_2). CC″ records the total slip at angle of dip $\theta_1 + \theta_2$, and ψ records the total angular shear strain.

Let us examine this deformation in more detail by considering two surfaces *ABC* and *A′B′C′* situated in a flexural-slip fold made up of two circular arcs, that between *A′* and *B′* having a radius of curvature r_1 and that between *B′C′*, a radius r_2 (Fig. 7-56). The distance between the two layers is δr; the dip of the surface at *B′* is θ_1 (radians) and at *C′* is $(\theta_1 + \theta_2)$. The amount of flexural slip between the two surfaces at *B′* is the difference in length of the two arcs *AB* and *AB″*.

$$BB'' = AB'' - AB = AB'' - A'B'$$

$$= \theta_1(r_1 + \delta r) - \theta_1 r_1 = \theta_1 \delta r \qquad (7\text{-}15)$$

The total slip at C' is BB'' plus that in the arc $B'C'$

$$CC'' = \theta_1 \delta r + \theta_2 \delta r = (\theta_1 + \theta_2) \, \delta r$$

The amount of slip therefore depends on the angle of dip of the folded layer and is independent of the changes in curvature of the surface in the fold. The actual slip is also directly proportional to the thickness of the layer. The finite shearing strain γ can be expressed in terms of the angle of dip of the layer; if ψ is the angular shearing strain,

$$\gamma = \tan \psi = \frac{CC''}{C'C''} = \theta_1 + \theta_2 \qquad (7\text{-}16)$$

This is graphically recorded in Fig. 7-57, and by using Figs. 3-21 and 3-22 it is possible to determine the values and orientations of the principal strains at any position in the fold (Fig. 7-54).

It is interesting to compare the variations in finite strain in different-shaped flexural-slip folds developed by the same amount of total shortening. The fold which gives the most uniform distribution of strain over the largest area in a cross section is that of the chevron-fold model, and the structure which gives the least strain over the largest area (but the greatest strain over a small area) is that of the box-fold model.

As a result of this flexural slip, various structures may develop in the rocks.

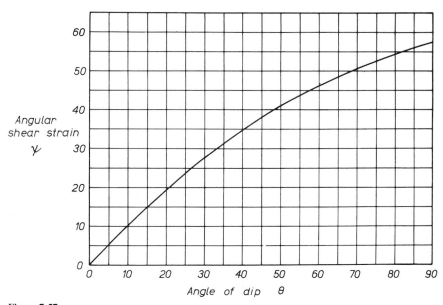

Figure 7-57

Angular shear strain in flexural folds.

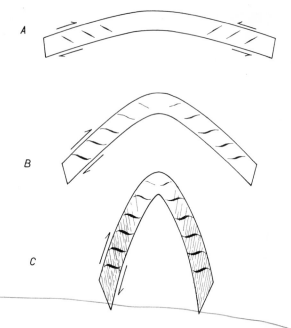

Figure 7-58

*Progressive development of
sigmoidal tension fissures and
slaty cleavage as a result of
progressive fold development with
internal deformation by
flexural flow.*

Zones of en echelon quartz- or calcite-filled tension fissures may begin to
form on the flanks of structures where the dips attain the highest value in
the fold (Fig. 7-58). As the fold develops, these fissures become progressively
deformed in the way described in Chap. 3. They always tend to dip at moderate
to low angles toward the antiformal axial surface, and if the shearing is inhomogeneous they may be sigmoidally shaped. Slaty cleavage may develop
on the fold limbs perpendicular to the maximum finite compressive strain
(Fig. 7-58C).

Particularly intense shear strains may be concentrated along the bounding
zones of the competent layers. This concentration frequently leads to the
development of *bedding plane thrusts* within the structure (Figs. 7-55, 7-59).
Because the amount of slip is greatest on the limbs of the fold and is directly
proportional to the thickness of the layer, these thrusts are best developed
in the fold limbs at the boundaries of the thickest and most competent rock
strata and die away toward the hinge of the structure. The thrust may show
slickensides which are often arranged perpendicular to the hinge lines of the
fold (Fig. 7-55) and are sometimes accompanied by crescentic grooves termed
lunules by Wegmann and Schaer (1957).

Although flexural-slip folding develops best in rocks which have a well-
marked stratification or layering, it may also be an important method of
internal deformation in rock sheets which have no layered structure. Kuenen
and de Sitter (1938) performed a series of experiments on homogeneous cakes

Figure 7-59

Tension fissures deformed by bedding plane slip in graywacke. Church Cove, South Cornwall, England.

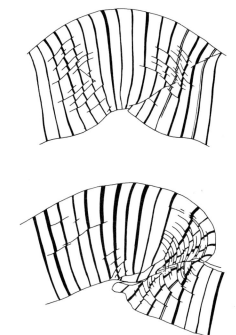

Figure 7-60

Development of flexural-slip thrusts in deformed, initially homogeneous slabs of clay. (After Kuenen and de Sitter, 1938.)

of modeling clay and proved that shear planes parallel to the upper and lower surfaces of the initially homogeneous clay sheet could be developed as a result of folding (Fig. 7-60). Ramberg (1961*a*) has suggested that shear planes formed in this way are most likely to be developed in folds where the ratio of thickness to initial wavelength is less than one to three.

In some flexural-slip folds in the Jura Mountains, Wegmann has noted that the slickensides are not perpendicular to the fold hinge, and he interpreted these observations as indicating that the movement history of the layers was more complex than that generally attributed to flexural-slip folds. If we consider a stack of competent layers of rock which are inclined obliquely to the principal axes of bulk rock strain, the folded structure can still be a true parallel type formed by flexural slip providing that the shear directions are always contained in the folded surfaces (Fig. 7-61).

Flexural slip is often said to produce small-scale asymmetric folds on the flanks of the major structure that are termed *drag folds* (Leith, 1923; Nevin, 1949; Billings, 1954). De Sitter (1958*a*) pointed out that drag folds should die out toward the fold hinges as the amount of simple shear decreases. In practice, however, this does not generally occur; the asymmetric "drag folds" on the fold limb pass into symmetric folds at the hinge zone. The development of these symmetric folds cannot be accounted for by slip along the layering for there is no shear at the fold hinge. De Sitter termed them *parasitic folds* and

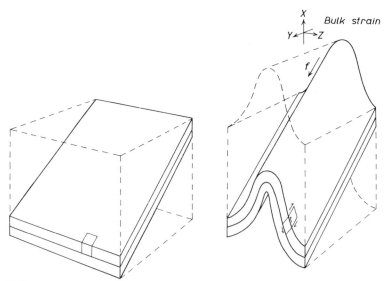

Figure 7-61

Development of an oblique flexural-slip fold, axis f, as a result of buckling a layer which is asymmetrically oriented with respect to the axes of bulk rock strain X, Y, and Z. Distortion occurs within the bedding surfaces of the structure as well as slip parallel to the layer boundaries.

suggested that the small-scale folds in the major structure on both the limbs and the hinge are most likely to be the result of a component of compressive strain acting along the competent layers. Simple shear on the bedding surfaces cannot lead to shortening of the layers, as these surfaces are planes of no infinitesimal (and no finite) longitudinal strain throughout the fold development. It therefore appears that the term "drag fold" has an incorrect genetic implication and does not provide a satisfactory description of these asymmetric small-scale folds. It seems likely that some folds of this type are the result of the modification of previously established symmetrical buckles by the late development of a larger wavelength structure (Fig. 7-62) (Ramberg, 1963a, 1964b); but it is also possible that some may develop during the formation of a larger fold in small competent layers inclined at a low angle to the principal contraction axis (de Sitter, 1958a).

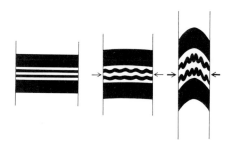

Figure 7-62

Development of parasitic folds.
(*After Ramberg, 1963b.*)

In flexural-slip folds it is possible to determine the amount of shortening of the layer by measuring along the length of *any* lithological unit on the boundary of, or inside, the competent layer. The flexural-slip-fold model is unique in this respect.

Longitudinal strains parallel to layer boundaries   The strains within buckled layers may be developed in ways other than those of flexural slip. *Tangential longitudinal strain* is a type of internal deformation commonly developed in buckled nonstratified sheets whereby the principal axes of strain are arranged tangentially and perpendicularly to the folded layers. It is possible to accommodate the internal strains of the layers entirely by this type of deformation, although in nature both tangential longitudinal strain and flexural slip generally proceed together. Lines drawn on a cross section perpendicular to the layering before deformation remain perpendicular (or nearly so) to the layering after deformation (Fig. 7-63). Within the buckled layer there is a surface known as the *neutral surface* (or *finite neutral surface*) along which the principal finite longitudinal strains are zero. This neutral surface need not be positioned exactly midway between the layer boundaries and it may, in practice, crosscut the internal layering at a low angle. On the outer arc of

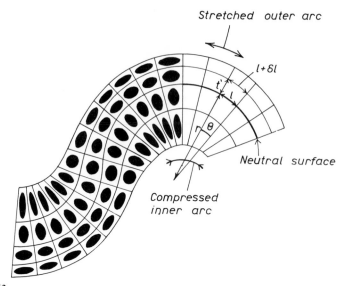

Figure 7-63

Buckled layer with internal deformation developed by tangential longitudinal strain.

the neutral surface the layers have been subjected to an extensive strain, while on the inner arc they have been compressed (Fig. 7-63). As a result of these strains the orthogonal thickness of the internal layers has been modified. During the development of the fold there is generally another surface, best termed the *infinitesimal neutral surface*, along which the infinitesimal longitudinal strains are zero. At any particular stage of fold formation the layers in the outer arc are being extended along their length, while those which lie inside the infinitesimal neutral surface are being contracted along their length. In general, the finite and infinitesimal neutral surfaces do not coincide, and those small-scale structures formed by an earlier period of compressive strain along the layering may become modified by later extensive strains (Fig. 7-65).

Consider a layer of original thickness t and length l which is folded into a sector of the fold so that one boundary lies on the finite neutral surface and the radius of curvature of the neutral surface in this fold sector is r (Fig. 7-63). The original orthogonal thickness t has become modified to t' in the fold. Assuming perfectly ductile deformation and plane strain, if we compare the areas of the original rectangle with that of the final folded segment, we find

$$lt = \frac{l}{2\pi r}\left[\pi(r + t')^2 - \pi r^2\right]$$

$$0 = t'^2 + 2rt' - 2rt$$

or

$$t' = -r + (r^2 + 2rt)^{\frac{1}{2}} \tag{7-17}$$

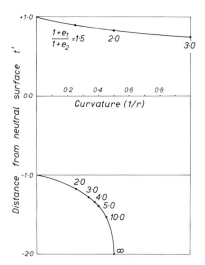

Figure 7-64

Modification of thickness in layers within a buckled layer deformed by tangential longitudinal strain, and strain ratios $(1 + e_1)/(1 + e_2)$ at these modified positions.

The values of the modified orthogonal thickness (originally unity) on either side of the neutral surface for fold sectors having different values of curvature ($c = 1/r$) are plotted in Fig. 7-64. As the curvature increases, the orthogonal thickness becomes either progressively decreased (if $t = +1$) or progressively increased (if $t = -1$). In the outer fold arcs this thinning of the layers near the fold hinge means that any one layer in this part of the fold has the form of a fold with strongly convergent dip isogons (Class 1A), while the layers of the inner fold arc have the form of a fold with weakly convergent dip isogons (Class 1C). When a particular band of material within a buckled layer is traced along a fold wave through a number of folds, the class of fold alternates between 1A and 1C in successive folds.

To determine the strains at different parts of the fold arc, assuming that there is no shearing parallel to the layer boundaries, consider the longitudinal strain on the surface at distance t' from the neutral surface (Fig. 7-63). If this arc sector was originally of length l, it is now expanded by an amount δl. Because $l = r\theta$ and $l + \delta l = (r + t')\theta$, it follows that

$$\delta l = \frac{lt'}{r} \qquad (7\text{-}18)$$

The extension e is given by $\delta l/l$ or

$$e = \frac{t'}{r} \qquad (7\text{-}19)$$

If t' is positive, this gives the greatest principal strain e_1; if negative, then this is the value of e_2. The value of the other principal strain can be calculated assuming plane strain from $(1 + e_1)(1 + e_2) = 1$, which gives e_2 (or e_1 on the inside of the neutral surface)

$$e = -\frac{t'}{r + t'} \qquad (7\text{-}20)$$

The ratios R of the principal strains at distance t' from the neutral surface where t was originally unit distance are obtained from (7-17), (7-19), and (7-20); on the outer arc

$$R = \frac{1 + e_1}{1 + e_2} = 1 + 2c \qquad (7\text{-}21)$$

and on the inner arc

$$R = \frac{1}{1 - 2c} \qquad (7\text{-}22)$$

where c is the curvature $(1/r)$ of the neutral surface. This geometrical investigation of internal deformation by tangential longitudinal strain alone shows that the principal finite extensive strains above the neutral surface increase in a linear fashion with the distance from the neutral surface, and that they also show a linear increase with the curvature. It therefore follows that the maximum strains of this type will always be set up at fold hinges on the boundaries of the layer, and that minimum strains occur throughout the layer in the fold limbs and at the neutral surface throughout the fold. This is the opposite arrangement of the position of the strongest strains developed by shear parallel to the boundary layers. The absolute maximum principal strains developed by tangential longitudinal strain always occur on the inner-arc layer boundary of a fold. If the internal deformation is accomplished entirely in this way, these maximum compressive strains (and strain ratio R) increase very rapidly as the curvature increases. Where the radius of curvature at the neutral surface reaches a value 2.5 times that of the original thickness of the material on the inside of the neutral surface, the finite-strain ratio attains a value of 5:1. Any further increase in the curvature results in an exceedingly large increase in the strain ratio $dR/dc = 2/(1 - 2c)^2$. This geometrical effect is extremely interesting, for the rheological properties of the material (or the rate of strain) might not allow ductile flow. There are several possibilities by which the further development of this fold may proceed.

1. The maximum curvature of the neutral surface at the fold hinge (which controls the maximum strain ratio) may keep a fairly constant value, and the fold develops further by an increase in the value of the curvature on the fold limbs until the neutral surface throughout the domain of the fold attains a fairly constant curvature. The fold becomes a true *concentric fold*.

2. The strain increments may be taken up by a change in the mechanism

of internal deformation, namely flexural slip (see p. 392). Although there is no absolute limit to the type of fold where this is likely to occur, those having an initial wavelength-thickness ratio of less than 4:1 are likely to build up maximum finite-strain ratios of more than 5:1 (see Fig. 7-64) and maximum strain-increment ratios of more than 3:1 so rapidly that they are likely to deform internally by flexural slip.

3. The neutral surface may alter its position and move nearer to the inner-arc boundary of the layer. This means that as the fold develops further the strain increments on the outer fold arc are increased, but those on the inner arc become less strongly marked. The tensional features associated with the outer arc become more pronounced and cut deeper into the layer. An experiment performed by Kuenen and de Sitter (1938, p. 236) by buckling hard clay rapidly shows the effect very clearly.

4. Conjugate shear planes may develop in the core of the structure, the strain increments being relieved by movements on local surfaces of discontinuity (Figs. 7-65, 7-66).

The nature of the small-scale structures which are developed as a result of longitudinal strain in the buckled layer depends to a large extent on the prop-

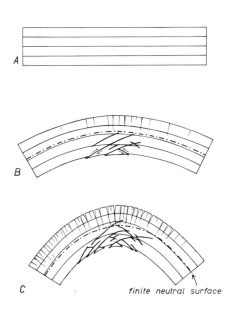

Figure 7-65

Structures developed in a layer progressively folded and deformed by tangential longitudinal strain.

Figure 7-66

Folded thrusts in deformed siltstones. Tarskavaig Bay, Isle of Skye, Scotland.

erties of the layer under a given strain rate (de/dt). In the inner arc of the fold the strong compressive strain perpendicular to the layering may lead to the development of slaty cleavage around the fold hinge and sometimes to the formation of tension fissures subparallel to the layering. If the layer is not completely ductile (perhaps during the closing stages of the fold history), the last strain increments may lead to the development of sets of conjugate shears (joints and faults) and to the formation of thrust faults which themselves may become involved in the folding (Figs. 7-65, 7-66). In the outer arc of the fold the principal strain along the layers is extensive and this may lead to the development of tension cracks and fissures crosscutting the layering (Fig. 7-65). Such fissures are often infilled with crystalline material that has migrated outward from the strongly compressed inner arc of the structure. Individual thin competent-layer components making up the buckled sheet may develop boudinage structures in the outer fold arc.

If a sheet of material of limited width is buckled, it will be found that in the outer arc of the fold the material tends to contract along the direction of the fold hinge, while in the inner arc of the fold the material tends to expand along the fold hinge. This gives rise to an additional curvature of the sheet in a direction perpendicular to the main fold axial direction (Fig. 7-67). This curvature is a well-known phenomenon studied by engineers investigating the stability of buckled plates and beams and is termed *anticlastic bending*. This process could be important in producing some of the variations in axial plunge seen in fold systems, but it should be remembered that although the strains set up in this way look impressive at the edges of a narrow sheet of folded material, they have less structural significance when the sheet is a competent layer of wide lateral extent.

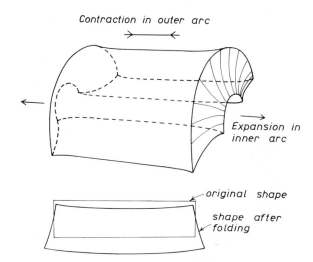

Figure 7-67

Anticlastic bending in a buckled layer.

If the position of the finite neutral surface within the fold can be established, measurement of the length of this surface will give the best possible estimate for the original length of the layer. It is not generally valid to measure along the layer boundaries to determine the original length because these surfaces have suffered both expansion and contraction in ways that do not always compensate one for the other. If the position of the finite neutral surface is unknown, the length of a surface midway between the boundaries of the layer is a first approximation to the original length.

Combinations of layer boundary slip and tangential longitudinal strain
Until now we have investigated the properties of internal distortion of the buckled layer in terms of simplified models of the deformation process. In general, it is much more common for the buckled layer to undergo internal distortions by a combination of flexural slip, tangential longitudinal strain, and other types of deformation.

Many of the variations in intensity and orientation of cleavage and other structures can be accounted for by various combinations of the several mechanisms of internal deformation. As an example, consider the effects of progressive deformation of two competent layers A and C separated by a thin incompetent layer B (Fig. 7-68), the three-layer sandwich enclosed in less

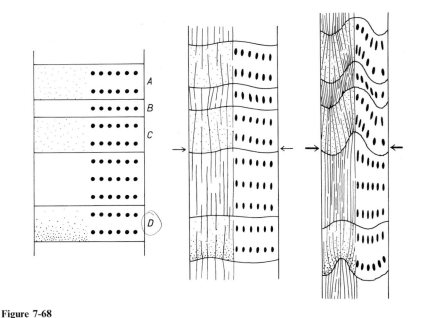

Figure 7-68

Progressive folding in a multilayered complex, the development of slaty cleavage, cleavage fans, and late shear along cleavage surfaces.

competent material. As the layers are progressively shortened along their length, they first contract approximately homogeneously, then a wavelike instability is set up which is amplified and the layer becomes buckled into a sinusoidal form with a characteristic wavelength depending on the viscosity ratios of the layers. The internal deformation of the competent layers is partly by shear parallel to their boundaries but mostly by tangential longitudinal strain. In the outer arcs of the folds the stretching parallel to the layers leads to a decrease in the ratios of the principal finite strain built up by the earlier part of the deformation, while in the inner arc the finite-strain ratios are increased. In layer B the dominant incremental distortions are those of simple shear on the layer boundaries, and the strain ellipsoids have their principal extensions dipping away from the axial surface of the anti-formal fold.

It has been noted previously (Chap. 4, Sec. 19) how the state of finite strain may be recorded in a rock as slaty cleavage or schistosity parallel to the XY plane of the finite-strain ellipsoid. If the strains are sufficiently large, cleavage will develop; but because the finite strains vary throughout the fold, the cleavage will be of variable orientation and intensity. Where internal defor-mation has been accomplished with a large component of simple shear on the layering (layer B), the XY strain planes and cleavage generally diverge downward from the axial surface of the antiformal folds; whereas, where tangential longitudinal strain is predominant (layers A and C), these planes

Figure 7-69

Convergent cleavage fan, siltstone. Spitsbergen.

converge downward on the axial surface of the antiform. The type of *cleavage fan* seen in layers *A* and *C* is that which has been described most frequently in geological literature (e.g., Dale, 1897; Leith, 1923; Wilson, 1946, 1961; Cloos, 1947; and many others). It is probably best termed a *normal cleavage fan* or, preferably, *convergent cleavage fan* (Fig. 7-69) (using the term "convergent" in the sense of tracing the cleavage from the outer to an inner arc of a particular fold). The fan structure of layer *B* is most conveniently referred to as a *reversed cleavage fan* or, preferably, *divergent cleavage fan*. In metamorphic environments and regions of strong deformation the fanlike form may be very weakly developed and the cleavage is parallel or subparallel to the axial surface of the fold. It is then termed an *axial plane cleavage* (Fig. 7-70). It will be seen from Fig. 7-68 that it is unwise to assume that any cleavage measured on the flank of a fold is parallel to the axial surface, although where the cleavage is perpendicular to the bedding surfaces it lies very close to the axial surface.

The change in orientation of the cleavage from layer to layer in the fold produced by the variation in strained state of the rock material is known as *cleavage refraction* (Fig. 7-71). Where lithological changes in a layer are of a transitional nature (e.g., graded units in turbidite deposits), the slaty cleavage develops a curved form through the bed (Fig. 7-68D). This curved cleavage may be mistaken for cross-bedding within the sedimentary layer; the truncated tops of the apparent "foreset" beds would appear at the base of the graded

Figure 7-70

Axial plane cleavage, Mesozoic marls. Helvetic Alps, Valais, Switzerland.
(*Photograph by J. McManus.*)

Figure 7-71

Slaty-cleavage refraction in alternating graywackes and slates. Rio Tinto, Spain.

unit where the cleavage refraction is greatest and thus the deductions made as to the "way-up" or direction of facing of the layer would be incorrect.

Some workers have suggested that the cleavage developed in the more competent layers has a different mechanical significance from that seen in the less competent strata. Because the cleavage developed in the competent layer often appears as closely spaced, clean-cut surfaces, it has been termed *fracture cleavage.* Billings (1954) and Wilson (1946) have suggested that fracture cleavage forms along the planes of no finite longitudinal strain, in contrast to the slaty cleavage which forms perpendicular to the maximum finite contraction (Fig. 7-72). This mechanical explanation is not convincing; observations of cleavage in graded units show that "fracture cleavage" passes without a break into slaty cleavage, thus indicating a common origin. Refracted slaty cleavage frequently undergoes modification during folding, and it seems that many of the special features of fracture cleavage may result from modification of slaty cleavage during the later stages of fold development.

It has been shown previously that, at the beginning of the buckling process, the layers tend to push against and displace the surrounding and less competent material, but that at a certain stage of fold development, the further strain increments require the hinges of adjacent folds to separate more than can be accomplished by simple buckling. Increasing separation of these folds can be assisted by movement along the weaknesses that have developed in the rock, namely, the previously formed slaty cleavage. If the hinges of adjacent folds are separated, the slaty cleavage provides a fabric which may be taken

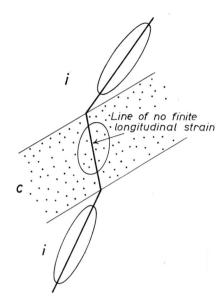

Figure 7-72

Interpretation of cleavage refraction. (After Billings, 1954 and Wilson, 1946.) In incompetent material (i), cleavage formed parallel to principal finite extension; in competent material (c), "fracture cleavage" formed parallel to a plane of no finite longitudinal strain.

over to assist the displacements. The slaty cleavage becomes a shear surface, and movements displace the contacts of the lithological layers into a series of steps (Figs. 7-73 and 7-68). If the slaty-cleavage planes show a refracted pattern, then slip on these surfaces will open up cracks and fissures in the competent layers wherever there is a convergent cleavage fan. These cracks develop the structure called fracture cleavage; if the displacements are large, openings will develop in the rock and become the site of deposition of quartz and carbonate material derived from the surrounding rocks. This crystalline material frequently has a fibrous structure with the fibers oriented subparallel to the shear direction in the cleavage of the less competent rocks. Although the total strain in a zone in which this process has been developed is altered by the shear, the strain state in the slivers of material positioned between the actual shear planes generally remains practically unmodified and is still related to the main slaty-cleavage formation (Fig. 7-74).

As a result of shearing, the fold shape will become modified. We can calcu-

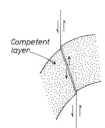

Figure 7-73

Movement along a refracted cleavage surface producing a dilation in the competent layer.

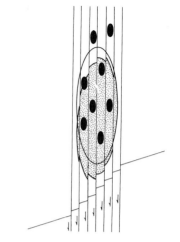

Figure 7-74

Comparison of state of strain inside slivers of sheared slate with the bulk strain modification caused by the slip. The solid ellipses show the strain within the cleavage slices.

late the geometrical effect of this modification as follows: consider a point P (dip α) situated on the arc of a fold cut by shear planes making an angle β with the axial trace (Fig. 7-75). Let the thickness of the layer at P measured parallel to the axial trace of the fold be $T = T_1 + T_2$. As a result of shear, the angle of dip at P is increased to a value $(\alpha + \theta)$; and if γ is the superimposed shear strain, then from (3-71) it follows that

$$\theta = \tan^{-1} \left[\tan (\alpha - \beta) + \gamma\right] - (\alpha - \beta) \tag{7-23}$$

or

$$\theta + \alpha = \tan^{-1} \left[\tan (\alpha - \beta) + \gamma\right] + \beta \tag{7-24}$$

Considering now the modification of thickness T to a new value T_m, then if d is the length across the layer measured in the shear direction,

$$T_m = T_1 + T_2 + T_3 = d \cos \beta + d \sin \beta \tan (\alpha + \theta) \tag{7-25}$$

But $T \cos \alpha = t$, and $d \cos (\alpha - \beta) = t$

therefore

$$d = \frac{T \cos \alpha}{\cos (\alpha - \beta)} \tag{7-26}$$

Combining (7-25) and (7-26),

$$T_m = \frac{T \cos \alpha}{\cos (\alpha - \beta)} \left[\cos \beta + \sin \beta \tan (\alpha + \theta)\right] \tag{7-27}$$

If we know the variation in γ and β around the fold, it is possible to compute the new dip α' and the thickness T_m for the modified structure.

The geometrical modification of folds in this way can produce some peculiar and complex fold styles. To take one example—the modification of a parallel fold by shear parallel to the axial surface—the angle β is zero throughout

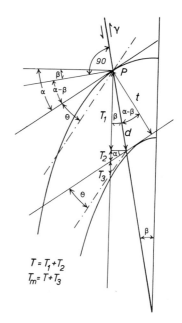

Figure 7-75

Modification of the thickness T in a fold to a new value T_m as the result of shear along a surface making an angle β with the axial surface of the fold.

$$T = T_1 + T_2$$
$$T_m = T + T_3$$

the structure and (7-27) and (7-24) simplify to the conditions $T_m = T$ and and $\alpha + \theta = \tan^{-1} (\tan \alpha + \gamma)$. If the hinge zone is sheared past the limbs such that the maximum shear strain is developed on the limbs and minimum shear strain at the hinge of the structure, the dip values T' and t' are modified as shown in Table 7-5. The graphical variations in T'_m and t'_m with α for the modified fold are shown in Fig. 7-76. The fold has a complex form; it has convergent dip isogons throughout and may be more exactly classified as Class $1C$ modified by Class $1A$.

Generally the variably oriented cleavage planes making up a cleavage fan are cylindrically disposed with respect to the fold axis, and the graphical construction of the line of intersection of any two nonparallel cleavage planes

Table 7-5 *Modification of T' and t' in a fold as a result of simple shear parallel to the axial surface*

α	0	10	20	30	40	50	60	70	80	90
γ	0.0	0.27	0.5	0.7	0.85	0.93	0.98	0.99	1.0	1.0
α'	0	24	41	52	60	65	70	75	82	90
T'_m	1.0	1.01	1.06	1.15	1.31	1.56	2.00	2.92	6.10	∞
t'_m	1.0	0.92	0.80	0.71	0.65	0.66	0.68	0.76	0.85	1.00

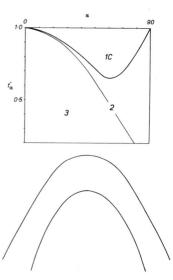

Figure 7-76

Parallel fold modified by heterogeneous shear parallel to the axial surface. The graph records the values of t'_α through the cross section.

gives the axial direction of the fold. If the cleavage is arranged in a fan or is an axial plane cleavage, a simple and very useful geometrical method may be used to determine the axial direction; the trace of a bedding surface on a cleavage plane, or vice versa, gives the axial direction at that locality (Leith, 1923).

The relationship between the orientations of cleavage and bedding surfaces within a structure is especially useful in determining the disposition of antiforms and synforms. Figure 7-77 shows two outcrops of dipping rock strata. The cleavage structures developed in the layers, however, are different at the two localities and it can be deduced that there is an antiformal fold between the two outcrops. The general rule for interpreting such cleavage-bedding relationships states " . . . where the schistosity and the beds dip in

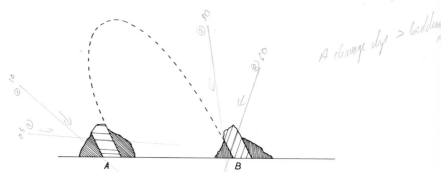

Figure 7-77

Relationship of slaty cleavage and bedding in an overturned limb (A) and normal limb (B) of a fold.

the same direction, if the schistosity is steeper than the stratification, the succession is in the correct order; if the schistosity dips less steeply than the bedding, then the beds are probably inverted." (Wilson, 1961, p. 466). Wilson emphasizes, however, that this technique *cannot be used* to determine the true stratigraphic order of the strata but only shows whether they are "normal" or "overturned" with respect to the fold in question. For example, if the rock layers have been previously inverted by some tectonic process (and this is not uncommon in some strongly folded zones of the crust) and then these beds were subsequently folded and cleaved, the strata on the "normal limbs" of the folds would be in inverted stratigraphic order, while those on the "overturned limbs" of the folds would have been reinverted and therefore arranged in their correct stratigraphic sequence. Although there is some initial problem in applying the above rules where cleavage and bedding dip in opposite directions (as in Fig. 7-77B), the geometry is generally easily resolved when the two possible structural interpretations are contrasted.

Modifications of shape by superimposed homogeneous strain   In environments where the rock materials are ductile during deformation, it is common for the shape of the buckled layers to become modified by the superposition of a fairly homogeneous strain. De Sitter (1958) has suggested that there is a value of maximum shortening of the competent layers by buckling alone. He also showed that with parallel concentric folds, which have their limbs parallel to the axial surfaces, the layer is unlikely to contract more than $\pi/2$ of its original length, that is, 36 percent (Fig. 7-48). Although the limitations that de Sitter imposed on the type of fold he investigated are not directly

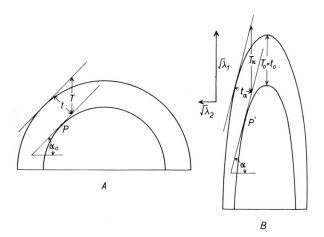

Figure 7-78

Modification of the shape of a parallel fold by a homogeneous strain. (After Ramsay, 1962a).

applicable to all parallel folds, it seems generally true that unlimited shortening of a competent layer by parallel folding alone is not possible. De Sitter suggested that further compressive strain leads to the modification of the parallel shape of the buckled layer by a process of flattening, and the folds formed in this way have been termed *flattened parallel folds* (Ramsay, 1962a).

Consider a point P on the lower surface of a folded layer (Fig. 7-78A) where the angle of inclination (with reference to the normal to a vertical axial surface) is α_o. Let the orthogonal thickness of the layer at this point be t and the thickness parallel to the axial surface, T. Then from (7-5)

$$T = t \sec \alpha_o \tag{7-28}$$

If this shape is modified by a superimposed homogeneous strain with principal quadratic extensions λ_1 parallel to the axial surface and λ_2 normal to this surface (Fig. 7-78B), the point P takes up a new position P'. The angle of dip at P' is α, and the orthogonal and axial-surface-direction thicknesses of the layer are t_α and T_α, respectively. From (7-5)

$$T_\alpha = T(\lambda_1)^{\frac{1}{2}} = t_\alpha \sec \alpha \tag{7-29}$$

Combining (7-28) and (7-29),

$$t_\alpha = \frac{t(\lambda_1)^{\frac{1}{2}} \sec \alpha_o}{\sec \alpha} \tag{7-30}$$

From (3-34),

$$\tan^2 \alpha = \frac{\lambda_1}{\lambda_2} \tan^2 \alpha_o$$

or

$$\frac{\lambda_2}{\lambda_1} \sec^2 \alpha + \left(1 - \frac{\lambda_2}{\lambda_1}\right) = \sec^2 \alpha_o \tag{7-31}$$

Combining (7-30) and (7-31),

$$t_\alpha{}^2 = t^2 \lambda_1 \left[\frac{\lambda_2}{\lambda_1} + \left(1 - \frac{\lambda_2}{\lambda_1}\right) \cos^2 \alpha\right] \tag{7-32}$$

This equation is true for any original fold shape, and the thickness ratio (t'_α) for any angle of dip α in the modified fold may be computed.

Consider the modification of an originally parallel fold. At the fold hinge, $\alpha = 0$ and $t_o^2 = t^2 \lambda_1$. We can now determine the orthogonal thickness ratio $t'\alpha$ for modified parallel folds, $t'_\alpha = t_\alpha/t_o$

$$t'_\alpha{}^2 = \frac{\lambda_2}{\lambda_1} + \left(1 - \frac{\lambda_2}{\lambda_1}\right) \cos^2 \alpha$$

or

$$t'_\alpha = \left(\cos^2 \alpha + \frac{\lambda_2}{\lambda_1} \sin^2 \alpha\right)^{\frac{1}{2}} \tag{7-33}$$

All the parallel folds modified in this way are folds with weakly convergent dip isogons (Class 1C). The curves which represent this family of folds with various values of λ_2/λ_1, the superimposed strain, are plotted in Figs. 7-79 and 7-80. The lowest curve of the family where $\lambda_2/\lambda_1 = 0$ gives the function $t'_\alpha = \cos \alpha$, that of the similar-fold model (Class 2). The stronger the compressive modification, the closer the flattened flexure folds approach the similar-fold model. Similar folds may be envisaged in mathematical terms as parallel folds which have been subjected to an infinite compressive strain. These graphs may be used to determine the amount of superimposed compressive strain in naturally formed folds of this type (Ramsay, 1962a, p. 314),

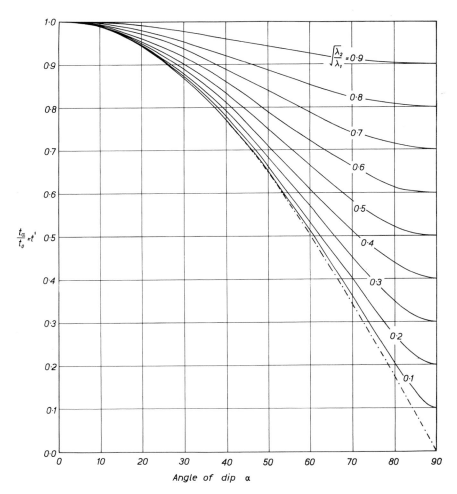

Figure 7-79

Values of t'_α in flattened parallel folds with variation in λ_2/λ_1.

and it is possible by graphical methods to restore the folds to an unmodified parallel shape, and to measure along the center of the parallel-fold layer to obtain a value for the original length of the layer. By this method it is possible to estimate the total compressive strain that has acted along this layer. For these calculations to be accurate, we must be sure that the original shape

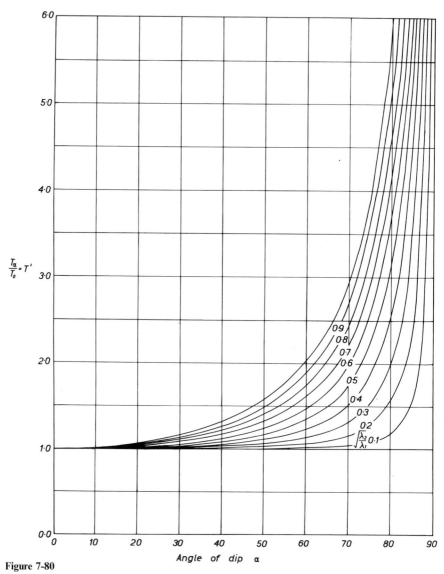

Figure 7-80

Values of T'_α in flattened parallel folds with variation in λ_2/λ_1.

was that of the parallel model. If the buckled layer has a well-developed lithological banding, we have a method of checking this assumption: the t'_α values for each component bed should fall along the same curve. If the original buckling led to the development of internal deformation by a mechanism of tangential longitudinal strain, we should be careful about interpreting the results. It may be that the changes of shape on either side of the finite neutral zone approximately compensated for each other and that the layer boundaries formed a parallel-fold structure. However, the tangential-longitudinal-strain modification on the inner arc of the fold may lead to an abnormal inhomogeneous thickening of the layer near the fold hinge. The t'_α values on the limbs of such a fold would be abnormally low, and if the structure subsequently suffered a homogeneous compressive strain, the computed λ_2/λ_1 ratios for the superimposed strain would be too small. Wherever possible, independent checks should be made of the component lithological bands making up the layer. If the t'_α values for each component vary greatly, the method for computing the superimposed strains should not be used.

Another problem in interpretation occurs where the original fold was of the parallel model but where the principal superimposed strain axes in the fold axial surface were oriented obliquely to the fold hinge. The measurements for computing the strain components by the method described above must be made on the profile section of the *original* fold, and yet this has come to lie in some position that is no longer perpendicular to the hinge line of the modified structure. The profile section of the *modified* fold was not originally a profile section of the original fold and hence not a true parallel-model cross section. This is the same effect as was encountered when determining the state of strain from initially cylindrical objects oriented at random in space (see Sec. 5-12).

The effects of homogeneous strain components on the other types of fold models may be summarized as follows:

Class 1A Generally modified to a complex compound form of the type
shown in Fig. 7-27, with strong deformation modified to Class 1*C*
Class 1B Modified to Class 1*C*
Class 1C Modified, approaching Class 2 but remaining Class 1*C*
Class 2 Modified, but remaining Class 2
Class 3 Modified, approaching Class 2 but remaining Class 3

7-9 STATES OF STRAIN OUTSIDE THE BUCKLED LAYERS

Where competent layers undergo buckling, the shape of the surrounding material has to be accommodated to that of the developing fold. The nature of this modification depends to a large extent on the proximity or otherwise of other buckling layers and the rheological state of the surrounding material.

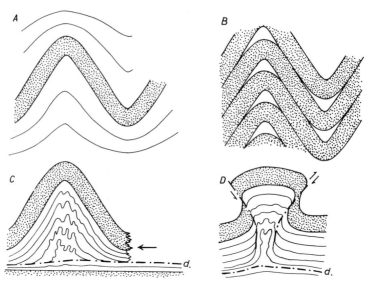

Figure 7-81

Behavior of rock material around buckling layers. A, layer surrounded by incompetent material; B, closely packed competent multilayers; C, décollement structure d; D, box-fold formation with limb faults and underlying décollement zone d.

If the buckled layer is surrounded by less viscous material, this material is displaced by the developing fold (Figs. 7-81A and 7-82). Some distance away from the buckled layer the banding in the material undergoes homogeneous shortening without folding, but near the layer the banding is subjected to inhomogeneous strain and takes up a folded form. This region of inhomogeneous strain set up around the buckled layer is known as the *zone-of-contact strain*. Theoretically, inhomogeneous strain should extend to an infinite distance from the contacts of the layer, but for geological considerations the effects of contact strain are negligible at distances greater than about one initial wavelength (W_i, see Fig. 7-82) on either side of the buckle (Kienow, 1942; Wunderlich, 1959a, 1960; and especially Ramberg, 1960, 1961b). The strain distributions in this zone give rise to two classes of fold. In the incompetent material on the inner arc of the buckle the folds which form are of Class 3, while in the outer arc they are of Class 1A. These two classes of fold alternate periodically along the length of a single band; on one side of the buckle, antiforms are all Class 3 and synforms are Class 1A, while on the other side antiforms are all Class 1A and synforms are Class 3.

The finite deformations give maximum strain ratios in the incompetent material situated on the inside of the fold arcs. At points situated on the outer fold arcs there may be no finite strain. These points will be called *finite neutral points* or f.n.p. (Fig. 7-82). The development of a finite neutral point depends on the amount of compressive strain taken up by both competent and in-

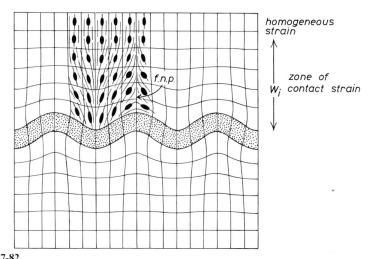

Figure 7-82

Zone of contact strain around a single buckled layer. (f.n.p. is the finite neutral point where the finite strains are zero.)

competent layers before the development of the buckles (e.g., in Fig. 7-68 there is no finite neutral point on the outer fold arcs). There is generally an *infinitesimal neutral point* somewhere on the outside of the outer arc of the buckling layer where the principal infinitesimal strains are zero. The principal axes of strain show a variation in orientation, and this variation is systematic with reference to the folds in the buckled layer (Fig. 7-82). The axes of principal extensive strain converge toward the inner arcs of the buckles (toward the region of greatest finite strain) and converge away from the outer arcs of the buckles. A rather peculiar pattern of principal strain axes is developed around the finite neutral point (Fig. 7-82). The material around the buckled layer may show a slaty-cleavage or schistosity distribution which reflects the variation in amount and in orientation of the finite strain. The characteristic type of cleavage distribution gives rise to a *divergent cleavage form*. In rocks which already possess a cleavage or schistosity before the development of the buckles it is common for crenulation cleavage to be developed in the manner previously described (p. 390). The shear discontinuities set up by these microfolds reflect the strain pattern in that the discontinuities or shear planes form normal to the maximum compressive strain (Fig. 7-83).

If the contact strain zones around two adjacent competent layers overlap, the folding developed in the two layers is related. Further, if the competent layers are of about the same viscosity and thickness, then *harmonic folds* are set up and the folds have the same wavelength. Where the thicknesses or viscosities of the competent layers are variable, the folds show a *compound wavelength of two or more orders* of size (Figs. 7-11, 7-33, 7-62). If, however, the layers are separated at a distance which is greater than their initial wave-

Figure 7-83A

Figure 7-83B

Figure 7-83

Strain-slip cleavage developed in incompetent mica-phyllites around buckled silty layers. A, appearance on outcrop surface; B, appearance in thin section. Compare orientation of cleavage with direction of maximum and minimum finite elongation in Fig. 7-82. Holy Island, North Wales.

lengths, then the buckles develop independently of one another to produce *disharmonic folds* (Fig. 7-84).

Figure 7-84

Harmonic and disharmonic folding in buckled calc-silicate layers enclosed in marble. Sokumfjell, Norway.

The competent layers may be close to each other in the fold (Fig. 7-81B). Because the shapes of the upper and lower boundaries of the layers are not the same, neighboring layers will not lie closely against each other throughout the structure. If the layers are in contact on the fold limbs, they cannot be so at the hinge zone of the folds. Where there are thin sheets of incompetent material between the competent layers, then this material is often squeezed out into these regions at the hinge zones. If incompetent material is not available, curved, open fissures may be developed in this region. These spaces often become infilled with quartz and carbonate material derived from the surrounding rocks or precipitated from primary solutions, and the curved veins so formed are known as *saddle reefs*. The best known examples of these structures occur at Bendigo, Australia (Stillwell, 1917), where gold deposits are found in the hinge regions of the folds.

The layers of material surrounding the buckling layer may resist the ductile shortening of the type shown in Fig. 7-81A. If the layers have a resistance, they will buckle and generally produce a complex mass of small crumples and contortions in the core of the structure (Fig. 7-81C). Sometimes adjacent thick competent layers do not shorten by the same amount and a discon-

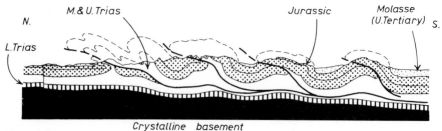

Figure 7-85

Décollement structure in the Jura Mountains of Switzerland. Jurassic limestones have become detached from the underlying crystalline basement and lower Triassic quartzites by sliding on the saliferous middle and upper Triassic strata. (After Heim, 1921.)

tinuity develops in the incompetent material which separates the layers (Fig. 7-81*C*). There is no completely satisfactory English term to describe this type of fold detachment and either the French term *décollement* or the German *abscherung* is generally used. As is well known, this phenomenon occurs in the Jura Mountains of Switzerland and France where the Mesozoic strata have been detached from the underlying basement as a type of thrust nappe and folded in an independent manner, with the upper Triassic saliferous marls acting as a lubricant at the base of the folded sheet (Fig. 7-85). Décollement

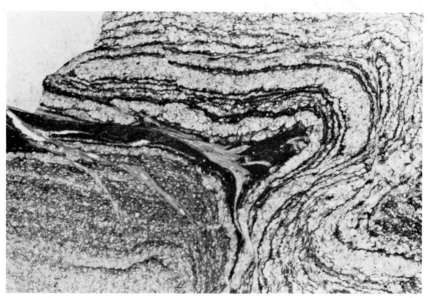

Figure 7-86

Décollement between folded sandstone layers. Lower Torridonian sediments, Tarskavaig Bay, Isle of Skye, Scotland.

may also occur between individual folded competent layers, and the fault surface itself becomes folded during its development (Fig. 7-86).

Where the material which surrounds the buckling layer will not deform viscously, a space problem develops inside the inner arc of the fold, especially if there is a plane of décollement beneath the structure. This material tends to form a resistant mass in the core of the structure which restricts and modifies development of the fold (Fig. 7-81*D*). In such circumstances the shape of the buckled layer may be modified by the development of *limb faults* on the flank of the structure and the formation of a *box fold*. The faults enable the hinge region of antiformal folds to be lifted and so partially allow for the accommodation of the material in the fold core.

7-10 SIMILAR FOLDS AND THE PROBLEM OF SHEAR FOLDING

Folds with a geometrical form which is that of the similar-fold model (Class 2), or which closely approximate it (Figs. 7-87, 7-88), are generally found in the central parts of belts of regional folding and metamorphism. The zones of the crust where these folds are most common were probably in a very ductile state during fold formation. The rocks were probably positioned at some considerable depth below the surface when folding took place and subjected to high temperatures and pressures (both hydrostatic and deviatoric). Similar folds are sometimes encountered outside this environment in materials which

Figure 7-87

Folds of similar style in metamorphosed feldspathic sandstone (Moine Series). Loch Monar, Scotland.

Figure 7-88

Folds which closely approximate to the similar fold model in Devonian phyllites. Tintagel, North Cornwall, England.

are known to have been very mobile during folding (e.g., salt deposits, muds, etc.). Biot (1965*a*) has put forward a theory of similar fold formation in viscous multilayers. He suggests that folds of this type only form where viscosity contrasts are very small and that if this condition does not hold, the embedding medium acts as a strong confinement which prevents similar folding and leads to internal buckling.

If perfect similar folds are developed in a material which originally had a uniform parallel layering, then the inhomogeneous strain conditions within that substance implied by the constancy of T_α throughout the fold can *only* be accounted for in two basic finite-strain plans (Fig. 7-89*A* and *B*). With plan *A* the variably oriented strain ellipses all have a direction of no finite longitudinal strain parallel to the axial surface, and with plan *B* they all have an identical strain (quadratic extension λ) parallel to the axial surface. If, at any one locality, there is a range of size of similar folds of various amplitudes and wavelengths that probably represent different stages in the development of the fold model (i.e., the various plans of finite-strain variation represent various stages in progressive deformation), then there are *only two* possible processes that can account for the formation of the folds: (1) progressive, inhomogeneous, simple shear; and (2) a uniform, homogeneous strain throughout the region together with a progressive inhomogeneous

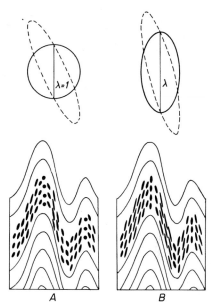

Figure 7-89

The two possible types of strain variation in ideal similar folds. The folds in model A are formed by heterogeneous simple shear and T_α is constant and of unchanged length ($\lambda = 1$). In model B the similar folds are formed by heterogeneous simple shear and a homogeneous strain so that T_α is constant with quadratic extension λ.

simple shear. The actual mechanism of fold formation is in both instances a variable component of shear acting across the layering and transposing the layers as passive markers; hence the folds are sometimes called *shear folds* or *slip folds*. Although this idea of shear folding has been criticized (Gonzalez-Bonorino, 1960; Ramberg, 1963c), the geometrical form of an ideal similar fold affecting several layers of material has never been satisfactorily explained in any other way.

The shear planes may be infinitely close (and need not be visible in naturally deformed rocks) and give rise to folds in which the lithological layers form continuously curving sheets (Fig. 7-88); but sometimes the shear displacements have steplike discontinuities, and the folded surface may be sliced up into a number of separate pieces (Fig. 7-90).

7-11 THE SHEAR COMPONENTS

The geometrical form of similar folds can be explained by a deformation scheme which can be analyzed into two components: a fold-forming inhomogeneous shear and a non-fold-forming homogeneous strain. We shall now examine the fold-forming shear component in more detail.

Consider a plane marker contained in a block of material (Fig. 7-91) and subject the block to simple shear so that the shear axes *a*, *b*, and *c* remain fixed but the amount of shear varies throughout the block. The upper and lower surfaces of the block and the plane marker inside it become folded into similar folds. The *axial surfaces* of the folds are always parallel to the *ab*

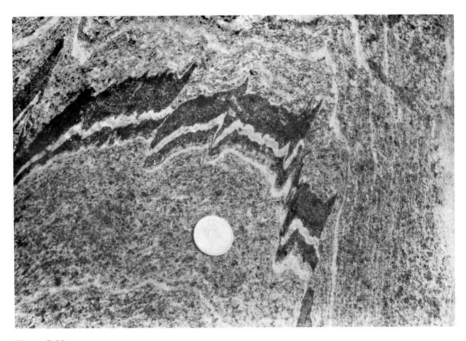

Figure 7-90

Similar-type folds in biotite gneisses with shear discontinuities parallel to the axial surfaces. Pennine Alps, Ticino, Switzerland.

or shear plane. The *hinge lines and fold axes* are always parallel to the line of intersection of the shear plane (*ab*) with the surface being folded (Borg and Turner, 1953, p. 1348). Where the *b* direction of the shear lies in the marker plane, the fold axis *f* is parallel to *b* (Fig. 7-91*B*); but this condition does not generally hold, and the fold axis *f* is generally oriented at some angle (β) to the *b* axis (Fig. 7-92*B*). With folds formed in this way, we should *never use the orientation of the fold axis to determine the shear displacement* (*a*). If the marker plane inside the block is inclined so that it contains the *a* direction (Fig. 7-93*A*), then, although the surface becomes internally deformed, *no folds are formed* (Fig. 7-93*B*). From Figs. 7-91, 7-92, and 7-93, it will be seen that the fold amplitude produced by a given shear is at a maximum where *f* is parallel to *b*, and a minimum where the intersection of the markers with the *ab* plane is parallel to *a* (Ramsay, 1960; 1963, p. 157). If A_{max} is the maximum amplitude of the shear movements, *f* the intersection of the marker plane with *ab*, and β the angle between *f* and *b*, then the actual fold amplitude *A* (measured in a profile section perpendicular to *f*) is given by the simple relationship

$$A = A_{max} \cos \beta \qquad (7\text{-}34)$$

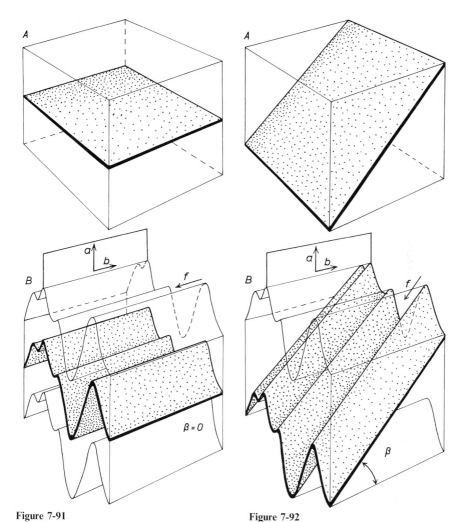

Figure 7-91

The geometric form of similar folds formed by heterogeneous simple shear where the slip direction a is perpendicular to the layers. The fold axes f form parallel to b.

Figure 7-92

The geometric form of similar folds formed by heterogeneous simple shear where a is not perpendicular to the layers. The fold axes f form at an angle β to the b axis.

The location of the hinge lines and axial surfaces of the folds in relation to the amount of shear displacement depends on the initial angle of intersection of the layering and the shear surfaces. If, for example, within the *bc* plane the layering is initially inclined at an angle to the *c* axis and is displaced by a symmetrical periodic shear motion (Fig. 7-94), then the hinge line of the fold

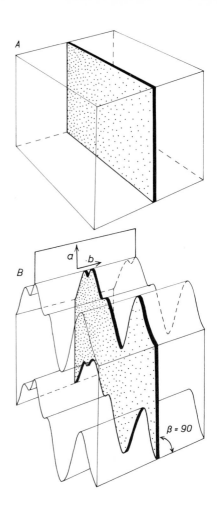

Figure 7-93

The deformation of layers by heterogeneous simple shear where a lies within the layers. No folds are formed.

will not coincide with the lines of maximum and minimum relative translation, and the cross-sectional shape of the folds that are produced will be asymmetric. This is because the hinge lines must occur where the curvature reaches a maximum value on the transposed surface given by $A \sin x + mx$, where m is the slope of the layering with the c axis.

The profile plane of a similar fold produced by the type of displacements of Figs. 7-91 and 7-92 is a plane of symmetry with reference to the folded surfaces and their pole distributions, and the folded surfaces show a monoclinic symmetry. It will be seen in Fig. 7-92B that the symmetry of the movement plan (monoclinic, with ac symmetry plane) does not generally coincide with that of the fold profile except in the special circumstances of Fig. 7-91 where $\beta = 0$. Similar folds generally have a total symmetry which is triclinic,

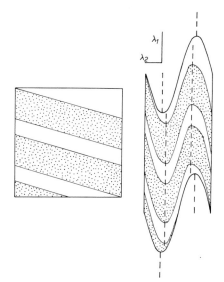

Figure 7-94

Position of the axial traces of similar folds formed by a symmetric periodic heterogeneous simple shear and a homogeneous strain. The axial traces are not symmetrically disposed to the surfaces and the folds are asymmetric.

and it follows that the symmetry and orientations of the movement axes cannot be deduced from the shape of the folded surface alone.

7-12 THE COMPRESSIVE STRAIN COMPONENTS

It has been shown how perfect similar folds must have a strain plan which can always be separated into shear components and homogeneous strain components. It is also true that variations in finite-strain state through folds of Class 3 and Class 1C which lie close to the form of the similar model can also be closely approximated in terms of these same two components. Although it is possible to consider the structural effects of similar folding in this way, it must be remembered that the separation of the two components is only a matter of analytical convenience. The two types of strain are probably developed simultaneously as part of a single deformation process, and throughout the section which follows, the term "superimposed strain" has no time significance. Although the variations in state of finite strain throughout the fold may be numerically described in terms of a specific simple shear strain followed by a certain homogeneous strain, it must be remembered that the order of superposition of these specific strains may not be reversed. The products of the two components are what is known as *noncommutative* (see Sec. 3-4).

Folds produced by inhomogeneous simple shear will always retain a similar fold form if homogeneously deformed by principal strains of any value and any orientations. However, the states of strain seen in naturally produced

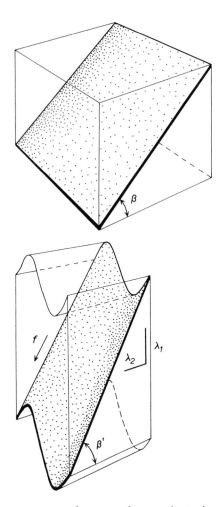

Figure 7-95

Change of plunge of a shear fold as the result of a superimposed homogeneous irrotational strain.

similar folds suggests that the maximum compressive superimposed strain component is generally perpendicular to the *ab* shear plane, and that probably the maximum superimposed extensive strain is parallel to the *a* direction (although this is not always so).

As a result of the superimposed strain component, the shapes and orientations of the component parts of shear folds are modified. If the maximum compressive superimposed strain is perpendicular to the axial surface (*ab*), then the amplitude of the folds is increased and their wavelength decreased, and the position of the axial surface is modified slightly (Fig. 7-94). The angle of plunge of the fold is also generally modified (Fig. 7-95); if λ_1 and λ_2 are the principal quadratic two-dimensional extensions within the axial surface, then the fold axis approaches the λ_1 axis (Ramsay, 1962*a*; Turner and Weiss,

1963) according to the relationship (3-34), that is,

$$\tan \beta' = \tan \beta \left(\frac{\lambda_1}{\lambda_2}\right)^{\frac{1}{2}} \tag{7-35}$$

7-13   THE STATES OF FINITE STRAIN WITHIN SIMILAR FOLDS

The inhomogeneous finite-strain states that are seen in similar folds should always be of the type shown in Fig. 7-89. The XY planes of the strain ellipsoids are always inclined at some angle to the layering (although this may be very small on the fold limbs) and dip away from the axial surfaces of antiforms and toward the axial surfaces of synforms. As the superimposed homogeneous strain component becomes larger, the XY planes lie closer to the axial surfaces of the folds. These principal XY planes of the finite-strain ellipsoid are often paralleled in the rocks by slaty cleavage or schistosity in the form of *divergent cleavage fans*. In similarly folded rocks the cleavage planes do not generally represent the traces of the fundamental flow plane which led to the formation of the fold. However, in highly metamorphosed environments where similar folds are often abundant and the superimposed strain components large, the

Figure 7-96

Similar fold in metamorphic quartzo-feldspathic sandstones with well-developed axial plane schistosity. Loch Monar, Scotland.

Figure 7-97

Thin section of the rock material in Fig. 7-96 to illustrate the preferred orientation of the muscovite and biotite which leads to the macroscopic schistosity.

schistosity is frequently subparallel to the axial surfaces of the folds (Figs. 7-96 and 7-97).

7-14 THE MECHANISM OF FORMATION OF SIMILAR FOLDS

The only mechanism that has been suggested to explain satisfactorily the geometrical form of *perfect* similar folds is that of slip, shear, or flow across a set of layers or beds. The layers act as reference surfaces or markers, and points on each layer became differentially displaced with respect to each other and so the layer takes up a folded form. Since the layers take no active part in the development of the structure, the folds have been termed *passive folds* (Donath, 1962) or *bending folds* (Ramberg, 1963c). If the flow planes remain parallel during the fold formation, the similar folds have a T value given by

$$T = t \csc \alpha \tag{7-36}$$

where t is the original orthogonal thickness of the layer and α the original angle between the layer and the plane of flow (Fig. 7-98). If the original distance between the flow planes is modified by a multiplying factor x (Fig. 7-99, where $x = d'/d$), then the similar folds have a constant T value given by

$$T = \frac{t \csc \alpha}{x} \tag{7-37}$$

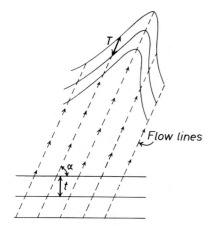

Figure 7-98

Relationship of original orthogonal thickness (t) of a layer and the thickness measured parallel to the axial surface of a similar fold (T) formed by heterogeneous simple shear.

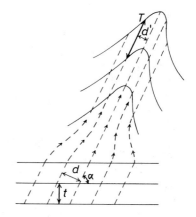

Figure 7-99

Modification of the value of T in Fig. 7-98 where the flow lines converge.

It has been suggested that similar folds can be formed by a dilation ($x > 1$) so that the distance between the flow planes increases (Carey, 1962). Although this is geometrically possible, no evidence has been found to suggest that it does occur in naturally produced folds.

The theory that similar folds are formed by a flow process raises a number of important mechanical problems. First, is the generating mechanism of the differential flow either (*a*) contained within the domain of the similar fold or (*b*) does it arise from outside this region? Second, if the differential flow is generated within the fold domain, how can it be explained in terms of material properties and stress distribution? Third, if (*b*) holds, why should the strain variations developed at the boundary of the similar fold domain be conveyed so uniformly through the material although minimum strain energy requirements necessitate that such strains should diminish in some exponential way from the generating point? These questions will now be examined further.

The similar-fold model is unique in an infinite range of possible fold shapes. Some investigations have been made to determine how closely naturally formed similar folds do accord with the Class 2 model. These show that many folds which superficially resemble similar folds show divergences from the perfect model (Ramsay, 1962*a*). Some of these variations will now be discussed to see if they cast any light on the problems listed above.

A number of similar folds when subjected to detailed morphological examination show forms which are more or less obviously the result of the buckling process.

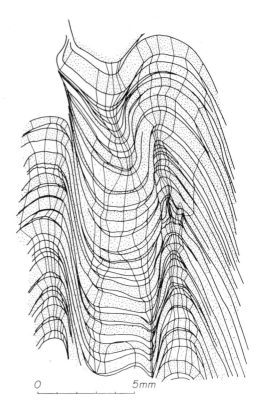

Figure 7-100

Dip isogons drawn on multilayered phyllite (Quartenschiefer, Brigels, Switzerland). The fold is composed of layers of alternating classes 1C and 3.

0   5mm

Figure 7-100 illustrates a number of folded layers superficially resembling the shape of a similar fold—the layer shapes at the upper part of the diagram crudely approximate those at the lower part, individual layers are thinned on the fold limbs, and in one zone of the structure there is a structural discontinuity or shear zone which runs approximately parallel to the axial surfaces of the folds. When dip isogons are drawn through the structure it is clearly seen that the resemblance to the Class 2 model is entirely the result of summation of the properties of two types of fold model: Class 1C in the (competent) silty layers and Class 3 in the (incompetent) argillaceous layers. This structural pattern is identical with that produced by buckling a multilayered complex with some modifications of the fold shapes by superimposed strain components (flattened flexure folds) and by shearing of the fold limbs in the manner described on p. 388 and in Fig. 7-50.

The folds shown in Fig. 7-101 are developed in alternating layers of calcite marble and calc-silicate material and appear to be very close to the similar-fold model. However, when measurements are made on the layers it is found that in each marble layer T'_α is slightly less than 1, the dip isogons diverge slightly, and the folds are of Class 3; while in the calc-silicate layers T'_α slightly exceeds

Figure 7-101

Folds of similar type in multilayered marble and calc-silicate rocks. The folded calc-silicate layers have slightly convergent dip isogons, and the marble layers have slightly divergent isogons. The similar folds are made up of alternations of folds of types 1C and 3. Sokumfjell, Norway.

1, the dip isogons converge slightly, and the folds are those of Class 1C. If a unit made up of a lithological pair is measured (i.e., from the top of one marble layer to the top of the next marble layer in the pile), the resulting form is a nearly perfect Class 2 fold. It therefore can be said that if we take a series of harmonically buckled multilayers and flatten them normal to their axial surfaces (Fig. 7-102), the styles of the folds in the individual layers come to lie close to that of the similar-fold model, while that of a combined incompetent and competent layer pair is almost exactly that of a similar fold. Where the ratio of the superimposed strains has a high value, the individual layer shapes approach more closely that of the similar-fold model, a geometrical fact that is true for *all* original fold shapes. Many of the apparently similar folds seen in naturally deformed rocks probably owe their initial development to the buckling mechanism (Flinn, 1962; Ramsay, 1962*a*; Ramberg, 1964*b*) and in consequence have some fixed wavelengths governed by the viscosities (or elasticities) of the layers during this early folding process. However, the folds produced in this way can never have perfect similar form and a careful investigation of the morphology of the two-component folded layer should reveal slight but significant divergence from the Class 2 model which shows

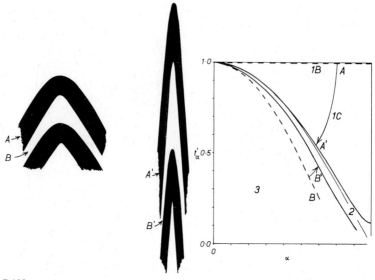

Figure 7-102

Folded layers A and B which have been subjected to a large homogeneous strain and their shapes modified to A' and B', respectively. Layers A and B originally had t'_α values as indicated in the graph, and as a result of the homogeneous strain the t'_α values are modified into curves A' and B' and approach the ideal similar-fold model (Class 2).

that they were not produced by axial-surface flow. Thus it is possible to produce folds which may closely approximate the similar-fold model without recourse to progressive flow or slip parallel to their axial surfaces.

Investigations into the geometry of naturally formed similar folds (Ramsay, 1962a) have shown another pattern of variation of fold style. Some folded layers within structures of approximately similar type show alternating maximum and minimum values of T in the axial surfaces of alternate antiforms and synforms, i.e., antiforms of Class 3 and synforms of Class 1. This is the pattern characteristic of zones-of-contact strain around buckled layers (p. 416); but in the example under investigation, the same pattern occurred in layers of quite different lithologies and no buckled layer could be detected. Fold patterns like this can be developed around regions of inhomogeneous finite strain caused perhaps either by initial inhomogeneities in the material properties or inhomogeneities that develop during metamorphic processes going on at the same time as the rock deformation (Fig. 7-103). Inhomogeneous strain must lead to differential flow and therefore to the production of folds (Fig. 7-104). This differential flow may be transmitted into regions outside that of the original inhomogeneity as shown in Fig. 7-104. Whether this takes place probably depends on the nature of the materials involved. If the material

Figure 7-103

Modifications of texture in amphibolitic rocks which have undergone similar folding. Sambuco, Ticino, Switzerland.

is a rather homogeneous substance, it seems likely that the differential displacements will decrease away from the region of inhomogeneity in some exponential manner in the way that the zone-of-constant strain dies away from the surface of a buckled layer. However, the finite strains may lead to the

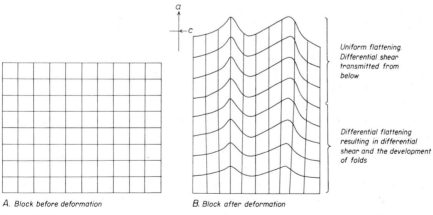

A. *Block before deformation*   B. *Block after deformation*

Figure 7-104

Deformation of the block A by inhomogeneous strain leading to the production of folds of general and ideal similar-fold type. (After Ramsay, 1962a.)

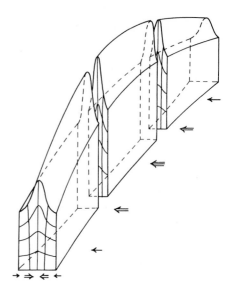

Figure 7-105

Variation in compressive strain along the length of the fold axial surface producing a fold culmination. (After Ramsay, 1962a.)

development of a rock fabric (cleavage or schistosity) which may well assist the translations into regions some distance from that where they were initiated. Although the "card pack" mechanism for producing similar folds is open to criticism and there seems little doubt that the primary origin of schistosity and cleavage is that of the XY strain ellipsoid plane and not that of a slip surface, this cleavage structure may be used as a slip surface of card-pack type during the late stages of deformation. The idea of material inhomogeneity leading to the production of folds has also been extended into three dimensions to account for some of the variations in plunge that can be seen within a single fold (Fig. 7-105) (Ramsay, 1962a).

7-15   KINK BANDS, CHEVRON FOLDS, AND CONJUGATE FOLDS

Certain types of straight-limbed angular folds with narrow hinge zones occur so frequently in some geological environments that they merit special description and examination. These folds may be either single hinge or multiple hinge types and characteristically form sharp, angular deflection of the layering. They may be developed as angular, steplike monoclines (known variously as *kink bands, knick zones,* or *joint drags*) which are formed on parallel-plane axial surfaces (Fig. 7-106) (Dewey, 1965). These terms are generally given to small-scale structures where the distance between the adjacent axial surfaces does not exceed about 10 cm; if the structure is larger it is termed a *zigzag fold* or *knee fold*. The monoclines may be arranged in two sets on axial surfaces

Figure 7-106A

Figure 7-106B

Figure 7-106

Characteristic appearance of kink bands. A is an example of a kink band in slates seen in cross section. West Cork, Ireland. (Photograph by K. Coe.) B shows the appearance of the hinge lines of a set of kink folds developed in a phyllitic schistosity. Tintagel, North Cornwall, England.

which are inclined toward one another (Fig. 7-107) and give rise to a polyclinal structure generally known as a *conjugate fold* (Johnson, 1956; Ramsay, 1962b). Where the folds are symmetric and the limbs are of equal length, the structure is generally termed a *chevron* or *concertina fold* (Fig. 7-108).

All these varieties are characteristically developed in materials which possess a well-developed layering, such as thin-bedded sediments or in rocks which have developed a planar tectonic anisotropy (slaty cleavage or schistosity) as a result of some earlier deformation. Folds of this type are also formed in some individual crystals in deformed rocks, especially those which have a well-developed cleavage or glide surfaces (micas, Fig. 7-109; enstatite, Fig. 7-110; kyanite, etc.) (Mügge 1898, Turner and Weiss, 1965).

The fold shape of different layers throughout the structure generally keeps fairly constant, and the overall geometric form of kink bands, chevron folds, or of the individual monoclinal parts of a conjugate fold is generally of the similar type (Class 2). When the individual layers are examined in more detail, however, it is often found that they keep a constant orthogonal thickness throughout the structure and are approximately of parallel form (Class 1*B*). The axial surface of each fold generally bisects or nearly bisects the fold limbs.

The internal deformation that goes on within the material folded by these

Figure 7-107

Conjugate kink bands in cross section. West Cork, Ireland. (Photograph by K. Coe.)

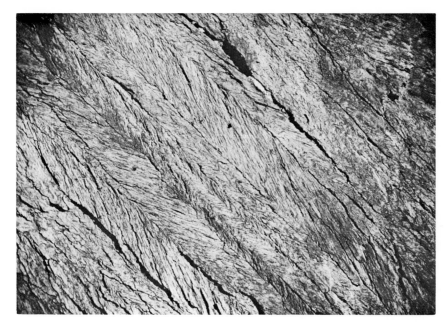

Figure 7-108

Chevron folds in Archean phyllites. Barberton, East Transvaal, South Africa.

Figure 7-109

Thin section of kink bands in micaceous phyllites. Anglesey, North Wales.

Figure 7-110

Thin section of kink bands in an enstatite crystal.

structures appears to be almost completely accomplished by slip on the layering as with the flexural-slip model, and the surface of the layers may show signs of slickensiding. The folds form as a response to a compressive stress acting along the layering, and folds of this type have been formed experimentally in this way by Paterson and Weiss (1962).

We shall now examine a model of a chevron fold to determine its geometrical properties at different stages of its development. Probably the most realistic model is that shown in Fig. 7-111. Although this model differs slightly from that chosen by de Sitter (1958*a*), the geometric properties of the two models are alike in their broad outline. The model consists of a number of layers each of thickness t folded to a symmetrical chevron structure so that the limbs of the structure make an angle $(90 - \alpha)$ degrees with the axial surface. The length of any layer measured from one fold hinge to the next hinge retains its original length l. The perpendicular distance between the axial surfaces of adjacent folds after folding is $l(1 + e)$, where e is the shortening measured along the original direction of the layering.

The total amount of slip between the upper surfaces of two adjacent layers is $t \tan \alpha$, and the total finite shear strain is $\gamma_t = \tan \alpha$. This total shear is compounded of two parts. One part occurs internally within the layer (s_i), and the amount of this shear across a layer of thickness t is

$$s_i = t\alpha \tag{7-38}$$

The other part is concentrated as slip along the boundary of two adjacent layers (s_b) and has a value

$$s_b = t(\tan \alpha - \alpha) \tag{7-39}$$

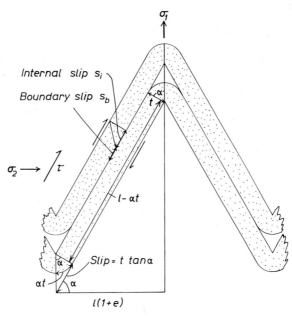

Figure 7-111

Chevron-fold model.

We can thus envisage these two components in terms of shear strain across the layer such that

$$\gamma_t = \gamma_i + \gamma_b \qquad (7\text{-}40)$$

where

$$\gamma_i = \alpha \qquad (7\text{-}41)$$

and

$$\gamma_b = \tan \alpha - \alpha \qquad (7\text{-}42)$$

De Sitter (1958*a*) has pointed out that as the thickness of the layers (t) is increased there is a corresponding linear increase of total slip, and this relationship may well account for the observed fact that folds of this type are always preferentially developed in rocks composed of thin layers where t is small.

From simple trigonometrical relationships it follows that

$$(l - \alpha t + \tan \alpha) \cos \alpha = l(1 + e) \qquad (7\text{-}43)$$

or

$$\frac{l - \alpha t}{l} \cos \alpha + \frac{t}{l} \sin \alpha = 1 + e \qquad (7\text{-}44)$$

From this expression the inclination angle α of the fold limbs can be calculated for any particular compressive strain e and t/l ratio (Fig. 7-112). It will be seen clearly from these curves that the angles of dip increase rapidly during the initial development of the fold, and that for any given finite shortening $(1 + e)$, a fold with a low t/l ratio can accommodate these strains with smaller limb dip than one with a high t/l value. This also implies that if the layers

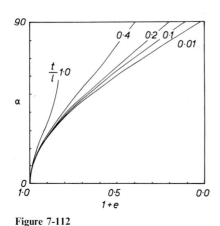

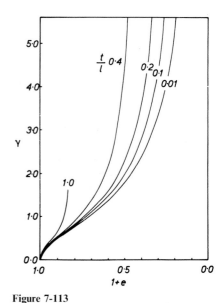

Figure 7-112

Angle of inclination α of the limbs of chevron folds (model Fig. 7-111).

Figure 7-113

Variation in total shearing strain γ developed in the limbs of chevron folds.

in a fold show a variety of t/l values, then the different layers try to take up fold shapes with different interlimb angles; or if this does not occur, either an extensive strain is developed along the layering of the thinner sheets within the fold or a compressive strain acts along the thicker sheets. There is a limit to the amount of shortening that can be taken up by folds formed according to this model, and larger compressive strains can be taken up more effectively by layers with low t/l values than those with high values.

The total finite shear strain ($\gamma_t = \tan \alpha$) developed on the fold limbs and expressed with respect to the shortening $(1 + e)$ has been calculated from (7-39) and is graphically illustrated in Fig. 7-113 for folds with layers of various t/l ratios. For any finite shortening these strains are largest for a fold with a large t/l ratio—another reason why such structures form most easily in thin-layered material. The rates at which the shear strain increases at different stages during the fold development show a pronounced variation and this is investigated further. Differentiating (7-44) with respect to α,

$$\frac{de}{d\alpha} = \sin \alpha \left(\frac{\alpha t}{l} - 1\right) \qquad (7\text{-}45)$$

Also, because $\gamma_t = \tan \alpha$,

$$\frac{d\gamma_t}{d\alpha} = \sec^2 \alpha \qquad (7\text{-}46)$$

$$\frac{d\gamma_t}{de} = \frac{d\gamma_t}{d\alpha}\frac{d\alpha}{de} = \frac{\sec^2 \alpha \csc \alpha}{\alpha t/l - 1} \tag{7-47}$$

This function is illustrated in Fig. 7-114 and it will be shown that it has properties which account for some of the morphological features of chevron folds. From (7-47) it follows that the shear-strain ratios are infinite when either $\alpha = 0°$, $\alpha = l/t$, or if $l/t > \pi/2$, when $\alpha = 90°$.

Let us now examine the shear stresses which act on the limbs of the fold at different stages of its development. Take the principal compressive stress σ_2 ($\sigma_1 > \sigma_2$ are principal tensile stresses) to act normal to the axial surface of the fold; then the shear stress τ acting along the fold limbs dipping at an angle α is given by (2-13), $\tau = [(\sigma_1 - \sigma_2)\sin 2\alpha]/2$. The values of this shear stress at different stages during the folding have been determined by combining (2-13) and (7-43) and are plotted in Fig. 7-115 in units of stress difference $(\sigma_1 - \sigma_2)/2$. At the start of the deformation where $e = 0$, $\alpha = 0$ and the incremental shear strain on these surfaces must be infinitely large. Because σ_2 is acting along the layer, no shear stress can act on these surfaces no matter what values are taken by σ_1 and σ_2. The fold cannot develop, therefore, unless either the layer is slightly curved or the stresses lead to the production of a wave instability and the formation of buckles. The second of these possibilities

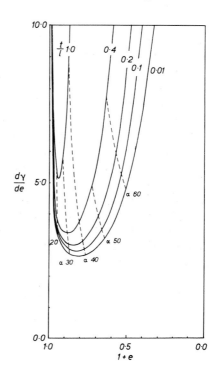

Figure 7-114

Variation in the shearing-strain increments $d\gamma/de$ with variation in total shortening and t/l ratio. The dashed curves show values of inclination α.

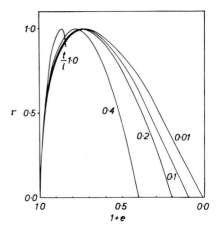

Figure 7-115

Variation in the shearing stresses on the limbs of chevron folds at various stages of development and at various t/l values.

would lead to the initiation of buckles with a specific wavelength and so incline the layers until the shear stresses can overcome the resistance of the material, accomplish the necessary strains, and overcome the threshold of folding. Once this threshold has been passed, the incremental shear strains on the layering become progressively diminished as the fold develops. Providing the external stresses σ_1 and σ_2 are not appreciably altered by the fold development, the actual shear stresses on the surface of the folded layers increase and the fold develops easily. The fold initiation may completely relieve the external stresses, however, and the fold will no longer develop because the shear stresses fall to a low value insufficient to overcome the material resistance and develop the necessary strain. Let us now investigate the possibility of further fold development where the stresses keep a high value.

Where the fold limbs are inclined at an angle $\alpha = 25$ to $35°$ (the exact value depends on the value of t/l), the shear strain increments attain a minimum value. Further (negative) increments in $(1 + e)$ now lead to a progressive increase in the infinitesimal shear strains on the layers, and the folding probably proceeds less easily as greater slip must be accomplished on the surfaces. Where the shear-strain rates reach an infinite value ($\alpha = l/t$ or, if $l/t > \pi/2$, where $\alpha = 90°$), the fold again cannot develop further. In practice, these increasing shear-strain increments on the layer surface give rise to a "lockingup" of the fold structure well before the absolute limiting threshold value is attained. De Sitter (1958a) was the first to suggest this as a reason for the commonly observed feature of natural chevron folds which generally show an interlimb angle of about 60°, and he further pointed out that this signifies that a total compressive shortening of the crust by more than 50 to 65 percent by this folding mechanism is unlikely (see Figs. 7-112 and 7-114). The actual position of the repression of fold development depends on a number of factors: the values of t/l and $\sigma_1 - \sigma_2$, the material properties of the rock layers, and the coefficient of friction at the contacts. Further research developing from a reappraisal of de Sitter's work on this subject confirms his deductions about the locking of chevron folds; but it is now apparent that, before the folding which he describes can begin to develop, there is an initial stress-strain threshold to be overcome. From Fig. 7-114 it will be seen that the folds with a large t/l ratio have a greater incremental shearing strain along the layers than those with a small t/l value, and are therefore likely to become locked earlier in the contraction history with a relatively small angle of dip on the fold limbs. Where $0.1 > t/l > 0.0$ the folding proceeds most easily, and variations of t/l within this range do not significantly alter $d\gamma/de$, the ease of formation of the structure, or the most probable limiting interlimb angle of the final fold.

The most likely reason why open chevron folds are rare in naturally formed folds is that, once the initial stress-strain threshold has been overcome and if $\sigma_1 - \sigma_2$ is not greatly altered, it actually becomes progressively easier for the fold to go on developing until the minimum value of $d\gamma_t/de$ is reached. If the strain rates are kept at a certain level during the formation of the folds, the structure will pass from an open to a close form with little resistance. When the fold has reached its "locked" state, further compression can lead to its modification by either a ductile change of shape or the development of rupture by faults (de Sitter, 1958a) depending on the rheological state of the rock materials and the strain rate.

We have seen that the total shear-strain increments on the limbs of the structure have maximum values at the start of the folding (threshold value) and at the end of the folding (locking value). The total shear increment is compounded of two parts, γ_i within the layer and γ_b on the layer boundaries. The relative importance of γ_i and γ_b will now be investigated. Differentiating (7-40) with respect to e, we find that the total shear increment at any stage during the folding represents the sum of the increments in γ_i and γ_b for

$$\frac{d\gamma_t}{de} = \frac{d\gamma_i}{de} + \frac{d\gamma_b}{de} \tag{7-48}$$

The partial increments can be obtained by differentiating (7-41 and 7-42) with respect to α and combining with (7-45)

$$\frac{d\gamma_i}{de} = \frac{d\gamma_i}{d\alpha} \frac{d\alpha}{de} = \frac{\csc \alpha}{\alpha t/l - 1} \tag{7-49}$$

$$\frac{d\gamma_b}{de} = \frac{d\gamma_b}{d\alpha} \frac{d\alpha}{de} = \frac{\sin \alpha \sec^2 \alpha}{\alpha t/l - 1} \tag{7-50}$$

From (7-49) and (7-50) it is possible to assess the relative importance of these two types of shear strain in controlling the development of the fold. The two functions are graphically recorded in Fig. 7-116 for different t/l values. From

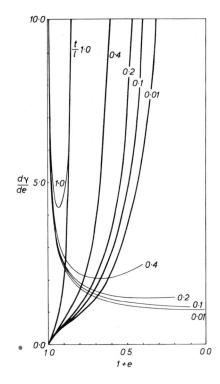

Figure 7-116

Variation in the shearing-strain increments on the layer boundaries (thick lines) and internally in the layers (thin lines) with development of chevron folds.

these curves it will be seen that, at the start of the folding, shear on the boundaries of the layer is of little or no importance, whereas the rates of internal shear within the layers are very large; the initial strain threshold, which must be overcome before folding can start, is localized within the layers. As folding proceeds, the shear-strain increments inside the layers become progressively smaller, while the incremental displacements on the layer boundaries take larger values. For any particular t/l ratio the folding takes place most easily where there is a minimum value in the sum of $d\gamma_i/de$ and $d\gamma_b/de$ (this gives the minimum value in the curves of $d\gamma_t/de$ shown in Fig. 7-114). As the shortening goes on and this minimum value is passed, folding becomes progressively more difficult as more boundary rearrangements have to be made in the material, and eventually these incremental strains build up to give an upper limiting value at which the fold becomes "locked." This limiting value is generally developed by the rapid rate of increase in shear along the contact surfaces of adjacent layers; but if the t/l ratio exceeds 0.4, it will be seen that there is a significant increase in the incremental shear strains inside the layer which also assists the "locking up" of the structure. These functions for internal and boundary shear increments are all lowest for folds with low t/l ratios, although the differences probably have no practical significance in the range

$0.1 > t/l > 0.0$. Again this supports the conclusion that chevron folds with low t/l ratios are likely to form more easily than those where the ratio is high.

Folding on the scheme of the model shown in Fig. 7-111 leads to an increase in the area of the profile section of the structure as the fold develops. Since the deformation is not a plane strain, shortening of the rocks along the fold axis must occur if the volume of the material is to keep constant. Another possibility is the formation of *saddle reefs* by the introduction of material into the open zones at the crests and troughs of the folds. The area change or dilation Δ in the profile plane as a result of folding is given by

$$\Delta = \frac{t(l - \alpha t + t \tan \alpha) - tl}{tl} = \frac{t}{l}\left(\tan \alpha - \alpha\right) \qquad (7\text{-}51)$$

From this function (Fig. 7-117) it follows that saddle reefs are most likely to develop in rocks where t/l takes a large value (> 0.1) and that they will be best developed where α is large and the interlimb angle small. Openings of this type are unlikely to be important in thinly laminated rock material where t/l takes a value less than 0.1.

Kink bands are geometrically related to chevron folds and are probably formed in a like manner. The internal deformation within kink zones is generally accomplished by flexural slip on the layering. The layers are bent at the hinge (Boschma, 1963) and the orthogonal thickness of the layers is generally nearly constant throughout the fold. The strain set up in the kinked zone by this flexural slip often leads to the production of zones of en echelon tension fissures perpendicular to the principal extension which crosscut the layering (Fig. 7-118). The axial surface of each fold usually lies close to the bisector of the angle between the straight limbs. Occasionally the axial surface makes a smaller angle (β_1) with the dominant trend of the layering than it does with the layering in the kink zones (β_2). If the layer thicknesses remain constant throughout the fold, an opening up of the rock material across the layers occurs within the kinked zone (Anderson, 1964). The fissures may remain open, but often they are filled with crystalline material segregated

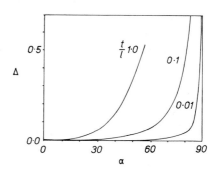

Figure 7-117

Dilation Δ in chevron folds with variation in t/l ratio and limb inclination α.

Figure 7-118

Development of sigmoidal en echelon quartz veins in kink bands in phyllite. Tintagel, North Cornwall, England.

Figure 7-119

Dilation veins in a kink band developed in calcareous slates. Lake District, England.

from the surrounding rock (Fig. 7-119). The amount of expansion δt made across any layer of thickness t within the zone is given by the relationship (see Fig. 7-120)

$$\frac{t}{\sin \beta_1} = \frac{t + \delta t}{\sin \beta_2} \tag{7-52}$$

The dilation Δ given by $\delta t/t$ from (7-52) is

$$\Delta = \frac{\sin \beta_2}{\sin \beta_1} - 1 \tag{7-53}$$

If the fissures contain economically valuable mineral deposits, this expression can be used to find the potential importance of the vein development. If $\beta_2 < \beta_1$, the layer suffers a contraction across the layering. This is generally uncommon in naturally formed kink bands, but where it is developed, the strains produced in this way often lead to the accentuation of the crosscutting zones of en echelon tension fissures.

It is possible to discuss the geometric properties of kink bands in the same way as was done previously with chevron folds. Consider a simple fold model (Fig. 7-121) made up of a number of initially horizontal layers each of thickness t folded by a kink band so that the dip of the layering in the kinked zone is α. Let the axial surface bisect the fold limbs and be inclined at an angle β. Angles α and β are dependent variables connected by the relationship

$$\beta + \frac{\alpha}{2} = 90° \tag{7-54}$$

The length of the layers in the kinked sector is of unchanged value l and the shortening along a direction parallel to the initial layering is $l(1 + e)$. It follows

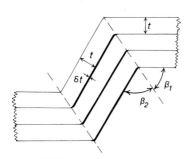

Figure 7-120

Amount of dilation δt in a kink band where β_1 and β_2 are of unequal value.

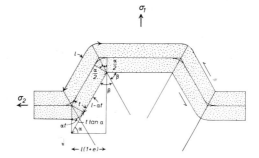

Figure 7-121

Conjugate kink-band model developed by principal tensile stresses σ_1 and σ_2 acting perpendicular and parallel to the layers.

that the total slip on the kinked surfaces is $2t \tan \alpha/2$ and the shear strain $\gamma_t = 2 \tan \alpha/2$ (cf. chevron folds where $\gamma_t = \tan \alpha$).

The relationships connecting l, t, α, and e are identical to those of the chevron-fold model

$$\left(\frac{l - \alpha t}{l}\right) \cos \alpha + \frac{t}{l} \sin \alpha = 1 + e \qquad (7\text{-}44)$$

and the relationship between the dip of the surfaces in the kinked zone and the amount of shortening are the same as those expressed in Fig. 7-112. The

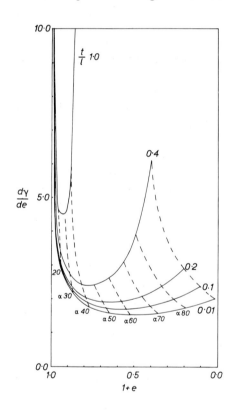

Figure 7-122

Variation in shearing strain increments in various conjugate kink-fold models. The dashed curves show values of inclination α.

finite shear strain on the surfaces of a kink band for any given amount of shortening $(1 + e)$ is less than that for the chevron-fold model. This leads to the important conclusion that less work is expended to shorten a layer by the production of kink bands (or paired conjugate kinks) than by the formation of symmetrical chevron folds. It seems likely that primary chevron folds only form where the fold hinges are initially fixed by a buckling process which produced sinusoidal fold waves.

The shear-strain increments on the surfaces of the layers within the kink zone can be determined in the same way as was done for chevron folds. Because $\gamma_t = 2 \tan \alpha/2$,

$$\frac{d\gamma_t}{de} = \frac{d\gamma_t}{d\alpha} \frac{d\alpha}{de} = \frac{\sec^2 \alpha/2 \csc \alpha}{\alpha t/l - 1} \tag{7-55}$$

This function has been evaluated for different values of compressive strain $(1 + e)$ and t/l ratios (Fig. 7-122) using the relationships of e and α expressed by (7-44).

The graphs show broadly similar relationships to those for chevron folds (Fig. 7-114). At the start of folding, the shear-strain increments are very large; but as folding progresses, they fall to a minimum value where the fold limbs are inclined at about 30° ($t/l = 1$) to 60° ($t/l = 0.01$) and then rise as the fold limbs become further inclined. Practically all the previous discussion on the development of chevron folds may therefore be applied to the development of kink bands. The initial stress-strain threshold must be overcome by the development of buckles, and the kink bands probably initiate about imperfections of structure in the rock layers. Once the threshold has been overcome, the folding proceeds easily provided that $\sigma_1 - \sigma_2$ maintains a sufficiently large value to give the required shearing stress on the fold limbs. Where the limbs are inclined at more than 45°, if the principal stress orientations and values are unchanged, the shear stresses begin to take lower values and eventually a position occurs where they are insufficient to develop the shearing strains necessary for the folding to go on and the folds become locked. The position where this occurs depends on the stress difference, $\sigma_1 - \sigma_2$, and the t/l values of the layers in the kinked zone; for a given stress difference, folds with low t/l values can take up more shortening than those with high t/l values. This locking of the fold is likely to occur somewhere in the range of limb dips where $\alpha = 60$ to 80°. If the axial surfaces of the kinks bisect the fold limbs, their dips are controlled by (7-54) and lie in the range $\beta = 50$ to 60°. It therefore follows that the intersecting axial surfaces of conjugate kink zones will have an obtuse angle between the planes of 100 to 120° which faces the direction of maximum shortening in the material.

The total shear γ_t at any stage during the folding may be separated into two components: an internal shear in the layer γ_i and a shear along the layer boundaries γ_b such that

$$\gamma_i = \alpha \tag{7-56}$$

$$\gamma_b = 2 \tan \frac{\alpha}{2} - \alpha \tag{7-57}$$

The rates of change of these shear components with respect to e are given by

$$\frac{d\gamma_i}{de} = \frac{\csc \alpha}{\alpha t/l - 1} \tag{7-58}$$

$$\frac{d\gamma_b}{de} = \frac{\tan^2 \alpha/2 \csc \alpha}{\alpha t/l - 1} \tag{7-59}$$

Equation (7-58) is identical to (7-49) derived for chevron folds. It is the internal shear which controls the strain threshold that has to be overcome before the folding may develop. The function expressing the rates of layer boundary slip (7-59) is somewhat similar to that (7-50) evaluated for chevron folds. It takes progressively larger values as the fold develops; but for any particular stage in the fold development, the layer boundary slip increment is less for kink bands than for chevron folds by an amount $(1 - \tan^2 \alpha/2)^2/4$.

The orientation of the axial surfaces of conjugate folds appears to be controlled by the dips of the fold limbs and these dips depend on the values of the applied principal stress difference, on the persistence of this stress system with fold development, and on the internal and boundary frictional properties of the rock layer. It seems unlikely that the axial surfaces originate as a primary

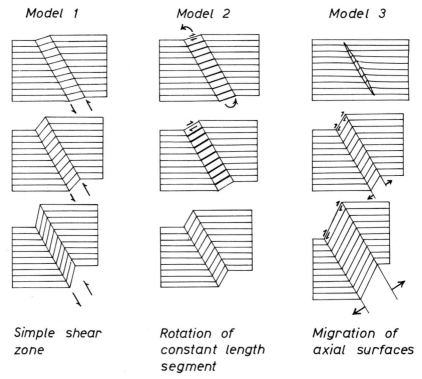

Model 1	Model 2	Model 3
Simple shear zone	Rotation of constant length segment	Migration of axial surfaces

Figure 7-123

Geometric models for the progressive development of a kink band. Model 3 is probably the most commonly found in naturally deformed rocks.

shear system as suggested by Johnson (1956), Ramsay (1962*b*), and Ramsay and Sturt (1963). The shearing hypotheses put forward by these authors have never satisfactorily accounted for the generally observed geometrical fact that it is the obtuse angle of the conjugate axial surface which faces the maximum shortening in the rock material. Primary conjugate shear systems developed in rocks always have the acute angle between the shear planes facing the direction of maximum shortening (and maximum compressive stress). It seems likely that the applied stresses develop an initial buckle, and this grows into a kink fold by extending the lengths of this kinked sector and progressively decreasing the t/l ratio for the structure. From the start of the folding the kinked layer acquires a specific limiting dip value and the fold grows by sideways migration of the axial surfaces (Figs. 7-123, model 3; 7-124). If the rock materials change their properties during deformation and become too brittle to take up the severe internal shear necessary for the migration of the axial surfaces, then faults and shears may develop along the axial surfaces and lead to the rupture (and perhaps brecciation) of the kink zone. Thus, although the axial surfaces are not primary shear planes generating the folds, they are commonly taken over by faulting at a later stage of fold development as the rocks pass from a ductile to a brittle state.

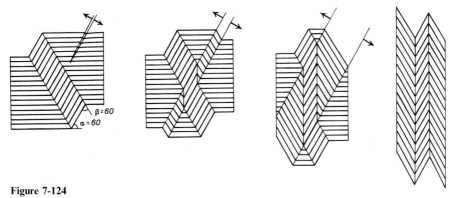

Figure 7-124

Development of chevron folds by the progressive development of one kink band across another. For this process to be possible, $\alpha = \beta = 60°$. (After the mechanism proposed by Paterson and Weiss, 1966.)

It is my experience that the total shortening by the formation of kink bands and conjugate folds rarely exceeds 25 percent. They are generally developed at the end of an orogenic deformation after the rocks have been previously cleaved and made schistose by large strains developed when the rocks were in a ductile state. Paterson and Weiss (1966) have suggested that it may be possible for the strong development of conjugate folds to lead to the formation of chevron folds (Figs. 7-124 and 7-125). This geometrical "working out" of con-

Figure 7-125

Natural development of conjugate kinks into chevron folds with independent fold development on either side of a surface of décollement. West Cork, Ireland. (Photograph by K. Coe.)

jugate structures can only occur where $\alpha = \beta = 60°$ and where the kink bands can mutually displace one another. Complete transformation of conjugate folds into chevron types requires a total compressive strain of 50 percent.

Single or conjugate kink bands developed according to the model of Fig. 7-121 are not formed under plane strain conditions; there is a dilation Δ given by

$$\Delta = \frac{t}{l}\left(2 \tan \frac{\alpha}{2} - \alpha\right) \qquad (7\text{-}60)$$

This may lead to the development of open fissures or saddle reefs along the axial trace, but the amount of dilation is always significantly less than that produced by the chevron-fold model (7-51).

The discussion of conjugate fold geometry has been developed around a two-dimensional model where the principal stresses σ_2 and σ_1 act along the layering and perpendicular to the layers, respectively. The kink zones which form have an orthorhombic symmetry which reflects that of the stress tensor (Johnson, 1956; Ramsay, 1962b). The conjugate kink zones intersect in a line parallel to the folded surface; as a result, the fold axes (Fig. 7-126, f_a, f_b) of all the kinked surfaces are parallel. Many naturally formed conjugate folds

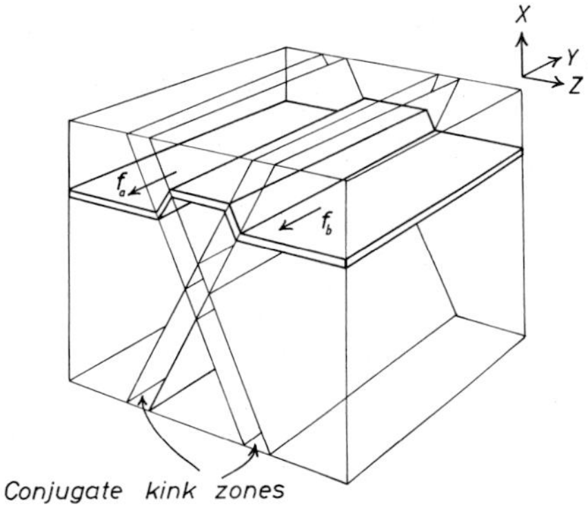

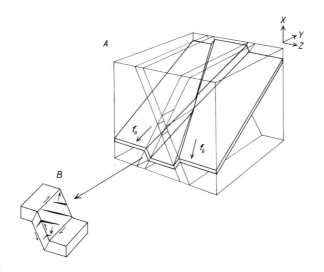

Figure 7-126

Conjugate kinks with overall orthorhombic symmetry where the axes of bulk strain are symmetrically disposed to the layering.

Conjugate kink zones

show a more complex arrangement of the fold axes (Fig. 7-127, f_a, f_b) (Ramsay, 1962b; Koopmans, 1962; Ramsay and Sturt, 1963). The line of intersections of the two kink zones passes obliquely through the folded surfaces; in consequence, the fold axes that developed in the differently inclined kink zones have different orientations (Fig. 7-128). The structure has an overall monoclinic or triclinic symmetry. It seems likely that structures of this type occur where the principal bulk strains of the material are not symmetrically disposed to the layering, and that this probably implies that the principal stresses which lead to kink-band formation were likewise asymmetrically inclined to the layers

Figure 7-127

Conjugate kinks of low symmetry. A shows conjugate kink bands with triclinic symmetry where the axes of bulk strain are asymmetrically disposed to the layering. B illustrates the type of strain in the kink zones and the possible development of en echelon tension veins.

Figure 7-128

Crossing fold axes in triclinic conjugate kink folds developed in mica schists. Ross of Mull, Scotland.

(Ramsay, 1962*b*). Where this type of deformation occurs, the distortion within the kink zones is rather complex. In each zone the distortions are generally accomplished by a combination of layer slip perpendicular to the hinge of the folds and a shear within the layering with slip direction parallel to the fold axes. This distortion sometimes leads to the formation of en echelon fissures and sometimes to the formation of subsidiary small folds obliquely inclined to the axes of the principal kink folds. These are second-order structures produced by the local strain states in the kinked zones.

REFERENCES AND SUGGESTED FURTHER READING

Anderson, T. B.: Kink Bands and Related Geological Structures, *Nature*, **202**: 272-274 (1964).

Bain, G. W.: Flowage Folding, *Am. J. Sci.*, **22**: 503-530 (1931).

Bayley, M. B.: A Theory of Similar Folding in Viscous Materials, *Am. J. Sci.*, **262**: 753-766 (1964).

Billings, M. P.: "Structural Geology," Prentice-Hall, Inc., Englewood Cliffs, N.J., 1954.

Biot, M. A.: Folding Instability of a Layered Viscoelastic Medium under Compression, *Proc. Roy. Soc. (London)*, Ser. *A*, **242**: 444-454 (1957).

Biot, M. A.: Theory of Folding of Stratified Viscoelastic Media and Its Implication in Tectonics and Orogenesis, *Geol. Soc. Am. Bull.*, **72**: 1595-1620 (1961).

Biot, M. A., H. Odé, and W. L. Roever: Experimental Verification of the Folding of Stratified Viscoelastic Media, *Geol. Soc. Am. Bull.*, **72**: 1621-1630 (1961).

Biot, M. A., and H. Odé: On the Folding of a Viscoelastic Medium with Adhering Layer under Compressive Initial Stress, *Appl. Math. Quart.*, **19**: 351-355 (1962).

Biot, M. A.: Internal Buckling under Initial Stress in Finite Elasticity, *Proc. Roy. Soc. (London)*, Ser. *A*, **273**: 306-328 (1963*a*).

Biot, M. A.: Theory of Stability of Multilayered Continua in Finite Anisotropic Elasticity, *J. Franklin Inst.*, **276**: 128-153 (1963*b*).

Biot, M. A.: Stability of Multilayered Continua Including the Effect of Gravity and Visco-elasticity, *J. Franklin Inst.*, **276**: 231-252 (1963*c*).

Biot, M. A.: Theory of Viscous Buckling of Multilayered Fluids Undergoing Finite Strain, *Am. Inst. Phys. Mag.*, **7**: 855-859 (1964*a*).

Biot, M. A.: Theory of Internal Buckling of a Confined Multilayered Structure, *Geol. Soc. Am. Bull.*, **75**: 563-568 (1964*b*).

Biot, M. A.: Theory of Similar Folding of the First and Second Kind, *Geol. Soc. Am. Bull.*, **76**: 251-258 (1965*a*).

Biot, M. A.: Theory of Viscous Buckling and Gravity Instability of Multilayers with Large Deformation, *Geol. Soc. Am. Bull.*, **76**: 371-378 (1965*b*).

Borg, I., and F. J. Turner: Deformation of Yule Marble, Part VI, *Geol. Soc. Am. Bull.*, **64**: 1343-1352 (1953).

Boschma, D.: Successive Hercynian Structures in Some Areas of the Central Pyrenees, *Leidse Geol. Mededel.*, **28**: 166-170 (1963).

Breddin, H., and H. Furtak: Zur geometrie asymmetrischer Falten, *Geol. Mitt. Aachen*, **3**: 197-219 (1962).

Busk, H. G.: "Earth Flexures," Cambridge University Press, London, 1929.

Campbell, J. D.: En Echelon Folding, *Econ. Geol.*, **53**: 448-472 (1958).

Campbell, J. W.: Some Aspects of Rock Folding by Shear Deformation, *Am. Jour. Sci.*, **249**: 625-639 (1951).

Carey, S. W.: The Rheid Concept in Geotectonics, *J. Geol. Soc. Austr.*, **1**: 67-117 (1954).

Carey, S. W.: Folding, *J. Alberta Soc. Petrol. Geologists*, **10**: 95-144 (1962).

Cloos, E.: Oolite Deformation in the South Mountain Fold, Maryland, *Geol. Soc. Am. Bull.*, **58**: 843-918 (1947).

Currie, J. B., H. W. Patnode, and R. P. Trump: Development of Folds in Sedimentary Strata, *Geol. Soc. Am. Bull.*, **73**: 655-674 (1962).

Dahlstrom, C. D. A.: "Statistical Analysis of Cylindrical Folds," *Trans. Can. Min. Inst.*, **57**: 140-145 (1954).

de Sitter, L. U.: The Principle of Concentric Folding and the Dependence of Tectonical Structure on Original Sedimentary Structure, *Proc. Koninkl. Ned. Akad. Wetenschap.*, **42**: 5 (1939).

de Sitter, L. U.: Schistosity and Shear in Micro and Macrofabrics, *Geol. Mijnbouw.*, **16**: 429-439 (1954).

de Sitter, L. U.: The Strain of Rock in Mountain Building Processes, *Am. J. Sci.*, **254**: 585-604 (1956).

de Sitter, L. U.: Cleavage Folding in Relation to Sedimentary Structures, *Intern. Geol. Congr.*, *20th session, Mexico*, Sec. V: 53-64 (1957).

de Sitter, L. U.: Boudins and Parasitic Folds in Relation to Cleavage and Folding, *Geol. Mijnbouw.*, **20**: 277-286 (1958*a*).

de Sitter, L. U.: "Structural Geology," McGraw-Hill Book Company, New York, 1958*b*.

Dewey, J. F.: Nature and Origin of Kink-bands, *Tectonophysics*, 1:459-494 (1965).

Donath, F. A.: Role of Layering in Geologic Deformation, *Trans. N.Y. Acad. Sci.*, Ser. 2, **24**: 236-249 (1962).

Donath, F. A.: Fundamental Problems in Dynamic Structural Geology, in T. W. Donnelly (ed,) "The Earth Sciences," pp. 83-103, University of Chicago Press, Chicago, Ill., 1963.

Donath, F. A., and R. B. Parker: Folds and Folding, *Geol. Soc. Am. Bull.*, **75**: 45-62 (1964).

Elliott, D.: The Quantitative Mapping of Directional Minor Structures, *J. Geol.*, **73**: 865-880 (1965).

Eskola, P. E.: The Problem of Mantled Gneiss Domes, *Quart. J. Geol. Soc.*, **104**: 461-476 (1949).

Evans, A. M.: Conical Folding and Oblique Structures in Charnwood Forest, Leicester, *Proc. Yorkshire Geol. Soc.*, **34**: 67-80 (1963).

Fitzgerald, E. L., and L. T. Braun: Disharmonic Folds in Besa River Formation, Northeastern British Columbia, Canada, *Bull. Am. Assoc. Petrol. Geologists*, **49**: 418-432 (1965).

Fleuty, M. J.: The Description of Folds, *Geol. Assoc. Proc.*, **75**: 461-492 (1964).

Flinn, D.: A Tectonic Analysis of the Muness Phyllite Block of Unst and Uyea, Shetland, *Geol. Mag.*, **89**: 263-272 (1952).

Flinn, D.: On Folding during Three Dimensional Progressive Deformation, *Quart. J. Geol. Soc.*, **118**: 385-433 (1962).

Fyson, W. K.: Folds in the Carboniferous Rocks near Walton, Nova Scotia, *Am. J. Sci.*, **262**: 513-522 (1964).

Goguel, J.: Introduction à l'étude méchanique des déformations de l'écorce terrestre, *Ser. Carte Geol. France Mem.*, 1-531 (1948).

Goldstein, S.: The Stability of a Strut under Thrust When Buckling is Resisted by a Force Proportional to the Displacement, *Cambridge Philos. Soc. Proc.*, **23**: 120-129 (1926).

Gonzalez-Bonorino, F.: The Mechanical Factor in the Formation of Schistosity, *21st Intern. Geol. Congr., Copenhagen*, Pt. 18: 303-316 (1960).

Gunn, R.: A Quantitative Study of Mountain Building on an Unsymmetrical Earth, *J. Franklin Inst.*, **224**: 19-53 (1937).

Heim, A.: "Geologie der Schweiz," Bernhard Tauchnitz Verlag Gmbh, Stuttgart, 1921.

Hills, E. S.: Examples of the Interpretation of Folding, *J. Geol.*, **53**: 47-57 (1945).

Hills, E. S.: "Elements of Structural Geology," especially pp. 211-286, John Wiley & Sons, Inc., New York, 1963.

Hoeppener, R.: Zum Problem der Bruchbildung, Schieferung und Faltung, *Geol. Rundschau*, **45**: 247-283 (1956).

Johnson, M. R. W.: Conjugate Fold Systems in the Moine Thrust Zone in the Lochcarron and Coulin Forest Areas of Wester Ross, *Geol. Mag.*, **93**: 345-350 (1956).

Kienow, S.: Grundzüge einer Theorie der Faltungs—und Schleiferungsvorgänge, *Fortschr. Geol. Palaeontol.*, **14**: 1-129 (1942).

King, B. C., and N. Rast: Tectonic Styles in the Dalradians and Moines of Part of the Central Highlands of Scotland, *Proc. Geologists' Assoc., (Engl.)*, **66**: 243-269 (1956).

King, B. C.: Problems of the Precambrian of Central and Western Uganda; Structure, Metamorphism and Granites, *Science Progr.*, **47**: 735-739 (1959).

Kleinsmiede, J. F.: The Geology of the Valle de Aran, Central Pyrenees, *Leidse Geol. Mededel.*, **25**: 129-246 (1960).

Knill, J. L., and D. C. Knill: Some Discordant Fold Structures from the Dalradian of Craignish, Argyll and Rosguill, Co. Donegal, *Geol. Mag.*, **95**: 497-510 (1958).

Knill, J. L.: The Tectonic Pattern in the Dalradian of the Craignish-Kimelfort District, Argyllshire, *Quart. J. Geol. Soc.*, **115**: 339-364 (1960).

Knill, J. L.: Joint Drags in Mid-Argyllshire, *Proc. Geologists' Assoc. (Engl.)*, **72**: 13-19 (1961).

Koopmans, B. N.: The Sedimentary and Structural History of the Valsurvio Dome, Cantabrian Mountains, Spain, *Leidse Geol. Mededel.*, **26**: 121-232 (1962).

Kuenen, P. H.: Observations and Experiments in Ptygmatic Folding, *Bull. Comm. Geol. Finlande*, **123**: 11-27 (1938).

Kuenen, P. H., and L. U. de Sitter: Experimental Investigation into the Mechanisms of Folding, *Leidsche Geol. Mag.*, **10**: 217-240 (1938).

Leith, C. K.: "Structural Geology," Henry Holt and Company, Inc., New York, 1923.

Love, A. E. H.: "The Mathematical Theory of Elasticity," Dover Publications, Inc., New York, 1944.

Lugeon, M.: Notice explicative de la feuille des Diablerets, Atlas géol. Suisse au 1/25,000, Comm. géol. Suisse (1940).

McBirney, A. R., and M. G. Best: Experimental Deformation of Viscous Layers in Oblique Stress Fields, *Geol. Soc. Am. Bull.*, **72**: 495-498 (1961).

Mathews, D. H.: Dimensions of Asymmetrical Folds, *Geol. Mag.*, **95**: 511-513 (1958).

Mead, W. J.: Folding, Rock Flowage and Foliate Structures, *J. Geol.*, **48**: 1007-1021 (1940).

Mendelsohn, F.: Structure of the Roan Antilope Deposit, *Trans. Inst. Mining Met.*, **68**: 229-263 (1959).

Mertie, J. B.: Classification, Delineation and Measurement of Non-parallel Folds, *U.S. Geol. Surv. Profess. Paper* 314 *E*: 91-124 (1959).

Muehlberger, W. R.: Conjugate Folds of Small Dihedral Angle, *J. Geol.*, **69**: 211-219 (1959).

Mügge, O.: Über Translationen und verwandte Erscheinungen in Krystallen, *Neues Jahrb. Mineral. Geol. Palaeontol., Abt. B*, **1**: 71-75 (1898).

Nevin, C. M.: "Principles of Structural Geology," John Wiley & Sons, Inc., New York, 1949.

Paterson, M. S., and L. E. Weiss: Experimental Folding in Rocks, *Nature*, **195**: 1046-1048 (1962).

Paterson, M. S., and L. E. Weiss: Experimental Deformation and Folding in Phyllite, *Bull. Geol. Soc. Am.*, **77**: 343-374 (1966).

Phillips, F. C.: The Study of Small Scale Structures in the Variscan Fold Belt, *9th Inter-University Geol. Congr.*, Manchester University Press: 109-128 (1961).

Pilger, A., and W. Schmidt: Definition des Begriffes "Mullion Struktur," *Neues Jahrb. Geol. Palaeontol.*, 24 (1957*a*).

Pilger, A., and W. Schmidt: Mullion-Strukturen in der Nord-Eifel, *Abhandl. Hess. Landesamtes Bodenforsch.*, **20**: 53 (1957*b*).

Ramberg, H.: Evolution of Ptygmatic Folding, *Norsk Geol. Tidsskr.*, **39**: 99-151 (1959).

Ramberg, H.: Relationships between Length of Arc and Thickness of Ptygmatically Folded Veins, *Am. J. Sci.*, **258**: 36-46 (1960).

Ramberg, H.: Relationship between Concentric Longitudinal Strain and Concentric Shearing Strain during Folding of Homogeneous Sheets of Rock, *Am. J. Sci.*, **259**: 382-390 (1961*a*).

Ramberg. H.: Contact Strain and Folding Instability of a Multilayered Body under Compression, *Geol. Rund.*, **51**: 405-439 (1961*b*).

Ramberg, H.: Evolution of Drag Folds, *Geol. Mag.*, **100**: 97-106 (1963*a*).

Ramberg, H.: Fluid Dynamics of Viscous Buckling Applicable to Folding of Layered Rocks, *Bull. Am. Assoc. Petrol. Geol.*, **47**: 484-515 (1963*b*).

Ramberg, H.: Strain Distribution and Geometry at Folds, *Bull. Geol. Inst. Univ. Upsala*, **42**: 1-20 (1963*c*).

Ramberg, H.: Note on Model Studies of Folding of Moraines in Piedmont Glaciers, *J. Glaciology*, **5**: 207-218 (1964*a*).

Ramberg, H.: Selective Buckling of Composite Layers with Contrasted Rheological Properties; a Theory for Simultaneous Formation of Several Orders of Folds, *Tectonophys.*, **1**: 307-341 (1964*b*).

Ramberg, H., and O. Stephansson: Compression of Floating Elastic and Viscous Plates Affected by Gravity, a Basis for Discussing Crustal Buckling, *Tectonophys.*, **1**: 101-120 (1964*c*).

Ramsay, D. M., and B. A. Sturt: A Study of Fold Styles, Their Association and Symmetry Relationships, from Sørøy, North Norway, *Norsk Geol. Tidsskr.*, **43**: 411-430 (1963).

Ramsay, J. G.: The Deformation of Early Linear Structures in Areas of Repeated Folding, *J. Geol.*, **68**: 75-93 (1960).

Ramsay, J. G.: The Geometry and Mechanics of Formation of "Similar" Type Folds, *J. Geol.*, **70**: 309-327 (1962*a*).

Ramsay, J. G.: The Geometry of Conjugate Fold Systems, *Geol. Mag.*, **99**: 516-526 (1962*b*).

Ramsay, J. G.: Structure, Stratigraphy and Metamorphism in the Western Alps, *Proc. Geologists' Assoc. (Engl.)*, **74**: 357-391 (1963*a*).

Ramsay, J. G.: Structure and Metamorphism of the Moine and Lewisian Rocks of the Northwest Caledonides, in Johnson and Stewart (eds.), "The British Caledonides," Oliver & Boyd Ltd, Edinburgh and London, 1963*b*.

Rast, N.: Morphology and Interpretation of Folds—a Critical Essay, *Geol. J.*, **4**: 177-188 (1964).

Richter, D.: Die δ-Achsen und ihre räumlich-geometrischen Beziehungen zu Faltenbau und Schiefrigkeit, *Geol. Mitt. Aachen*, **2**: 1-35 (1961).

Schmidt, W.: "Tektonik und Verformungslehre," Berlin, 1932.

Sederholm, J. J.: Uber ptygmatische Faltungen, *Neues Jahrb. Min. Petrol.*, **36**: 491 (1913).

Smoluchowski, M.: Versuche über Faltungserscheinungen schwimmender elastischen Platten, *Anz. Akad. Wiss., Krakau, Math. Phys. Kl.*, 727-734 (1909).

Stauffer, M. R.: The Geometry of Conical Folds, *New Zealand J. Geol. Geophys.*, **7**: 340-347 (1964).

Stillwell, F. L.: The Factors Influencing Gold Deposition in the Bendigo Goldfield, *Advisory Council Sci. Ind. Bull.* 4, 8, and 16, Melbourne, 1917, 1918, 1919.

Stillwell, F. L.: Replacement in the Bendigo Quartz Vein, *Econ. Geol.*, **13**: 108-111 (1918).

Stockwell, C. H.: The Use of Plunge in the Construction of Cross Sections of Folds, *Proc. Geologists Assoc. Can.*, **3**: 97-121 (1950).

Tromp, S. W.: "The Mechanism of the Geological Undulation Phenomena," A. W. Sijthoff, Leiden, 1937.

Turner, F. J., and L. E. Weiss: "Structural Analysis of Metamorphic Tectonites," especially pp. 104-125 and 469-494, McGraw-Hill Book Company, New York, 1963.

Turner, F. J., and L. E. Weiss: Deformational Kinks in Brucite and Gypsum, *Proc. Nat. Acad. Sci.*, **54**: 359-364 (1965).

van Hise, C. R.: Principles of North American Pre-Cambrian Geology, *U.S. Geol. Surv. 16th Ann. Rept.*, 581-843 (1894).

Wegmann, E., and J. P. Schaer: Lunules tectoniques et traces de movements dans les plis du Jura, *Eclogae Geol. Helv.*, **50**: 491-496 (1957).

Weiss, L. E.: Fabric Analysis of a Triclinic Tectonite, and Its Bearing on the Geometry of Flow in Rocks, *Am. J. Sci.*, **253**: 225-236 (1955).

Williams, E.: The Deformation of Confined Incompetent Layers in Folding, *Geol. Mag.*, **98**: 317-323 (1961).

Williams, E.: The Deformation of Competent Granular Layers in Folding, *Am. J. Sci.*, **263**: 229-237 (1965).

Wilson, G.: The Relationship of Slaty Cleavage and Kindred Structures to Tectonics, *Proc. Geologists Assoc.*, **57**: 263-302 (1946).

Wilson, G.: The Geology of the Tintagel Area, North Cornwall, *Quart. J. Geol. Soc.*, **106**: 393-432 (1951).

Wilson, G.: Mullion and Rodding Structure in the Moine Series of Scotland, *Proc. Geologists Assoc.*, **64**: 118-151 (1953).

Wilson, G.: The Tectonic Significance of Small Scale Structures, *Ann. Soc. Geol. Belg. Bull.*, **84**:423-548 (1961).

Wunderlich, H. G.: Zur Gesteinsdeformation im Einflussbereich gefalteter Bänke, *Deut. Geol. Ges. Z.*, **111**: 525-542 (1959*a*).

Wunderlich, H. G.: Erzeugung engständiger Scherflächen in plastischen Material, *Neues Jahrb. Geol. Pal.*, **1**:34-44 (1959*b*).

Wunderlich, H. G.: Gesteinsdeformation im Einflussbereich von Falten mit geneigter Achsenebene, *Deut. Geol. Ges., Z.*, **111**: 599-611 (1960).

Wynne-Edwards, H. R.: Flow Folding, *Am. J. Sci.*, **261**: 793-814 (1963).

Zandvliet, J.: The Geology of the Upper Salat and Pallaresca Valleys, Central Pyrenees, *Leidse Geol. Mededel.*, **25**: 1-127 (1960).

8 ⦀ Deformation of linear structures

It was shown in the last chapter how rock strata may take up a great variety of folded forms depending on the rheological state of the layers and their environmental conditions. Consider an originally straight, linear (rectilinear) feature contained on the plane surface of a rock layer before folding (Fig. 8-2A). The aim of this chapter is to investigate the form taken up by this line when the plane surface is folded (Figs. 8-1, 8-2B). The deformed line traces out a curved locus in space whose form depends entirely on the states of finite strain in the folded layers. When the fold axis is constantly oriented, the locus can be defined by describing the angular relationships of the moving line to the fixed fold axis: a relationship which is most conveniently represented on a projection where the curving line in space appears as a linear locus (Fig. 8-2C).

The study of the geometry of deformed lines has a number of very important practical applications:

1. VARIATION IN FINITE STRAIN IN A FOLD It has been shown that any one particular fold shape may be developed in an infinite number of different arrangements of finite-strain variation within the folded layer. The degree of freedom of this variety can be reduced by a study of the orientations of deformed lineations on the folded surfaces. For example, it is possible to discover whether identical parallel folds have been formed by simple flexural slip (Figs. 7-54, 7-55) or oblique flexural slip (Fig. 7-61) if the fold surfaces contain deformed lineations since the lineation loci of these two types of parallel fold are distinctive.

2. FLOW DIRECTION OF SIMILAR FOLDS   In Sec. 7-11 it was shown that movement directions of the rock material in similar folds could not be determined from observation of the orientation of the fold axes. However, if the folds were developed in previously lineated rocks, the direction of flow of material can be calculated from the deformed lineation loci.

Figure 8-1

Deformed linear structures. Pennine Alps, Ticino, Switzerland.

3. INITIAL ORIENTATION OF DEFORMED LINEATIONS   The geologist may wish to determine the initial orientation of a deformed linear structure situated on the limb of a fold. For example, by establishing the initial orientation of a particular sedimentation structure in a specific layer, he could determine the original direction of the currents (paleocurrents) which brought in the sedimentary material making up that particular bed of rock. As another example, it may be necessary to correct paleomagnetic observations obtained from folded and tilted strata, or to determine the original orientations of early fold axes in regions of superimposed folding. In all these problems a technique is used to remove the effects of the rock strain on the linear feature. The techniques employed depend on what is known about (1) the state of strain in the fold, (2) the style of folding of the rock layers, and (3) the mechanism of formation of the fold. The main types of fold-forming mechanisms will now be examined to determine the behavior of rectilinear features in folds.

8-1 FOLDS FORMED BY BUCKLING

Flexural-slip folds In these folds all the internal deformation is accomplished by simple shear parallel to the layer boundaries. Folds of this type are always of the parallel model (Class 1*B*), and no distortion occurs within the curved surfaces of the folded layers. The folded surfaces undergo only a rotation about the fold axis, and the angle (α) between the linear structure and the fold axis remains constant at all points on the fold surface (Fig. 8-2). The locus of the deformed line is part of a helix, and, because α remains constant, the deformed lineations plot on an equal-area projection to give a locus which is a partial small circle (Fig. 8-2C) (Sander, 1948, pp. 174-178; Phillips, 1954; Weiss and McIntyre, 1957; Weiss, 1959a, pp. 99-100; 1959b, p. 32; Ramsay, 1958a, 1960, 1963; Wilson, 1961). Where $\alpha = 90°$ the locus becomes a great circle representing a plane normal to the fold axis. Although the small-circle patterns have been widely discussed from a theoretical viewpoint, they seem to be rather uncommon in regions where naturally deformed lineations have been observed.

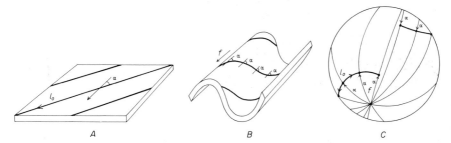

Figure 8-2

Linear structures deformed by flexural-slip folding.

Buckle folds with internal deformation by tangential longitudinal strain
It is possible to accommodate the strains within a buckled layer by variation in tangential longitudinal strain (Sec. 7-8, p. 397; Fig. 7-63). On the convex side of the finite neutral surface, the layers are subjected to a tangential extension; whereas on the concave side, they undergo tangential contraction. The finite neutral surface is unstrained, and here linear structures will be deformed in exactly the same way as with the flexural-slip model described above. In the material on the outer arc of the neutral surface, the extensive strain reaches a maximum value where the curvature is greatest, that is, around the hinge zone of the structure. On surfaces in this part of the fold, the angle between the deformed lineation and the fold axis is increased by an amount which depends on the amount of the extension. If there is no strain parallel to the fold hinge, the relationship of the original angle (α)

between the lineation and the fold axis and this angle after deformation (α') may be found by combining (7-19) and (3-34):

$$\tan \alpha' = \tan \alpha \, (1 + t'c) \tag{8-1}$$

where t' is the distance above the neutral surface and c the curvature of the neutral surface. By similar arguments, the linear structures on the folded surfaces which lie on the concave side of the curving neutral surface at a distance t' from it are deformed so that they come to lie closer to the fold axis by an amount given by

$$\tan \alpha' = \tan \alpha \, (1 - t'c) \tag{8-2}$$

The loci of the deformed lineations on different surfaces in the fold will vary depending on the way the curvature and the position of the finite neutral surface vary throughout the structure. The type of deformed lineation loci seen in structures of this form are shown in Fig. 8-3. At the lines of inflection between adjacent folds, $c = 0$; therefore $\alpha' = \alpha$. If the amount of tangential longitudinal strain is not known, the linear structures situated on the fold limbs are often the simplest to choose for restoration to their original unfolded orientation.

The layers in many buckle folds show internal deformation by a combination of flexural slip and tangential longitudinal strain. Where combined deformations of this type are present, the angles between the deformed linear structures and the fold axis will be altered by an amount less than that given by (8-1) and (8-2). The loci of deformed lineations recorded on a projection will now show the same general forms as those shown in Fig. 8-3, but they

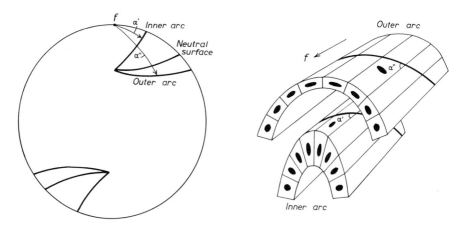

Figure 8-3

Linear structures deformed where the planes on which they lie are subjected to buckling and internal deformation by the process of tangential longitudinal strain.

will not diverge so much from the locus on the finite neutral surface of the new structure (where $\alpha' = \alpha$). The deviations of observed values of α' from those given by (8-1) and (8-2) might be used to determine the relative amounts of deformation by tangential longitudinal strain and flexural slip.

Oblique flexural-slip folds Although buckled stratified layers often show internal deformation by flexural slip, the a directions of simple shear deformation are not always perpendicular to the fold axis and may vary around the folded surface (Sec. 7-8, p. 396). If this occurs, the component of slip in the folded surface parallel to the fold axis varies and the deformed lineation locus will depart from that of the simple helix. As an example of the effects of this type of deformation, the amount of shear parallel to the fold axis has been computed (Table 8-1) for the oblique flexural-slip fold illustrated in Fig. 7-61. In this fold the angle between the fold axis f and the intermediate bulk strain axis Y is $30°$ ($\lambda = 1$ along f).

Table 8-1   *Variation in shearing strain within the bedding surface of the oblique flexural-slip fold of Fig. 7.60*

Angle between axial plane and folded surface	0	20	40	60	80	90
Shear strain γ in folded surface measured in direction of fold axis	± 0.58	± 0.55	± 0.44	± 0.29	± 0.11	0

These values have been used with (3-71) to calculate the lineation distortions and deflections of the angle between the lineations and the fold axis (α')

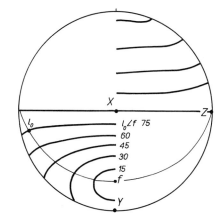

Figure 8-4

Loci of lineations deformed in an oblique flexural-slip fold, axis f, with different initial angles $\alpha = l_0 f$. X, Y, and Z are the principal axes of bulk finite strain.

from that of the original angle (α). The loci for a number of original values of α are plotted in Fig. 8-4. In my experience this type of deformation is common in many naturally formed flexural-slip folds, especially those in metamorphic environments. The extent of the oblique slip can be determined by measuring the amount by which the deformed lineations depart from the constant-angle helix locus of the simple flexural-slip model.

Flattened parallel (flexural-slip) folds   It was shown in Sec. 7-8, p. 411, how the geometric form of folds may be modified by the superposition of a homogeneous strain or a flattening process. Linear structures deformed by flexural-slip folding will take up a new locus depending on the ratios and orientations of the principal superimposed strains. Deformed linear structures near the hinge zone of the fold come to lie closer to the fold axis, while those situated on the fold limb generally come to lie at a higher angle to the fold axis.

The symmetry of the deformed lineation locus depends on the orientation of the superimposed strains with reference to the axis of flexural-slip folding. If one of the superimposed strain axes coincides with the fold axis f (generally $f = Y$), the lineation locus is symmetric and forms part of an elliptical helix. The angular relationship of the lineations when plotted on a projection are identical to those of a cone with elliptical base (Fig. 8-5) (Ramsay, 1963, p. 159). From this locus it is sometimes possible to determine the principal strain ratios. If the maximum and minimum angles between the deformed

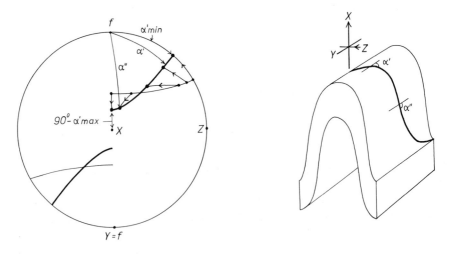

Figure 8-5

Lineations deformed in a flattened flexural-slip fold where X, Y, and Z are the axes of flattening and where the intermediate strain axis Y coincides with the axis of folding f.

lineations and the fold axis are determined (Fig. 8-5, α'_{max}, α'_{min}), then it follows from (4-21b, c) that

$$\tan (90 - \alpha'_{max}) = \left(\frac{\lambda_2}{\lambda_1}\right)^{\frac{1}{2}} \tan (90 - \alpha) \tag{8-3}$$

$$\tan \alpha'_{min} = \left(\frac{\lambda_3}{\lambda_2}\right)^{\frac{1}{2}} \tan \alpha \tag{8-4}$$

Multiplying together,

$$\tan \alpha'_{min} \cot \alpha'_{max} = \left(\frac{\lambda_3}{\lambda_1}\right)^{\frac{1}{2}} \tag{8-5}$$

This method offers a check on the technique of computing λ_3/λ_1 by measuring the orthogonal thickness of layers in profile section (p. 413).

If the principal superimposed strain axis Y does not coincide with the axis of flexural-slip folding f, then the deformed lineation locus is complex and asymmetric, i.e., the geometric form on one side of the axial surface is not identical to that on the other side. In structures of this type it is not possible to determine the strain ratios using variations in orthogonal thickness in a profile section (see discussion on p. 415). If the positions of the principal strain axes are known and the complete lineation locus has been found or can be extrapolated, then it is possible to compute the three principal strain ratios. Considering the original flexural-slip fold (Fig. 8-6A) and its flattened modification (Fig. 8-6B), then from (4-21b) the following relationships hold:

$$\tan x' = \left(\frac{\lambda_2}{\lambda_1}\right)^{\frac{1}{2}} \tan x \tag{8-6}$$

$$\tan x'' = \left(\frac{\lambda_2}{\lambda_1}\right)^{\frac{1}{2}} \tan (x - \alpha) \tag{8-7}$$

$$\tan x''' = \left(\frac{\lambda_2}{\lambda_1}\right)^{\frac{1}{2}} \tan (x + \alpha) \tag{8-8}$$

These three equations can be solved for the three unknowns, α, $(\lambda_2/\lambda_1)^{\frac{1}{2}}$, and x.

The various types of deformed lineation loci are all distinctive and they have one feature in common: where the locus is completely represented (from an isoclinal fold), the direction of the deformed lineations on one limb of the isocline is not parallel to that on the other limb. On a stereogram these angular relationships appear as a *discontinuity* in the position of the lineation locus across the axial surface of the fold. This discontinuity is characteristic of folds which contain a component of body rotation of the layers around the fold axis.

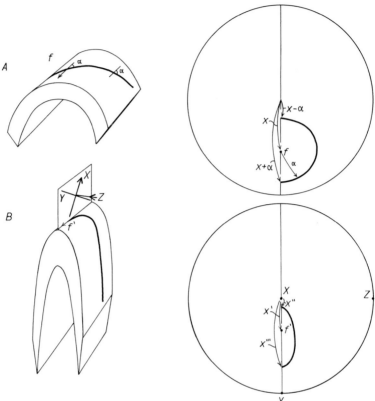

Figure 8-6

Lineations deformed in a flattened flexural-slip fold where the intermediate strain axis Y does not coincide with the initial axis of the flexural-slip fold f. Angle α does not remain constant and f changes its spatial position to f'.

8-2 CONICAL FOLDS

Conical folds have been divided into two types, depending on whether the cross section of the structure perpendicular to the cone axis is a part of a circle or an ellipse. The two types of fold have been designated *circular conical folds* and *elliptical conical folds*, respectively (Haman, 1961; Stauffer, 1964). Stauffer has further suggested that elliptical conical folds might be formed from circular conical folds by the process of flattening (superimposed homogeneous strain).

The deformation of early linear structures on surfaces subjected to conical folding has been discussed by Stauffer. Consider a plane surface *PQR* containing rectilinear structures (Fig. 8-7*A*) and fold this into a circular cone so that *PQ* coincides with *RQ* (Fig. 8-7*B*). The apical angle of the cone is

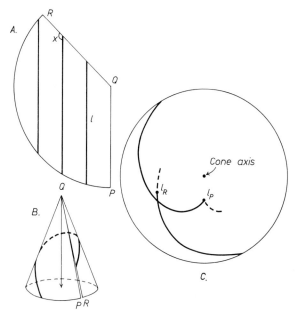

Figure 8-7

Deformation of lineations by formation of a circular conical fold. A shows the surface on which the initial rectilinear structure was developed, and B shows the deformation of this surface into a cone. C illustrates the plot of the positions of the deformed lineations in B.

$2 \sin^{-1} x/360$, where x is the angle between the linear structure and PQ (Fig. 8-7A). If no internal deformation goes on within the surface of the cone, the angle between the original rectilinear structure and straight lines drawn from the apex of the cone down the side of the conic surface change regularly through the structure. The deformed lines on the cone surface trace out a complex spatial locus. This locus has been plotted on a projection (Fig. 8-7C) with the cone axis oriented vertically for convenience of representation; it forms a complex spiral. Although only one cone structure has been examined in this diagram, all other cones show similar geometric features. In geological deformations producing conic structures of this type, only a part of the spiral locus would be represented and the cone axis would probably be less steeply inclined than that of Fig. 8-7C.

8-3 SIMILAR FOLDS

Similar folds are common in the most strongly deformed zones of orogenic belts. In these regions the total deformation is most commonly built up by a number of separate movement pulses leading to the formation of several generations of folds. The linear features related to one set of folds are commonly deformed by the development of later folds. The patterns of the deformed lineations in these zones of similar folding are quite unlike those produced by the buckling process. The geometry of lines deformed by shear was discussed by Weiss (1955, p. 229); but the application of these principles to the study of linear structures deformed by similar folding was not developed until later (Weiss, 1959a, p. 100; Ramsay, 1960).

1. Similar folds formed by inhomogeneous simple shear with shear axes *a*, *b*, **and** *c* **constantly oriented**   Consider a plane surface *ABCD* containing a rectilinear structure *AOPQRC*, and deform this plane by shear on constantly oriented planes (*ab*) with variable displacement in the direction of slip (*a*) (Figs. 8-8, 8-9). Points *KLMN* on the plane take up new positions *klmn* and the surface becomes folded with an axial surface parallel to *ab* and a fold axis *f* parallel to the intersection of *ab* with the original plane surface. Points *OPQR* on the rectilinear feature are transported variable distances along *a* and come to lie at *opqr* on the fold surface such that they are contained in a plane controlled by the orientation of the original lineation and *a*. The angle between the fold axis *f* and the deformed linear structure on the fold's surface (*α′*) varies throughout the structure and departs from that of the original (*α*); *α′* may be greater or smaller than *α* and zones of constant *α′* run parallel to the axial surfaces of the fold (Fig. 8-16). The lineation locus of Fig. 8-8 is

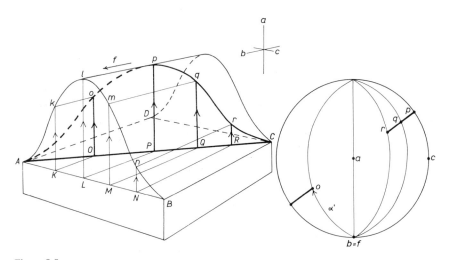

Figure 8-8

Deformation of a linear structure by heterogeneous simple shear where the b direction coincides with the axis of folding.

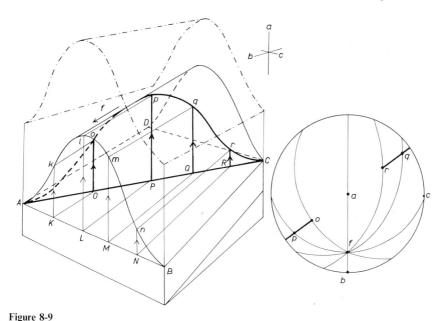

Figure 8-9

Deformation of a linear structure by heterogeneous simple shear where the b direction does not coincide with the axis of folding.

symmetric; but in general the lineation locus is asymmetric (Fig. 8-9) because *b* and the fold axis *f* do not necessarily coincide and the structure as a whole has a triclinic symmetry.

If the plane *ABCD* is inclined so that the movement direction *a* lies on the surface, variable translation of *KLMN* to *klmn* does not lead to the formation of folds in the surface (Figs. 8-10, 8-11). Points *OPQR* on the linear structures, however, are translated to *opqr*, the deformed lineation is curved, and the variation in attitude indicates the relative displacements of the particles. If the *a* direction lies at a high angle to the original lineation, then the changes in direction of the deformed linear structure tend to be very large (Fig. 8-12*A*); whereas if the *a* direction is close to the direction of the original lineation, the same shear variations lead only to small changes in the orientation of the deformed linear structures (Fig. 8-12*B*). If the dihedral angle between *ab* and a number of parallel original surfaces, *r,s,t,* and *u,* was small (Fig. 8-13) and *a* is contained in the surfaces, then the deformed lineation curves on one surface differ from those immediately next to it (cf. lineation directions l_r, l_s, l_t, and l_u). This produces the peculiar effect of deformed lineations on one surface having a different direction from those on the surface adjacent to it (Fig. 8-13, cf. l_r and l_s; Fig. 8-14).

These geometric principles have been found to hold in a number of naturally

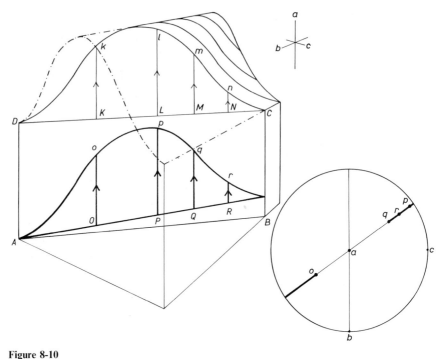

Figure 8-10

Deformation of a linear structure by heterogeneous simple shear, where the a direction lies in the surface on which the linear structure is situated.

folded rock materials (Ramsay, 1960, 1965; Best, 1963) over a considerable range of scale. Lineation patterns which show extremely complex variations over large areas (Fig. 8-15) may appear more orderly when the data are subjected to geometrical analysis (Figs. 8-16, 8-17).

The variations in plunge and direction of the linear structures shown on the map in Fig. 8-15 can be seen to be systematically arranged when plotted onto a projection (Fig. 8-17). The deformed linear structures all lie about a great circle zone, and so appear to conform to the general geometric rule established in Fig. 8-9.

It is possible to use the plane lineation locus in a similar fold to compute the *a* direction of shear. The lineation locus is first determined; this may be done by making a number of separate observations of linear directions at different points on the folded surface, plotting the data, and finding the great circle on which they lie. If the folds occur within a single outcrop it is sometimes possible to make direct measurements of the locus of deformation; the individual folds are viewed from different directions until one deformed linear feature passing over the surface appears aligned on a plane (Figs. 8-18,

Figure 8-11

Linear structures deformed so that the plane on which they lie is unfolded. Monar, Inverness-shire, Scotland.

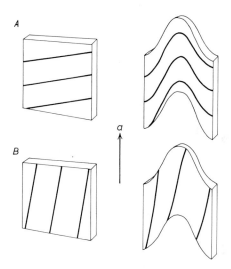

Figure 8-12

Variation in orientation of deformed lineation. A, where the angle between the original lineation and a is large; B, where the angle between the original lineation and a is small.

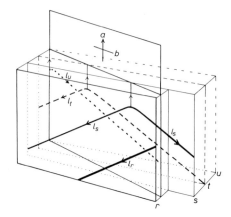

Figure 8-13

Variation in pattern of deformed lineations on surfaces r, s, t, and u resulting from the oblique intersection of the shear plane ab with these surfaces.

8-19). At this position the observer is viewing the linear structure along the plane locus, and the orientation of this surface may be directly measured. This lineation locus is controlled by the original direction and *a* (Fig. 8-9). The axial surface (*ab*) can be determined, and the intersection of this with the

Figure 8-14

Lineations l_s and l_r on adjacent surfaces deformed so that they are not parallel. Loch Hourn, Scotland.

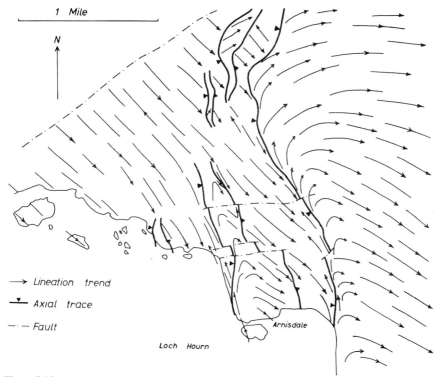

Figure 8-15

Traces of deformed lineations at Arnisdale, Western Highlands of Scotland. (After Ramsay, 1960.)

lineation locus gives the position of the only line common to both surfaces, the *a* direction. The application of this technique often produces valuable structural information in regions of complex folding (Ramsay, 1960, 1965). As an example of the use of the technique, the structural variation seen in a region of complexly folded phyllites from Anglesey (North Wales) has been chosen (Fig. 8-20). The fold axes of the most dominant folds show considerable plunge variations. These folds were developed in rocks which had been previously folded, lineated and metamorphosed. The deformed linear structures lie on plane loci and the slip directions (*a*) have been computed (Figs. 8-20, 8-21). These directions and the axial surfaces of the folds show a fairly constant orientation, indicating that the orientation of the fold-forming translations were fairly constant over quite a large region. It is also interesting to note that the *a* directions show slight but systematic changes in orientation through the area (cf. Figs. 8-21*A*, 8-21*B*). In contrast to this the fold axes are very variable, either because the surfaces on which the folds were developed were themselves variably oriented, or because the amount of compressive

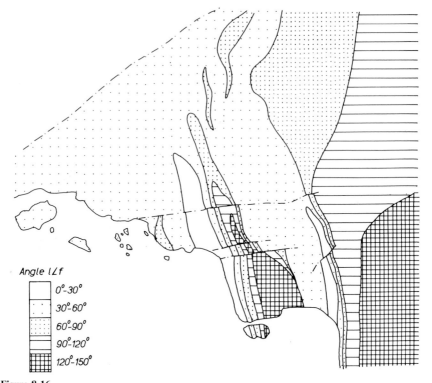

Figure 8-16

Contoured map of angles between deformed lineations and axes of late folds in Fig. 8-15.

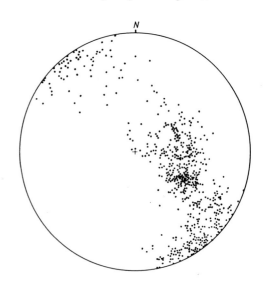

Figure 8-17

Equal-area plot of the deformed lineations of Fig. 8-15.

Figure 8-18

Lineations deformed over the hinge of a fold. Anglesey, North Wales.

Figure 8-19

Same outcrop as Fig. 8-18 viewed so that the deformed lineations are aligned in plane surfaces.

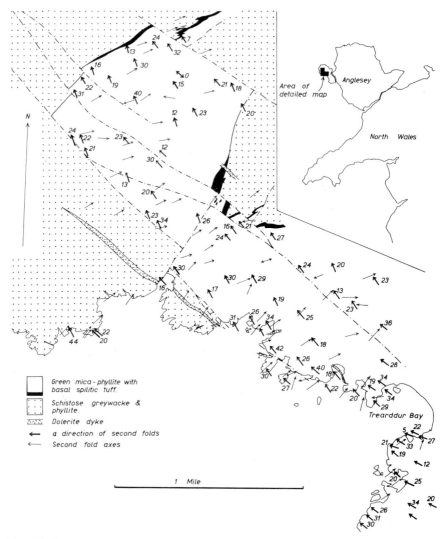

Figure 8-20

Map of a directions in a part of Holy Island, North Wales.

strain accompanying the folding was variable (see below), or, as seems most likely, a combination of these factors.

Studies of deformed linear structures may give information about the fold-forming processes that cannot be obtained from a study of the fold surfaces alone. For example, Clifford (1960) has described a region of multiple folding in the Northern Highlands of Scotland in which two large, steeply plunging

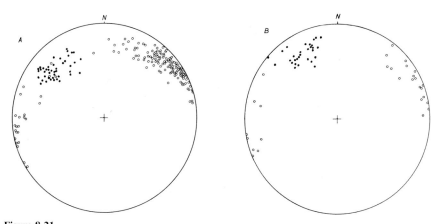

Figure 8-21

Projection of a directions (filled circles) and minor fold axes (open circles) from the area shown in Fig. 8-20. A from the southeast, B from the northwest of the area.

late folds deform early linear structures in the manner illustrated in Fig. 8-22. A plot of these deformed lineations shows that they lie about a plane surface which dips to the south less steeply than do the beds involved in the fold. Using the technique described above to determine the *a* direction of the movement, it is apparent that the *a* direction plunges less steeply southward than do the axes of the late folds, and that as a result the rock material in the synformal structure must have suffered a greater upward translation relative

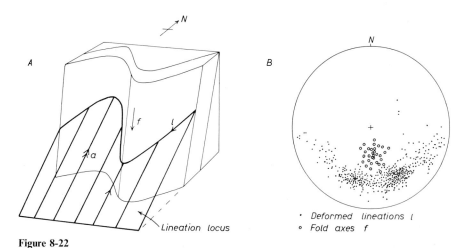

Figure 8-22

Plots of deformed lineations and fold axes from the Loch Luichart region, Northern Highlands of Scotland, and their interpretation. (After Clifford, 1960.)

to that in the adjacent antiform. Generally one tends to assume that antiformal folds coincide with zones where the upward translation of material has been greatest; but if the *a* direction were less steeply inclined than the surface undergoing folding, this would be an incorrect deduction.

Another common feature of deformed linear structures in metamorphic terrains is that they may not be uniformly well preserved in different parts of the later folds. At Monar (Northern Highlands of Scotland) the intensity of development of early linear structures is quite different on the opposite limbs of small-scale folds (Fig. 8-23). The most likely explanation of this feature is that one of the fold limbs suffered greater internal deformation than the other. This explanation is further strengthened by the presence of a much stronger axial plane fabric (schistosity) on the limb where the early linear structures are weakest, and also by the development of sheets of migmatitic granite cutting through this fold limb. It appears that recrystallization accompanying the fold development was concentrated in the zone of maximum deformation. Similar arguments based on the degree of preservation of old linear structures and the degree of development of late structures have been used to compare the relative amounts of deformation in different parts of major folds in other areas of the Scottish Highlands (Ramsay, 1958*b*, p. 512).

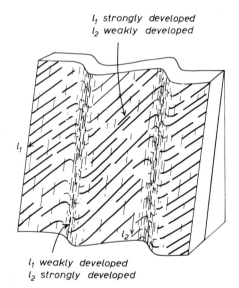

l_1 strongly developed
l_2 weakly developed

l_1

l_2

l_1 weakly developed
l_2 strongly developed

Figure 8-23

Comparative development of two sets of linear structures l_1 and l_2 from Loch Monar, Scotland, to illustrate that the short limbs of the folds have been deformed more than the long limbs. (After Ramsay, 1958a.)

2. Similar folds formed by shear with a component of homogeneous strain

If the shear fold structure described above has a component of homogeneous strain, various geometrical features of the structure are changed but the structure remains a similar fold. Generally it appears that the homogeneous strain components are arranged so that the maximum shortening is perpendicular to the shear plane (axial surface) *ab*, but a more general relationship will be considered below. If the fold is subjected to a homogeneous strain with axes $X > Y > Z$ (Fig. 8-24), the following structural modifications occur:

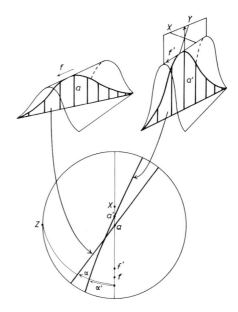

Figure 8-24

Deformed lineations produced by subjecting a shear fold to a homogeneous strain with principal strains $X > Y > Z$.

A. The fold axis and axial surface change position in a way which depends on the orientation and ratios of the principal strains. The fold axis moves toward the X axis but the structure remains cylindrical, while the axial surface comes to lie closer to the XY plane of the strain ellipsoid. If the XY plane coincides with the axial surface, the fold axis changes position according to the relationship $\tan \beta' = (\lambda_2/\lambda_1)^{\frac{1}{2}} \tan \beta$, where β and β' are the angles between the fold axis and the X axis before and after deformation, respectively.

B. The plane locus of deformed lineations changes orientation; but, if the strain is homogeneous, it still remains planar. The intersection of this modified lineation with the axial surface of the fold gives the modified position a' of the transport direction. This means that it is still valid to use the construction described above to determine the movement direction of the flow which produced the folded form of the marker layers. If the XY principal plane coincides with the axial surface, the plane lineation locus moves to a new position so that the angle α (measured between the line of intersection of the lineation locus and YZ plane, and the Y axis) is changed to α' by an amount indicated by $\tan \alpha' = (\lambda_3/\lambda_2)^{\frac{1}{2}} \tan \alpha$.

If the a direction coincides with one of the principal superimposed strain axes, and both are perpendicular to the fold axis, the lineation locus is symmetric about the axial surface of the fold. Generally these conditions do not hold; the lineation locus is then asymmetric and the fold has a triclinic symmetry. With truly similar folds, where the deformed lineation locus is completely represented (from an isoclinal fold), the linear directions on the fold limbs are parallel, and on a plot the locus is *continuous*.

3. Folds of Classes 1C and 3 which closely approximate the similar-fold model (Class 2)   In folds of this type the deformed lineation locus is generally not contained exactly in a plane surface. There are a number of possible explanations for this feature, and the significance of the geometric form of the deformed lineations is best considered together with that of the shape of the folded layers. Some of the commonest deformed lineation loci of this type will now be discussed.

A. FOLDS OF SIMILAR TYPE WITH BUCKLING COMPONENTS   Many folds approximating similar form show a component of buckling. This component can sometimes be detected by careful examination of the profile shape, for the structures generally fall in the types previously described as folds with weakly convergent dip isogons (Class 1C), or they may be made up of complex combinations of classes.

Many fold structures are probably initiated by buckling instability within the competent layers and the shape of the initial buckles becomes modified by the process of flattening and shear parallel to the axial surface. The locus of deformed lineations produced by an initial buckling and a later shear may be rather complex (Ramsay, 1960, p. 92). Consider a lineation with initial orientation l_o positioned on a horizontal plane surface. Fold the plane (flexural slip) about an axis f; the lineation takes up various orientations l' on the fold surface so that the locus is now that of a small circle (Fig. 8-25). If folding now proceeds by shear parallel to the axial plane, the lineations on each part of the fold are deformed so that they move on a great-circle locus connecting l' and the shear direction a, and so come to lie at a new position l''. The resulting locus has a complex form depending on the comparative amounts of folding by the two mechanisms (Fig. 8-25). If the shear component is strongly developed and entirely later than the buckling component, the deformed lineation locus develops a *continuous* geometric form characteristic of true similar folds. This locus should not, however, coincide exactly with that of a plane, although in practice it may lie so close to one that the lineation locus is indistinguishable from that of the perfect similar-fold model.

Probably many natural folds formed by a combination of buckling and shear mechanisms do not develop in such clear-cut stages. However, the lineation pattern is likely to be close to that of the simple two-stage model described above. If buckling and axial plane shear proceeded together (Fig. 8-26), the locus will be *slightly discontinuous*, the amount of discontinuity

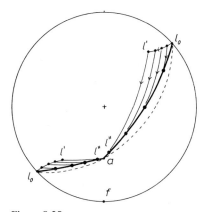

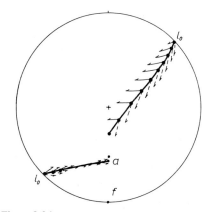

Figure 8-25

Deformed lineation locus formed by buckling (flexural-slip mechanism) followed by simple shear.

Figure 8-26

Deformed lineation locus formed by flexural slip and simple shear proceeding together. During the development of the fold the lineation is deformed partly by rotation about the fold axis (thin lines with arrows), and partly by shear towards the a direction (dotted lines). It therefore adopts a compromise locus between these two trends.

depending on the amount of the buckling component during the fold development.

In many naturally formed folds in metamorphic complexes it appears that similar-type structures formed by translation of passive layers have their geometry modified as environmental conditions change during the process of deformation. If the metamorphic intensity becomes less, the mechanical properties of the layers may no longer be identical, the layers are no longer translated in a homogeneous fashion, and the more competent ones may be buckled. The type of lineation pattern produced by the buckling of previously established similar folds is illustrated in Fig. 8-27. The lineation initially positioned at l_o is deformed by shear to position l' on the fold surface, and these lines are then rotated to new positions l'' about the fold axis such that for any one direction angle $l'f =$ angle $l''f$. The locus of deformed linear structures produced in this way generally has a considerable *discontinuity* across the axial surface of the fold (Fig. 8-27). (See also Ramsay, 1960, pp. 90-91; 1963, p. 159.)

If the lineation loci produced by various combinations of rotation and axial plane shear are compared (Fig. 8-28), it will be seen that the linear locus is generally asymmetric and discontinuous. The amount of discontinuity depends on three main factors: (1) the angle between l_o and the fold axis, (2) the amount of rotation produced by buckling, and (3) the relative proportions of deforma-

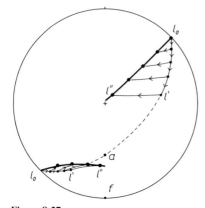

Figure 8-27

Deformed lineation locus formed by heterogeneous simple shear followed by flexural-slip folding.

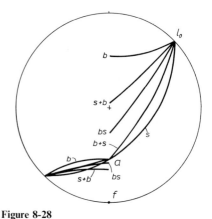

Figure 8-28

Comparison of the deformed lineation loci formed by flexural slip (b), simple shear followed by flexural slip (s + b), shear and flexural slip together (sb), flexural slip followed by shear (b + s) and simple shear alone (s).

tion by rotation and axial plane shear at different stages in the fold development. If the value of the buckling component is identical in two folds, then the lineation locus discontinuity in that fold where buckling developed early during folding (Fig. 8-28, locus $b + s$) will be less marked than that found where the buckling component was most active at a later stage in the folding (Fig. 8-28, locus $s + b$).

Compressive strains acting with the other deformation components will modify these loci; the deformed lineations will approach more closely the XY plane of the superimposed strain and will move toward the X axis (especially if the deformation is of a constriction type, $X \gg Y > Z$). Any of those linear loci which showed a discontinuity in Fig. 8-28 will continue to do so, but the amount of discontinuity will be decreased.

B. FOLDS OF SIMILAR TYPE WITH INHOMOGENEOUS COMPRESSIVE STRAIN
It has been suggested (Sec. 7-14) that the geometry of some Class 3 folds which approximate that of the similar-fold model may have developed by a process of progressive inhomogeneous contraction. This process leads to the development of differential translation of the rock particles and the production of folds in the layers. Deformed linear structures in folds of this type generally show a complex continuous lineation locus which departs from that of the plane locus of true similar folds by an amount which depends on the compressive strain variations and the orientations of the principal strain axes.

For the purpose of analysis, the deformation at any point on the fold shown in Fig. 8-29*A* can be divided into two parts: first, a component of shear parallel

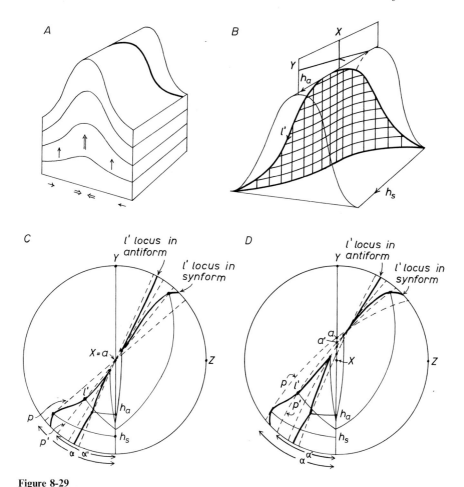

Figure 8-29

Complex lineation loci in folds of Class 3 and Class 1C close to the Class 2 model. A and B, deformed lineation locus in a fold developed by heterogeneous compressive strain. C, the plot of loci where the intermediate strain Y is always perpendicular to the shear axis a, and D is where this does not hold. The lineation locus in the antiform is different from that in the adjacent synform.

to the axial surface of the fold, and second, a strain with principal compression perpendicular to the axial surface. Generally the slip direction *a* of the shearing component will coincide with, or lie close to, the principal elongation X of the compressive component because the two components of deformation act simultaneously and the shear component is, in fact, the result of the inhomogeneity in shortening of the layer.

Figure 8-29B illustrates the geometrical form of the deformed lineation in a fold structure of this type where *a* and X coincide. Because the amount of compressive strain varies through the fold, the orientation of the line of intersection of the shear surface *ab* and the folded layering will vary from place to place. The fold is not cylindrical and the hinge lines of adjacent antiformal and synformal folds, h_a, h_s, will in general be nonparallel (Ramsay, 1962, pp. 325-326). For analytical convenience the effects of the total deformation of the deformed linear structure will be considered in two parts. The shear component deforms the initial lineation l_o to new positions such that they are contained in a plane, locus *p* (Fig. 8-29C). The inhomogeneous compressive strain component modifies the orientation of this plane *p* to various new orientations *p'* such that the angle α between the lineation plane *p* and the *Y* axis before compression is modified to α' depending on the strain ratio $Z/Y = (\lambda_3/\lambda_2)^{\frac{1}{2}}$ such that tan $\alpha' = (\lambda_3/\lambda_2)^{\frac{1}{2}}$ tan α. Where *X* and *a* coincide, these new positions *p'* all contain *a*, and they form a cylindrically folded surface about axis *a*. The deformed lineations after shear take up new positions *l'* in the fold and fall on a continuous locus where the cylindrically folded lineation planes *p'* intersect the noncylindrically folded surfaces of the primary layering. The lineation locus in the antiformal folds does not coincide with that in the synformal folds.

If *X* and *a* are not coincident, then the lineation plane *p* changes position to *p'* (Fig. 8-29D), and the *a* direction changes position to *a'*. This means that *p'* is noncylindrically folded. The final lineation locus will be found at the intersection of the noncylindrically folded surface *p'* with the noncylindrically folded primary layering. If the directions of the strain axes *X*, *Y*, and *Z* are not known, this lineation pattern will be practically indistinguishable from that where *X* and *a* do coincide.

C. VARIATION IN THE *a* DIRECTION THROUGH THE FOLD   If the direction of slip *a* varies through the fold but the slip surface *ab* remains constantly oriented, the deformed linear structures trace out a similar type of complex locus to those described in subsection *B*, above; but, because the orientation of the slip planes and folded layering remains constant, the fold is cylindrical. If the slip directions have different orientations on the two fold limbs, the lineation locus will be discontinuous.

If the orientation of the slip plane *ab* is also variable throughout the structure, the deformed lineation locus will be a discontinuous, complex curve and the folded surfaces will be noncylindrically disposed. If the slip planes converge toward the axial surface when traced toward the concavity of the curved fold surface, the folded structure will be of Class 1*B*; if they diverge, the structure will be Class 3.

8-4   RESTORATION OF LINEAR STRUCTURES TO THEIR UNDEFORMED STATE

The technique that is used for the restoration of deformed lineations positioned in a folded structure depends on the information available on the state of finite strain, or on some knowledge of the mechanism which led to the fold formation.

In Sec. 4-2 it was shown how any state of strain can be analyzed into two parts, and how the finite-strain tensor can be considered as being made up of two components, an irrotational part and a rigid-body rotation. Many reconstructions of the initial orientations of deformed sedimentation structures which have been made in paleocurrent studies assume that only the rigid-body rotation is important and that distortion produced by the irrotational part of the strain tensor can be ignored. This is true for slightly deformed sedimentary strata and for those folded by the mechanism of flexural slip, but for all other fold types this assumption may lead to serious error in the position of the unfolded linear structure.

I. Restoring linear structures deformed only by rotation (tilting or flexural-slip folding)

A. HORIZONTAL FOLD AXIS   Consider a steeply inclined (and inverted) layer within a flexural-slip fold which has a horizontal fold axis f. This stratum has a linear feature on it which is plotted on a projection at l (Fig. 8-30A). To restore this lineation to its unfolded position on a horizontal surface, rotate l along a small circle about the fold axis f to a position l_o on the circumference of the net so that angle $lf =$ angle $l_o f$.

B. INCLINED FOLD AXIS   If the axis of flexural-slip folding is inclined, it is not valid to restore the surface to a horizontal position by direct rotation about the strike of the bed. The axis of rotation to remove the effect of folding is the fold axis f. The restoration proceeds in two parts: first, rotate the fold

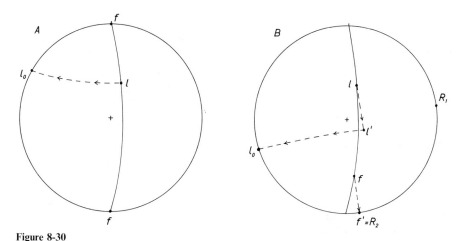

Figure 8-30

Method for removing the effects of deformation of lineation in flexural-slip folds. A, where the fold axis is horizontal; B, where the fold axis is inclined.

axis to a horizontal position, and second, remove the effects of folding. The axis of rotation R_1 about which the fold axis is rotated into a horizontal position is situated on the circumference of the net at an angle of 90° to the plunge *direction* of the actual fold (Fig. 8-30B). The fold axis f is rotated through an angle equal to the angle of fold plunge to the periphery of the net (f'), and the lineation l is moved to a position l' along a small-circle locus about R_1 and also through an angle equal to the plunge of the fold such that angle $lR_1 = $ angle $l'R_1$. The effects of the folding are now removed by rotation about axis $R_2 = f'$ of the surface to a horizontal position and of the contained lineation to a position l_o so that angle $l'R_2 = $ angle l_oR_2. Comparing Fig. 8-30A with 8-30B it will be seen that it is of the utmost importance to establish the correct procedure for unfolding plunging flexural-slip folds, or large errors in the position of l_o will result (see Ten Haaf, 1959; Ramsay, 1961, for a discussion of the amount of angular error produced by the use of incorrect techniques).

II. Restoring linear structures deformed by internal strain and rotation
From a study of the shapes of distorted objects (ooids, fossils, etc.) in the rock material, it may be possible to determine the state of strain in the rock material. In flexural-slip folds the two-dimensional strain ratios within the folded surface are everywhere unity, but in all other types of fold this does not hold. The method of restoring the structure depends on how much is known of the ratios of the principal strains.

A. PRINCIPAL STRAIN RATIOS WITHIN THE SURFACE (X/Y) KNOWN   The restoration technique has three parts: (1) removal of the strain within the folded surface, (2) removal of the fold plunge, and (3) the unfolding of the rotational component of folding. The effects of surface strain on the linear structure are removed utilizing Eq. (3-34) in the graphical form (Fig. 4-6). For example, in Fig. 8-31, if $X/Y = 0.4$, then angle $Xl = 30°$, angle $Xf = 70°$; these are restored to positions l' and f' such that angle $Xl' = 55°$ and angle $Xf' = 82°$. The second part of the construction proceeds in exactly the same manner as *IB* above by making the fold axis f' horizontal (new position f'') and moving the lineation l' to a new position l''; and finally the rotation is removed about f'', with the lineation taking up a final position l_o.

B. PRINCIPAL STRAIN RATIOS OF THE FINITE-STRAIN ELLIPSOID KNOWN WITH THEIR ORIENTATIONS   Again the technique for restoration of the lineation has three stages. The first is the removal of the finite irrotational strain. This is done using the method described in Sec. 4-5. In Fig. 8-32 where $Z/X = 0.4$ and $Y/X = 0.7$, l moves to l' and f to f'. The rotational parts of the strain are now removed, and the lineation is successively transferred to positions l'' and l_o using identical techniques to those of method *IB*. This technique is the most accurate method of general strain and it is considerably more accurate than that of *IIA* used when only a limited amount of information is available.

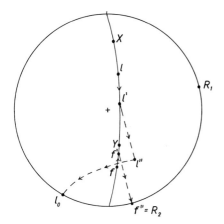

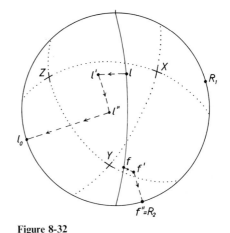

Figure 8-31

Method for removing the effects of deformation on a lineation in a fold where the state of finite strain on the surface is known (principal strains X and Y).

Figure 8-32

Method for removing the effects of deformation on a lineation in a fold where the three-dimensional finite-strain state at that point is known.

Unfortunately, the type of data required is not always available in naturally deformed rocks.

Comparison of the final positions of l_o in Figs. 8-30, 8-31, and 8-32 should make it clear how important it is to use the correct method for removing the effects of folding in order to restore the linear structure to its original orientation.

Often the data available from a field investigation do not enable an accurate restoration of the linear structures to be made. If the folding is not of the parallel type and if nothing is known about the states of strain throughout the fold, it is probably best to attempt to establish the deformation lineation locus throughout the fold. This may be difficult if the fold is large for it may be an unjustifiable assumption to presume that the linear structure was originally rectilinear through the domain now occupied by the fold. If the geometrical form of the deformed lineation locus can be established and the shapes of the layers in a cross section of the fold determined, it may be possible to make some deductions about the most likely mechanisms of formation. It may then be possible to reverse this process to establish the original orientation of the deformed lineations. However, it must be acknowledged that the restoration of lineations using such techniques is likely to give only approximate answers to the problem.

REFERENCES AND FURTHER READING

Best, M. G.: Petrology and Structural Analysis of Metamorphic Rocks in the Southwestern Sierra Nevada Foothills, California, *Univ. Calif. (Berkeley) Publ. Geol. Sci.*, **42**: 111-158 (1963).

Borg, I., and F. J. Turner: Deformation of Yule Marble, Part VI, *Geol. Soc. Am. Bull.*, **64**: 1343-1352 (1953).

Clifford, P.: The Geological Structure of the Loch Luichart Area, Ross-shire, *Quart. J. Geol. Soc.*, **115**: 365-388 (1960).

Clifford, P. et al.: The Development of Lineation in Complex Fold Systems, *Geol. Mag.*, **94**: 1-24 (1957).

Cummins, W. A.: Current Directions from Folded Strata, *Geol. Mag.*, **101**: 169-173 (1964).

Haman, P. J.: Manual of the Stereographic Projection, *West Can. Res. Publ.*, Ser. 1, no. 1, 1961.

Norman, T. N.: Azimuths of Primary Linear Structures in Folded Strata, *Geol. Mag.*, **97**: 338 (1960).

Phillips, F. C.: "The Use of Stereographic Projection in Structural Geology," p. 86, Methuen & Co., Ltd., London, 1954.

Ramsay, J. G.: Superimposed Folding at Loch Monar, *Quart. J. Geol. Soc.*, **113**: 271-307 (1958*a*).

Ramsay, J. G.: Moine-Lewisian Relations at Glenelg, *Quart. J. Geol. Soc.*, **113**: 487-523 (1958*b*).

Ramsay, J. G.: The Deformation of Early Linear Structures in Areas of Repeated Folding, *J. Geol.*, **68**: 75-93 (1960).

Ramsay, J. G.: The Effects of Folding on the Orientation of Sedimentation Structures, *J. Geol.*, **69**: 84-100 (1961).

Ramsay, J. G.: The Geometry and Mechanics of Formation of Similar Type Folds, *J. Geol.*, **70**: 309-327 (1962).

Ramsay, J. G.: Structure and Metamorphism of the Moine and Lewisian Rocks of the North West Caledonides, in "The British Calidonides," M. R. Johnson and F. H. Stewart (eds.) pp. 143-175, Oliver & Boyd Ltd., Edinburgh and London, 1963.

Ramsay, J. G.: Structural Investigations in the Barberton Mountain Land, Eastern Transvaal, *Geol. Soc. S. Africa Trans.*, **66**: 353-401 (1965).

Sander, B.: "Einführung in die Gefügekunde der geologischen Körper," Pt. 1, Springer-Verlag OHG, Berlin, 1948.

Stauffer, M. R.: The Geometry of Conical Folds, *New Zealand J. Geol. Geophys.*, **7**: 340-347 (1964).

Turner, F. J., and L. E. Weiss: "Structural Analysis of Metamorphic Tectonites," McGraw-Hill Book Company, New York, 1963.

Ten Haaf, E.: Graded Bedding of the Northern Apennines, Thesis, Rijksuniversiteit te Groningen, 1959.

Weiss, L. E.: Fabric Analysis of a Triclinic Tectonite and Its Bearing on the Geometry of Flow in Rocks, *Am. J. Sci.*, **253**: 223-236 (1955).

Weiss, L. E., and D. B. McIntyre: Structural Geometry of Dalradian Rocks at Loch Leven, Scottish Highlands, *J. Geol.*, **65**: 575-602 (1957).

Weiss, L. E.: Geometry of Superimposed Folding, *Geol. Soc. Am. Bull.*, **70**: 91-106 (1959*a*).

Weiss, L. E.: Structural Analysis of the Basement System at Turoka, Kenya, *Overseas Geol. Mineral Resources (Gt. Brit.)*, **7**: 3-35 (1959*b*).

Wilson, G.: The Tectonic Significance of Small Scale Structures and Their Importance to the Geologist in the Field, *Ann. Soc. Geol. Belg.*, **84**: 423-548 (1961).

9 ⫴ Folding of obliquely inclined surfaces

THE discussion of the geometrical forms taken up by folded rock layers developed in Chap. 7 considered only those structures formed in rocks containing a single set of initially plane-parallel surfaces. Many of the structural problems encountered in naturally deformed rocks involve the folding together of groups of surfaces which are initially inclined to each other. For example, sedimentary strata often show angular discordances over a wide range of scales, from large unconformities and disconformities between parts of the stratigraphic succession to small sedimentation structures such as cross-bedding or "washout" structures. In rocks which have been folded more than once, the disposition of the layering after the first phase of folding may be highly variable, and the later folds therefore have to form themselves on a whole series of variously inclined layers. If the development of the earlier folds was accompanied by the formation of schistosity or cleavage crosscutting the lithological layers, then the superimposed folds are formed in rocks containing mutually intersecting surfaces.

The aim of this chapter is to examine the geometric properties of rock complexes made up of folded, inclined, discordant, or intersecting surfaces. The geometric theory will be built up in the following way: a layer of material bounded by two initially horizontal plane surfaces (planes A) containing an inclined plane (plane B) will be subjected to folding. The material between the planes A at the layer boundaries will be subjected to deformation according to the various mechanisms discussed in Chap. 7 (flexural slip, shear folding, etc.), a cylindrical fold will be formed, and the behavior of plane B will be computed. This is most conveniently calculated graphically by using a equal-area net; the poles of plane A will trace out a great-circle locus, and then, by allowing the internal deformation to go on in some known manner, the position of the locus of the poles of plane B will be determined (Ramsay, 1963a).

9-1   FLEXURAL-SLIP FOLDING

Consider a plane A horizontal (Fig. 9-1) and a plane B inclined to plane A so that they intersect in a line l_o. When plotted onto an equal-area net, plane A_o (initial position) will occupy the periphery of the net, and its pole falls at the central point (Fig. 9-3, πA_o). The pole of the original position of plane B_o will fall at πB_o. The initial dihedral angle between these two planes is measured by the angle δ_o. Plane A will now be cylindrically folded about an axis f_A, and the internal deformation will be accomplished by flexural slip (simple shear) on plane A (Fig. 9-2).

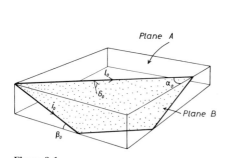

Figure 9-1

Two planes A and B intersecting in direction l_o and with dihedral angle δ_o.

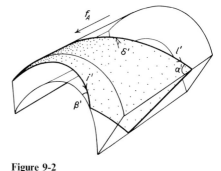

Figure 9-2

Geometry of the fold formed by deforming planes A of Fig. 9-1 by the mechanism of flexural slip about an axis f_A.

Fold plane A about f_A to a new position A' so its angle of dip is θ; its pole takes up a new position $\pi A'$ (Fig. 9-4). The linear intersection of planes A and B initially situated at l_o will change position to l' according to the principles established in Sec. 8-1, so that angle $l_o f_A$ = angle $l' f_A$ = α_o. This line l' must be positioned on the deformed surface of plane B (that is, B'), and if we can determine one other line on this plane it is possible to fix the position of B'.

Consider the profile plane of the fold normal to f_A (Fig. 9-1). Planes A_o and B_o both initially intersect this surface in lines which make an angle β_o on the profile. As a result of folding (Fig. 9-2), the line i_o changes its position to i' at an angle of $\beta' + \theta$ from the horizontal. Because the internal deformation

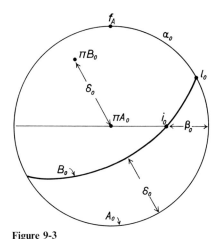

Figure 9-3

Equal-area plot of the geometric features of Fig. 9-1.

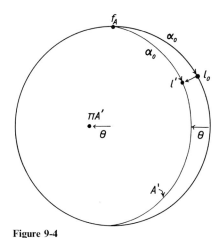

Figure 9-4

Deformation of surface A_o to a new position A (pole $\pi A'$) with angle of dip θ.

is accomplished by simple shear on A', we can compute the value of β' from the relationships expressed in Eqs. (7-16) and (3-71):

$$\cot \beta' = \cot \beta + \theta \qquad (9\text{-}1)$$

where θ is measured in radians. On one side of the fold hinge, β' will exceed β; on the other it will be less than β. Determine the position of i' on the net (Fig. 9-5). Because plane B' contains both l' and i', the great circle passing through these points represents the position of B'. The dihedral angle between A' and B' (δ') can be measured between the poles $\pi A'$ and $\pi B'$. As a result of the internal deformation, δ' generally differs from δ_o.

By repeating this construction with varying values of θ, it is possible to determine completely the spatial locus of πB (Fig. 9-6). In general, this has a rather complex form signifying that surface B is folded into a complex, non-cylindrical surface. The geometrical meaning of this surface can be established by considering the effects of the internal deformation from a different viewpoint. At any point in the fold the deformation is made up of two parts: (1) a simple shear on surface A, and (2) a rigid-body rotation about the fold axis f_A. If the two surfaces A_o and B_o are sheared (Fig. 9-7), then the direction of l_o does not change; plane B' must always contain l_o, and hence it is cylindrically folded about an axis l_o. The actual position of B' can be computed by determining the angle β' from (9-1). The poles of B' (Fig. 9-8, $\pi B'$) fall on part of a great-circle locus with normal l_o. The second component of the deformation is a rigid-body rotation about axis f_A, the poles $\pi B'$ move through the angle of dip θ of the surface A' (Fig. 9-9). Thus the poles of surface B are folded into a complex helicoidal form diagrammatically illustrated in Figs. 9-11 and 9-12.

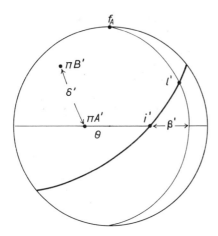

Figure 9-5

Calculation of the position of surface B′ from the position of 1′ and i′. The pole of surface B′ is positioned at πB′, and the dihedral angle between the surfaces A′ and B′ is modified to δ′.

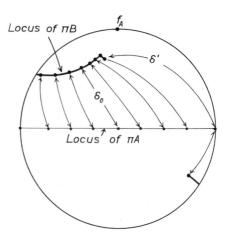

Figure 9-6

Plot of the complete spatial locus of deformed plane B′ and variation in dihedral angle δ.

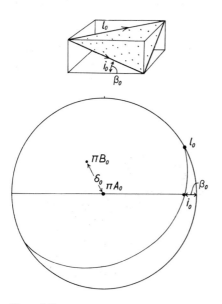

Figure 9-7

Initial orientations of surfaces A_o and B_o (poles πA_o, πB_o) of Fig. 9-1.

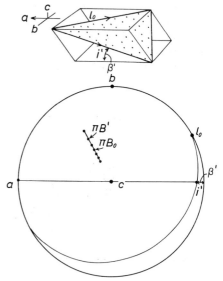

Figure 9-8

Shear component of the flexural-slip deformation. πB_o is deformed to new position $\pi B′$.

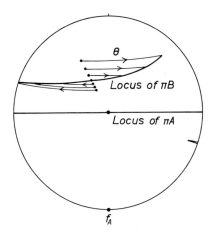

Figure 9-9

Effect of the rotational component (θ) of the folding to establish the complete locus of πB′.

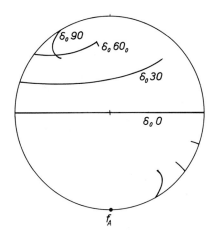

Figure 9-10

Various loci of πB′ with variation in initial dihedral angle δ_o.

Whenever δ' exceeds $90°$ the resulting locus of πB shows a cusplike form. Figure 9-10 illustrates a number of different loci of πB developed by flexural slip of two surfaces, all of which have the same angle $\alpha_o (l_o f_A = 60°)$ but variable initial dihedral angles δ_o.

Two special initial arrangements of the planes A and B give rise to rather simple geometrical forms. The first (Fig. 9-13) is where the linear intersection of A_o and B_o is parallel to the axis of folding f_A ($\alpha_o = 0$). The complex folded

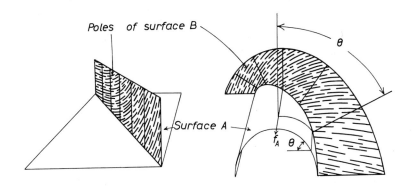

Figure 9-11

Three-dimensional form of the poles of surface B (πB′) of Fig. 9-8.

Figure 9-12

Completed form of the poles of surface B (πB′) of Fig. 9-9 after rotation through the angle of dip θ of surface A_1.

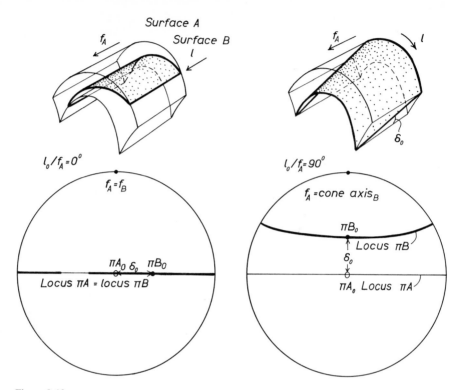

Figure 9-13

Special cylindrical locus of surface B where angle $l_o f_A = 0°$.

Figure 9-14

Special conic locus of surface πB where angle $l_o f_A = 90°$.

surface of B degenerates into part of a simple cylinder that is coaxial with f_A. The second special arrangement (Fig. 9-14) is where the linear intersection of A_o and B_o is perpendicular to f_A ($\alpha_o = 90°$). With this arrangement the dihedral angle δ' remains constant at all points in the fold, and therefore surface B takes up a simple conical form, with the cone axis parallel to f_A.

9-2   VARIATIONS IN DIHEDRAL ANGLE PRODUCED BY FLEXURAL SLIP

As a result of the internal deformation, the original angle δ_o is modified except in the special instance mentioned above where $\alpha_o = 90°$. These modified dihedral angles δ' have been computed for a number of folds with various initial values of δ_o and α_o (angle $l_o f_A$) and are expressed graphically in Fig. 9-18. On one fold limb the dihedral angle is always decreased, and on the

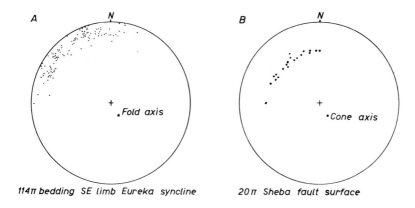

114π bedding SE limb Eureka syncline   20π Sheba fault surface

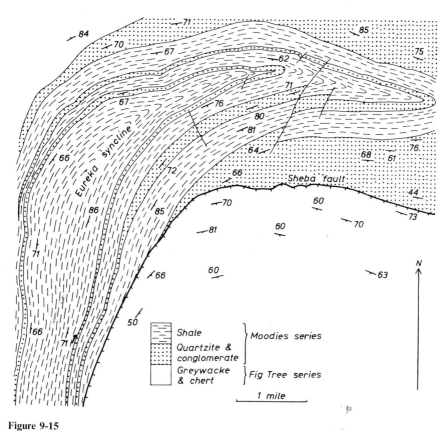

Figure 9-15

Deformation of two inclined surfaces, a fold limb and a fault surface, by flexural-slip folding. Bar-berton region, East Transvaal, South Africa. (After Ramsay, 1965.)

other it is increased. The amount of these changes depends on the value of α_o, and the changes become most strongly marked as α_o approaches a zero value and become least marked as α_o approaches 90°.

Deformation of this type has been suggested to explain the phenomenon of oversteepened cross-lamination in false-bedded sediments. It is generally accepted that the original angle of repose of sands in cross-laminations cannot exceed 35°, yet in rocks which have been folded the foreset beds may be inclined at much higher angles to the truncating surfaces (Brett, 1955; Pettijohn, 1957; Ramsay, 1961). Brett and Pettijohn have shown how the mean inclinations of cross-laminations in folded pre-Cambrian quartzites south of Lake Superior (Canada) show systematic variations that appear to accord with the flexural-slip mechanism of deformation. Brett found that the mean inclination in the northern limb of the Baraboo syncline was 20°, whereas on the southern limb it was 24.5°; and Pettijohn's investigations in the more tightly folded Lacloche syncline showed means of 17.9° and 32.0° on the northern and southern limbs, respectively.

Figure 9-15 is an example which illustrates the behavior of inclined surfaces folded together by flexural-slip folding. The region is a part of the Barberton Mountain Land of the East Transvaal (South Africa) and is made up of extensively folded and faulted Archaen sediments. Massive quartzites and incompetent shales have been folded into a great arcuate syncline, the Eureka syncline. This structure has been refolded so that its axial surface dips steeply to the south, southeast, or east. The synclinal hinge plunges WSW, SSW, or SSE, respectively, on this curving surface. The beds on the southeastern limb of the Eureka syncline are cut across by a steeply inclined reversed fault, the Sheba fault. This fault surface has been folded by the same deformation which refolded the Eureka syncline, and it now dips to the southeast, south, or east. An analysis of the inclinations of the massive quartzite horizons situated on the southeastern limb of the Eureka syncline shows that these beds have been cylindrically refolded about an axis which plunges steeply to the southeast (Fig. 9-15A). The poles to the curving surface of the Sheba fault do not, however, fall on a great circle. They appear to lie on a small circle which represents a cone of apical angle 70°, and a cone axis coinciding exactly with the cylindrical axis of refolding of the southeastern limbs of the Eureka syncline (Fig. 9-15B). This geometric form appears to accord with that established in Fig. 9-14 and has resulted from flexural slip of the lithological layering of the southeastern limb of the Eureka syncline (Ramsay, 1965).

9-3   FLATTENED FLEXURAL-SLIP FOLDS

In Sec. 9-1 it was shown how, if plane A was folded by flexural slip, the form of plane B could be determined. It is now a simple matter to determine how these surfaces change their orientations as a result of a superimposed homogeneous (or inhomogeneous) strain through the structure by using the projection techniques described in Sec. 4-5.

It is not feasible here to examine the effects of all the various possible orientations and values of superimposed strain on the flexural-slip fold, but as an illustration of the methods of determining these modifications we shall examine the effect of superimposing a homogeneous strain with ratios $X:Y:Z = 4:2:1$ on a flexural fold. The XY plane of the superimposed strain will be arranged so that it coincides with the axial surface of the fold of surface A, and the Y direction coincides with f_A (Fig. 9-16B and C).

In Fig. 9-17, the poles of plane B' ($\pi B'$), after flexure folding, have been moved according to the geometric conditions of homogeneous strain of the structure. They take up new positions $\pi B''$ on a new locus. One particularly interesting feature of this new arrangement is that it leads to considerable changes in the dihedral angle between planes A and B through the fold. These have been graphically recorded for flattened modifications of all the flexural-slip folds discussed previously (Fig. 9-19). The variations in dihedral angle are rather unlike those of the simple flexural-slip model (cf. Figs. 9-18, 9-19), and some show an approach to the type of variation that is seen in similar folds and which will be discussed below. In general, there is a much greater variation in dihedral angle in a flattened flexural-slip fold than in the simple flexural-slip fold. This variation is greatest where α_o (angle $l_o f_A$) is small,

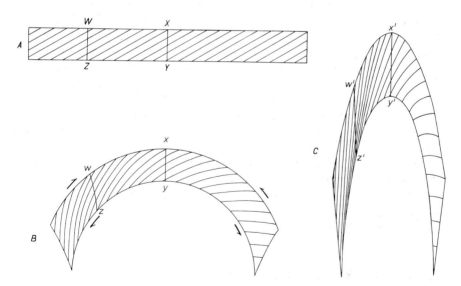

Figure 9-16

Modification of two inclined surfaces. A, undeformed condition; B, deformed by flexural slip; C, effect of an additional superimposed homogeneous strain. (After Ramsay, 1961.)

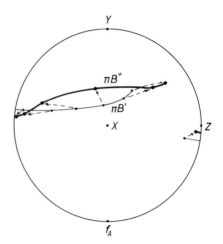

Figure 9-17

Modification of the locus of the poles of surface B after flexure folding ($\pi B'$) by a superimposed homogeneous strain ($\pi B''$) with principal strains X, Y, and Z.

while on the limbs of tight flattened flexural-slip folds the angle δ is either considerably reduced or oversteepened, both effects leading to the blurring of the original angular discordance. However, the dihedral angle always has a finite value even in isoclinal flattened flexural-slip folds and if, for example, the layers were originally cross-bedded, it should always be possible to detect the cross-bedding on the fold limbs. Another very interesting feature in folds where the original dihedral angle between planes A and B was less than about 40° is the general development of a maximum value in the measured dihedral angle situated just to one side of the hinge zone of the structure. In this position, original angular discordances are actually made clearer by the deformation brought about with the folding, and it is here that the field geologist most profitably searches for cross-bedding, unconformities, etc., in the structure (Fig. 9-16C).

9-4   BUCKLE FOLDS WITH INTERNAL DEFORMATION BY TANGENTIAL LONGITUDINAL STRAIN

The geometric form assumed by surface B, if surface A is folded cylindrically with internal deformation only by tangential longitudinal strain, is most readily determined in two stages. Each position in the fold will undergo (1) a homogeneous strain with principal axes parallel and perpendicular to the layering of surface A, and (2) a rigid-body rotation about the fold axis f_A. The nature of the folded form of surface B depends on the amount and variation in the amount of tangential longitudinal strain around the fold, and this depends on the curvature of the finite neutral surface and the distance of the layer from this surface (Fig. 9-20). There is therefore no fixed solution to the problem, for these variables depend on the shape of the fold and the

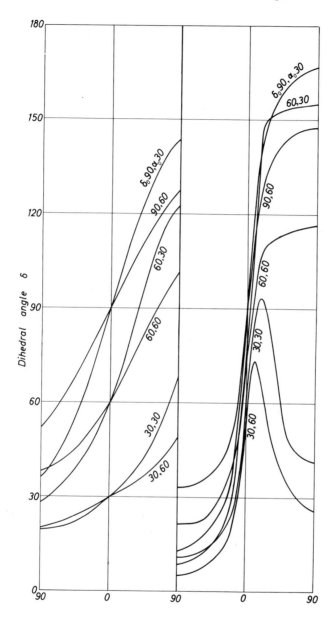

Figure 9-18

Modification of dihedral angle δ between two inclined surfaces as a result of flexural-slip folding. Abscissa records the dip of surface A.

Figure 9-19

Modification of dihedral angle δ between two inclined surfaces as a result of flexural-slip and a homogeneous strain.

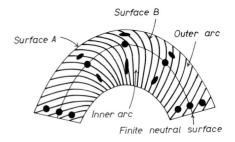

Figure 9-20

*Distortion of surface B when surface A is
buckled and the strains are accommodated
by tangential longitudinal strain.*

thickness of the layer; however, we can determine the form of the general
solution.

On the finite neutral surface the tangential longitudinal strains are zero,
and so the layers here will suffer only a body rotation. The poles of plane B
will therefore fall on a small-circle locus (Fig. 9-21).

On the concave side of the finite neutral surface, tangential compressive
strains will attain a maximum value near the fold hinge and decrease to a
zero value on the fold limb where the curvature is zero. We can therefore
determine the type of locus of the poles of surface B by taking the original
pole (Fig. 9-22, πB_o) and then (1) superimposing a variable compressive plane
strain on this to determine the change of position ($\pi B'$), and finally (2) rotating
these poles about the fold axis f_A through the angle of dip of plane A (θ)
to a new position $\pi B''$ (Fig. 9-22). Similarly, on the convex side of the finite
neutral surface, greatest tangential extensive strain occurs at the fold hinge,
and a similar construction allows the position of the poles of πB_o after straining

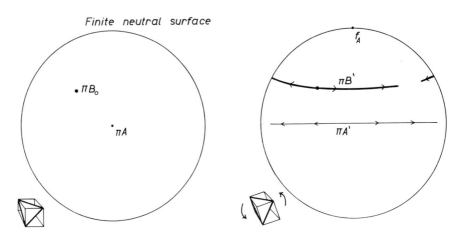

Figure 9-21

Distortion of πB_o on the neutral surface of the fold in Fig. 9-20.

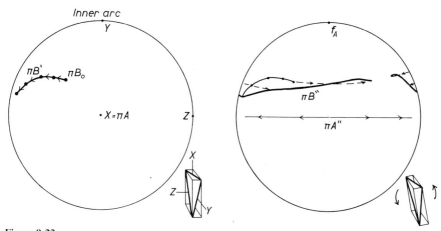

Figure 9-22

Distortion of πB_o on the inner arc of the fold in Fig. 9-20.

($\pi B'$) to be computed and then these poles are rotated to new positions in the fold ($\pi B''$) (Fig. 9-23).

The form taken up by surface B varies at different levels in the structure. If the initial intersection of planes A_o and B_o was parallel to the fold axis f_A ($\alpha_o = 0$), then B is cylindrically folded about f_A; but for all other relationships the surface form of B is complex and noncylindrical.

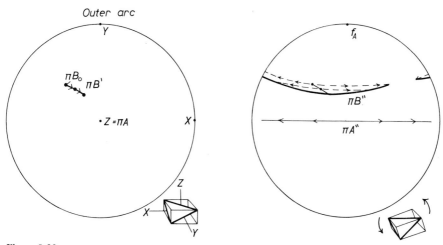

Figure 9-23

Distortion of πB_o on outer fold arc in Fig. 9-20, first to $\pi B'$ by strain components X, Y, and Z and then to $\pi B''$ by rotation about the fold axis f_A.

The dihedral angles between surfaces A and B vary throughout the fold. On the neutral surface they are unchanged through the fold. On the inner (concave) side of the neutral surface they increase toward the hinge zone and approach $90°$; whereas on the outer (convex) side the dihedral angles are always reduced, reaching a minimum value at the fold hinge (Fig. 9-20).

9-5 SIMILAR FOLDS FORMED BY SHEAR

Consider a layer of rock bounded by two parallel-plane surfaces (Fig. 9-24, plane A) containing another plane (B) inclined obliquely to A and intersecting in a line l_o (Fig. 9-26). A similar fold can be made from this layer by heterogeneous simple shear according to the method of shear fold formation described in Sec. 7-11 (Fig. 9-25). The slip direction a is arranged normal to surface A which is cylindrically folded about an axis f_A and this coincides with b. We now wish to investigate the changes in orientation undergone by surface B during the production of this fold.

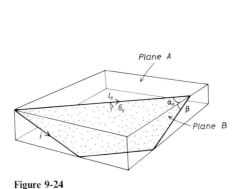

Figure 9-24

Undeformed orientation of surfaces A and B.

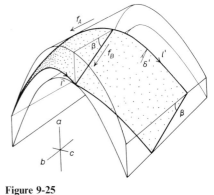

Figure 9-25

Surfaces A and B of Fig. 9-24 deformed by heterogeneous simple shear.

Consider a surface A' in this fold with its pole positioned at $\pi A'$ (Fig. 9-27). The original intersection of planes A_o and B_o (l_o) is deformed by the folding to a position l' on A'. The angle α_o (angle $f_A l_o$) is modified to α' according to the principles established in Sec. 8-3 so that the lineation lies on a plane containing l_o and the slip direction a. The line l' must lie on the deformed surface B'; if we can determine one other line on this surface we can exactly locate its position in space. This could be accomplished in a similar way to that undertaken for flexural-slip folding, by determining the displacement of i to i' (Figs. 9-24, 9-25), but one simplifying feature of the geometry of the shear fold makes this unnecessary. It will be seen in Figs. 9-24 and 9-25 that the

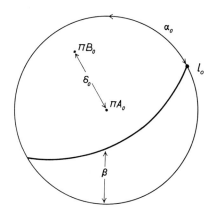

Figure 9-26

Angular relationships of Fig. 9-24.

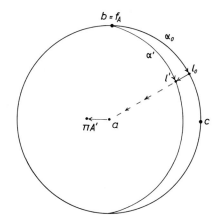

Figure 9-27

Deformation of surface A so that its pole is moved to πA'. Lineation l_o is deformed to l' on this surface.

angle between the lines of intersection of planes A and B with the shear plane ab (β) is fixed at the start of the deformation and remains unaltered throughout the deformation. This implies that, as ab and $f_A = b$ do not change orientation throughout the fold formation, the fixed point f_B must always be located on surface B throughout the deformation. The only way

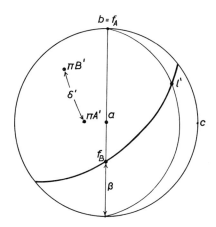

Figure 9-28

Calculation of the position of pole to surface B ($\pi B'$). The dihedral angle between $\pi A'$ and B' is δ'.

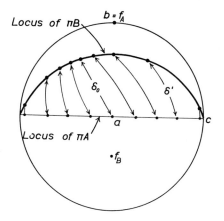

Figure 9-29

Complete locus of poles of surface B when surface A is folded by heterogeneous simple shear.

this can happen is for surface *B* to be folded cylindrically about the axis f_B, and this greatly simplifies the construction to find the position of *B'*. The great circle passing through *l'* and f_B represents surface *B'*, with the pole situated at $\pi B'$ (Fig. 9-28). By repeating this construction throughout the fold, the complete great-circle locus of $\pi B'$ can be computed; and by measuring the angular distance between $\pi A'$ and $\pi B'$, it is possible to determine the variation in dihedral angle δ through the fold (Fig. 9-29).

The most important geometrical rule we have established is that when discordant surfaces are subjected to shear folding each surface takes up a cylindrical fold form about an axis which is positioned at the intersection of the surface with the shear plane *ab*.

9-6   VARIATION IN DIHEDRAL ANGLE IN SHEAR FOLDS

The variations in dihedral angle δ have been calculated for a number of shear folds with different initial values of δ_o and α_o, and these are graphically recorded in Fig. 9-30. The pattern of variation of δ is the same for all shear folds except those where one of the initial plane surfaces contains the slip direction *a*. The characteristic features of this dominant pattern are:

1. On one of the fold limbs the dihedral angle is everywhere decreased; if $\alpha_o = 90°$, it is decreased throughout the fold.
2. On the other fold limb, δ reaches a maximum value just to one side of the fold hinge (providing $\alpha_o \neq 90°$) and then decreases with increasing inclination of the surface *A*.
3. On both fold limbs, δ approaches a zero value as the angle between the fold limbs and the axial surface tends toward zero.

If one of the surfaces contains the slip direction *a*, then the dihedral angle varies in a slightly different manner: on one limb it progressively increases, on the other it progressively decreases.

The variations in dihedral angle are greatest where the initial angle α_o between l_o and $f_A = b$ has a small value.

9-7   SIMILAR FOLDS FORMED BY SHEAR
AND A HOMOGENEOUS STRAIN

In Sec. 7-12 it was shown how the shape of a shear fold was modified by the superposition of a homogeneous strain but that the fold still retained its similar form (Class 2). If a homogeneous compressive strain is superimposed on the shear fold illustrated in Fig. 9-32B, the geometric shapes of the sets of obliquely inclined surfaces are modified (Fig. 9-32C), but both retain their similar form. These changes in geometry can be investigated in the same way

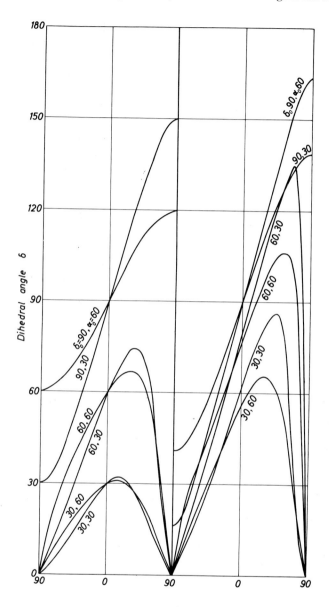

Figure 9-30

Modification of dihedral angle δ between two inclined surfaces as a result of similar folding by heterogeneous simple shear. Abscissa records the dip of surface A.

Figure 9-31

Modification of dihedral angle δ between two inclined surfaces as a result of similar folding by heterogeneous simple shear and a homogeneous strain.

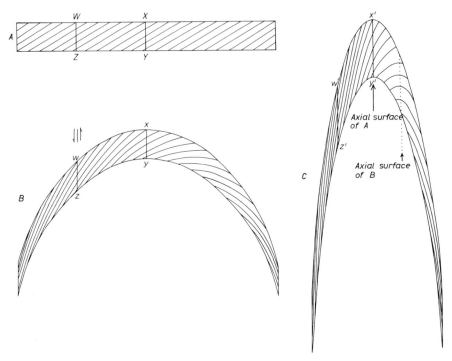

Figure 9-32

Modification of two inclined surfaces. A, undeformed condition; B, deformed by heterogeneous simple shear; C, effect of a superimposed homogeneous strain on B. (After Ramsay, 1961. Cf. Fig. 9-16.)

as was done in Sec. 9-3 for flattened flexural-slip folds. As an example, consider the superposition of a homogeneous strain on the fold illustrated in Fig. 9-29 such that the superimposed strains have ratios $X:Y:Z = 4:2:1$. The XY principal plane coincides with the shear plane ab and Y coincides with the fold axis on surface A, f_A (Fig. 9-33). The fold axis of surface B changes to f_B' and the poles of surface B move onto a new great circle $(\pi B'')$. The extent of these changes has been computed using the methods set out in Sec. 4-5. As a result of this additional deformation, the dihedral angles δ between surfaces A and B through the fold suffer further modification. These changes have been computed for various initial values of α_o and δ_o and are graphically recorded in Fig. 9-31. The general form of these variations is very similar to that established for shear folding alone, but the actual values of variations in δ are much greater and increases occur in the vicinity of the hinge zone on *both* sides of the hinge line.

In Sec. 7-11 (and Fig. 7-94) it was shown that the position of the axial surface in a similar fold depends on the angle between the surfaces being

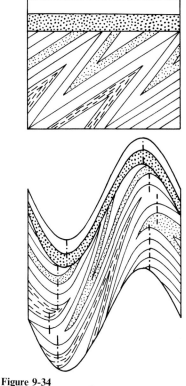

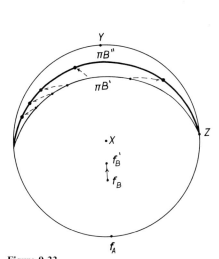

Figure 9-33

Effect of a homogeneous strain on the geometric features of Fig. 9-29. The fold axis of surface B′ is modified to position f′$_B$.

Figure 9-34

Deformation of pre-existing folds and an angular unconformity by similar folding. The axial surfaces (dot and dash lines) of the new folds are arranged in an en echelon manner across the structures.

folded and the shear plane *ab*. It therefore follows that where the folded surfaces initially had a variable inclination the axial surfaces of the folds will not coincide. This is a particularly important phenomenon in regions of superimposed folding where the axial surfaces of the later folds are often arranged in an "en echelon" manner across the previously formed fold structures (Fig. 9-34) (O'Driscoll, 1964). If the two sets of surfaces (e.g., cleavage and bedding surfaces) intersect one another, then each surface develops its own axial surface (Fig. 9-32*B* and *C*, Ramsay, 1961).

Geological examples of these phenomena will now be described as illustrations of the practical importance of the various theoretical principles that have been established. The first example of similarly folded discordant surfaces (Fig. 9-35*A*) comes from the Glenelg region situated on the western side of the Caledonian mountain chain of the Scottish Highlands. (Clifford et al.,

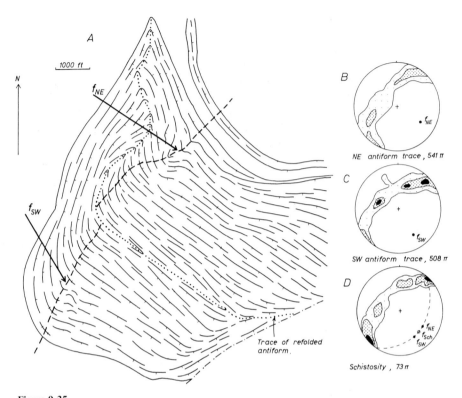

Figure 9-35

A map of a refolded antiformal fold in the Glenelg region, Northwest Scottish Highlands. B and C show plots of bedding poles from the refolded structure, and D shows plots of the refolded early schistosity. Each set of surfaces is cylindrically folded about different axes f_{NE}, F_{SW}, and f_{sch}.

1957; Ramsay, 1958*b*). In this region the cover metasediments (Moine series) form the core of a large, tightly folded antiform and are surrounded by gneisses of the basement (Lewisian complex). This antiformal structure is itself a superimposed fold which was developed in rocks which had previously been overturned. A well-developed schistosity runs parallel to the axial surface of the fold throughout all the structure. The two limbs of this antiform and the crosscutting schistosity have all suffered refolding by a reclined fold whose axial surface dips toward the southeast. The axial traces of this later fold are arranged in an en echelon manner on the two refolded antiform limbs in the manner shown in Fig. 9-34. This latest folding closely approximates the similar-fold model. It has folded three initially inclined sets of surfaces, namely, the two fold limbs and schistosity of the earlier antiform. Analysis of the data from refolded surfaces (Fig. 9-35*B*, *C*, *D*) shows very clearly the following features:

1. Each surface is refolded cylindrically.

2. The axis of each cylindrical fold lies on planes parallel to the axial surfaces of the common folds.

3. The axis of refolding of the northeastern limb of the antiform has a more easterly component of plunge direction than the axis of refolding of the southwestern limb of this fold. Thus the divergent plunge directions of these new folds clearly reflect the initial divergent dips of the antiformal fold limbs.

4. The axis of refolding of the schistosity surfaces lies midway between the axes of refolding on the antiformal limbs.

These features of cylindrical refolding accord with the principles established from a theoretical viewpoint. They have been established in many regions where superimposed folding is developed (Weiss, 1955; Clifford and others, 1956; Weiss and McIntyre, 1957; Ramsay, 1958a; Turner and Weiss, 1963).

The changes in dihedral angle between folded inclined surfaces can be studied in regions of superimposed folding. Figure 9-36 illustrates one such example; the dihedral angle between the limbs of the first fold varies through

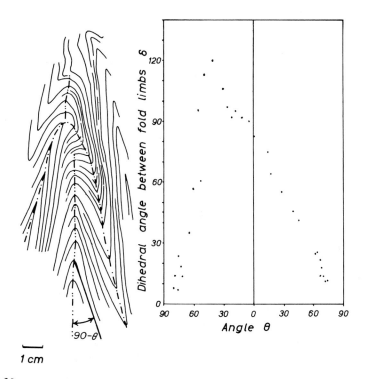

Figure 9-36

Plots of the dihedral angle between the limbs of a refolded fold. Holy Island, North Wales.

the structure depending on the position within the superimposed structure. Maximum values occur around the hinge zone of the late fold, and the high values are asymmetrically disposed in the structure in a similar way to that shown in the theoretical examples graphed in Figs. 9-30 and 9-31.

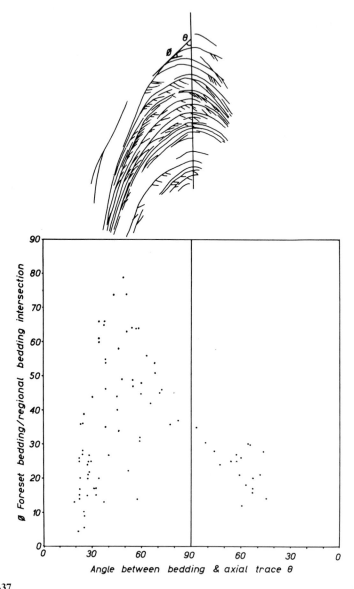

Figure 9-37

Plots of the angles between the truncated foreset beds and truncating surfaces in cross-bedded Moine sediments. Northern Highlands of Scotland. (Data collected by J. Soper.)

Similarly folded cross-bedded sediments often show variations in the inclinations of the foreset beds in different parts of the fold (Ramsay, 1961). Figure 9-37 is drawn from unpublished work of Dr. J. Soper and shows the exposed part of a profile section of a small fold in cross-bedded sandstones from the Moine series of Scotland. The angle of inclination of the cross laminations on this section has been measured (ϕ) and this has been plotted against the angle between the truncating surface (regional bedding) and axial surface of the fold (θ). The systematic, asymmetric variations in ϕ through this fold clearly produced by deformation and the plots accord well with the theoretical curves. Some of the variations in ϕ must be attributed to initial variation in the dihedral angle and in the orientation of the cross-bedding laminae.

In tight similar folds the dihedral angle between oblique surfaces is severely reduced, and in truly isoclinal folds no angular discordance can be detected (Figs. 9-30, 9-31). These geometrical facts appear to account for the progressive obliteration of angular discordances within the central part of the fold belts where similar folds (and flattened flexural-slip folds) are so common. As an example of this phenomenon three areas from the foreland, marginal zone, and strongly deformed internal zone of the Caledonian mountain chain of Scotland will be contrasted (Figs. 9-38 through 9-41) (Ramsay, 1963b).

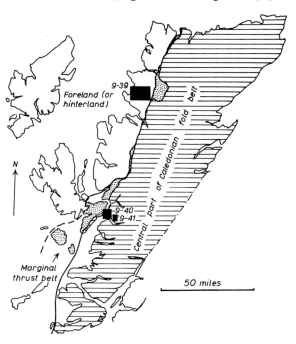

Figure 9-38

Foreland, marginal, and central parts of the Caledonian fold belt of the Northwest Highlands of Scotland showing the location of the areas illustrated in Figs. 9-39, 9-40, and 9-41.

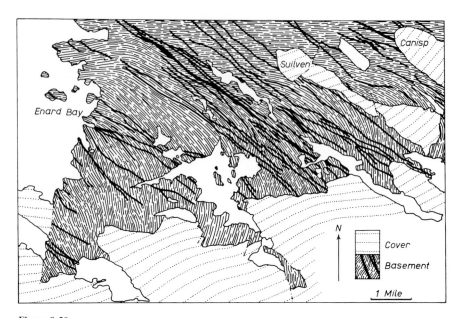

Figure 9-39

The foreland (or more probably hinterland) region of the Caledonian belt. Gently inclined Tor-ridonian sediments rest unconformably on the gneissic Lewisian basement containing crosscutting basic dikes.

In the foreland region (Fig. 9-39) the cover rocks consist of horizontal or very gently inclined, strongly cross-bedded feldspathic sandstone (Torridon-ian) resting with a marked angular unconformity on the underlying basement gneisses and schists and crosscutting dikes (Lewisian complex). Four marked angular discordances exist here:

1. The crosscutting dikes in the basement (δ up to $90°$)
2. The angular unconformity between gneissose layering in the basement and cover sediments (δ up to $90°$)
3. The angular uncomformity between dikes in the basement and the cover sediments (δ up to $90°$)
4. The cross laminations in the Torridonian sediments (δ up to $33°$)

In the marginal zone of the deformed mountain belt (Fig. 9-40), it is possible to recognize each of these angular discordances, but the dihedral angles have all been reduced. Thus in the region just southeast of the Sound of Sleat the dikes crosscut the basement gneisses at an angle of not more than $20°$, the cover rocks (here metamorphosed to garnet grade and known as the Moine series) show an angular unconformity with the (remetamorphosed) basement with $\delta < 15°$, while the current bedding inclination never exceeds $10°$. Farther

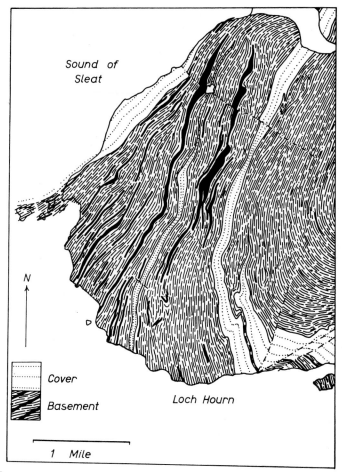

Figure 9-40

Relationships of Lewisian basement and cover (Moine series, metamorphic equivalent of a Torridon-ian sediment) in the west-central part of the mountain belt north of Loch Hourn. Unconformable basement-cover relations still exist, and the basic dikes crosscut the composition layering in the Lewisian gneisses.

east in regions where the basement and cover have been severely interfolded, the discordant features previously recognizable become progressively more difficult to find (Fig. 9-41). Although the basement and cover rock types are still distinguishable, it is only by tracing individual bands within the Lewisian complex for a considerable distance along the strike that it becomes apparent that there is some slight discontinuity between the two series of rocks. Similar obliterations of angular unconformities are known in many mountain chains (e.g., Heim, 1921, in the Swiss Alps; Reece, 1960, in pre-Cambrian orogenic

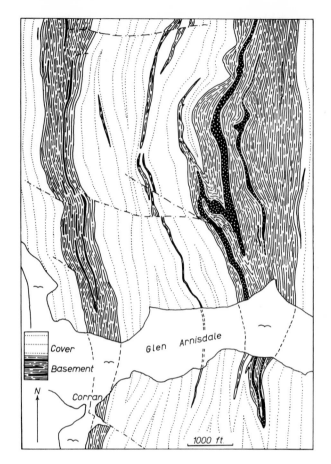

Figure 9-41

Basement-cover relationships in the Arnisdale region. Discordant relationships have been almost completely effaced.

belts in Uganda). This process does not require any special deformation along the uncomformable contacts; it is the inevitable result of internal deformation of the rocks by shear and flattening.

Although discordances may be difficult to detect with certainty on the limbs of similar folds, there is one position in the structure just to one side of the hinge line where careful observation may lead to an identification of the original discordance. It is here that the dihedral angles between the surfaces may be increased and the discordance be more marked than in the undeformed original.

REFERENCES AND SUGGESTED FURTHER READING

Brett, G. W.: Cross Bedding in the Baraboo Quartzite of Wisconsin, *J. Geol.*, **63**: 143-148 (1955).

Clifford, P. et al.: The Development of Lineation in Complex Fold Systems, *Geol. Mag.*, **94**: 1-23 (1957).

Heim, A.: "Geologie der Schweiz," 2, Tauchnitz, Leipzig, 1921.

O'Driscoll, E. S.: Cross Fold Deformation by Simple Shear, *Econ. Geol.*, **59**: 1061-1093 (1964).

Pettijohn, F. J.: Palaeocurrents of Lake Superior Pre-Cambrian Quartzites, *Geol. Soc. Am. Bull.*, **68**: 469-480 (1957).

Ramsay, J. G.: Superimposed Folding at Loch Monar, *Quart. J. Geol. Soc.*, **113**: 271-307 (1958*a*).

Ramsay, J. G.: Moine-Lewisian Relations at Glenelg, *Quart. J. Geol. Soc.*, **113**: 487-523 (1958*b*).

Ramsay, J. G.: The Effects of Folding on the Orientation of Sedimentation Structures, *J. Geol.*, **69**: 84-100 (1961).

Ramsay, J. G.: The Folding of Angular Unconformable Sequences, *J. Geol.*, **71**: 397-400 (1963*a*).

Ramsay, J. G.: Structure and Metamorphism of the Moine and Lewisian Rocks of the North West Caledonides, in "The British Caledonides," Oliver & Boyd Ltd., Edinburgh and London, pp. 143-175 (1963*b*).

Ramsay, J. G.: Structural Investigations in the Barberton Mountain Land, Eastern Transvaal, *Trans. Geol. Soc. S. Africa*, **66**: 353-401 (1965).

Reece, A.: The Stratigraphy, Structure and Metamorphism of the Pre-Cambrian Rocks of NW Ankole, Uganda, *Quart. J. Geol. Soc.*, **115**: 389-416 (1960).

Turner, F. J., and L. E. Weiss: "Structural Analysis of Metamorphic Tectonites," McGraw-Hill Book Company, New York, 1963.

Weiss, L. E.: Fabric Analysis of a Triclinic Tectonite, *Am. J. Sci.*, **253**: 225-236 (1955).

Weiss, L. E., and D. B. McIntyre: Structural Geology of Dalradian Rocks at Loch Leven, *J. Geol.*, **65**: 575-602 (1957).

10 Superimposed folding

WHEN layered materials are subjected to an inhomogeneous strain, folds are developed in the layers. If the nature and orientation of the strain increments change during the history of the deformation, any previously folded rock layers may be subjected to refolding about different axial directions, and the axial surfaces of the original folds may take up some folded form. The finite strain produced by this process is markedly heterogeneous and the initial layering, therefore, takes up a complex three-dimensional shape. The aim of this chapter is to discuss the interpretation and significance of the geometric form of these complex compounded structures.

10-1 ENVIRONMENTS OF SUPERIMPOSED FOLDING

Superimposed folding is common in a number of geological environments; it may result from a number of different types of deformation processes.

1. Crossing orogenic belts Superimposed folds may develop where layered rocks are subjected to successive deformations separated by a very large time interval, as, for example, when a folded basement complex is subjected to a later orogenic deformation. Oulianoff (1953) has described examples of structures generated in this way in the Swiss and French Alps, where a folded basement made up of pre-Alpine structural components was subjected to a further deformation during

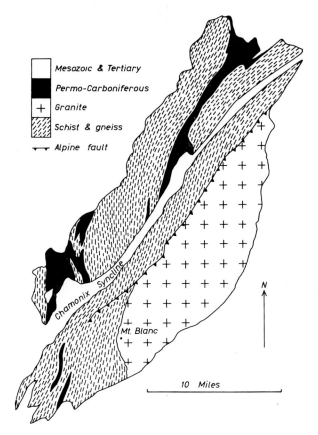

Figure 10-1

Geological map of the Aiguilles Rouges and Mont Blanc massifs (Switzerland and France) illustrating how the trends of the Alpine fold structures (e.g., Chamonix syncline) have been superimposed obliquely across the folded structures in the pre-Triassic basement.

the Alpine orogenic movements (Fig. 10-1). Under these circumstances the strain plans of the successive foldings are completely unrelated.

2. Successive deformation phases in one orogenic cycle This very common phenomenon probably accounts for the most frequently developed examples of superimposed folding. The total deformation within an orogenic cycle is generally built up of a series of separate pulsatory, compressive deformations. Under these circumstances the successive folds that may be produced often have very different styles depending on the strain rates and physical states of the rock during deformation; but there may be some common geometric feature of the various sets of folds. For example, in the Caledonian orogenic belt of the Scottish Highlands the folds show a great range of styles and axial orientations but their axial surfaces generally have a regional trend between NNE and ENE (Sutton, 1960*a*, 1960*b*; Rast, 1963; Ramsay, 1963).

3. Successive folding during a single progressive deformation   During the progress of a single, continuous deformation it is general for there to be a change in the orientation of the layering relative to the axes of the incremental · strain ellipsoid. This change leads to the oblique straining of any originally formed folds and perhaps to the development of superimposed fold structures with new axial directions and axial surfaces (Sec. 4-18).

4. Simultaneous folding in several directions during one deformation   If a competent layer is subjected to a constrictive type of two-dimensional strain as a result of a constrictive type of three-dimensional strain, it is possible for folding to be developed in several directions simultaneously (Secs. 3-17, 4-9).

The determination of the timing relationships of the development of folds formed in several directions in the same rock generally has to be accomplished by combining geometric criteria with other geological evidence. Synchronous folding may usually be distinguished from successive folding by studying the relationship of the axial surfaces of the two (or more) sets of folds. If these mutually fold each other, a synchronous development is likely (but not inevitable, see Sec. 10-6). If the axial surfaces of one fold set have a folded form with the axes of folding contained in constantly oriented axial surfaces of the other fold set, then these features generally imply a successive development of the folds. In rocks which have undergone metamorphic reconstruction, it is sometimes possible to recognize separate development of the folds by observing the relationships of individual folds to crystal growth and rock fabric, e.g., one set of folds earlier than a certain episode of metamorphism, the other set later than the metamorphic event.

10-2   INTERFERENCE PATTERNS PRODUCED BY TWO SUCCESSIVE FOLDINGS

The three-dimensional form taken up by originally planar surfaces as a result of the superposition of two sets of folds may be likened to the pattern of interference caused by the intersection of two sets of waves. The nature of these interference patterns depends on the orientation relationships of the component fold systems, and there appear to be only three types of basic pattern. These types, which are described in detail in the following sections, are characterized by certain conditions (Fig. 10-2):

Type 1   Flow direction of the superimposed movement (a_2) oriented close to the axial surfaces of the first folds (α takes any value except zero, $\beta > 70°$)

Type 2 Flow direction of the superimposed movements (a_2) oriented at a high angle to the axial surface of the first folds, and the axis of first folds positioned at a moderate or high angle to the b_2 direction ($\alpha > 20°$, $\beta < 70°$)

Type 3 Flow direction of the superimposed movement (a_2) oriented at a high angle to the axial surface of the first folds, and the axis of first folds positioned close to the b_2 direction (α close to zero, $\beta < 70°$)

Consider a set of folded surfaces in a rock and superimpose a new periodic flow motion such as would produce a regular set of homogeneous folds in unfolded strata (Fig. 10-2). As a result of this new flow of the material, the previously folded layers take up a complex three-dimensional form, the exact nature of which depends on the values taken by angles α and β in Fig. 10-2. The geometric form of the final pattern may be modified further by other types of homogeneous strain and by buckling. These modifications do not fundamentally affect the type of interference form, and their effects will be discussed later.

Figure 10-2

Relationship between the two wave forms in a superimposed fold system. As the angles α and β vary, so different types of three-dimensional interference patterns result.

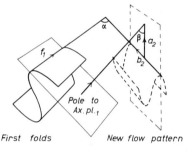

Type 1 interference pattern   Figure 10-3*A* shows a set of homoaxial cylindrical folds whose axes trend E-W horizontally across the block and Fig. 10-3*B* shows the result of superimposing on these folds a wave form produced by heterogeneous simple shear with shear plane a_2b_2, and flow direction a_2. As a result of this flow of material, the first folds become distorted and their limbs become refolded about new axial directions. Wherever the antiformal second folds are superimposed across a first-fold antiform there is a mutal culmination in both sets of folds resulting in the formation of a domelike structure. Similarly, crossing synforms produce mutal depressions in both sets of fold axes and the development of a basin structure. Where antiforms of one fold set cross synforms of another set, the antiformal hinges are depressed while the synformal hinges show a culmination. Where this occurs a saddle or col structure develops in the surfaces. As a result of this interaction the surfaces are folded into a series of alternating domes and basins, each basin surrounded by four domes, each dome by four basins—a structure which may be likened to an egg carton (see Moore and Trueman, 1939; de Sitter, 1952; Weiss 1959*a*; Ramsay, 1962*b*; Carey, 1962). Structural patterns of this type have been

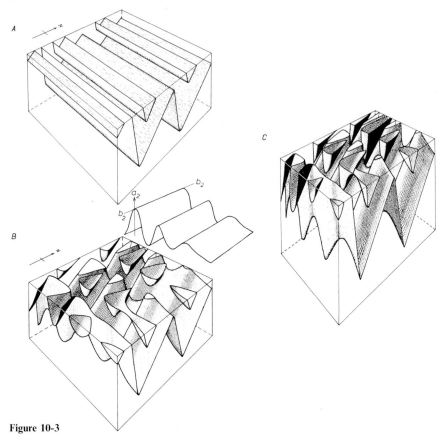

Figure 10-3

Interference pattern of type 1: A illustrates the original form of the first folds, B shows the inter-ference pattern which results from the superposition of a heterogeneous simple-shear deformation on the folds of A. a_2b_2 represents the slip surface of the second deformation, and a_2 the slip direction. C shows the modifications of the form of B by the superposition of a homogeneous strain with max-imum shortening perpendicular to a_2b_2.

developed experimentally in models deformed by lamellar gliding by O'Driscoll (1962*a*, 1962*b*, 1964*a*, 1964*b*).

Interference structures of this type developed over a region several hundred square kilometers in extent have been described by de Sitter (1952) in non-metamorphic Mesozoic sediments in the Atlas Mountains of Morocco (Fig. 10-4). A set of NE-SW directed folds were developed during a period of post-Eocene deformation (Pyreneic phase), and these were refolded in post-Miocene times by another set of folds whose axial traces trend E-W (Vindobonic phase). The Jurassic strata were detached from the underlying basement, sliding on a group of saliferous Triassic sediments, so that now diapiric Triassic material

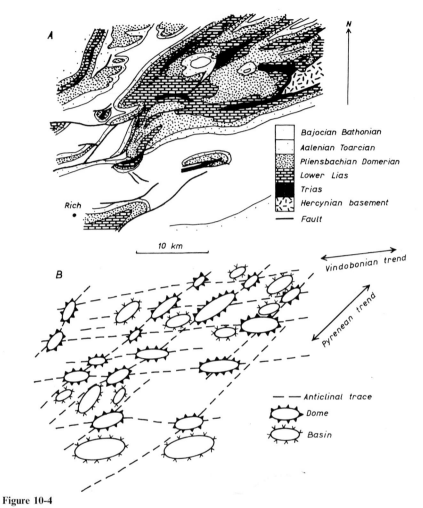

Figure 10-4

Patterns of interfering systems at Rich, Morocco. (After de Sitter, 1952.) A shows the disposition of the rock types, and B the interpretation of the structural pattern.

appears in domelike culminations situated where the two sets of anticlines cross. Basinlike depressions of Jurassic strata occupy the regions between the culmination points. The adaptation of curvature of these strata into the interference forms led to the setting up of a complex stress system in the layers and to the brittle fracture of the rocks in a complex series of faults.

In environments where metamorphic processes are active during the formation of the superimposed folds, ductile behavior of the rock masses often prevails and interference patterns involving complex three-dimensional curvature of the layers can be developed without the formation of faults and fractures. An example of such a pattern is shown in Figs. 10-5 and 10-6. These structures were developed in layered quartz-feldspathic sandstones which were being metamorphosed to the almandine facies at the same time as the fold development. The pattern on the outcrop surface represents a two-dimensional cross section through domes (black area in Fig. 10-6*B*) or basins (stippled area). By joining the lines of culmination and depression, the traces of the two sets of folds can be found.

Large-scale development of dome and basin patterns has been described in the central parts of orogenic belts, and from regions where basement rocks have been reactivated during later orogenic events (King, 1959; Fig. 7-44), and perhaps some of the forms known as *mantled gneiss domes* (Eskola, 1949) might be the result of superimposed folding. One especially good example of these patterns developed in metasediments in the Caledonian fold belt of Scotland has been described by Tobisch (1966). This region is made up mostly of metasedimentary rocks (Fig. 10-7) which have been repeatedly folded during the Caledonian mountain-building process. The outcrop pattern of the layers is very complex and Tobisch shows that the closed forms seen on the map represent cross sections through very pointed domes and basins of considerable size.

A homogeneous strain superimposed on an interference pattern which was itself produced by heterogeneous flow will modify the pattern but not alter

Figure 10-5

Dome and basin forms of type 1 interference pattern. Monar, Northern Highlands of Scotland. (After Ramsay, 1962a.)

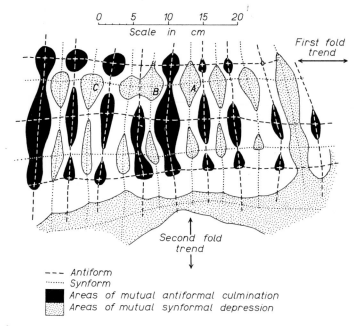

Figure 10-6

Structural interpretation of the outcrop of Fig. 10-5.

its fundamental geometric form (Fig. 10-3C). A compressive strain which acts across the axial surfaces of the superimposed strain will cause a decrease in the wavelengths of both the original and the superimposed folds and an increase in their amplitudes. Comparison of Fig. 10-3B with 10-3C indicates how the compressive strain makes the domes and basins take up a more strongly pointed form; a common feature when this type of interference pattern is developed in metamorphic terrains (Ramsay, 1962b; Tobisch, 1966).

Type 2 interference pattern   With this geometric arrangement the axial surfaces of the first folds become folded with the limbs of the first folds. Figure 10-8B illustrates the three-dimensional morphology of the compound structure developed by superimposing a heterogeneous flow on the folds illustrated in Fig. 10-8A. The first-fold hinges are bowed upward or downward by the heterogeneous flow and both limbs of any one first fold become refolded upward into a common antiform, or downward into a common synform. The second folds, which develop on the limbs of the first folds, plunge in approximately the same direction (cf. Type 1). If this complex three-dimensional form is progressively unroofed, the successively exposed two-dimensional outcrop patterns show a procession of systematically changing forms. The patterns have rather surprising shapes; they have been produced experimentally in

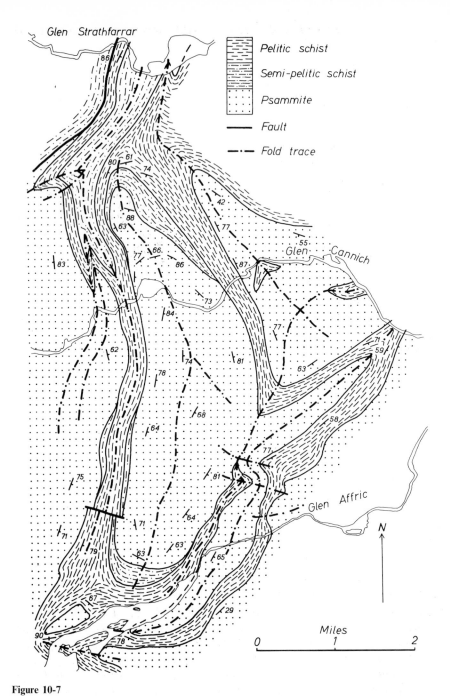

Figure 10-7

Large-scale interference pattern in the Moine metasediments of Glen Cannich, Northern Highlands of Scotland. (After Tobisch, 1966.)

526

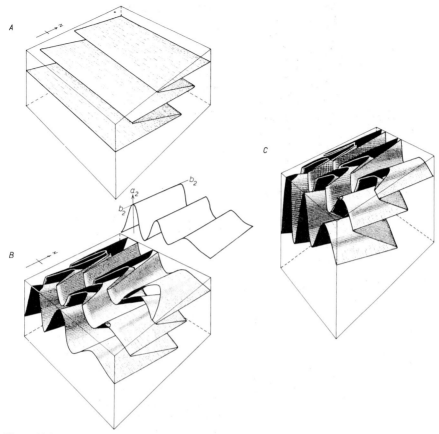

Figure 10-8

Interference pattern of type 2. A, original form of first folds; B, interference pattern produced by superposition of heterogeneous simple-shear flow on A; C, modification of B by a homogeneous finite strain.

layered clay models by Holmes and Reynolds (1954). The first form is nearly circular and represents the cross section through the culmination point of the antiformal first-fold hinge. In deeper sections these circular patterns become asymmetric and eventually crescent shaped (Fig. 10-9). These represent sections through arched first folds (Fig. 10-8B). If the number of deformed first folds are present in the region, their erosion to lower levels eventually leads to progressive enlargement and joining up of adjacent crescent cross sections and mushroom-like patterns, or more complex multilobed forms appear (Figs. 10-8B, 10-13G and H). Although these peculiar shapes look to be of limited relevance to geological problems, they are in fact fairly common in certain environments. In parts of the Pennine zones of the Swiss Alps and in some localities in the

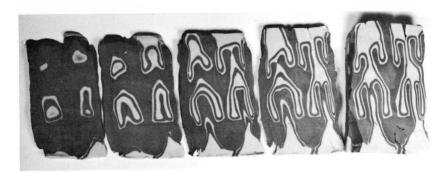

Figure 10-9

Type 2 interference pattern made experimentally in layered clays.

Figure 10-10

Characteristic "mushroom" outcrop pattern in gneissic rocks from the Lower Pennine nappes of Switzerland. Cristallina region, Ticino.

Caledonian fold belt of the Scottish Highlands, recumbent folds have been refolded by new structures with steeply inclined axial surfaces, and here these types of interference patterns are abundant (Figs. 10-10, 10-11, and 10-12). If the two sets of folds cross one another at right angles ($\alpha = 90°$, $\beta = 0$), the pattern that is produced shows perfectly symmetric mirror-image forms on either side of the axial trace of the second-fold structures (Fig. 10-13G). If the two sets are not superimposed exactly in this way, the two-dimensional interference patterns of crescents and mushroomlike forms are asymmetric (Fig. 10-13H). Holmes and Reynolds pointed out that the mean trend of the first-fold hinges can be determined by finding the direction of the join of the

Figure 10-11

Lobate forms of type 2 interference patterns in Moine metasediments, Loch Hourn, Western Highlands of Scotland. (Photo by G. Tanner.)

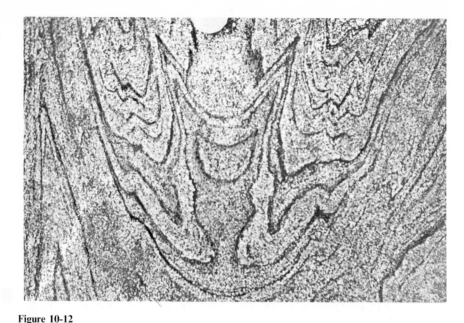

Figure 10-12

Type 2 interference patterns in Moine rocks of Beinn Clachach, Western Scottish Highlands.

ends of the "prongs" of the lobate outcrop shapes where they occur at the same topographic level.

White and Jahns (1950) have described the development of fold interference of this type over an area of about 60 square miles in Vermont. The outcrop patterns of the Gile Mountain quartz-mica schists and the Waits River calcareous schists have a peculiar form on the map (Fig. 10-14). In this region two distinct styles of minor folds exist; the earlier folds are tight or isoclinal and show a strongly developed axial plane schistosity, while the later folds are open structures with an associated strain-slip cleavage deforming the earlier schistosity. The major first folds appear to have been arched over a later antiformal structure, and erosion has cut through the upper arched parts of these folds. On the eastern side of the later fold arch the individual first-fold hinges plunge to the northeast (Fig. 10-14, traces *A*, *B*, and *C*), while on the western side of the arch the counterparts of these folds (traces *A'*, *B'*, and *C'*) plunge to the west-southwest.

Compressive strain components accompanying the flow deformation do not alter the fundamental type of interference pattern (Fig. 10-8*C*). The pattern is, however, modified to some extent, the crescentic forms tending to become more strongly curved.

Type 3 interference pattern   This pattern of superimposed structures, like

Figure 10-13

Various types of two-dimensional interference patterns and their dependence on the angles α and β (Fig. 10-2). Where α = 0 and β = 90° (C), the two sets of folds have parallel axial surfaces and axes, and there is no characteristic pattern of interference. All other variations of α and β produce patterns of types 1, 2, or 3, or transitional forms (D and E).

Type 2 discussed above, is commonly developed where recumbent folds are refolded by new structures with steeply inclined axial surfaces. Its development in the initially recumbent folds of Fig. 10-15A is illustrated in Fig. 10-15B. Because of the relative orientations of the two sets of fold waves, the axes of the new folds coincide, or nearly coincide, with the axial direction of the first folds. The two-dimensional patterns produced by this arrangement of the waves do not show the closed outcrop shapes characteristic of the previous types of pattern because the periodic undulations of the first-fold hinges are not well developed. The outcrop traces of the first folds show continuously converging or diverging forms (Fig. 10-13F and I). Homogeneous compressive

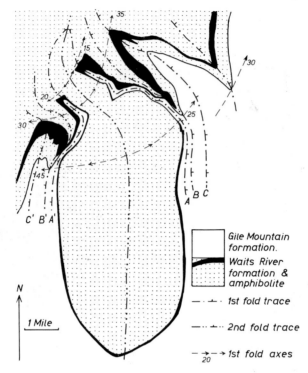

Gile Mountain formation.

Waits River formation & amphibolite

— · —·— 1st fold trace

— ·· —··— 2nd fold trace

— →—→ 1st fold axes
        20

N

1 Mile

Figure 10-14

Geologic map of the type 2 interference patterns seen in the folded Paleozoic schists of the Strafford Village area, Vermont. (After White and Jahns, 1950.)

strain components acting with the fold-forming shear components of the superimposed deformation modify the basic pattern but do not alter its fundamental structure (Fig. 10-15C).

Small-scale structures of this type developed across single first folds are illustrated in Figs. 10-16 and 10-17; more complex patterns where several first folds are present are shown in Fig. 10-18. The refolded limbs of the first folds may develop an almost similar waveform as a result of strongly developed refolding. If this type of structure is formed in tight or isoclinal first folds, it is sometimes possible for the hinge zones of the first folds to remain un-detected. The order of the rock layers in the superimposed fold might be assumed to represent the true stratigraphic succession. This would be an incorrect assumption, however, and careful study of Fig. 10-18 should make the dangers apparent. The many quartzo-feldspathic layers that occur in the left of this figure beneath the coin probably represent repeated parts of only three primary layers.

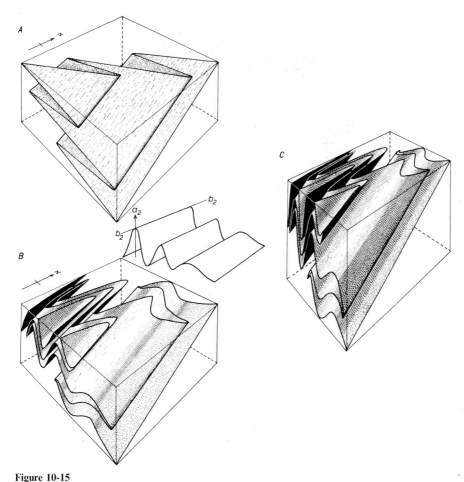

Figure 10-15

Interference pattern of type 3. A, original form of first folds; B, interference pattern produced by superposition of heterogeneous simple-shear flow on A; C, modification of B by a homogeneous finite strain.

Large-scale examples of this pattern of structure are well known in the Swiss Alps where recumbent fold nappes are common. They were illustrated in some of the first cross sections to be drawn across this fold belt (e.g., Schardt, 1894; Lugeon, 1901). Refolding patterns of this type are less commonly seen on maps than in cross section; but one example from the basement rocks of South Africa is illustrated in Fig. 9-15. It is interesting to compare the form of this structure with the smaller scale folds in Fig. 10-16: the structural forms are much alike, yet one is 3.5×10^4 the scale of the other.

Figure 10-16

Type 3 interference pattern in Moine metasediments, Loch Hourn, Western Highlands of Scotland.

Figure 10-17

Type 3 interference pattern in banded pre-Triassic gneisses in the core of a Lower Pennine fold nappe. Cristallina, Ticino, Switzerland.

Figure 10-18

Complex multifolded layers showing type 3 interference in banded biotie gneiss. Cristallina, Ticino, Switzerland.

Techniques for mapping interference patterns   In Sec. 7-2 (p. 351) it was shown how the shapes and symmetry of the small-scale folds at different positions in a large fold could be recorded on a map by employing those letters of the alphabet which most closely resemble the profile pattern of the fold when viewed down its axis (see also Fleuty, 1961, p. 475). The letters S and Z convey the form of asymmetric fold forms, and the letter M is used for symmetric forms. This method of recording fold symmetry can now be extended to cover the pattern of superimposed fold forms (Ramsay, 1958*a*, 1962; Turner and Weiss, 1963). Both the new and old folds in a compound system of super-imposed structures may have their own congruous minor folds, and the combined superimposed minor-fold patterns at any point in the larger structure can always be represented in terms of the nine possible symmetry combinations of the individual fold sets, that is,

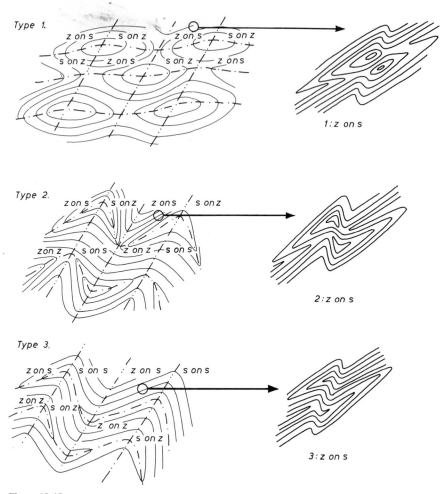

Figure 10-19

Methods of mapping interference patterns. The three diagrams on the left show the disposition of layers and axial traces of major structures of differing interference types, and the characteristic varieties of symmetry combinations. The figures on the right show the actual appearance of minor folds in certain zones of the major structures.

S on Z	M on Z	Z on Z
S on M	M on M	Z on M
S on S	M on S	Z on S

The actual appearance of the minor-fold interference structure depends on the type of interference pattern involved according to the principles established above. Figure 10-19 shows the distribution of the various combinations of minor folds in major structures interfering to produce the three basic types

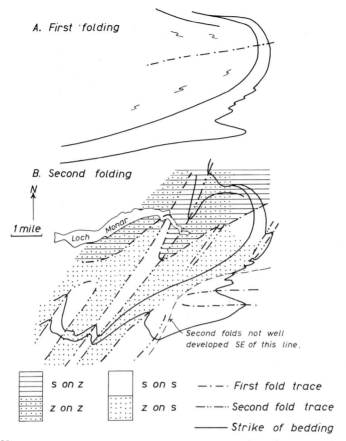

A. First folding

B. Second folding

N

1 mile

Loch Monar

Second folds not well
developed SE of this line.

☰ s on z	▢ s on s	— · — · First fold trace
▦ z on z	⸫ z on s	— ·· — ·· Second fold trace
		——— Strike of bedding

Figure 10-20

An example of the method of mapping interference structures as applied to refolded folds in the Loch Monar region, Northern Highlands of Scotland. A illustrates the probable form of the Loch Monar synform before refolding, and B shows the various types of interference symmetry as it is seen now. (After Ramsay, 1958a.)

of pattern. The method of recording the minor structure form depends on recording both the interference type and the form of the fold combination, for example, $2:S$ on Z, or $3:Z$ on M. If these observations are systematically made from outcrop to outcrop in an area of superimposed folding and recorded on a map, their distribution offers a powerful tool for locating the positions of the traces of both sets of major folds (e.g., Fig. 10-20).

10-3 STRATIGRAPHIC ORDER

Methods for reconstructing the cross-sectional shapes of folded rock strata (structural profiles) have been described by Wegmann (1929), and in more detail by Weiss (1959b) and Turner and Weiss (1963). These techniques can be employed where the folded forms have a monoclinic symmetry, and using them, the order of the layers in the folds (known as the structural succession) may be found. Where superimposed folding is developed in layered rocks, however, the symmetry of the compound structure generally has a triclinic form; under these conditions, application of the techniques of profile construction is not valid. Because individual layers may recur several times in a single fold of the superimposed set, both in correct and inverted order (see Fig. 10-17), the original order of the strata may be difficult to determine, particularly where only a small part of a large structure has been mapped.

10-4   GEOMETRIC CONTROL OF NEW FOLDS IN SUPERIMPOSED FOLD SYSTEMS

In any region of superimposed folding the structures that are seen in the rocks may be separated into two groups: first, those formed synchronously with the development of the new or superimposed deformation, and second, those geometrically related to the early formed folds, and which have been distorted as a result of the later rock deformation. The first of these groups will now be discussed, and the dependence of the morphology of the later folds on that of the pre-existing folds and structures will now be examined.

Axial surfaces of the new folds   The axial surfaces of the new folds generally have a fairly constant orientation even when other surfaces and linear directions are variable. The significance of this fact is open to a number of interpretations. If the new folds are formed by heterogeneous simple shear, then the constancy of the axial surfaces implies that the orientation of the shear surfaces is independent of that of the surface being folded. Where the new folds are developed by buckling of the competent rock layers in the limbs of the old folds, the axial surfaces of the new folds might be expected to have a variable orientation. During progressive deformation, however, these axial surfaces should rotate toward the XY plane of the strain (Flinn, 1962, p. 424), and so they might be expected to become more closely alined as deformation proceeds.

Because the surfaces on which the new folds are developed have variable orientations due to the previous folding, the position of the axial surfaces of the new folds generally varies according to the principles established in Secs. 7-11 and 9-7 (see Fig. 9-34). When individual superimposed folds are traced across the early structures, they may show an "en echelon" arrangement where they cross the axial surface of the first folds.

The symmetry of the major superimposed folds depends on the angular relationships that exist between the previously folded layers and the axial surfaces of the new folds. If the sheet-dip of the layers is perpendicular to the axial surfaces of the new folds, symmetric new folds are formed (Fig. 10-21, fold limb *A*); whereas if these features are not perpendicular, the new structures take up an asymmetric form (Fig. 10-21, fold limb *B*) (Ramsay 1958*a*, p. 288; Turner and Weiss, 1963, p. 141).

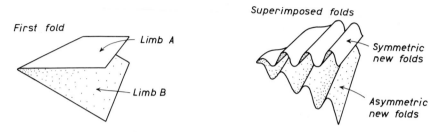

Figure 10-21

The formation of symmetric and asymmetric superimposed folds depending on whether the axial surfaces of the new folds are subperpendicular to the first-fold limb, or if they make some other angle with the fold limb.

Axial directions of the new folds   Because the new folds are generated on variably inclined and previously folded surfaces, it follows that, in general, they have variably oriented axial directions (Fig. 10-22). If the superimposed folds are formed by deformation processes of heterogeneous flow by simple shear, or heterogeneous simple shear together with a homogeneous compressive strain, they will be perfectly cylindrical folds. If the folds are produced by the process of buckling of the more competent layers, with internal distortions within the layers accomplished by true flexural slip, or by tan-

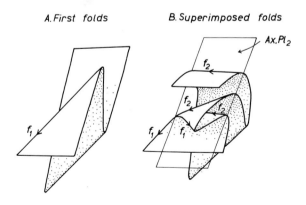

Figure 10-22

The orientation of the axes of superimposed folds. A shows a simple antiformal and synformal first-fold pair, B illustrates how the axial direction of the second folds (f_2) are controlled by the orientations of the limbs of the first fold.

gential longitudinal strain, it is possible for these folds to have a perfectly cylindrical form provided that some additional strains accompany their development (e.g., slip on the axial surface of the first fold—see Sec. 10-5).

In systems where the new folds are of a similar type and produced by flow and heterogeneous simple shear, the axial directions of the new folds are situated at the intersection of the shear planes *ab* and the surface undergoing folding (see Sec. 9-5 for the geometric proof of this fact). Where additional components of compressive strain are present, these axial directions are modified, and the fold axes come to lie closer to the axis of greatest extension (Sec. 7-12). If these additional strain components are homogeneous through the structure, the folds produced are cylindrical; but if these strains are not homogeneous, the folds take up a noncylindrical form (Ramsay, 1962*a*, p. 325; see also Sec. 8-3).

Figure 10-23 illustrates the application of these principles and shows how the pattern of axial variation of the new folds may depend on the degree of tightness and size of hinge zone of the old folds. If the surfaces within the first folds show a considerable range of orientation, then the axial directions of the folds produced by refolding this structure will show a considerable spread. This variation is generally greatest where the axial surfaces of the new folds make high angles with the first fold axes, and least where this angle is small (Fig. 10-23, cf. *A* and *B*). If the distribution of the surfaces in the first fold does not show a uniformly large range, and the first folds have well-developed limbs and narrow hinge zones, the superimposed folds tend to have two dominant directions controlled by the limbs of the first folds (Fig. 10-23*C*). If the first folds are isoclinal, or nearly so, with narrow hinge zones, then only one dominant new-fold axial direction will develop.

Direction stability of the new-fold axes   It has been shown how variations in the attitudes of the surfaces on which the new folds are developed are inherited into these new folds as a variation in their axial directions. If these surfaces have a variable orientation, the attitudes of the axes of the new folds will be great and their *axial direction stability* will be low. Other geometric features influence the direction stability, and the way in which new folds with a low axial direction stability can be formed on surfaces which only show slight initial variations will now be investigated. Where the axial directions of the new folds make high angles with the surfaces undergoing folding, any initial variation in the orientation in the surface being folded gives rise to a change in pitch of the fold which is equal to (Fig. 10-24*A*) or less than (Fig. 10-24*B*) that of the variation of the initial surfaces. In consequence, the variation in the orientation of the new-fold axes is less strongly marked than that of the surfaces, and the folds have a high axial direction stability. If, however, the initial angle between the axial surfaces of new folds and the surface being folded is small, any slight variation in these surfaces undergoing

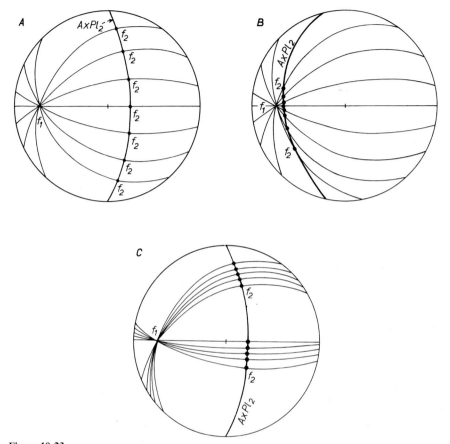

Figure 10-23

Projections showing the relationships of second-fold axes to the shape of the first-fold structure and the orientation of the axial surfaces of the second folds ($AxPl_2$). A and B show first folds where the folded surfaces have a wide range of orientation, C where the first folds have well-developed limbs. In A and C the second-fold axial planes make a high angle with the first-fold axis, while in B this angle is small.

folding can be inherited into new-fold geometry as a great variation in axial direction (Fig. 10-24C). Under such circumstances the axial direction stability is low. This large variation in axial orientation is somewhat analogous to the large errors in positioning of β points when the β-diagram method of fold analysis was applied to tight or open folds (Sec. 1-9).

Figure 10-25 illustrates the application of the principles of fold symmetry, axial orientation control, and direction stability of the later folds in a region of superimposed folding. The map shows a region made up of metamorphosed sedimentary rocks (Moine series) and gneissic basement rocks (Lewisian

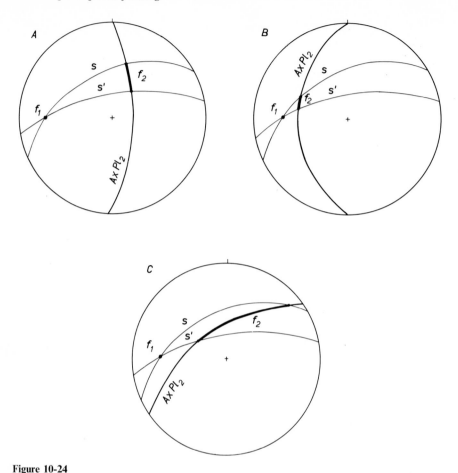

Figure 10-24

Axial direction stability of the second folds developed on curving surfaces varying in orientation from s to s'. f_2 shows the range of orientation of the second-fold axes.

complex) in part of the Northern Highlands of Scotland (Clifford and others, 1957; Ramsay, 1958a, 1963; Fleuty, 1961). Two synformal folds (Glen Orrin and Loch Monar) are separated by a neutral fold (Sec. 7-4, p. 358) of the same generation. The shapes of the two synforms are not quite the same—both have axial surfaces dipping to the southeast; but in the Loch Monar synform this surface has been strongly folded by later deformations. The limbs of the Glen Orrin synform are subparallel and show a sheet-dip of 60 to 75° to the southeast, whereas the limbs of the Loch Monar synform show a more diverse orientation; the northern limb has a sheet dip of 40 to 50° to the south and

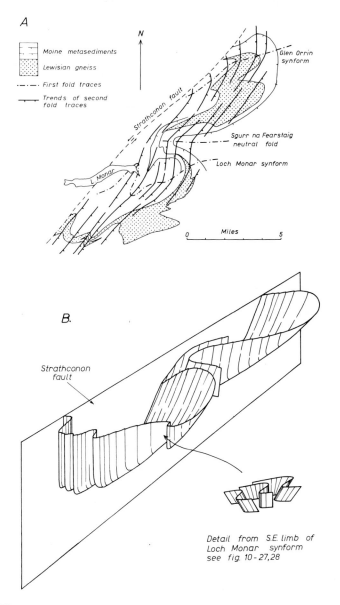

Figure 10-25

Superimposed fold structures in the Glen Orrin and Loch Monar regions of the Northern Highlands of Scotland. The Lewisian rocks probably occupy nappe cores which predate the formation of the two sets of folds illustrated here. For further details see text. (After Ramsay, 1958a, and Fleuty, 1961.) B shows a simplified diagram of the form of the folded bedding surfaces.

the southern limb has a sheet-dip of 80 to 90° to the southeast. These folds have been subjected to refolding by structures whose steeply inclined axial surfaces trend approximately NE. The axial directions of the later folds show an ordered systematic variation from place to place, depending on which limb of the earlier folds they are situated:

NW limb, Glen Orrin synform, new folds plunge 45 to 50° to SW
SE limb, Glen Orrin synform, new folds plunge 50 to 60° to SSW
NW limb, Loch Monar synform, new folds plunge 30 to 40° to SW
SE limb, Loch Monar synform, new folds plunge 75 to 90° to SSE

The initial variation in dip of the limbs of the Loch Monar synform controlled the axial orientations of the new folds. On the gently inclined NW fold limb, the folds plunge at moderate or low angles; whereas those developed on the steeply inclined SE limb are steeply or vertically plunging over large areas (Fig. 10-26). The limbs of the Glen Orrin synform are subparallel, and the new folds developed there do not show such great plunge variation as those on the Loch Monar fold. The direction stability of the new folds is fairly high on both limbs of the Loch Monar synform and it is greater on the more gently inclined NW fold limb than on the steeply inclined SE limb. This can be seen in the relative spreads of the lineation fields of new minor folds in Fig. 10-26B. Associated with this feature is the fact that the forms of the major new folds on the NW limb are generally symmetric, while those on the steeply inclined limb are asymmetric and Z-shaped. Thus, where the

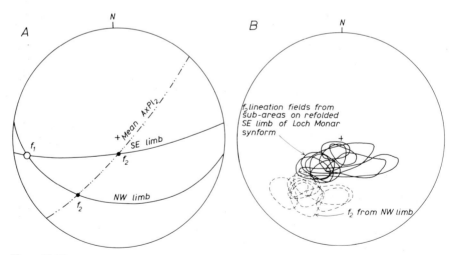

Figure 10-26

Analysis of the orientation of axes of superimposed folds in the Loch Monar syncline. B is a synoptic diagram showing the actual lineation fields in this region. (After Ramsay, 1958a.)

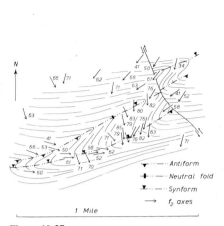

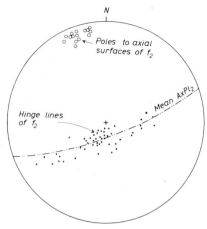

Figure 10-27

A pair of superimposed folds on the SE limb
of the Loch Monar synform showing a low
axial direction stability.

Figure 10-28

Projection of the data from the area of
Fig. 10-27 to show the great variation in
axial direction of the f_2 folds even though
their axial surfaces are fairly constantly
oriented.

angular relationships of fold limb to new axial surfaces was subperpendicular;
the new folds are symmetric with a high degree of directional stability;
whereas on the steeply inclined first-fold limb, the oblique relationships of
the new-fold axial surfaces to the limb produced asymmetric folds with a
lower degree of directional stability. At some localities along this steeply
inclined fold limb the axial direction stability of the new folds is exceedingly
low, even though the axial surfaces of the new folds and the sheet-dip of the
limb both show a fairly constant orientation. The geometric form of a pair
of new folds in this region is illustrated in Fig. 10-27. The axial directions of
the large- and small-scale new folds in this area change their angle of pitch
by about $120°$ over a distance of about a mile (Fig. 10-28). This systematic
and very pronounced change in orientation of the plunge direction results
in an "inversion" of fold plunge in the large structures; when the antiformal
structure exposed in the east of the region is followed along its axial trace
toward the WSW, the fold form passes through a neutral shape into a synformal structure (Figs. 10-25 and 10-27).

Amplitude of new folds In regions of superimposed folding it is common
for the amplitude of the new folds to change from locality to locality. This
may be attributed to variation in the intensity of the deformation that produced the structure. Moreover, where folds are formed by a heterogeneous
simple shear, another factor has some influence on the amplitude: this is the
angular relationship that exists between the surfaces being folded and the
movement direction a of the shear system (Ramsay, 1960, p. 90; 1963, p. 157;

and Sec. 7-11). If the flow direction is positioned at a high angle to the surfaces being folded, the folds have a larger amplitude than where this angle is small. Because these angles will vary from place to place, depending mainly on the attitude of the layers in the first folds, the fold amplitude of the structures is likely to vary also.

10-5   GEOMETRIC FORMS OF DEFORMED FIRST FOLDS

In regions of superimposed folding, the various component parts of the first-fold structures are distorted and take up new orientations in a manner which depends on the mechanism and intensity of the new deformations.

Linear structures   As a result of the first-fold development, the rocks may acquire a variety of differently oriented linear features, some parallel to the axial directions of the folds, others related to the states of finite strain and orientation of the finite-strain ellipsoid. These linear features take up some curved form in space according to the principles discussed in Chap. 8.

Planar structures   When a first-fold structure is refolded, its form is modified; in particular, its shape and the dihedral angle between the fold limbs are altered. If the refolding is accomplished by a heterogeneous shear, the dihedral angles are modified according to the principles established in Secs. 9-5 to 9-7. Small first folds refolded into a position on the limbs of the new folds have a much tighter form than the original shape, whereas those near to the hinge zones of the new folds may be more open than the original (Fig. 9-36). After refolding, the first folds show a variable dihedral angle between their limbs, and it is therefore not advisable in regions of super-imposed folding to correlate minor folds of various generations by means of their tightness.

If components of flexural-slip folding enter into the deformation process, the first folds may undergo rupture. To illustrate how this may develop, consider an antiformal first fold whose axis (f_1) plunges to the south (Fig. 10-29A). This fold is refolded about an E-W trending axial surface so that each limb is folded independently by buckling and flexural slip on the bedding surfaces. The east and west limbs of the original antiform are refolded about axes f_{2E} and f_{2W}, respectively. As the deformation is by flexural slip, any linear features originally positioned on the first-fold limbs parallel to the axis f_1 take up a new spatial orientation on two small-circle loci, l_{1E} at an angle α from f_{2E}, and l_{1W} at an angle β from f_{2W} (Fig. 10-29B) according to the principles described in Sec. 8-1. Because these two loci do not coincide (unless the first fold is an isoclinal structure and f_{2E} and f_{2W} are parallel), it follows that the lineation l_1 on adjacent limbs of the refolded first fold can no longer be parallel. This implies that a rotating shearing motion must be

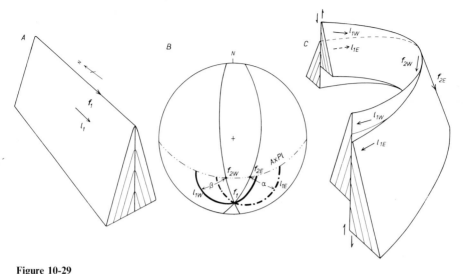

Figure 10-29

Refolding of an antiformal first fold (A) by the mechanism of flexural slip leading to the production of a slide along its axial surface.

set up on the surface separating the two first-fold limbs, and that the folded axial surface of the first must develop into a slide (Fig. 10-29C; Ramsay, 1963, p. 158). If the first fold has an axial plane schistosity, sliding is probably easily accomplished; but if this is not so, there may be a definite resistance to slide formation. If the slide movement is restricted, the strains must be adjusted in other ways within the structure.

First, the effect of resisting the slide development will be examined, keeping the two lineation loci l_{1E} and l_{1W} coincident in some intermediate position l'_1, yet retaining a true flexural-slip folding by retaining the constancy of angles α and β. The only way in which this may be geometrically accomplished is by moving the positions of the fold axes f_{2E} and f_{2W} along their axial surface toward the intersection point of the first- and second-fold axial surfaces (Fig. 10-30). In order to do this, the positions of the first-fold limbs must be altered so that the dihedral angle between them decreases. This implies that restriction of the sliding motion necessitates a tightening of the first fold by rotation of the fold limbs through f_1. The mechanical effect of this deformation can easily be seen by refolding a previously creased sheet of paper; as the paper is folded, the initial creases become more sharply plicated until they become isoclinal folds.

A second possibility of adjusting the strains within the deforming first fold if the axial plane sliding is restricted is by relaxing the condition of constancy of the two angles α and β. If the deformed first lineation locus l'_1 is the same

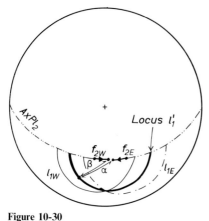

Figure 10-30

Geometric effects of restricting development of the axial surface slide shown in Fig. 10-29, keeping angles α and β constant.

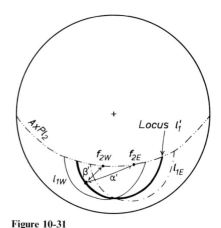

Figure 10-31

Slide restriction, allowing α and β to vary.

for both refolded limbs, and the positions of f_{2E} and f_{1E} remain static, then the angles α and β change to α' and β' (Fig. 10-31). For this to occur, the second folds can no longer be simple flexural-slip folds. There must be developed additional shearing strains within the folded surface so that the formation goes on by the mechanism of oblique flexural slip (p. 396).

It seems likely that during refolding of previously formed folds by flexural-slip mechanism, the internal strains will be developed in part by all three mechanisms of axial plane sliding, tightening of the first fold and development of oblique flexural-slip folds. All of these mechanisms require more energy than that needed for refolding the two first-fold limbs in isolation, and this means that rocks which already contain folded structures are much more difficult to buckle than those where the layers are all unfolded.

10-6   REACTIVATION OF OLD FOLDS

It has been shown how first folds may be tightened when their limbs are refolded. As this process continues, it is possible for the orientation of the superimposed structures to be modified. The structures in the Glen Orrin and Loch Monar regions of Scotland illustrated in Fig. 10-25 will be used to show an example of this type of process. In this area the axial surfaces of the superimposed new folds vary in both their dip and strike in a manner which is symmetrically related to their location on the limbs of the first folds. On the NW limbs of the Glen Orrin and Loch Monar synforms, these axial surfaces generally dip steeply to the west; whereas on the SE limbs of these folds, the axial surfaces dip to the SE. Figure 10-32 shows an equal-area plot

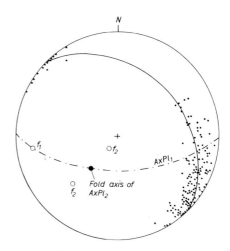

Figure 10-32

Projection of orientations of second-fold axial surface poles from the Loch Monar syncline.

of the poles of these axial surfaces as found in the eastern part of the Loch Monar synform. The poles lie about a part of a great circle showing that they have been cylindrically folded (Fig. 10-32). The axial direction of these folds is not parallel to any of the first- or second-fold axial directions, but it lies on the axial surface of the Loch Monar first-fold synform. There is no doubt whatsoever that these new axial surfaces are folded by the axial surfaces of the older structure. It therefore would appear that the development of the new folds led to the reactivation of the Loch Monar synform and to the warping of their axial surfaces (Ramsay, 1958a, p. 292; Fleuty, 1961, pp. 465-468).

10-7   INITIAL ORIENTATION OF FIRST-FOLD AXES

There are some precise methods available for the determination of the original orientations of the axial directions of first folds in superimposed systems, but these methods require a considerable knowledge of the refolding mechanisms and also rather special geometric circumstances.

If the refolding is developed by the production of true flexural-slip folds and with sliding on the axial surface of the first fold, the various deformation loci of the first-fold lineation on the refolded limbs of the first structure are found. They will intersect in one point, and this point will give the original axial direction of the first folds (Fig. 10-33A).

If the new folds are formed by heterogeneous simple shear, and if the movement directions of the shear vary from place to place, then, providing that the early fold axis was originally rectilinear, the loci of deformed first-fold lineations should all pass through a single point representing the initial position of the first-fold hinge (Fig. 10-33B).

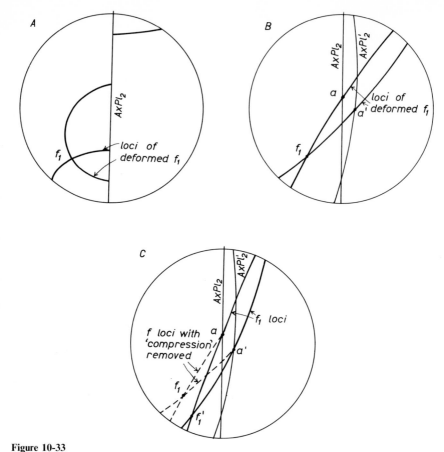

Figure 10-33

Methods for determination of the original orientation of first-fold axes. A, flexural-slip refolding; B, shear folding; C, shear folding with compressive strain components.

Most shear folds have components of compressive strain which modify the positions of the deformed lineation loci away from the af_1 plane. This means that the amount of this strain must be determined before the original first-fold direction can be found (Fig. 10-33C).

A more practical way for finding the approximate original orientation of the first-fold axes uses a comparison of the relative intensities of the minor structures related to the two sets of folds. An actual example of this method applied to folded folds in the Glenelg region of the Northwest Highlands of Scotland is illustrated in Fig. 10-34. The axial trace of a refolded antiform varies from N20°E to E-W. Where the fold trends N to N20°E, the structures associated with it (minor folds, schistosity) are very well developed in the

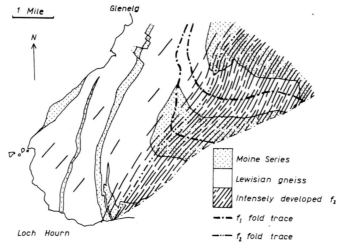

Figure 10-34

Intensity of various sets of structures in the Glenelg region, Western Highlands of Scotland.

rocks, whereas minor structures related to the later folds are rather weakly developed. In contrast, in the region where the first-fold trace is E-W, the minor folds of the later generation are very well developed. Comparison of the relative intensity of development of the two sets of structures therefore suggests that the original trend of the first-fold trace was about N-S.

None of these methods described here determine the amount of body rotation suffered by the whole block of material; they are essentially methods of comparison of the amounts of deformation on the various limbs of the later folds. The determination of the body-strain rotation components is usually impossible, because the true spatial orientation of lines before and after deformation is generally unknown.

10-8 GEOMETRIC ANALYSIS OF SUPERIMPOSED FOLDS USING PROJECTION TECHNIQUES

The analysis of the axial orientation of the various fold components in a superimposed fold system by projection methods using only orientation data of the folded surface is usually a rather complex procedure because of the great heterogeneity in these directions. The application of the β-diagram technique is unwise because of the possible appearance of many spurious β points (Turner and Weiss, 1963, pp. 160-163; Ramsay, 1964, pp. 442-447). An analysis is best accomplished by subdividing the whole structure into small homogeneous domains and then plotting the data onto a π diagram as described by Turner and Weiss (1963, pp. 144-193).

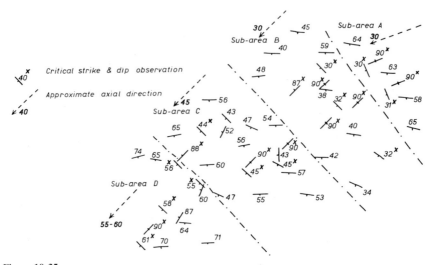

Figure 10-35

Map inspection methods of using dip variations to determine the approximate limits of domains of homogeneity.

The subdivision of a heterogeneous folded region into relatively homogeneous subareas is often a tricky problem. It may be carried out by trial and error, plotting data situated at greater and greater distances from a single point onto the projection until the pole plots begin to depart from any initially established π circle, but this is an unsatisfactory procedure. It is best to try to determine the degree of homogeneity of axial orientation of the folds by inspection before any data are plotted. This can be done sometimes by the application of two simple rules:

1. The strike of vertical beds is always parallel to the trend of the axial direction of the folds.

2. If beds have a dip direction which is parallel to the strike of adjacent vertical beds, then this angle of dip is equal in value to the angle of plunge of the fold axes.

Figure 10-35 illustrates the application of these rules to give approximate determinations of the axial orientations of the folds which enable the area to be subdivided into smaller domains for more precise analysis.

If the axial traces of the first folds can be determined on a map from a study of the major fold interference patterns, the general principles of axial control and direction stability established in Sec. 10-4 can be employed profitably to delimit the areas which are most likely to have homogeneous axial orientations. In Fig. 10-36 the folded axial surfaces divide the region into three parts. New folds developed in regions *A*, *B*, and *C* that developed on the limbs of the first fold are likely to have a high degree of axial stability.

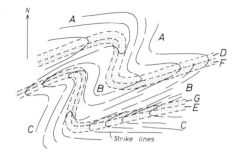

Figure 10-36

Division of a region of superimposed folding into subareas for detailed structural analysis.

Those in zones *D* and *E* near the hinges of the first folds, and those in *F* and *G* where the axial surfaces of the new folds lie close to the layering, are likely to have a low axial direction stability and the data from these zones should not be combined with those from *A*, *B*, and *C*.

In all these problems of geometric analysis it is generally more useful to discover the axial directions of the folds directly in the field by recording the orientation of minor-fold structures and congruent linear features.

REFERENCES AND FURTHER READING

Berthelsen, A.: Geology of Tovqussap Nuna, *Medd. Groenland*, 123, No. 1 (1960).

Best, M. G.: Petrology and Structural Analysis of Metamorphic Rocks, in the Southwestern Sierra Nevada Foothills, California, *Univ. Calif. (Berkeley) Publ. Geol. Sci.*, **42**: 111-158 (1963).

Bhattacharji, S.: Theoretical and Experimental Investigations on Cross-Folding, *J. Geol.*, **66**: 625-667 (1958).

Carey, S. W.: Folding, *J. Alberta Soc. Petrol. Geologists*, **10**: 95-144 (1962).

Choubert, B.: Le problème des structures tectoniques superposées en Guyane francaise, *Bull. Soc. Geol. France, Ser.* 7, **2**: 855-861 (1960).

Clifford, P.: The Geological Structure of the Loch Luichart Area, Ross-shire, *Quart. J. Geol. Soc.*, **115**: 365-388 (1960).

Clifford, P., M. J. Fleuty, J. G. Ramsay, J. Sutton, and J. Watson: The Development of Lineation in Complex Fold Systems, *Geol. Mag.*, **94**: 1-23 (1957).

Clough, C. T.: The Geology of Cowal, *Mem. Geol. Surv. Scotland*, 1897.

Derry, D. R.: Detailed Structure in Early Pre-Cambrian Rocks of Canada, *Quart. J. Geol. Soc.*, **95**: 109-133 (1939).

de Sitter, L. U.: Plissement croisé dans le Haut Atlas, *Geol. Mijnbouw*, **14**: 277-282 (1952).

de Sitter, L. U.: Cross Folding in Non Metamorphic of the Cantabrian Mountains and in the Pyrenees, *Geol. Mijnbouw*, **39**: 189-194 (1960).

Eskola, P. E.: The Problem of Mantled Gneiss Domes, *Quart. J. Geol. Soc.*, **104**: 461-476 (1949).

Fleuty, M. J.: The Three Fold System in the Metamorphic Rocks of Upper Glen Orrin, Ross-shire, *Quart. J. Geol. Soc.*, **117**: 447-479 (1961).

Haller, J.: Gekreuzte Faltensysteme in Orogenzonen, *Schweiz. Mineral. Petrog. Mitt.*, **37**: 11-30 (1957).

Holmes, A., and D. L. Reynolds: The Superposition of Caledonian Folds on Older Fold Systems in the Dalradians of Malin Head, Co. Donegal, *Geol. Mag.*, **91**: 417-444 (1954).

Johnson, M. R. W.: The Structural Geology of the Moine Thrust Zone in the Coulin Forest, Wester Ross, *Quart. J. Geol. Soc.*, **113**: 241-270 (1958).

Johnson, M. R. W.: The Structural History of the Moine Thrust Zone at Lochcarron, Wester Ross, *Trans. Roy. Soc. Edinburgh*, **64**: 139-168 (1960).

Johnson, M. R. W.: Polymetamorphism in Movement Zones in the Caledonian Thrust Belt of Northwest Scotland, *J. Geol.*, **69**: 417-432 (1961).

Johnson, M. R. W.: Relations of Movement and Metamorphism in the Dalradians of Banff-shire, *Trans. Edinburgh Geol. Soc.*, **19**: 26-64 (1962).

King, B. C.: Problems of the Precambrian of Central and Western Uganda; Structure, Meta-morphism and Granites, *Sci. Progr.* (*London*), **47**: 735-739 (1959).

Lambert, J. L. M.: Cross-folding in the Gramscatho Beds at Helford River, Cornwall, *Geol. Mag.*, **96**: 489-486 (1959).

Lugeon, M.: Les grandes nappes de recouvrement des Alpes du Chablais et de la Suisse, *Bull. Soc. Geol. France*, Ser. 4, **1**: 723-823 (1901).

Moore, L. R., and A. E. Trueman: The Structure of the Bristol and Somerset Coalfields, *Proc. Geologists Assoc.* (*Eng.*), **50**: 46-67 (1939).

Nicholson, R.: Eyed Folds and Interference Patterns in the Sokumfjell Marble Group, Northern Norway, *Geol. Mag.*, **100**: 59-68 (1963).

O'Driscoll, E. S.: Models for Superposed Laminar Flow Folding, *Nature*, **196**: 1146-1148 (1962*a*).

O'Driscoll, E. S.: Experimental Patterns in Superimposed Similar Folding, *J. Alberta Soc. Petrol. Geologists*, **10**: 145-167 (1962*b*).

O'Driscoll, E. S.: Interference Patterns from Inclined Shear Fold Systems, *Bull. Can. Pet. Geol.*, **12**: 279-310 (1964*a*).

O'Driscoll, E. S.: Cross Fold Deformation by Simple Shear, *Econ. Geol.*, **59**: 1061-1093 (1964*b*).

Oulianoff, N.: Superposition successive des chaînes de montagnes, *Scientia*, **47**: 1-5 (1953).

Ramsay, J. G.: Superimposed Folding at Loch Monar, Inverness-shire and Ross-shire, *Quart. J. Geol. Soc.*, **113**: 271-307 (1958*a*).

Ramsay, J. G.: Moine-Lewisian Relations at Glenelg, *Quart. J. Geol. Soc.*, **113**: 487-523 (1958*b*).

Ramsay, J. G.: The Deformation of Early Linear Structures in Areas of Repeated Folding, *J. Geol.*, **68**: 75-93 (1960).

Ramsay, J. G.: The Geometry and Mechanics of Formation of "Similar" Type Folds, *J. Geol.*, **70**: 309-327 (1962*a*).

Ramsay, J. G.: Interference Patterns Produced by the Superposition of Folds of "Similar" Type, *J. Geol.*, **60**: 466-481 (1962*b*).

Ramsay, J. G.: Structure and Metamorphism of the Moine and Lewisian Rocks of the North West Caledonides, in "The British Caledonides," Oliver & Boyd Ltd., Edinburgh and London, pp. 143-175, 1963.

Ramsay, J. G.: The Uses and Limitations of Beta and Pi-diagrams in the Geometrical Analysis of Folds, *Quart. J. Geol. Soc.*, **120**: 435-454 (1964).

Ramsay, J. G.: Structural Investigations in the Barberton Mountain Land, Eastern Transvaal, *Geol. Soc. S. Africa Trans.*, **66**: 353-401 (1965).

Rast, N.: Tectonics of the Schichallian Complex, *Quart. J. Geol. Soc.*, **114**: 25-46 (1958).

Rast, N., and J. I. Platt: Cross Folds, *Geol. Mag.*, **94**: 159-167 (1957).

Rast, N.: Structure and Metamorphism of the Dalradian Rocks of Scotland, in "The British Caledonides," pp. 123-143, Oliver & Boyd Ltd., Edinburgh and London, 1963.

Schardt, H.: Profil du Simplon, *Livret guide geol. de la Suisse*, 1894.

Simpson, A.: Stratigraphy and Tectonics of Manx Slate Series, Isle of Man, *Quart. J. Geol. Soc.*, **119**: 367-397 (1963).

Sturt, B. A.: The Geological Structure of the Area South of Loch Tummel, *Quart. J. Geol. Soc.*, **117**: 131-156 (1961).

Sturt, B. A., and A. L. Harris: The Metamorphic History of the Loch Tummel Area, Central Perthshire, *Liverpool Manchester Geol. J.*, **2**: 689-711 (1961).

Sutton, J.: Some Cross Folds and Related Structures in Northern Scotland, *Geol. Mijnbouw.*, **39**: 149-162 (1960*a*).

Sutton, J.: Some Structural Problems in the Scottish Highlands, *21st Intern. Geol. Congr., Copenhagen*, **18**: 371-383 (1960*b*).

Sutton, J., and J. Watson: The Structural and Stratigraphic Succession of the Moines of Fannich Forest and Strath Bran, Ross-shire, *Quart. J. Geol. Soc.*, **110**: 21-53 (1955).

Sutton, J., and J. Watson: Structure in the Caledonides between Loch Duich and Glenelg, Northwest Highlands, *Quart. J. Geol. Soc.*, **114**: 231-257 (1959).

Tobisch, O. T.: A Large Scale Basin and Dome Pattern Resulting from the Interference of Major Folds, *Geol. Soc. Am. Bull.*, **77**: 393-408 (1966).

Turner, F. J., and L. E. Weiss: "Structural Analysis of Metamorphic Tectonites," McGraw-Hill Book Company, New York, 1963.

Wegmann, O. E.: Beispiele Tektonischer Analysen des Grundgebirges in Finnland, *Comm. Geol. Finland Bull.*, **87**: 98-127 (1929).

Weiss, L. E., and D. B. McIntyre: Structural Geology of Dalradian Rocks at Loch Leven, Scottish Highlands, *J. Geol.*, **65**: 575-602 (1957).

Weiss, L. E.: Geometry of Superimposed Folding, *Geol. Soc. Am. Bull.*, **70**: 91-106 (1959*a*).

Weiss, L. E.: Structural Analysis of the Basement System at Turoka, Kenya, *Overseas Geol. Mineral Resources*, **7**: 3-35 and 123-153 (1959*b*).

White, W. S., and R. H. Jahns: The Structure of Central and East-Central Vermont, *J. Geol.*, **58**: 179-220 (1950).

Whitten, E. H. T.: A Study of Two Directions of Folding: The Structural Geology of Mona-dhliath and Mid-Strathspey, *J. Geol.*, **67**: 14-47 (1959).

Zwart, H. J.: Relations between Folding and Metamorphism in the Central Pyrenees, and Their Chronological Succession, *Geol. Mijnbouw.*, **22**: 163-180 (1960).

Author index

Adams, F. D., 338
Allen, P., 253
Anderson, E. M., 183, 293, 294, 298, 301–302, 334, 338, 340, 447
Anderson, T. B., 456
Andrade, E. N. da C., 335
Arges, K. P., 48
Arslan, A., 48

Badgley, P. C., 22
Badoux, H., 253
Bain, G. W., 456
Balk, R., 183
Balmer, G., 338
Bancroft, J. A., 338
Barrett, C. S., 265, 339
Bates, T. F., 183
Baumberger, E., 253
Bayley, M. B., 139, 183, 456
Becker, G. F., 183
Benioff, H., 337
Berthelsen, A., 553
Best, M. G., 376, 459, 472, 490, 553
Bhattacharji, S., 553
Billings, M. P., 396, 406–407, 456
Binder, R. C., 252

Biot, M. A., 373–380, 422, 456, 457
Birch, F., 340
Bland, D. R., 268, 335
Blyth, F. G. H., 184
Boltzmann, L., 273–274, 335
Boozer, G. D., 340
Borg, I., 424, 457, 490
Boschma, D., 447–457
Brace, W. F., 70, 183, 252, 253, 338
Bragg, W. H., 339
Bragg, W. L., 339
Braun, L. T., 458
Breddin, H., 253, 457
Bretherton, F. P., 252
Brett, G. W., 498, 517
Broughton, J. G., 183
Bryan, W. H., 253
Bucher, W. H., 22, 251, 338
Buessen, W. R., 183
Bunzl, M., 252
Burckhardt, C. E., 252
Busk, H. G., 457

Campbell, J. D., 457
Campbell, J. W., 457

Carey, S. W., 340, 431, 457, 521, 553
Chapman, F., 251
Charlsworth, H. A. K., 183
Chayes, F., 22
Chinnery, M. A., 294, 338
Choubert, B., 553
Clifford, P., 478–479, 490, 509, 511, 517, 542, 553
Cloos, E., 141, 180, 183, 184, 187, 191, 193, 202–203, 208–209, 251, 253, 338, 405, 457
Clough, C. T., 553
Coe, K., 184
Cohen, M., 339
Coker, E. G., 47, 49
Colette, B. J., 183
Conway, H. D., 48
Cottrell, A. H., 265, 339
Crandall, S. H., 48, 335
Crook, K. A. W., 183
Currie, J. B., 47, 49, 376, 457
Cummins, W. A., 490

Dahl, N. C., 48, 335
Dahlstrom, C. D. A., 22, 457
Dale, T. N., 183, 254, 405
Daubrée, G. A., 187, 254
Derry, D. R., 553
De Sitter, L. U., 338, 394–397, 401, 411, 440–441, 444–445, 457, 458, 521–523, 553
Dewey, J. F., 436, 457
Deyarmond, A., 48
Donath, F. A., 338, 430, 457
Donn, W. L., 22
Donner, J. J., 252
Drescher, F. K., 252
Drysdale, C. V., 335
Dufet, H., 254
Durelli, A. J., 48
Dziedzie, K., 252

Eberly, S. W., 22
Eirich, F., 252, 268, 335
Elliott, D., 363, 458

Elwell, R. W. D., 252
Engels, B., 254
Eskola, P. E., 386, 458, 524, 553
Evans, A. M., 458
Evans, C. R., 183
Eyring, H., 265, 338

Fairbairn, H. W., 22, 183, 220, 252
Fearnsides, W. G., 184
Ferguson, L., 254
Filon, L. N. G., 47, 49
Finnie, J., 337
Fitzgerald, E. L., 458
Fleuty, M. J., 349, 358–360, 458, 535, 542–543, 549, 553
Flinn, D., 22, 137, 154–155, 164, 184, 252, 329, 335, 340, 433, 458
Ford, H., 48, 318, 336
Fourmarier, P., 184
Frankel, J. P., 336
Fried, B., 49
Frocht, M. M., 47, 49
Furtak, H., 183, 254, 457
Fyson, W. K., 458

Garland, S. J., 22
Garson, M. S., 338
Glenn, J. W., 252
Goguel, J., 183, 254, 340, 341, 458
Goldschmidt, V. M., 252
Goldstein, S., 336, 373, 458
Gonzalez-Bonorino, F., 423, 458
Gräf, I., 254
Green, A. E., 318, 336
Green, A. P., 336
Green, J. F. N., 251
Griffith, A. A., 292–293, 338
Griffith, B. A., 48
Griggs, D. T., 259–261, 264, 275, 337, 339
Gunn, R., 373, 458
Gunthert, A., 252

Hafner, W., 298, 301–306, 338

Hager, A. D., 252
Hager, C. H., 252
Hager, R. C., 260–261, 340
Haller, J., 554
Haman, P. J., 22, 468, 490
Handin, J., 260–261, 340
Hardy, H. R., 337
Harker, A., 183, 187, 254
Harland, E. B., 139, 183
Harris, A. L., 555
Harris, F. C., 22
Haughton, S., 180, 183, 186, 254
Heard, H. C., 259–261, 264–265, 340
Heim, A., 180, 183, 187, 251, 254, 333, 340, 458, 515, 517
Heller, W. R., 337
Hellermann, E., 254
Hellmers, H., 254
Hencky, H., 289, 317, 336
Hetényi, M., 48–49
Hietenan, A., 251
Hill, M. J., 294, 338
Hill, R., 313, 322, 336
Hills, E. S., 254, 388, 458
Hitchcock, E., 252
Hodge, P. G., 322, 336
Hodgson, J. H., 338
Hoeppener, R., 183, 458
Holmes, A., 527, 554
Holmes, C. D., 252
Hubbert, M. K., 291, 295–298, 338, 339
Hundy, B. B., 318, 336

Ingerson, E., 8, 22, 183

Jaeger, J. C., 48, 268, 293, 336, 338
Jahns, R. H., 530–532, 555
Jameson, A. H., 336
Jannettaz, E., 254
Jeffery, G. B., 222–226, 252
Jeffreys, H., 48, 341
Jessop, H. T., 49
Johnson, M. R. W., 438, 453–454, 458, 554
Johnson, W., 336

Johnson, W. A., 252
Jones, O. A., 253
Jones, O. T., 251

Kamb, W. B., 22
Kehle, R. O., 319, 336
Kennedy, A. J., 264, 337
Kienow, S., 373, 416, 458
King, B. C., 458, 524, 554
Kleinsmiede, J. F., 458
Knill, D. C., 458
Knill, J. L., 184, 458
Knopf, A., 48
Knopf, E. B., 8, 22, 183, 340
Koerber, G. G., 336
Koopmans, B. N., 455, 458
Krumbein, W. C., 252
Kuenen, P. H., 394–395, 401, 458
Kvale, A., 252, 340

Ladurner, J., 254
Lake, P., 254
Lamb, H., 336
Lambert, H. L. M., 554
Lee, G. H., 48
Leith, C. K., 184, 396, 405, 458
Levy, M., 319, 336
Lianis, G., 318, 336
Lohest, M., 184
Lomnitz, C., 337
Long, R. R., 48, 336
Love, A. E. H., 336, 387, 458
Lugeon, M., 459, 533, 554

McBirney, A. R., 376, 459
McCallie, S. W., 252
MacDonald, G. J. F., 339
McIntyre, D. B., 463, 490, 511, 517, 555
McKinstry, H. E., 294, 338
Margaretha, H., 252
Mason, S. G., 224–225, 252
Mathews, D. H., 459
Matisto, A. S. I., 251
Maxwell, J. C., 184, 273–277, 336

Mead, W. J., 459
Mehnert, K. R., 221, 252
Mellor, P. B., 336
Mendelsohn, F., 459
Mertie, J. B., 459
Michelson, A. A., 336
Mises, R. von, 316–319, 336
Misra, A. K., 337
Mohr, O., 69
Moore, L. R., 521, 554
Morris, T. O., 184
Muehlberger, W. R., 459
Mügge, O., 438, 459
Murrell, S. A. F., 337

Nadai, A., 48, 317, 336
Nagy, B., 183
Neuvonen, K. J., 251
Nevin, C. M., 396, 459
Nicholson, R., 554
Noble, D. C., 22
Norman, T. N., 490

Odé, H., 45–46, 48, 293, 298, 322, 339, 374, 376, 378, 456, 457
O'Driscoll, E. S., 509, 517, 522, 554
Oertel, G., 340
Oftedahl, C., 252
Olszak, W., 313, 336
Orowan, E., 337, 339
Oulianoff, N., 518, 554

Palmer, A. E., 48
Parker, J. M., 339
Parker, R. B., 457
Paterson, M. S., 333, 336, 340, 440, 453, 459
Patnode, H. W., 47, 49, 376, 457
Peach, B., 252
Pearson, C. E., 336
Pettijohn, F. J., 252, 498, 517
Phillips, A., 336
Phillips, E. A., 48
Phillips, F. C., 22, 459, 463, 490

Phillips, J., 180; 183, 184, 186, 254
Pilger, A., 459
Pincus, H. J., 22
Platt, J. I., 554
Popov, E. P., 336
Prager, W., 322, 336
Prescott, J., 336
Price, N. J., 268, 275–277, 285, 337, 339
Pugh, W. J., 251

Quinney, H., 318, 337
Quirke, T. T., 184

Raleigh, G. B., 339
Ramberg, H., 113, 154–155, 184, 313, 336, 374–382, 397, 416, 423, 430, 433, 459
Ramsay, A. C., 252
Ramsay, D. M., 253, 453, 455, 459
Ramsay, J. G., 14, 22, 212, 253, 330, 335, 337, 340, 383, 411–413, 424, 428, 431–436, 453–456, 459, 463, 470–475, 480–483, 490, 492, 497–499, 509–515, 517, 520–524, 539–544, 553, 554
Rast, N., 184, 458, 460, 520, 554
Read, W. T., 265, 339
Ree, F. H., 265, 338
Ree, T., 265, 338
Reece, A., 515, 517
Reiner, M., 268, 337
Renevier, E., 187, 254
Reynolds, D. L., 527, 554
Richter, D., 460
Rickard, M. J., 184
Riedel, W., 184
Roark, R. J., 48
Robertson, E. C., 338–340
Robinson, P., 22
Robinson, R., 22
Roever, W. L., 374, 376, 378, 456
Rogers, G. L., 337
Rosenhain, W., 339
Rosenthal, D., 337
Rouse, H., 337
Rubery, W. W., 295–298, 338, 339

Runner, J. J., 220, 253
Rutherford, D. E., 308, 337
Rutsch, R. F., 254

Sander, B., 22, 164, 184, 220, 253, 333, 340, 463, 490
Sanford, A. R., 298, 306, 339
Schaer, J. P., 394, 460
Schardt, H., 533, 554
Scheidegger, A. E., 341
Scheumann, K. K., 253
Schmidt, W., 333, 340, 388, 459, 460
Sdzuy, K., 254
Sederholm, J. J., 460
Serdengecti, S., 340
Shainin, V. E., 184
Sharpe, D., 180, 183, 184, 186, 254
Shiner, J. A., 22
Siebel, M. P. L., 318, 337
Simpson, A., 555
Smoluchowski, M., 373, 460
Sokolnikoff, I. S., 337
Sommerfeld, A., 337
Soper, N. J., 512–513
Sorby, H. C., 180, 183, 184, 186, 251, 254
Sperry, W. C., 268, 337
Stanier, X., 184
Stauffer, M. R., 460, 468, 490
Stephansson, O., 459
Stillwell, F. L., 419, 460
Stockwell, C. H., 460
Strand, T., 253
Streeter, V. C., 337
Sturt, B. A., 453, 455, 459, 555
Sutton, J., 520, 553, 555
Swanson, C. O., 184
Synge, J. L., 48

Tanner, G., 529
Tavener-Smith, R., 253
Taylor, G. I., 253, 318, 337
Ten Haf, E., 490
den Tex, E., 22
Thomas, D. E., 254
Thompson, W., 271, 337

Timoshenko, S. P., 48, 301, 337
Tobisch, O. T., 524–526, 555
Tocher, D., 338
Tresca, M. H., 315–316, 337
Trevelyan, B. J., 224, 253
Tromp, S. W., 460
Trueman, A. E., 521, 554
Trump, R. P., 47, 49, 376, 457
Tsao, C. H., 48
Turner, F. J., 8, 22, 181, 183, 184, 253, 259–261, 335, 340, 351, 424, 428, 438, 457, 460, 490, 511, 517, 535, 538–539, 551, 555
Twenhofel, W. H., 253

Urbanowski, T., 313, 336

Van Hise, C. R., 460
Verma, A. R., 339
Voight, W., 270–274, 337
Voll, G., 254, 266, 339

Walsh, J. B., 339
Watson, J., 553, 555
Weaver, S. H., 264, 338
Wegman, O. E., 394, 396, 460, 537, 555
Weiss, L. E., 8, 22, 183, 333–335, 340, 351, 428, 440, 453, 459, 460, 463, 470, 490, 511, 517, 521, 535, 538–539, 551, 555
Weller, R., 49
Wellman, H. W., 254
West, R. G., 252
Westergaard, H. M., 337
Wettstein, A., 187, 254
White, W. S., 184, 530–532, 555
Whitten, C. A., 265, 341
Whitten, E. H. T., 555
Williams, E., 460
Williams, H., 251
Wilson, G., 177, 183, 184, 251, 386, 391, 405–407, 460, 463, 490
Winchell, H., 22

Wunderlich, H. G., 46, 460
Wynne-Edwards, H. R., 460

Young, J. F., 337

Zandvliet, J., 460
Zener, C. M., 337
Zerna, W., 336
Zingg, T., 137, 183, 253
Zwart, H. J., 55

Subject index

Abscherung, 420
Aggregate formation, 225
Amygdules, deformation of, 190
Annealing, 261
Anticlastic bending, 402
Axes, of folds, 12, 348, 356, 424
    of ooids, fluctuation in, 202–209
    of simple shear, 334–335
    of symmetry, 333–335

β axis, 12–14
    errors in position, 13–14
    spurious, 13
β diagram, 12–14, 551
Basin, 346
    resulting from superposed
        folding, 521–525
Biharmonic equation, 300, 312
Billing's net, 3
Boltzmann's principle, 273–274
Boudinage, 103–109, 112–116, 184
    folded, 115
    measurement of strain from,
        250
Brittle fracture, 257–258, 289–297
    Coulomb criteria for, 291–292
    Griffith criteria for, 292–293

Brittle strength, 258, 289
Buckling, 109, 372–386

Chalazoidites, deformation of,
    188–189
Chocolate-tablet structure,
    112–113
Cleavage, axial plane, 405
    crenulation, 177–179, 390,
        417–418
    fracture, 177–178, 406–407
    shear, 389
    slaty, 89, 93–94, 177–181,
        183–184, 403
    strain-slip, 177, 322, 389, 418
Cleavage-bedding intersection,
    410–411
Cleavage fan, convergent, 403–405
    divergent, 403–405, 417, 429
Cleavage refraction, 405–406
Cold working, 261
Compaction, 228, 229
Compatibility equations, 99
Competence, 103
Concretions, deformation of, 188
Conglomerates, deformation of,
    201–202, 211–228

Conglomerates, pressure solution
    in, 226–228
Continuity equation, 308
Creep, 262–264
    equations for, 264, 279–281
Cross bedding (*see* False bedding)

Decollement, 416, 420
Deformation field, 141
Deformation path, 96, 140, 326–332
Dilation, in elastic deformation, 284–285
    in finite strain, 65, 123, 161–162
    in flow, 308
    in folding, 447–449, 454
    in infinitesimal strain, 103, 174, 282
Dip, apparent and true, 6
Dip isogons, 363–369, 432
Dislocations in crystals, 261, 267–268
Dome, 346
    mantled gneiss, 524
    resulting from superposed folding,
        521–525

Elastic deformation, 258, 266, 283–289
Elastic limit, 258
Elastic moduli, definitions of, 283–287
    relationships between, 286–287
Elasticas, 349, 387–388
Ellipse, frequency of section sizes, 192
Ellipsoidal objects, deformation of,
    211–221
Elliptic orbits, 222–225
Elliptical objects, deformation of,
    202–211
Equal-area net, 3
Equilibrium equations, 41–43
Experiments, long-term rock deformation,
    262–265
    model, boudinage, 103–109
        folding, 110, 372, 374, 377, 380, 382,
            383, 396
        superposed deformation, 94
        superposed folding, 527–528
    short-term rock deformation, 256–261
Extension, 52

False bedding, deformation of, 498,
    512–516
Fault, bedding plane thrust, 394
    deformation during folding, 401,
        497–498
    limb, 416, 421
    movement direction of, 6
    normal, 293
    second-order, 294
    splay, 295
    strike-slip, 293
    thrust, 293, 295–296, 304–306, 401
Flame structure, 386
Flow, cataclastic, 261
    ductile, 103–113, 256
Flow direction in similar folds, 423–426,
    462, 470–480
Fluid pressure, 295–297
Folds, accordion, 350
    amplitude of, 352–355, 424
    anticlinal, 358
    antiformal, 358
    axes of, 12, 348, 356, 424
    axial direction of, 10, 18, 348, 410
    axial surfaces of, 10, 355–356, 423,
        426–427
    axial traces of, 10, 356–357, 427
    bending, 430
    boudinaged, 115–120
    box, 357–358, 416, 421
    chevron, 350, 388, 438–447
    concentric, 110, 387, 400
    concertina, 438
    conical, 10, 20–22, 349, 496–498
        deformation of lineations in, 468–469
    conjugate kink, 357–358, 438, 449–456
    convergent dip isogon, 365–367, 413
    crest lines of, 345–346
    crest surfaces of, 355–356
    crestal traces of, 356–357
    culmination points of, 345–346, 436
    curvature of surfaces in, 345–347, 350,
        363–370
    cylindrical, 348
        analysis of, 9–14
    depression points of, 345–346
    disharmonic, 418–419

Folds, divergent dip isogon, 365–368
drag, 396–397
enveloping surfaces of, 351–353
finite neutral points of, 416–417
finite neutral surfaces of, 397–398
flattened parallel, 387, 411–415, 432–434
deformation of lineation in, 466–467
flexural, 110, 391–396
deformation of lineation in, 463
shortening in, 397, 403
formed by buckling, 372–421
geometric analysis of, 9–14, 18–22
geometric classification of, 359–372
harmonic, 387–417
hinge lines of, 347, 356, 424
hinge zones of, 11, 348–349
inclined, 359–360
infinitesimal neutral points of, 417
infinitesimal neutral surfaces of, 398
inflection lines of, 346–347
inflection surfaces of, 355–357
interlimb angles of, 349–350
knee, 436
limbs of, 11, 348–349
mechanisms of formation of, 372–456
median surface of, 351–353
monoclinal, 436
multiple hinge, 347–348, 357–358
neutral, 358
noncylindrical, 9, 348–349, 540
analysis of, 14
oblique flexural slip, 396, 465–466
orthogonal thickness in, 360–362
overturned, 359
parallel, 110, 372
parallel dip isogon, 365–367, 421
parasitic, 396–397
passive, 430
polyclinal, 358
profile section of, 347
reclined, 359–360
rectilinear generator of, 9, 12
recumbent, 358–360
second-order, 456
shear, 423.
sheet-dip in, 351–353
similar, 365–367, 413, 421–436

Folds, similar, deformation of
lineations in, 469–482
simultaneous development in several
directions, 112–113, 161, 520
single hinge, 347–348
slip, 423
superposed, amplitude of, 545–546
axial direction in, 509–511, 539–545
axial direction stability in, 540–545
axial surfaces of, 509, 538–539
development of slides in, 546–548
geometric analysis of, 551–553
initial orientation of first folds,
549–551
interference patterns in, 520–537
profile construction in, 538
reactivation of first folds, 548–549
stratigraphic order in, 532, 537–538
symmetry of, 535–539
symmetry of, 351–354, 357, 535
synclinal, 358
synformal, 358
trough lines of, 345–346
trough surfaces of, 355–356
unfolded, 115–118
upright, 359–360
variant and invariant features of,
344–358
wavelength of, 110, 352–355, 373–383
zigzag, 350, 436
zone of contact strain around, 416–417
Force, body, 23
surface, 23
Fossil deformation, 186–189, 228–250
Friction, internal, 291

Gleitbretter, 388-390
Graben, 306

Hooke's law, 283–284
Horst, 306
Hot working, 261

Invariants, of deviatoric stress tensor, 41

Invariants, of finite-strain tensor, 81–83, 145–146
of infinitesimal strain tensor, 172
of strain, 81–83, 145–146, 172
of stress tensor, 28–29, 34–35
and yield criteria, 314–316
Isoclinic line, 47
Isotropic point, 43–44

Joint, 91, 113, 277, 402
Joint drag, 436

Kink band, 436–440, 447–456

Lambert net, 3
Linear fabric, 213, 220–221, 225–226
Linear structures, angle between, 6, 31–32
deformation of, 461–489
direction significance of, 5
initial orientation of, 462, 486–489
plotting methods, 4–5
Lineation, 183–184
Load casting, 386
Lunules, 394

Mantled gneiss dome, 386, 524
Mechanical analogs for rock behavior, Bingham model, 280–281
elastic model, 268–269
elastoviscous models, 269–270, 274–281
Kelvin-Voigt model, 270–274, 279–280
Maxwell model, 273–277, 279–280
St. Venant model, 268–269
Standard linear solid, 277–279
viscoelastic models, 268–274, 277–279
viscous model, 268–269
Microlithon, 322, 388–390
Mohr diagram, for strain, 69–81, 102, 149–153, 173–174
for stress, 151, 289–291
Mohr envelope, 290–291
Mud pellets, deformation of, 187–188

Mullion structure, 385–386

Nappe, 51, 295, 420
Necking, 104, 112

Oblique surfaces, deformation, by flattened flexural-slip folds, 498–501
by flexural-slip folds, 492–496
by similar folds, 504–516
Ooids, deformation of, 187–188, 193–197, 202–211
fluctuation in axes of, 202–209
Orogenic belts, crossing, 518–519
Orogenic cycles, 519–520, 524

π axis, 10
π circle, 10
π diagram, 9–12, 551–552
Photoelasticity, 47–48
Pinch-and-swell structure, 103–104
Pisolites, deformation of, 187–189
Planar fabric, 225–226
Planar structures, angle between, 7
direction significance of, 5
intersection of, 6
modification of angle between, 496–501, 504, 506–507
plotting methods, 4
Plastic deformation, 258–261, 266–268, 313–322
Plastic flow, proportionality factor of, 320–321
Polycrystalline aggregates, deformation of, 265–268
Polygonization, 261
Pressure shadow, 181–182
Pressure solution, 195–197, 226–228, 249–250
Progressive deformation, definition of, 55
folds development during, 114–120, 177, 343, 520
strain states during, 139–140, 322–332
structures developed during, 114–120, 174–177

Projection, contouring methods, 8–9
    density distribution, 8–9
    equal-area, 3
    for presenting structural data, 7–8
    spherical, 2
    stereographic, 1–4
Ptygmatic structure, 110–111, 119–120,
        387–388
    measurement of strain from, 250

Quadratic elongation, 52
Quartz rods, 391

Reduction spots, deformation of, 188–189
Relaxation, 269, 275–279
Retardation, 269, 272
Riecke's principle, 195
Rigid particles, deformation of, 221–226
Rods, 391

Saddle Reef, 419, 447
Schistosity, 177–181, 436
Schmidt net, 3
Singular point, 43–44
Slickensides, 394
Slide, 546–548
Slip lines, 321–322
Spherulites, deformation of, 187, 191,
        197–199
Spotted rocks, deformation of, 190
Strain, analysis in three dimensions,
        121–134
    analysis in two dimensions, 55–69
    compatibility equations for, 99
    definition of, 51
    finite, 55, 114
    finite angular changes, 67, 129–134
    finite length changes, 65–67, 124–128
    graphic representation of, 94–96,
        134–141
    homogeneous, 53–54
    infinitesimal, 55, 96–103, 114, 167–174
    infinitesimal length changes, 99–100,
        172–173

Strain, infinitesimal shear, 173
    inhomogeneous, 53–54
    invariants of, 81–83, 145–146, 172
    logarithmic, 52–53
    longitudinal, 52
    Mohr diagrams for, 69–81, 102,
        149–153, 173–174
    natural, 52–53
    parameters of, 52–53
    permanent, 258
    plane, 135, 137
    principal axes of, 59–60, 100, 122
    principal value of, 61–63, 122
    pure shear, 56, 115–119
    rotational and irrotational, 60, 100–101,
        114, 120, 169, 324–325
    shear, 53, 67–69, 102–103, 128–129
    simple shear, 56, 83–91, 119–120, 184,
        324, 334–335
    superimposed, 91–94, 165–167
    tangential longitudinal, 397–400,
        463–465
    true, 52–53
Strain ellipse, 58–59, 94–96, 99, 111–114
    measurement of, 193–199
    strain components from, 199–200
Strain ellipsoid, 122, 134–141, 186
    circular sections of, 127–128
    determined from two-dimensional
        data, 142–149
    infinitesimal, 171
    k value of, 137
    measurement of, 200–201
    types of, 135–139, 154–161
Strain energy in elastic deformation,
        287–289
Strain field, 111–113
Strain hardening, 260, 265, 268
Strain state, in folds, 343, 391–393,
        397–401, 403, 415–417, 422–423, 461
    during progressive deformation,
        139–140, 322–332
Strain tensor, 63–64, 123–124, 170,
        281–283
Stream function, 312–313
Strength, brittle, 258
    creep, 263–264

Strength, ultimate, 261
Stress, acting on a plane, 29–31, 35–36
  cohesive, 289–290
  components of, 24–26
  definition of, 23
  deviatoric, 39–41, 283
  effective, 295
  equilibrium equations of, 41–43
  hydrostatic, 39–40, 283
  invariants of, 28–29, 34–35, 41
  Mohr diagram for, 151, 289–291
  normal, 23–24
  octahedral, 317
  plane, 25–27
  principal, 27–28, 32–35, 40–41
  principal planes of, 32
  shear, 23–25, 36, 289
    maximum values of, 31, 36–38, 289
  superposed systems of, 44–46
  yield, 39–40, 258–261, 280, 314
Stress function, 297–306
Stress-strain relations, elastic solids,
    283–289
  general theory of tensors, 281–283
  plastic flow, 307, 313–322
  viscoelastic and elastoviscous materials,
    270–279
  viscous fluids, 306–313
Stress tensor, 281–283
  invariants of, 28–29, 34–35
Stress trajectories, 43–46
Stretching fiber, 89–91
Stretching lineation, 248
Structural analysis, conical folds, 10,
    20–22
  curving surfaces, 9–14
  cylindrical folds, 18–20
  mean vector, 15–17
  presentation of results, 7–8

Structural analysis, statistical
    techniques, 14–22
  subareas of, 11, 13
Symmetry, axes of, 333–335
  concept of, 333–335
  of folds, 351–354, 357, 535

Tectonites, linear, 164
  planar, 165
Tension fissures, 88–91, 394–395, 401, 455
Tensor, asymmetric, 63–64, 123–124, 282
  skew symmetric, 124, 282–283
  strain, 63, 123–124, 281–283
  strain rate, 307–308
  stress, 281–283
  symmetric, 63–64, 123–124, 282–283
Thixotropy, 307
Translation, 51, 334
Triaxial rig, 257

Unconformity, deformation of, 500, 509,
    513–516

Viscous moduli, 306, 309–311
Volume change (*see* Dilation)
Vorticity, 308

Wulff net, 2

Yield, 39–40
Yield criteria, 315–319
Yield point, 258
Yield stress, 39–40, 258–261, 280, 314
Yield surface, 314–319